# COMPUTING IN EUCLIDEAN GEOMETRY

**LECTURE NOTES SERIES ON COMPUTING**
Editor-in-Chief: D T Lee (*Northwestern Univ., USA*)

---

Vol. 1: Computing in Euclidean Geometry
*Eds. D-Z Du & F Hwang*

*Forthcoming*

Vol. 2: Algorithmic Aspects of VLSI Layout
*Eds. D T Lee & M Sarrafzadeh*

Lecture Notes Series on Computing – Vol. 1

# COMPUTING IN EUCLIDEAN GEOMETRY

Edited by

**Ding-Zhu Du**

*Department of Computer Science*
*University of Minnesota, USA*
*and Institute of Applied Mathematics*
*Academia Sinica, P R China*

**Frank Hwang**

*Mathematical Sciences Research Center*
*AT&T Bell Laboratories, USA*

**World Scientific**
*Singapore • New Jersey • London • Hong Kong*

*Published by*

World Scientific Publishing Co. Pte. Ltd.

P O Box 128, Farrer Road, Singapore 9128

*USA office:* Suite 1B, 1060 Main Street, River Edge, NJ 07661

*UK office:* 73 Lynton Mead, Totteridge, London N20 8DH

Library of Congress Cataloging-in-Publication data is available.

**COMPUTING IN EUCLIDEAN GEOMETRY**



ISBN 981-02-0966-5

Printed in Singapore by World Publications Printers Pte. Ltd.

# PREFACE

This volume begins with the article by Hsiang that traces the evolution of Euclidean geometry from Pythagoras to its modern setting of analytic geometry. It provides a perfect background for the seven survey papers on computational geometry that follow. Each of these articles provides a thorough survey of a major topic within computational geometry. We believe that this volume provides the reader with a good perspective on computational geometry and brings the reader to the forefront of research in some important areas of this field.

We briefly introduce the topics of the seven surveys. Bern and Eppstein survey finite-element mesh generation, which is indispensible for physical simulation. Here the emphasis is on optimal triangulations of geometry domains in two- and three-dimensions. An optimal triangulation is a partition of the domain into triangles or tetrahedras. Chou and Rathi survey algebraic methods for machine proofs of geometric theorems which have been highly successful in the past decade. They concentrate on Wu's method and its variants, and provide detailed examples for illustrations. Clarkson surveys randomized geometry algorithms, where the focus is not on random input, but on algorithms which use a source of random numbers. Randomization gives a general way to divide and conquer geometric problems, and thus often leads to either simple or asymptotically faster algorithms. Du and Hwang survey the two recent major results in Steiner minimal trees, i.e. shortest networks spanning given sets of points. The first is a proof of the Gilbert–Pollak $\sqrt{3}/2$ conjecture of the Steiner ratio on the Euclidean plane. The second is an approach to construct heuristic Steiner minimal trees which for the first time are probably shorter than minimum spanning trees. Fortune surveys a central subject in computational geometry: Voronoi diagrams, and their geometric duals, Delaunay triangulations. The goal is to demonstrate that there are practical algorithms for the construction of Voronoi diagrams in Euclidean planes and higher dimensions. Four fundamental algorithms are discussed in some detail, with a brief comparison included. Seidel looks into a practical topic "computer aided geometric designs", which is concerned with the construction of complex three-dimensional objects on a computer. Most of these objects are best described in terms of their surfaces, and the most popular class of surfaces is the class of parametric piecewise polynomials. Seidel surveys a class of such surfaces whose development is based on the study of polar forms. Finally, Smith and Winter survey topological network designs which concern the identification, number, and location of nodes and their interconnections in the design, analysis and synthesis of communication, transportation and distribution networks. Geometric topological network design problems along with their data structures and paradigms are classified into a complete and exhaustive morphology which could prove useful in future research.

Computational geometry aspires to two goals, solving practical problems and creating beautiful mathematics. We hope that this volume succeeds in meeting these two challenges. Some surveys in this volume may have emphasized the mathematical aspects and others the practical aspects. Our vision of computational geometry, which hopefully is evidenced by this volume, is of a field where computer science meets mathematics and where the real world problems which nurture geometry find solutions.

*D.-Z. Du and F. K. Hwang*

## CONTENTS

*Computing in Euclidean Geometry*, pp. 1–21
Eds. D.-Z. Du and F. K. Hwang


# On the Development of Quantitative Geometry from Pythagoras to Grassmann

BY
WU-YI HSIANG
DEPARTMENT OF MATHEMATICS
UNIVERSITY OF CALIFORNIA
BERKELEY, CA 94720

## §1. Introduction.

In this survey article, we shall trace the evolutionary development of the classical euclidean geometry from its ancient origin to its modern setting of analytic geometry. Roughly speaking, geometry studies the concept of *space* and the space is the *totality of all possible locations*. It is the space in which we live and share with everything else in the universe. Therefore, a good understanding of the properties of the space is obviously fundamental and indispensable for both science and technology. It is rather natural that geometry is, indeed, the number one science: the most fundamental and oldest one among all sciences of human civilization. The development of geometry is actually one of the most glorious chapters of the entire history of the rational mind. It can be roughly summarized into the following four stages, namely, (i) experimental geometry, (ii) qualitative geometry, (iii) quantitative geometry and (iv) analytic (vector) geometry.

EXPERIMENTAL GEOMETRY: Although geometry had already attained the stature of a highly deductive science even at the time of Euclid, its origin and its foundation were *(of necessity) rooted in* basic *experimental facts* (sometimes called geometric intuition) on the fundamental properties of the space in which we live. For examples, the side-angle-side (S.A.S.) congruence condition of triangles and the experimental fact that the sum of the three inner angles of *all* triangles are always equal to a *flat angle.*

QUALITATIVE GEOMETRY: In the early stage of geometry, one seeks some basic *qualitative* understanding to begin with. The *congruences* of some basic geometric configurations such as triangles, tetrahedrons, circles etc. and the *parallelism* are the two types of basic *geometric equalities* which naturally constitute the central topic of the qualitative study of geometry. In the study of congruences, the *S.A.S. congruence condition* of triangles provides the foundational basic property while the logical equivalences between several *characteristic properties* of *isosceles triangles* constitute the sole fundamental lemma that all kinds of proofs of congruences will be based upon. In the study of parallelism, the *flatness of the sum of the three inner angles* of an arbitrary triangle provides the foundational basic property while the logical equivalences between several *characteristic properties* of *parallelograms* constitute the sole fundamental lemma that all kinds of proofs involving parallelism will be based upon.

QUANTITATIVE GEOMETRY: After achieving a solid foundation of qualitative geometry, it is rather natural to upgrade the study of geometry to the level of *quantitative correlations* among various geometric quantities such as length, angle, area, volume etc. For examples, the area of a rectangle is equal to the product of its length and its width, the famous Pythagoras Theorem and the powerful theorem of similar triangles. The latter are the two fundamental theorems in quantitative geometry and they can easily be deduced from the former. Therefore, this very simple area formula of rectangles already constitutes the logical foundation of the entire quantitative geometry (cf. §2).

ANALYTIC (VECTOR) GEOMETRY: Historically, the transition from deductive geometry to analytic geometry was first achieved by the method of coordinization by Descartes and Fermat in the seventeenth century. The concept of vector was introduced by Hamilton and Grassmann in the nineteenth century. The *Grassmann algebra* constitutes a systematic algebraization of the entire space structure thus providing a concise efficient setting for the study of $n$-dimensional analytic geometry. In the setting of Grassmann algebra, a *decomposable $k$-vector*, $\mathbf{a}_1 \wedge \cdots \wedge \mathbf{a}_k$, is simply the algebraic representation of the geometric quantity of the *$k$-dimensional parallelotope* spanned by $\mathbf{a}_1, \ldots, \mathbf{a}_k$, $//(\mathbf{a}_1, \ldots, \mathbf{a}_k)$, and all the fundamental theorems of quantitative geometry are simply transformed into the basic laws of operation (cf. §3). For example, the similar triangle theorem is transformed into the distributive law of *scalar product*, i.e., $k\cdot(\mathbf{a}+\mathbf{b}) = k\cdot\mathbf{a}+k\cdot\mathbf{b}$, and the Pythagoras theorem is transformed into the distributive law of *inner product*, namely, $\mathbf{a}\cdot(\mathbf{b}+\mathbf{c}) = \mathbf{a}\cdot\mathbf{b}+\mathbf{a}\cdot\mathbf{c}$. Moreover, one of the most fundamental and most useful laws of operation in the Grassmann algebra is the following formula of *high dimensional innere product*, namely

$$(1) \qquad \langle \mathbf{a}_1 \wedge \cdots \wedge \mathbf{a}_k, \mathbf{b}_1 \wedge \cdots \wedge \mathbf{b}_k \rangle = \det(\mathbf{a}_i \cdot \mathbf{b}_j).$$

As one shall see later, the above $k$-dimensional inner product formula is exactly the $k$-dimensional generalization of the classical Pythagoras theorem! In retrospect, the classical Pythagoras theorem and its modern $k$-dimensional generalizations already constitutes a wholesome set of basic theorems as well as a set of powerful tools for the study of high dimensional analytic geometry. It is amusing to note that it took a span of two thousand years to progress from the Pythagoras theorem to its high dimensional generalizations and it will probably take at least several more decades to fully develop their vast potential in applications.

## §2. Foundation of quantitative geometry.

In qualitative geometry, one studies equalities and inequalities among geometric configurations, e.g., congruence among triangles and parallelism among directions. For example, one only qualifies the relationship between two given intervals to be either equal, shorter or longer, but no quantitative ratio is attaching to them to measure their difference in length in the latter two cases. In quantitative geometry, one introduces precise measurements of various kinds of geometric quantities such as length, angle, area, volume etc. and then one studies the quantitative correlations among various geometric quantities of some basic geometric configurations. Therefore, the first foundational problem is the *concept of measurement* itself, e.g., the measurement of the length of intervals.

### (2.1) Commensurability and non-commensurability.

Since the length of an interval is clearly the most fundamental geometric quantity, let us consider the concept of the measurement of the lengths of intervals as a typical example. Recall that two intervals $a$ and $b$ are defined to be *commensurable* if there exists a "common yardstick" $c$ such that both $a$ and $b$ are *integral multiples* of $c$, say $a = m \cdot c$ and $b = n \cdot c$; otherwise, $a$ and $b$ are said to be *non-commensurable*. One of the fundamental theoretical problems that the geometers of ancient Greece encountered in their first attempt to build a foundation of quantitative geometry was the following problem on commensurability, namely

COMMENSURABILITY PROBLEM: Is it true that any pair of intervals $a, b$ are always commensurable?

If the answer to the above problem is affirmative, then the concept of the measurement of the lengths of intervals is quite simple and the ratio between the lengths of two intervals will always be a rational number. In fact, geometers of the Pythagorean era took the "universal validity of commensurability" as an *axiom* and then erected a rather impressive fundamental theory of quantitative geometry based upon such an axiom, only to discover later the invalidity of such an axiom! Initially, the discovery of *non-commensurable pairs* of intervals was, of course, a terrible shock to them, but it also turned out to be the most important critical junction which uplifted Greek geometry to a much higher plane.

### (2.2) The area formula of rectangles, Pythagoras theorem and the similar triangle theorem.

The well-known area formula of rectangles asserts that the area of a rectangle is equal to the product of its length and its width, namely

$$A = l \cdot w. \tag{2}$$

Suppose $u$ is the chosen unit of length and the length and the width of a given rectangle is equal to $l$ and $w$ times the unit $u$ respectively. Then the area of such a rectangle is equal to $l \cdot w$ times the area of the square of unit side-length which is the natural unit of area corresponding to the chosen unit of length.

REMARKS: (i) The above seemingly obvious formula actually needs some proof; the special case that both $l$ and $w$ are integral (or rational) can be proved by parallel subdivision of the given rectangles into small squares (or rectangles) while the proof of the general case (i.e. $l$ and/or $w$ are irrational) is directly correlated with the definition of the product of two real numbers.

(ii) The above simple area formula is not only the logical basis of all the other area formulas in geometry, but it is also the logical foundation of the two most fundamental theorems of quantitative geometry, namely, Pythagoras theorem and the similar triangle theorem. The following are the proofs of the above two theorems by straightforward area formula computations:

PROOF OF THE PYTHAGORAS THEOREM: The well-known chinese proof shows that a square of side-length $a+b$ can be cut into four right-angled triangles of side lengths $a$, $b$ and $c$ and a square of side-length $c$. Therefore

$$(a+b)^2 = c^2 + 4 \cdot \frac{1}{2}ab \Rightarrow a^2 + b^2 = c^2. \tag{3}$$

PROOF OF THE SIMILAR TRIANGLE THEOREM: Let $\Delta ABC$ and $\Delta A'B'C'$ be two triangles with equal corresponding angles. We may assume that they are already in the relative position as indicated in Figure 1.

Following the tradition of chinese geometry, let us compute the area of the trapezoid $BCC'B'$ in two different ways, namely

$$\begin{aligned}\text{trapezoid } BCC'B' &= \Delta ABC - \Delta A'B'C' = \frac{1}{2}ah - \frac{1}{2}a'h' \\ &= \Delta BCC' + \Delta BC'B' = \frac{1}{2}a(h-h') + \frac{1}{2}a'(h-h'),\end{aligned} \tag{4}$$

thus obtaining

$$ah' = a'h \text{ or } \frac{a}{a'} = \frac{h}{h'}. \tag{4'}$$

Therefore

$$\frac{\Delta ABC}{\Delta A'B'C'} = \frac{\frac{1}{2}ah}{\frac{1}{2}a'h'} = \left(\frac{a}{a'}\right)^2 \tag{5}$$

and hence

$$\begin{aligned}&\left(\frac{a}{a'}\right)^2 = \left(\frac{b}{b'}\right)^2 = \left(\frac{c}{c'}\right)^2 = \frac{\Delta ABC}{\Delta' ABC} \\ &\Rightarrow \frac{a}{a'} = \frac{b}{b'} = \frac{c}{c'}.\end{aligned} \tag{5'}$$

□

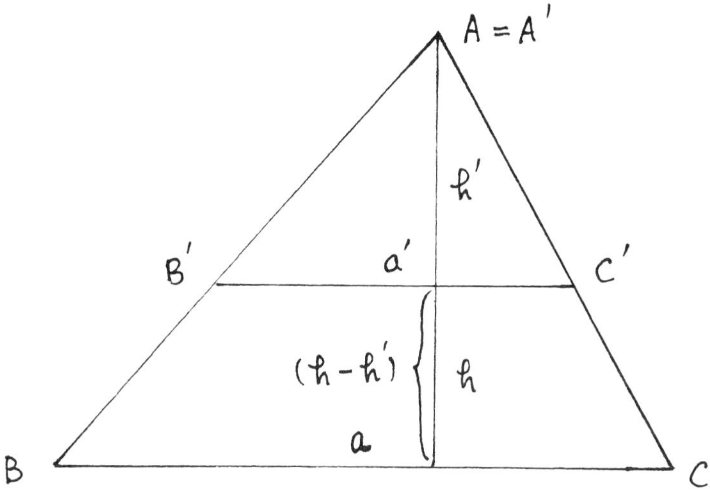


Figure 1

**(2.3) Euclidean trigonometry.**

Triangles are the simplest and also the most basic type of geometric configurations. Therefore, trigonometry, i.e. the quantitative geometry of triangles, is naturally of fundamental importance in the entire quantitative study of geometry, especially the following two basic laws of trigonometry, namely

LAWS OF SINE:

$$\frac{\sin A}{a} = \frac{\sin B}{b} = \frac{\sin C}{c} = \frac{2\Delta}{abc} = \frac{1}{2R} \tag{6}$$

where $\Delta$ is the area and $R$ is the circumradius of $\Delta ABC$.

LAWS OF COSINE:

$$\begin{cases} a\cos B + b\cos A = c \\ a\cos C + c\cos A = b \\ b\cos C + c\cos B = a \end{cases} \Rightarrow (7') \quad \begin{cases} \cos A = \frac{b^2+c^2-a^2}{2bc} \\ \cos B = \frac{a^2+c^2-b^2}{2ac} \\ \cos C = \frac{a^2+b^2-c^2}{2ab} \end{cases} \tag{7}$$

It is straightforward to use $(7')$ to show that

$$\frac{\sin^2 A}{a^2} = \frac{16s(s-a)(s-b)(s-c)}{4a^2b^2c^2}, \quad s = \frac{1}{2}(a+b+c) \tag{8}$$

and hence (by (8) and (6))

$$\Delta = \sqrt{s(s-a)(s-b)(s-c)}. \tag{9}$$

## §3. Foundation of analytic (vector) geometry.

In this section, we shall briefly describe a systematic algebraization of the entire structure of the space. The algebraic structure that it naturally leads to is the *Grassmann algebra* which is the foundation of high dimensional analytic euclidean geometry.

### (3.1) Displacement vectors.

The space is, in essence, the *totality of all possible locations* of the universe in which we live. Thus, locations are the most primitive type of geometric concept and the *difference in locations* will be the most fundamental type of geometric quantity, usually called *displacement vectors*. Recall that the most fundamental structure of the space is exactly the *unique existence* of the *shortest pathway* between an arbitrary pair of locations which is the *interval* joining the pair of points representing the given pair of locations. Therefore, the interval with direction, $\overrightarrow{AB}$, is a vivid representation of the difference in locations between $A$ and $B$. Hence, it is quite natural to *quantify* the difference in location by the following definition of the displacement vector.

DEFINITION: The differences in locations between $A, B$ and between $A', B'$ are defined to be equal if $\overrightarrow{AB}$ and $\overrightarrow{A'B'}$ have the same length and the same direction. A mapping $\tau$ of the entire space $E$ into itself is caled a *translation* if the differences in locations between $A$ and $\tau(A)$ are the same for all points $A$ in the space, namely, a translation is a motion of the entire space that all points are moved by the same amount of displacement. A translation is also called a *displacement vector*.

LEMMA 1. *The composition of two translations $\tau_1, \tau_2$ is again a translation.*

PROOF: Let $A$ and $B$ be two arbitrary points in the space and set $A' = \tau_1(A)$, $B' = \tau_1(B)$, $A'' = \tau_2(A')$, $B'' = \tau_2(B')$. Then $A'' = \tau_2 \circ \tau_1(A)$, $B'' = \tau_2 \circ \tau_1(B)$ and $\overrightarrow{AA'}$, $\overrightarrow{BB'}$ (resp. $\overrightarrow{A'A''}$, $\overrightarrow{B'B''}$) have the same length and the same direction. Therefore both $\square ABB'A'$ and $\square A'B'B''A''$ are parallelograms and hence $\overrightarrow{AB}$, $\overrightarrow{A'B'}$ and $\overrightarrow{A''B''}$ also have the same length and the same direction. This proves that $\square ABB''A''$ is also a parallelogram and $\overrightarrow{AA''}$, $\overrightarrow{BB''}$ have the same length and the same direction, thus verifying that $\tau_2 \circ \tau_1$ is again a translation. $\square$

REMARKS: (i) To any given pair of points $A, B$, there exists a unique translation $\tau$ with $\tau(A) = B$, it is a convenient notation to use $\overrightarrow{AB}$ to denote such a unique translation, i.e., displacement vector.

(ii) It is easy to verify that $\tau_1 \circ \tau_2 = \tau_2 \circ \tau_1$ and we shall define it to be the sum of $\tau_1$ and $\tau_2$, namely

DEFINITION: The sum of two translations $\tau_1, \tau_2$ is defined to be their composition, i.e., $\tau_1 + \tau_2 = \tau_1 \circ \tau_2 = \tau_2 \circ \tau_1$. For example, $\overrightarrow{AB} + \overrightarrow{BC} = \overrightarrow{AC}$.

NOTATION: From now on, we shall always use boldface latin latters such as $\mathbf{a}, \mathbf{b}$ etc. or arrow-intervals such as $\overrightarrow{AB}$, $\overrightarrow{BC}$, etc. to denote displacement vectors. Moreover, $n \cdot \mathbf{a}$ means the sum of $n$ $\mathbf{a}$'s and $\frac{m}{n} \cdot \mathbf{a}$ is the unique solution of the equation

$$n \cdot \mathbf{x} = m \cdot \mathbf{a}. \tag{10}$$

It is easy to verify that

$$\begin{cases} \frac{m}{n} \cdot \mathbf{a} + \frac{p}{q} \cdot \mathbf{a} = \left(\frac{m}{n} + \frac{p}{q}\right) \cdot \mathbf{a} \\ \frac{m}{n} \cdot \left(\frac{p}{q} \cdot \mathbf{a}\right) = \left(\frac{m}{n} \cdot \frac{p}{q}\right) \cdot \mathbf{a} \\ \frac{m}{n}(\mathbf{a} + \mathbf{b}) = \frac{m}{n} \cdot \mathbf{a} + \frac{m}{n} \cdot \mathbf{b} \end{cases} \tag{11}$$

for all fractional number $\frac{m}{n}, \frac{p}{q}$ and all vectors $\mathbf{a}, \mathbf{b}$. Therefore, it is quite natural to extend the meaning of the above fractional multiple of a vector by the following definition of *scalar product*.

DEFINITION: Let $\mathbf{a} = \overrightarrow{AB}$ be an arbitrary non-zero vector and $\lambda$ be an arbitrary positive real number. The scalar product $\lambda \cdot \mathbf{a}$ is defined to be the vector with the same direction of $\overrightarrow{AB}$ whose length is equal to $\lambda$ times of the length of $\overline{AB}$, namely, $\lambda \cdot \mathbf{a} = \overrightarrow{AC}$ where $C$ is located on the ray of $\overrightarrow{AB}$ and with $\overline{AC} : \overline{AB} = \lambda$. Moreover, $(-\lambda) \cdot \mathbf{a}$ is defined by $\lambda \cdot (-\mathbf{a}) = \lambda \cdot \overrightarrow{BA}$ and $0 \cdot \mathbf{a}$ or $\lambda \cdot \mathbf{o}$ are defined to be the zero vector $\mathbf{o}$.

It follows from the above definition and (11) that the scalar product has the following powerful laws of operation, namely

$$\begin{cases} (\lambda + \mu) \cdot \mathbf{a} = \lambda \cdot \mathbf{a} + \mu \cdot \mathbf{a} \\ \lambda \cdot (\mu \cdot \mathbf{a}) = (\lambda \cdot \mu) \cdot \mathbf{a} \\ \lambda \cdot (\mathbf{a} + \mathbf{b}) = \lambda \cdot \mathbf{a} + \lambda \cdot \mathbf{b}. \end{cases} \tag{12}$$

REMARK: The above scalar product is exactly the algebraization of the homothety magnification applied to the basic object of directed intervals, and the above laws of operation is the algebraization of the basic properties (or theorems) on homothety magnification. For

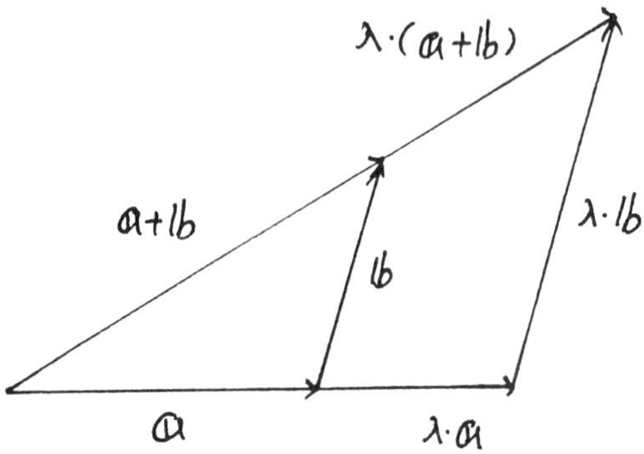


Figure 2

example, the distributive law $\lambda \cdot (\mathbf{a} + \mathbf{b}) = \lambda \cdot \mathbf{a} + \lambda \cdot \mathbf{b}$ is exactly the algebraization of the fundamental similar triangle theorem (cf. Figure 2)

Recall that a displacement vector $\mathbf{a}$ contains both a length and a direction. We shall use the notation $|\mathbf{a}|$ to denote its length and the notation $\angle\mathbf{a}, \mathbf{b})$ to denote the angle which measures the difference between their two directions. In terms of vectors, the fundamental theorem of Pythagoras asserts that

$$\angle\mathbf{a}, \mathbf{b}) = \pi/2 \Rightarrow |\mathbf{a} \pm \mathbf{b}|^2 = |\mathbf{a}|^2 + |\mathbf{b}|^2. \tag{13}$$

LEMMA 2. *For any pair of vectors* $\mathbf{a}, \mathbf{b}$

$$|\mathbf{a} + \mathbf{b}|^2 + |\mathbf{a} - \mathbf{b}|^2 = 2|\mathbf{a}|^2 + 2|\mathbf{b}|^2. \tag{14}$$

PROOF: It is clear that (14) holds for the special cases that $\angle\mathbf{a}, \mathbf{b}) = 0$, $\pi/2$ or $\pi$. In the general case, set $\mathbf{b} = \mathbf{b}_1 + \mathbf{b}_2$ such that $\angle\mathbf{a}, \mathbf{b}_1) = 0$ or $\pi$ and $\angle\mathbf{a}, \mathbf{b}_2) = \pi/2$. Then

$$\begin{aligned} |\mathbf{a} + \mathbf{b}^2| + |\mathbf{a} - \mathbf{b}|^2 &= |\mathbf{a} + \mathbf{b}_1|^2 + |\mathbf{a} - \mathbf{b}_1|^2 + 2|\mathbf{b}_2|^2 \\ &= 2|\mathbf{a}|^2 + 2|\mathbf{b}_1|^2 + 2|\mathbf{b}_2|^2 \\ &= 2|\mathbf{a}|^2 + 2|\mathbf{b}|^2. \end{aligned} \tag{15}$$

□

LEMMA 2′. *For any triple of vectors* $\mathbf{a}, \mathbf{b}, \mathbf{c}$

$$|\mathbf{a} + \mathbf{b} + \mathbf{c}|^2 - |\mathbf{a} + \mathbf{b}|^2 - |\mathbf{b} + \mathbf{c}|^2 - |\mathbf{c} + \mathbf{a}|^2 + |\mathbf{a}|^2 + |\mathbf{b}|^2 + |\mathbf{c}|^2 = 0. \tag{16}$$

PROOF: It follows directly from (14) that

$$\left\{\begin{array}{l} |\mathbf{a}+\mathbf{b}+\mathbf{c}|^2+|\mathbf{a}+\mathbf{b}-\mathbf{c}|^2-2|\mathbf{a}+\mathbf{b}|^2-2|\mathbf{c}|^2=0, \\ -|\mathbf{a}-\mathbf{b}+\mathbf{c}|^2-|\mathbf{a}+\mathbf{b}-\mathbf{c}|^2+2|\mathbf{a}|^2+2|\mathbf{b}-\mathbf{c}|^2=0, \\ |\mathbf{a}-\mathbf{b}+\mathbf{c}|^2+|\mathbf{a}+\mathbf{b}+\mathbf{c}|^2-2|\mathbf{a}+\mathbf{c}|^2-2|\mathbf{b}|^2=0, \\ -2|\mathbf{b}+\mathbf{c}|^2-2|\mathbf{b}-\mathbf{c}|^2+4|\mathbf{b}|^2+4|\mathbf{c}|^2=0. \end{array}\right. \tag{17}$$

The sum of the above four identities is exactly twice of (16). □

DEFINITION: Set

$$\mathbf{a}\cdot\mathbf{b}=\frac{1}{2}\{|\mathbf{a}+\mathbf{b}|^2-|\mathbf{a}|^2-|\mathbf{b}|^2\} \tag{18}$$

and call it the *inner product* of **a** and **b**.

The above inner product satisfies the following basic laws of operation, namely

$$\left\{\begin{array}{l} \mathbf{a}\cdot\mathbf{b}=\mathbf{b}\cdot\mathbf{a},\ \mathbf{a}\cdot\mathbf{a}=|\mathbf{a}|^2 \\ \mathbf{a}\cdot(\mathbf{b}+\mathbf{c})=\mathbf{a}\cdot\mathbf{b}+\mathbf{a}\cdot\mathbf{c} \\ \mathbf{a}\cdot(\lambda\cdot\mathbf{b})=\lambda\cdot(\mathbf{a}\cdot\mathbf{b}). \end{array}\right. \tag{19}$$

REMARKS: (i) Lemma 2 is usually called the generalized Pythagoras theorem.

(ii) The above distributive law of inner product can be considered as the optimal algebraization of the generalized Pythagoras theorem.

(iii) $\mathbf{a}\cdot\mathbf{b}=0$ if and only if $\angle\mathbf{a},\mathbf{b})=\pi/2$, and $\mathbf{a}\cdot\mathbf{b}=|\mathbf{a}|\cdot|\mathbf{b}|$ (resp. $-|\mathbf{a}|\cdot|\mathbf{b}|$) if $\angle\mathbf{a},\mathbf{b})=0$ (resp. $\pi$).

(iv) In the general case, one decomposes **b** into $\mathbf{b}_1+\mathbf{b}_2$ by the orthogonal projection used in the proof of Lemma 2 and then

$$\mathbf{a}\cdot\mathbf{b}=\mathbf{a}\cdot(\mathbf{b}_1+\mathbf{b}_2)=\mathbf{a}\cdot\mathbf{b}_1+\mathbf{a}\cdot\mathbf{b}_2=\mathbf{a}\cdot\mathbf{b}_1. \tag{20}$$

Summarizing the discussion of this section, the set of displacement vectors together with the above three basic operations, namely, addition, scalar product and inner product, constitute a type of algebraic structure called vector spaces with inner product (or inner product spaces). The displacement vectors quantify the *differences in locations* while the three basic operations algebraize the fundamental structures of the space. Such a systematic algebraization enables us to transform the fundamental theorems in quantitative geometry such as the similar triangle theorem and Pythagoras theorem into simple and extremely powerful laws of operations such as the distributive laws of the scalar product and the inner product.

**(3.2) Oriented volume and high dimensional inner product.**

Geometrically, the displacement vectors are the one-dimensional basic objects represented by intervals with directions. Of course, we shall also need to study higher dimensional basic geometric objects such as parallelograms or parallelotopes with directions. Let us begin with a brief analysis of the area of parallelograms in a given euclidean plane.

It is natural to consider a parallelogram $\square ABCD$ to be spanned by the pair of vectors $\overrightarrow{AB}$ and $\overrightarrow{AD}$ and its area as a function of its generating pair of vectors. However, there is a choice between the *absolute area* which is always positive and the *oriented area* which has a sign depending on the orientation of the parallelogram, namely, the orientation of $//(\mathbf{a},\mathbf{b})$ is positive (resp. negative) if $\angle \mathbf{a},\mathbf{b})$ is positive (resp. negative). Set $A(\mathbf{a},\mathbf{b})$ (resp. $A'(\mathbf{a},\mathbf{b})$) to be the oriented (resp. absolute) area of the parallelogram $//(\mathbf{a},\mathbf{b})$. Then the following is a brief comparison between their basic properties:

(20)
(i) $A(\mathbf{a},\mathbf{b}) = -A(\mathbf{b},\mathbf{a})$; $A'(\mathbf{a},\mathbf{b}) = A'(\mathbf{b},\mathbf{a})$
(ii) $A(\mathbf{a},\lambda\cdot\mathbf{b}) = \lambda A(\mathbf{a},\mathbf{b})$; $A'(\mathbf{a},\lambda\cdot\mathbf{b}) = |\lambda|\cdot A(\mathbf{a},\mathbf{b})$
(iii) **Shear invariance:** $A(\mathbf{a},\mathbf{b}+\mu\mathbf{a}) = A(\mathbf{a},\mathbf{b})$; $A'(\mathbf{a},\mathbf{b}+\mu\mathbf{a}) = A'(\mathbf{a},\mathbf{b}+\mu\mathbf{a})$

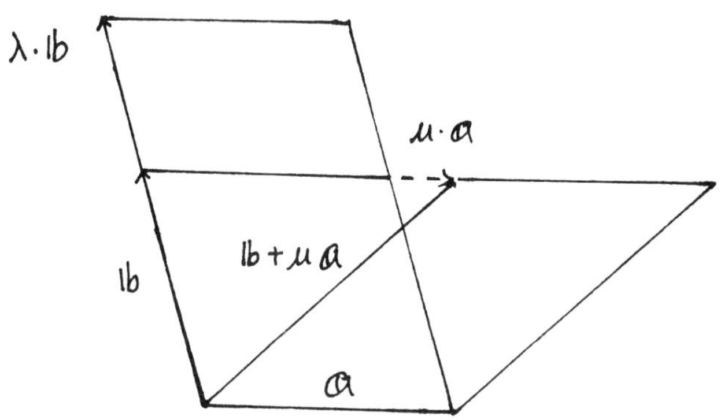


Figure 3

At first glance, the absolute area is slightly better in (i) while the oriented area is slightly better in (ii) than the other one, and both are shear invariant. However, a deeper analysis will show that the oriented area will have an overwhelming advantage as compared to the absolute area, namely, it follows from (20)-(ii),(iii) of $A$ that

$$A(\mathbf{a},\mathbf{b}+\mathbf{c}) = A(\mathbf{a},\mathbf{b}) + A(\mathbf{a},\mathbf{c}). \tag{20'}$$

and hence $A$ is a *bilinear* function of $\mathbf{a}$ and $\mathbf{b}$!

Similar discussion will show that the *oriented* $n$-dimensional volume of an $n$-dimensional parallelotope $//(\mathbf{a}_1, \ldots, \mathbf{a}_n)$ also has this overwhelming advantage of multi-linearity. We state it as the following Lemma 3.

LEMMA 3. *Let $\mathbf{a}_1, \ldots, \mathbf{a}_n$ be $n$ vectors in an $n$-dimensional euclidean space and $\mathbf{e}_1, \ldots, \mathbf{e}_n$ be an orthonormal basis of positive orientation. Then the oriented $n$-dimensional volume of the $n$-parallelotope $//(\mathbf{a}_1, \ldots, \mathbf{a}_n)$ is a multilinear alternating function of $(\mathbf{a}_1, \ldots, \mathbf{a}_n)$ whose value at $(\mathbf{e}_1, \ldots, \mathbf{e}_n)$ is 1, namely*

(i) *$v(\mathbf{a}_1, \ldots, \mathbf{a}_n)$ is linear with respect to each vector variable $\mathbf{a}_j$, $1 \leq j \leq n$,*
(ii) *$v(\mathbf{a}_1, \ldots, \mathbf{a}_n)$ is skew symmetric,*
(iii) *$v(\mathbf{e}_1, \ldots, \mathbf{e}_n) = 1$.*

REMARK: It is well-known that the above three properties have already constituted a set of *characteristic properties of the determinant function.* Therefore, the oriented $n$-dimensional volume of $//(\mathbf{a}_1, \ldots, \mathbf{a}_n)$ is, in fact, just equal to $\det(\mathbf{a}_1, \ldots, \mathbf{a}_n)$.

Next let us consider the problem of how to generalize the definition of inner product of 1-dimensional vectors to higher dimensional parallelotopes. Recall that there are essentially two different ways of defining the inner product for the 1-dimensional case, namely

(i) $\mathbf{a} \cdot \mathbf{b} = \frac{1}{2}\{\|\mathbf{a} + \mathbf{b}|^2 - |\mathbf{a}|^2 - |\mathbf{b}|^2\}$
(ii) $\mathbf{a} \cdot \mathbf{b} = \mathbf{a} \cdot \mathbf{b}_1 =$ the product of the oriented lengths of $\mathbf{a}$ and $\mathbf{b}_1$ where $\mathbf{b}_1$ is the orthogonal projection of $\mathbf{b}$ on to the line generated by $\mathbf{a}$ (cf. remark (iv) following (19)).

Since the concept of the sum of two $k$-dimensional parallelotope is still missing, it is difficult to generlize the first one to higher dimensional case. Therefore, it is only natural to try the second approach as follows.

Let $//(\mathbf{a}_1, \ldots, \mathbf{a}_k)$ and $//(\mathbf{b}_1, \ldots, \mathbf{b}_k)$ be two given $k$-dimensional parallelotopes in an $n$-dimensional euclidean space. Choose an orthonormal basis $\{\mathbf{e}_1, \ldots, \mathbf{e}_n\}$ such that the span of $\mathbf{e}_1, \ldots, \mathbf{e}_k$ is the same as the span of $\mathbf{a}_1, \ldots, \mathbf{a}_k$.

Set

$$\mathbf{b}'_j = \sum_{i=1}^{k} (\mathbf{b}_j \cdot \mathbf{e}_i)\mathbf{e}_i, \ \mathbf{b}''_j = \mathbf{b}_j - \mathbf{b}'_j, \ 1 \leq j \leq k. \tag{21}$$

Then $\mathbf{b}'_j$ is exactly the *orthogonal projection* of $\mathbf{b}_j$ and $//(\mathbf{b}'_1, \ldots, \mathbf{b}'_k)$ is exactly the *orthogonal projection* of $//(\mathbf{b}_1, \ldots, \mathbf{b}_k)$ onto the subspace (i.e. $k$-plane) spanned by $//(\mathbf{a}_1, \ldots, \mathbf{a}_k)$. In particular

$$\mathbf{a}_i \cdot \mathbf{b}''_j = 0, \ \mathbf{a}_i \cdot \mathbf{b}_j = \mathbf{a}_i \cdot \mathbf{b}'_j \tag{22}$$

for all $1 \leq i, j \leq k$.

DEFINITION: The inner product of $//(\mathbf{a}_1, \ldots, \mathbf{a}_k)$ and $//(\mathbf{b}_1, \ldots, \mathbf{b}_k)$ is defined to be the product of the oriented $k$-dimensional volume of $//(\mathbf{a}_1, \ldots, \mathbf{a}_k)$ and $//(\mathbf{b}'_1, \ldots, \mathbf{b}'_k)$ which are two $k$-dimensional parallelotopes belonging to the same $k$-plane. We shall denote it by $\langle //(\mathbf{a}_1, \ldots, \mathbf{a}_k), //(\mathbf{b}_1, \ldots, \mathbf{b}_k)\rangle$.

THEOREM 1. *The above inner product of two $k$-dimensional parallelotopes can be computed by the following formula*

$$\langle //(\mathbf{a}_1, \ldots, \mathbf{a}_k), //(\mathbf{b}_1, \ldots, \mathbf{b}_k)\rangle = \det(\mathbf{a}_i \cdot \mathbf{b}_j).$$

PROOF: Let $U$ be the $k$-plane spanned by $//(\mathbf{a}_1, \ldots, \mathbf{a}_k)$ and $(\mathbf{e}_1, \ldots, \mathbf{e}_k)$ be an arbitrarily chosen orthonormal basis of $U$. Set

(23)
$$a_{ij} = \mathbf{a}_i \cdot \mathbf{e}_j,\ b_{ij} = \mathbf{b}_i \cdot \mathbf{e}_j,\ 1 \leq i, j \leq k$$
$$\mathbf{b}'_i = \sum_{j=1}^{k} b_{ij}\mathbf{e}_j,\ \mathbf{b}''_i = \mathbf{b}_i - \mathbf{b}'_i.$$

Then $//(\mathbf{b}'_1, \ldots, \mathbf{b}'_k)$ is the orthogonal projection of $//(\mathbf{b}_1, \ldots, \mathbf{b}_k)$ onto $U$ and the inner product of $//(\mathbf{a}_1, \ldots, \mathbf{a}_k)$ and $//(\mathbf{b}_1, \ldots, \mathbf{b}_k)$ is defined to be the product of the oriented $k$-dimensional volume of $//(\mathbf{a}_1, \ldots, \mathbf{a}_k)$ and $//(\mathbf{b}'_1, \ldots, \mathbf{b}'_k)$, namely

(24)
$$\langle //(\mathbf{a}_1, \ldots, \mathbf{a}_k), //(\mathbf{b}_1, \ldots, \mathbf{b}_k)\rangle = |a_{ij}| \cdot |b_{ij}| = |c_{ij}|$$

where

(24′)
$$c_{ij} = \sum_{l=1}^{k} a_{il} \cdot b_{jl} = \mathbf{a}_i \cdot \mathbf{b}'_j = \mathbf{a}_i \cdot \mathbf{b}_j.$$

□

REMARK: Recall that the *bilinearity* of the inner product is essentially the optimal algebraic reformulation of the Pythagoras theorem. Therefore, the above theorem is actually a far-reaching generalization of the Pythagoras theorem, which is already in its optimal algebraic form of a concise multilinear formula.

### (3.3) The Grassmann algebra.

With the above fundamental inner product formula at hand, it is then quite straightforward to *formalize* the algebra of $k$ dimensional parallelotopes, $1 \leq k \leq n =$ dimension of the space, into a comprehensive, concise algebraic system, namely, the *Grassmann algebra*.

DEFINITION: Two $k$-dimensional parallelotopes $//(\mathbf{a}_1, \ldots, \mathbf{a}_k)$ and $//(\mathbf{b}_1, \ldots, \mathbf{b}_k)$ are defined to be equivalent if they are *parallel* (i.e. belong to the same $k$-plane up to a translation) and of the same oriented $k$-dimensional volume. We shall denote the *equivalence class* of $//(\mathbf{a}_1, \ldots, \mathbf{a}_k)$ by $\mathbf{a}_1 \wedge \cdots \wedge \mathbf{a}_k$ and call it the *exterior product* of $\mathbf{a}_1, \ldots, \mathbf{a}_k$.

LEMMA 4. (i) *Every $k$-dimensional parallelotope $//(\mathbf{a}_1, \ldots, \mathbf{a}_k)$ is equivalent to a rectangular one, namely, $//(\bar{\mathbf{a}}_1, \ldots, \bar{\mathbf{a}}_k)$ with $\bar{\mathbf{a}}_i \cdot \bar{\mathbf{a}}_j = 0$ for $1 \leq i < j \leq k$.*

(ii) *Two $k$-dimensional parallelotopes $//(\mathbf{a}_1, \ldots, \mathbf{a}_k)$ and $//(\mathbf{b}_1, \ldots, \mathbf{b}_k)$ are equivalent if and only if*

$$\langle //(\mathbf{a}_1, \ldots, \mathbf{a}_k), //(\mathbf{x}_1, \ldots, \mathbf{x}_k) \rangle = \langle //(\mathbf{b}_1, \ldots, \mathbf{b}_k), //(\mathbf{x}_1, \ldots, \mathbf{x}_k) \rangle \tag{25}$$

*for all possible $k$-dimensional parallelotopes $//(\mathbf{x}_1, \ldots, \mathbf{x}_k)$.*

PROOF: The first assertion follows easily from the *shear invariance* of $k$-dimensional oriented volume and the well-known Gram–Schmidt orthonormalization process.

It follows from the definition that (25) is clearly a necessary condition for the equivalence of $//(\mathbf{a}_1, \ldots, \mathbf{a}_k)$ and $//(\mathbf{b}_1, \ldots, \mathbf{b}_k)$. Therefore, one needs only to prove that it is also sufficient. We shall assume without loss of generality that both of them are already *rectangular* to begin with. Then, it follows from (25) that

$$\begin{aligned} \prod_{i=1}^{k} |\mathbf{a}_i|^2 &= \langle //(\mathbf{a}_1, \ldots, \mathbf{a}_k), //(\mathbf{a}_1, \ldots, \mathbf{a}_k) \rangle \\ &= \langle //(\mathbf{b}_1, \ldots, \mathbf{b}_k), //(\mathbf{a}_1, \ldots, \mathbf{a}_k) \rangle \\ &= \langle //(\mathbf{b}_1, \ldots, \mathbf{b}_k), //(\mathbf{b}_1, \ldots, \mathbf{b}_k) \rangle \\ &= \prod_{i=1}^{k} |\mathbf{b}_i|^2. \end{aligned} \tag{26}$$

Therefore

$$\prod_{i=1}^{k} |\mathbf{a}_i| = \prod_{i=1}^{k} |\mathbf{b}_i|. \tag{26$'$}$$

On the other hand, it is easy to see that

$$\langle //(\mathbf{a}_1, \ldots, \mathbf{a}_k), //(\mathbf{b}_1, \ldots, \mathbf{b}_k) \rangle \leq \prod_{i=1}^{k} |\mathbf{a}_i| \cdot \prod_{i=1}^{k} |\mathbf{b}'_i| \tag{27}$$

where

$$|\mathbf{b}_i|^2 = |\mathbf{b}'_i|^2 + |\mathbf{b}''_i|^2, \; 1 \leq i \leq k. \tag{27$'$}$$

Therefore, (26), (26′), (27) and (27′) imply that

$$\mathbf{b}_i = \mathbf{b}'_i \text{ (i.e., } \mathbf{b}''_i = 0), \; 1 \leq i \leq k$$

and hence

$$//(\mathbf{a}_1, \ldots, \mathbf{a}_k) \sim //(\mathbf{b}_1, \ldots, \mathbf{b}_k).$$

□

LEMMA 5. *Let* $\{\mathbf{e}_1, \ldots, \mathbf{e}_n\}$ *be an orthonormal basis. Then* (i) $//(\mathbf{a}_1, \ldots, \mathbf{a}_k) \sim //(\mathbf{b}_1, \ldots, \mathbf{b}_k)$ *if and only if*

$$\langle //(\mathbf{a}_1, \ldots, \mathbf{a}_k), //(\mathbf{e}_{i_1}, \ldots, \mathbf{e}_{i_k})\rangle = \langle //(\mathbf{b}_1, \ldots, \mathbf{b}_k), //(\mathbf{e}_{i_1}, \ldots, \mathbf{e}_{i_k})\rangle \tag{28}$$

*for* $1 \le i_1 < \cdots < i_k \le n$.

((ii))
$$\langle //(\mathbf{a}_1, \ldots, \mathbf{a}_k), //(\mathbf{c}_1, \ldots, \mathbf{c}_k)\rangle$$

$$= \sum_{1 \le i_1 < \cdots < i_k \le n} \langle //(\mathbf{a}_1, \ldots, \mathbf{a}_k), //(\mathbf{e}_{i_1}, \ldots, \mathbf{e}_{i_k})\rangle \langle //(\mathbf{c}_1, \ldots, \mathbf{c}_k), //(\mathbf{e}_{i_1}, \ldots, \mathbf{e}_{i_k})\rangle \tag{29}$$

PROOF: Direct consequence of Theorem 1 and Lemma 4. □

REMARK: (i) $//(\mathbf{e}_{i_1}, \ldots, \mathbf{e}_{i_k})$ is the "unit $k$-cube" in the *coordinate* $k$-plane spanned by $\{\mathbf{e}_{i_1}, \ldots, \mathbf{e}_{i_k}\}$ and

$$\langle //(\mathbf{a}_1, \ldots, \mathbf{a}_k), //(\mathbf{e}_{i_1}, \ldots, \mathbf{e}_{i_k})\rangle$$

is the oriented $k$-dimensional volume of the orthogonal projection of $//(\mathbf{a}_1, \ldots, \mathbf{a}_k)$ onto that coordinate $k$-plane.

(ii) The above lemma clearly suggests that the following $\binom{n}{k}$ values, namely

$$\{\langle //(\mathbf{a}_1, \ldots, \mathbf{a}_k), //(\mathbf{e}_{i_1}, \ldots, \mathbf{e}_{i_k})\rangle,\ 1 \le i_1 < \cdots < i_k \le n\} \tag{30}$$

can be considered as the components of $//(\mathbf{a}_1, \ldots, \mathbf{a}_k)$ with respect to the chosen coordinate system $\{\mathbf{e}_1, \ldots, \mathbf{e}_n\}$, and the formula (29) can be considered as the usual formula for computing inner product in terms of the components of the given pair of vectors.

(iii) From the formal algebraic point of view, the lack of "addition" among the equivalence classes of $k$-dimensional parallelotopes is one of the major "*imperfections*" that one would like to remedy if it is at all possible. The above lemma also provides a natural way of remedying such an imperfection, namely, formula (29) enables us to extend the definition of inner product to include *formal sum* of a finite collection of $k$-dimensional parallelotopes.

DEFINITION: Two *formal sums* of $k$-dimensional parallelotopes are defined to be equivalent if and only if their inner products with $//(\mathbf{e}_{i_1}, \ldots, \mathbf{e}_{i_k})$ are respectively equal for all $1 \le i_1 < \cdots < i_k \le 1$. The set of all *equivalence classes of such formal sums* forms a inner product space of dimension $\binom{n}{k}$, called the *space of $k$-vectors.*

DEFINITION: Let $V$ be the vector space of all displacement vectors in a given $n$-dimensional euclidean space and $\Lambda^k(V)$ be the vector space of $k$-vectors, $k = 2, \ldots, n$. Then

$$\Lambda(V) = \mathbb{R} + V + \Lambda^2(V) + \cdots + \Lambda^n(V) \tag{31}$$

together with the addition, scalar product, inner product and exterior product introduced in the above discussion forms an algebraic system called the *Grassmann algebra* of $n$-dimensional euclidean geometry. It is one of the most fundamental algebraic systems which is the natural outcome of a systematic algebraization of the basic geometric structures of high dimensional euclidean geometry.

REMARK: Usually, the above Grassmann algebra, $\Lambda(V)$, is introduced in the approach of the multilinear algebra formalism, namely, it is constructed as the *universal alternating graded algebra* generated by a given vector space $V$, and the above Theorem 1 is treated as a natural formula which enables us to extend the given inner product of $V$ to $\Lambda^k(V)$, $2 \leq k \leq n = \dim V$. However, the approach we outline in this section is a quite different one: we pay much more attention to the geometric origin and the geometric meaning of the four algebraic operations, i.e. addition, scalar product, inner product and exterior product, and the high dimensional Pythagoras theorem, namely, Theorem 1, plays a central role in the organization of the geometry of parallelotopes into the Grassmann algebra.

## §4. Some applications.

The applications of Gassmann algebra are almost boundless both in scope and in depth. We shall conclude this short survey article by mentioniong some examples of immediate applications in the realm of euclidean or spherical geometry.

### (4.1) Spherical trigonometry and spherical geometry.

The $(n-1)$-dimensional sphere, $S^{n-1}(1)$, can naturally be identified with the set of *unit vectors* in the $n$-dimensional inner product space $V$. Therefore, the Grassmann algebra is a powerful set of tools in the study of spherical geometry.

EXAMPLE 1: SPHERICAL COSINE LAW: Let $\widehat{\Delta ABC}$ be a given spherical triangle, $a, b, c$ be the lengths of the opposite sides of $A, B, C$ and $\mathbf{a}, \mathbf{b}, \mathbf{c}$ be the position vectors of $A, B, C$ respectively. Then

$$\begin{gathered} \mathbf{a}\cdot\mathbf{b} = \cos c, \;\; \mathbf{b}\cdot\mathbf{c} = \cos a, \;\; \mathbf{c}\cdot\mathbf{a} = \cos b, \\ |\mathbf{a}\wedge\mathbf{b}| = \sin c, \;\; |\mathbf{b}\wedge\mathbf{c}| = \sin a, \;\; |\mathbf{c}\wedge\mathbf{a}| = \sin b. \end{gathered} \tag{32}$$

Therefore, it follows from Theorem 1 that

$$\begin{aligned} \sin b \sin c \cos A = \langle \mathbf{a}\wedge\mathbf{c}, \mathbf{a}\wedge\mathbf{b}\rangle &= \begin{vmatrix} 1 & \cos c \\ \cos b & \cos a \end{vmatrix} \\ &= \cos a - \cos b \cos c \end{aligned} \tag{33}$$

and similarly

$$\begin{aligned}\sin a \sin c \cos B &= \cos b - \cos a \cos c,\\ \sin a \sin b \cos C &= \cos c - \cos a \cos b.\end{aligned} \tag{33'}$$

Thus the above spherical cosine law is, in fact, a direct consequence of Theorem 1 (in the special case of $k = 2$ and $n = 3$).

EXAMPLE 2: SPHERICAL SINE LAW: Straightforward algebraic computation will show that

$$(33) \Rightarrow \frac{\sin^2 A}{\sin^2 a} = \frac{1 + 2\cos a \cos b \cos c - \sum \cos^2 a}{\sin^2 a \sin^2 b \sin^2 c}.$$

Therefore, it follows from (33), (33′) and Theorem 1 that

$$\frac{\sin A}{\sin a} = \frac{\sin B}{\sin b} = \frac{\sin C}{\sin c} = \frac{|\mathbf{a} \wedge \mathbf{b} \wedge \mathbf{c}|}{\sin a \sin b \sin c} \tag{34}$$

this is a useful refinement of the spherical sine law.

EXAMPLE 3: Let $\Delta$ be the area of $\widehat{\Delta ABC}$. Then

$$\tan \frac{\Delta}{2} = \frac{|\mathbf{a} \wedge \mathbf{b} \wedge \mathbf{c}|}{1 + \sum \cos a}. \tag{35}$$

PROOF: Set $D = |\mathbf{a} \wedge \mathbf{b} \wedge \mathbf{c}|$ and $u = 1 + \sum \cos a$. Then it follows from (34) and the well-known fact $\Delta = A + B + C - \pi$ that

$$\begin{aligned}\sin \Delta &= -\sin(A + B + C)\\ &= \frac{D}{\prod \sin^2 a}\{D^2 + \sum \cos a \cos^2 b - \sum \cos a \cos b - \prod \cos a \sum \cos a\}\\ &= \frac{u \cdot D}{\prod (1 + \cos a)}\\ &= \frac{2uD}{u^2 + D^2}.\end{aligned} \tag{36}$$

Hence,

$$\tan \frac{\Delta}{2} = \frac{D}{u}. \tag{37}$$

EXAMPLE 4: THE COSINE LAW OF SPHERICAL SIMPLICES: Let $\sigma(A, B, C, D)$ be a spherical simplex (in $S^3(1)$) with $A, B, C, D$ as its vertices. Set

$$a = \widehat{AB},\ b = \widehat{AC},\ c = \widehat{AD};\ \bar{a} = \widehat{CD},\ \bar{b} = \widehat{BD},\ \bar{c} = \widehat{BC}.$$

Let $\alpha, \beta, \gamma$; $\bar{\alpha}, \bar{\beta}, \bar{\gamma}$ be the dihedral angles at the edges $\bar{a}, \bar{b}, \bar{c}$; $a, b, c$ respectively. The cosine law of spherical simplices enable us to express the cosines of dihedral angles in terms of the cosines of its edges.

Set $\mathbf{a}, \mathbf{b}, \mathbf{c}, \mathbf{d}$ to be the position vector of $A, B, C, D$ respectively. Then the dihedral angle $\alpha$ is exactly the angle between $\mathbf{a} \wedge \mathbf{c} \wedge \mathbf{d}$ and $\mathbf{b} \wedge \mathbf{c} \wedge \mathbf{d}$. Therefore

$$(38) \qquad |\mathbf{a} \wedge \mathbf{c} \wedge \mathbf{d}| \cdot |\mathbf{b} \wedge \mathbf{c} \wedge \mathbf{d}| \cdot \cos\alpha = \langle \mathbf{a} \wedge \mathbf{c} \wedge \mathbf{d}, \mathbf{b} \wedge \mathbf{c} \wedge \mathbf{d} \rangle$$

namely

$$(38') \qquad \begin{vmatrix} 1 & \cos b & \cos c \\ \cos b & 1 & \cos\bar{a} \\ \cos c & \cos\bar{a} & 1 \end{vmatrix}^{\frac{1}{2}} \cdot \begin{vmatrix} 1 & \cos\bar{c} & \cos\bar{b} \\ \cos\bar{c} & 1 & \cos\bar{a} \\ \cos\bar{b} & \cos\bar{a} & 1 \end{vmatrix}^{\frac{1}{2}} \cdot \cos\alpha = \begin{vmatrix} \cos a & \cos b & \cos c \\ \cos\bar{c} & 1 & \cos\bar{a} \\ \cos\bar{b} & \cos\bar{a} & 1 \end{vmatrix}.$$

Set

$$(39) \qquad A = \begin{bmatrix} 1 & \cos a & \cos b & \cos c \\ \cos a & 1 & \cos\bar{c} & \cos\bar{b} \\ \cos b & \cos\bar{c} & 1 & \cos\bar{a} \\ \cos c & \cos\bar{b} & \cos\bar{a} & 1 \end{bmatrix}.$$

Then the above formula can be written as

$$(40) \qquad \left.\begin{array}{l} A_{11}^{\frac{1}{2}} \cdot A_{22}^{\frac{1}{2}} \cos\alpha = -A_{12} \\ \text{and similarly, one has} \\ A_{11}^{\frac{1}{2}} \cdot A_{33}^{\frac{1}{2}} \cos\beta = -A_{13} \\ A_{11}^{\frac{1}{2}} \cdot A_{44}^{\frac{1}{2}} \cos\gamma = -A_{14} \\ A_{22}^{\frac{1}{2}} \cdot A_{33}^{\frac{1}{2}} \cos\bar{\gamma} = -A_{23} \\ A_{22}^{\frac{1}{2}} \cdot A_{44}^{\frac{1}{2}} \cos\bar{\beta} = -A_{24} \\ A_{33}^{\frac{1}{2}} \cdot A_{44}^{\frac{1}{2}} \cos\bar{\alpha} = -A_{34} \end{array}\right\}$$

where $A_{ij}$ is the $(i,j)$-algebraic cofactor of $A$.

EXAMPLE 5: THE DUAL COSINE LAW OF SPHERICAL SIMPLICES: Let us first consider the polar-duality among simplices of $S^3(1)$. Let $\pi_A, \pi_B, \pi_C, \pi_D$ be the equatorial-planes with $A, B, C, D$ as the poles respectively. Set $\sigma(A', B', C', D')$ to be the intersection of hemi-spheres containing the center of $\sigma(A, B, C, D)$. $\sigma(A', B', C', D')$ is called the polar dual of $\sigma(A, B, C, D)$ and viz versa. Set $a', b', c'$; $\bar{a}', \bar{b}', \bar{c}'$ and $\alpha', \beta', \gamma'$; $\bar{\alpha}', \bar{\beta}', \bar{\gamma}'$ to be the corresponding edge lengths and the corresponding dihedral angles. Observe that $\alpha'$ is the dihedral angle between the equatorial planes $\pi_A$ and $\pi_B$. Hence, it is easy to see that

$$(41) \qquad \begin{array}{l} \alpha' + a = \beta' + b = \gamma' + c = \bar{\alpha}' + \bar{a} = \bar{\beta}' + \bar{b} = \bar{\gamma}' + \bar{c} = \pi \\ a' + \alpha = b' + \beta = c' + \gamma = \bar{a}' + \bar{\alpha} = \bar{b}' + \bar{\beta} = \bar{c}' + \bar{\gamma} = \pi. \end{array}$$

Set

$$(42)\quad A' = \begin{bmatrix} 1 & \cos a' & \cos b' & \cos c' \\ \cos a' & 1 & \cos \bar{c}' & \cos \bar{b}' \\ \cos b' & \cos \bar{c}' & 1 & \cos \bar{a}' \\ \cos c' & \cos \bar{b}' & \cos \bar{a}' & 1 \end{bmatrix} = \begin{bmatrix} 1 & -\cos\alpha & -\cos\beta & -\cos\gamma \\ -\cos\alpha & 1 & -\cos\bar{\gamma} & -\cos\bar{\beta} \\ -\cos\beta & -\cos\bar{\gamma} & 1 & -\cos\bar{\alpha} \\ -\cos\gamma & -\cos\bar{\beta} & -\cos\bar{\alpha} & 1 \end{bmatrix}.$$

Then, it follows from (41) and the cosine law applying to $\sigma(A', B', C', D')$:

$$(43)\quad \begin{cases} {A'_{11}}^{\frac{1}{2}} \cdot {A'_{22}}^{\frac{1}{2}} \cos a = A'_{12} \\ {A'_{11}}^{\frac{1}{2}} \cdot {A'_{33}}^{\frac{1}{2}} \cos b = A'_{13} \\ {A'_{11}}^{\frac{1}{2}} \cdot {A'_{44}}^{\frac{1}{2}} \cos c = A'_{14} \\ {A'_{22}}^{\frac{1}{2}} \cdot {A'_{33}}^{\frac{1}{2}} \cos \bar{c} = A'_{23} \\ {A'_{22}}^{\frac{1}{2}} \cdot {A'_{44}}^{\frac{1}{2}} \cos \bar{b} = A'_{24} \\ {A'_{33}}^{\frac{1}{2}} \cdot {A'_{44}}^{\frac{1}{2}} \cos \bar{a} = A'_{34} \end{cases}$$

which express the cosines of the edges in terms of the cosines of the dihedral angles.

EXAMPLE 6: THE HEIGHT DECOMPOSITION OF A SPHERICAL SIMPLEX: Let $\overset{\frown}{AH}$ be the perpendicular geodesic interval from the vertex $A$ to its opposite face $\overset{\frown}{\Delta BCD}$ and $h$ be the length of $\overset{\frown}{AH}$, i.e., the height from the vertex $A$ to the base $\overset{\frown}{\Delta BCD}$. It is often needed to compute $\cos h$ in terms of the cosines of the six edges in the study of the geometry of spherical simplexes. The following method of applying Theorem 1 provides the most clean-cut solution of the above problem:

Let $A^*$ be the reflectional symmetric point of $A$ with respect to the equatorial subspace spanned by $\overset{\frown}{\Delta BCD}$, and $\mathbf{a}^*$ be its position vector. Then

$$(44)\quad \begin{aligned} \mathbf{a} \cdot \mathbf{a}^* &= \cos 2h, \\ \det(\mathbf{a}, \mathbf{b}, \mathbf{c}, \mathbf{d}) &= \det(\mathbf{a}^*, \mathbf{c}, \mathbf{b}, \mathbf{d}). \end{aligned}$$

Therefore

$$(45)\quad \begin{aligned} \langle \mathbf{a} \wedge \mathbf{b} \wedge \mathbf{c} \wedge \mathbf{d}, \mathbf{a}^* \wedge \mathbf{c} \wedge \mathbf{b} \wedge \mathbf{d} \rangle &= \begin{vmatrix} \cos 2h & \cos a & \cos b & \cos c \\ \cos b & \cos \bar{c} & 1 & \cos \bar{a} \\ \cos a & 1 & \cos \bar{c} & \cos \bar{b} \\ \cos c & \cos \bar{b} & \cos \bar{a} & 1 \end{vmatrix} \\ = \langle \mathbf{a} \wedge \mathbf{b} \wedge \mathbf{c} \wedge \mathbf{d}, \mathbf{a} \wedge \mathbf{b} \wedge \mathbf{c} \wedge \mathbf{d} \rangle &= \begin{vmatrix} 1 & \cos a & \cos b & \cos c \\ \cos a & 1 & \cos \bar{c} & \cos \bar{b} \\ \cos b & \cos \bar{c} & 1 & \cos \bar{a} \\ \cos c & \cos \bar{b} & \cos \bar{a} & 1 \end{vmatrix}, \end{aligned}$$

and hence, it is straightforward to solve $\cos 2h$ as a rational polynomial of the six cosines of its edges.

**(4.2) A volume formula of superscribing polyhedrons.**

In the study of sphere packing problems in the three-dimensional euclidean space, one needs to estimate the lower bound of the volumes of superscribing polyhedrons of the unit sphere $S^2(1)$ with a fixed number of faces, say eleven, twelve, thirteen etc. Let $\Sigma$ be a given number of points on $S^2(1)$ and $T(\Sigma)$ be the superscribing polyhedron bounded by the tangent planes of $S^2(1)$ at points of $\Sigma$. A basic geometric problem in the study of sphere packing problem is how to express the volume of $T(\Sigma)$ by a clean-cut formula which is advantageous for the above problem of lower bound estimation.

Let $D(\Sigma)$ be the *convex hull* of $\Sigma$. It is exactly the inscribing polyhedron of $S^2(1)$ with $\Sigma$ as its set of vertices and it is also the dual polyhedron of $T(\Sigma)$ with respect to the pole-polar duality of $S^2(1)$. Let $\mathcal{S}(\Sigma)$ be the radial projection of the faces of $D(\Sigma)$ onto $S^2(1)$ which provides a canonical subdivision of the total solid angle into the solid angles spanned by the faces of $D(\Sigma)$. Let us assume that all the faces of $D(\Sigma)$ are triangles (which is the generic case) and each of them contains its own circumcenter. Then the above solid angle subdivision cuts $T(\Sigma)$ into $(2n-4)$ convex polyhedrons of the following type.

Let $\tau = \overset{\frown}{\Delta ABC}$ be a spherical triangle containing its own circumcenter and $\mathbf{a}, \mathbf{b}, \mathbf{c}$ be the position vectors of $A, B, C$. Set $T(\tau)$ to be the intersection of the following six half-spaces, namely

$$\begin{aligned} &\overrightarrow{OX}\cdot\mathbf{a} \le 1,\ \overrightarrow{OX}\cdot\mathbf{b} \le 1,\ \overrightarrow{OX}\cdot\mathbf{c} \le 1, \\ &\overrightarrow{OX}\cdot\mathbf{a}\times\mathbf{b} \ge 0,\ \overrightarrow{OX}\cdot\mathbf{b}\times\mathbf{c} \ge 0,\ \overrightarrow{OX}\cdot\mathbf{c}\times\mathbf{a} \ge 0. \end{aligned} \tag{46}$$

(assume that $\mathbf{a}\cdot\mathbf{b}\times\mathbf{c} > 0$). Then, one has the following volume formula which plays a fundamental role in the solution of the classical sphere packing problem in [**Hsi-2**], namely

$$\operatorname{vol}(T(\tau)) = \frac{1}{3D}\left\{8 + u - \frac{4u(1+2u+v)}{u^2+D^2}\right\} \tag{47}$$

where

$$\begin{aligned} &D = \mathbf{a}\cdot\mathbf{b}\times\mathbf{c},\ u = 1 + \mathbf{a}\cdot\mathbf{b} + \mathbf{b}\cdot\mathbf{c} + \mathbf{c}\cdot\mathbf{a}, \\ &v = (\mathbf{a}\cdot\mathbf{b})(\mathbf{b}\cdot\mathbf{c}) + (\mathbf{a}\cdot\mathbf{b})\cdot(\mathbf{c}\cdot\mathbf{a}) + (\mathbf{b}\cdot\mathbf{c})\cdot(\mathbf{c}\cdot\mathbf{a}). \end{aligned} \tag{47'}$$

PROOF: Set $\mathbf{v}, \hat{\mathbf{a}}, \hat{\mathbf{b}}$ and $\hat{\mathbf{c}}$ to be the solution of the following sets of equations, namely

$$\begin{aligned} \mathbf{v}: &\quad \mathbf{a}\cdot\mathbf{v} = \mathbf{b}\cdot\mathbf{v} = \mathbf{c}\cdot\mathbf{v} = 1 \\ \hat{\mathbf{a}}: &\quad \hat{\mathbf{a}}\cdot\mathbf{b}\times\mathbf{c} = 0,\ \hat{\mathbf{a}}\cdot\mathbf{b} = \hat{\mathbf{a}}\cdot\mathbf{c} = 1 \\ \hat{\mathbf{b}}: &\quad \hat{\mathbf{b}}\cdot\mathbf{c}\times\mathbf{a} = 0,\ \hat{\mathbf{b}}\cdot\mathbf{c} = \hat{\mathbf{b}}\cdot\mathbf{a} = 1 \\ \hat{\mathbf{c}}: &\quad \hat{\mathbf{c}}\cdot\mathbf{a}\times\mathbf{b} = 0,\ \hat{\mathbf{c}}\cdot\mathbf{a} = \hat{\mathbf{c}}\cdot\mathbf{b} = 1. \end{aligned} \tag{48}$$

Then it is not difficult to see that

$$\begin{aligned} \mathbf{v} &= \frac{1}{D}\{\mathbf{a}\times\mathbf{b}+\mathbf{b}\times\mathbf{c}+\mathbf{c}\times\mathbf{a}),\\ \hat{\mathbf{a}} &= \frac{\mathbf{b}+\mathbf{c}}{1+\mathbf{b}\cdot\mathbf{c}},\ \hat{\mathbf{b}} = \frac{\mathbf{c}+\mathbf{a}}{1+\mathbf{c}\cdot\mathbf{a}},\ \hat{\mathbf{c}} = \frac{\mathbf{a}+\mathbf{b}}{1+\mathbf{a}\cdot\mathbf{b}}. \end{aligned} \tag{49}$$

Let $V, \hat{A}, \hat{B}$ and $\hat{C}$ be the points with $\mathbf{v}, \hat{\mathbf{a}}, \hat{\mathbf{b}}$ and $\hat{\mathbf{c}}$ as their position vector. Then $P(\tau)$ is the union of three cones with $V$ as their vertices and the quadrilaterals $OAB\hat{C}$, $OBC\hat{A}$ and $OCA\hat{B}$ as their basis respectively. In terms of Grassmann algebra, or equivalently, the three-dimensional vector algebra, the above three quadrilaterals can be represented respectively by

$$\begin{aligned} &\frac{\mathbf{a}\wedge\mathbf{b}}{1+\mathbf{a}\cdot\mathbf{b}},\ \frac{\mathbf{b}\wedge\mathbf{c}}{1+\mathbf{b}\cdot\mathbf{c}},\ \frac{\mathbf{c}\wedge\mathbf{a}}{1+\mathbf{c}\cdot\mathbf{a}}\\ &(\text{or equivalently, } \frac{\mathbf{a}\times\mathbf{b}}{1+\mathbf{a}\cdot\mathbf{b}},\ \frac{\mathbf{b}\times\mathbf{c}}{1+\mathbf{b}\cdot\mathbf{c}},\ \frac{\mathbf{c}\times\mathbf{a}}{1+\mathbf{c}\cdot\mathbf{a}}). \end{aligned} \tag{50}$$

Therefore

$$\begin{aligned} \operatorname{vol} T(\tau) &= \frac{1}{3}\mathbf{v}\cdot\left\{\frac{\mathbf{a}\times\mathbf{b}}{1+\mathbf{a}\cdot\mathbf{b}}+\frac{\mathbf{b}\times\mathbf{c}}{1+\mathbf{b}\cdot\mathbf{c}}+\frac{\mathbf{c}\times\mathbf{a}}{1+\mathbf{c}\cdot\mathbf{a}}\right\}\\ &= \frac{1}{3D}\{\mathbf{a}\times\mathbf{b}+\mathbf{b}\times\mathbf{c}+\mathbf{c}\times\mathbf{a}\}\cdot\left\{\frac{\mathbf{a}\times\mathbf{b}}{1+\mathbf{a}\cdot\mathbf{b}}+\frac{\mathbf{b}\times\mathbf{c}}{1+\mathbf{b}\cdot\mathbf{c}}+\frac{\mathbf{c}\times\mathbf{a}}{1+\mathbf{c}\cdot\mathbf{a}}\right\}\\ &= \frac{1}{3D}\left\{8+u-2u\cdot\left[\frac{1}{1+\mathbf{a}\cdot\mathbf{b}}+\frac{1}{1+\mathbf{b}\cdot\mathbf{c}}+\frac{1}{1+\mathbf{c}\cdot\mathbf{a}}\right]\right\}\\ &= \frac{1}{3D}\left\{8+u-\frac{4u(1+2u+v)}{u^2+D^2}\right\}. \end{aligned} \tag{41}$$

Hence

$$\operatorname{vol} T(\Sigma) = \sum\{\operatorname{vol} T(\tau);\ \tau\in\mathcal{S}(\Sigma)\} \tag{52}$$

can be effectively computed in terms of the cosines of the edges of $\mathcal{S}(\Sigma)$. We refer to [**Hsi-2**] for its application to the study of the sphere packing problem.

## References


[**Gra-1**] H. G. Grassmann, *Die lineale Ausdehnungslehre*, (1844), Chelsea (reprint) 1969.

[**Gra-2**] ————————, *Gesammelte mathematische und physikalische Werke*, 3 vols, B. G. Teubner, 1894-1911; Vol 1, Part I, 1–319; Vol 1, Part II, 1–383.

[**Ham-1**] Wm. R. Hamilton, *Elements of Quaternions*, 2 Vols, 1866, 2nd ed., 1899–1901, Chelsea (reprint) 1969.

[**Ham-2**] ______________, "Mathematical Papers," Cambridge Univ. Press, Vol 3, 1967.

[**Hsi-1**] W. Y. Hsiang, *On infinitesimal symmetrization and volume formula for spherical or hyperbolic tetrahedrons*, Quart. J. Math. Oxford (2) **39** (1988), 463–468.

[**Hsi-2**] ____________, *On the density of sphere packings in* $E^3$, *I and II*, Preprints PAM-530 and 535, Univ. of Calif., Berkeley (1991).

*Computing in Euclidean Geometry*, pp. 23–90
Eds. D.-Z. Du and F. K. Hwang


# MESH GENERATION AND OPTIMAL TRIANGULATION

MARSHALL BERN
*Xerox Palo Alto Research Center, Palo Alto, California 94304, U.S.A.*

and

DAVID EPPSTEIN
*Dept. of Information and Computer Science, University of California
Irvine, California 92717-3425, U.S.A.*

ABSTRACT

We survey the computational geometry relevant to finite-element mesh generation. We especially focus on optimal triangulations of geometric domains in two- and three-dimensions. An optimal triangulation is a partition of the domain into triangles or tetrahedra, that is best according to some criterion that measures the size, shape, or number of triangles. We discuss algorithms both for the optimization of triangulations on a fixed set of vertices and for the placement of new vertices (Steiner points). We briefly survey the heuristic algorithms used in some practical mesh generators.

## 1. Introduction

Computational geometry claims the two aims of solving practical problems and producing beautiful mathematics. There is a natural tension between these goals: the most elegant formulation of a problem rarely occurs in practice. But surprisingly often the aims complement each other. This chapter discusses the interplay between an important practical problem—finite element mesh generation, and a flourishing theoretical area—optimal triangulation algorithms.

Finite element methods have proved indispensable for physical simulation. These methods typically assume that the simulated domain—for example, the air around a wing—is divided into many small "elements", triangles or quadrilaterals in two dimensions and tetrahedra or hexahedra in three. The complex of elements is the mesh.

A triangulation is a partition of a geometric input, typically the region defined by a point set or a polytope, into simplices that meet only at shared faces. So in two dimensions, a triangulation consists of triangles that intersect only at shared edges and vertices. An optimal triangulation is one that is best according to some criterion that measures the size, shape, or number of simplices.

An example illustrates the complementary relationship of mesh generation and optimal triangulation. Numerical-analysis folklore had long held that finite element meshes should avoid elements with sharp angles. Remarkably, Lawson found that the Delaunay triangulation, a geometric construction with a long history, maximizes the minimum angle in a two-dimensional mesh [102]. At about the same time, work by other numerical analysts implicated the flattest—not the sharpest—angle as the critical factor in convergence [3, 90]. This problem did not have a solution waiting for it, and in developing an algorithm to minimize the maximum angle [64], computational geometers discovered a very interesting algorithmic paradigm (discussed in Section 2.2.2).

### *1.1. Background*

The differential equations arising in physical simulation require numerical solution. Most numerical methods—there are exceptions—assume that the domain of interest is divided into a mesh of small, simple elements. There are two major types of meshes: structured and unstructured. This chapter studies only unstructured meshes, but here we briefly sketch the larger context.

A *structured mesh* in two dimensions is most often simply a square grid deformed by some coordinate transformation. Each vertex of the mesh, except at the boundaries, has an isomorphic local neighborhood. In three dimensions, a structured mesh is usually a deformed cubical grid. An *unstructured mesh* is most often a triangulation with arbitrarily varying local neighborhoods.

Structured meshes offer certain advantages over unstructured. They are simpler, and also more convenient for use in the simpler finite difference methods. They require less computer memory, as their coordinates can be calculated, rather than explicitly stored. Finally, structured meshes offer more direct control over the sizes and shapes of elements.

The big disadvantage of a structured mesh is its lack of flexibility in fitting a domain with a complicated shape. A number of techniques have been developed to find appropriate coordinate transformations: conformal mapping, algebraic methods, and numerical methods that themselves solve differential equations [32, 167, 168]. Even armed with these techniques, it may be impossible to find a transformation that fits a complicated domain acceptably well. Faced with this problem, some practitioners cut out a region of the grid, without any transformation, to give a "stair-case approximation" to the domain. But then the computed solution will be quite inaccurate near the boundary of the domain, an area that is often of special interest. Other practitioners break up the domain into simpler regions, perhaps overlapping, each of which can be more nearly matched by a deformed grid. This method and its associated numerical analysis make up "domain decomposition", a large area of study in its own right.

Because of the need to fit complicated domains, such as aircraft and machine parts, the trend in simulation has been towards unstructured meshes [6, 117, 166], although both types of meshes will continue to be important for some time to come. Indeed, there are methods that combine many small structured meshes into an overall unstructured mesh [173].

As we shall see in this chapter, unstructured meshes can fit arbitrarily complicated domains. Simply fitting the domain, however, is not enough. A finite element mesh must also use elements of appropriate size and shape, and these quantities may vary over the domain. Multiple requirements lead to many interesting and difficult triangulation problems; these computational geometry problems define the subject of this chapter.

### *1.2. Formulating the problems*

In the formulation of triangulation problems, we see the tension between applicability and elegance. This tension pervades the major choices in the formulation: the type of input assumed, the type of triangulation desired, and the optimality criteria.

The most theoretically attractive inputs are polygonal regions in two dimensions and polyhedral regions in three, without any auxiliary information. Curved domains occur in practice, but usually the assumption of a flat-sided domain is not too limiting. The more severe restriction is the assumption of no auxiliary information. In practical mesh generation, some foreknowledge of the solution to the finite element computation usually guides the choice of element size, shape, and orientation.

A major binary choice arises in determining the type of triangulation. The vertices of the triangulation may be exactly the vertices of the input, or extra vertices—called *Steiner points*—may be allowed. In practice, mesh generation invariably allows Steiner points, although the placement of Steiner points is often separate from the subsequent triangulation process. Computational geometers have traditionally disallowed Steiner points, and thus their theorems have relevance only for the second stage of practical mesh generation, but recently a small number of geometers (especially ourselves) have considered problems allowing Steiner points. Steiner points change the character of optimal triangulation problems. Where previously the goal was an exact algorithm, it now must be an approximation algorithm that uses a modest number of Steiner points.

For practitioners, the ultimate optimality criteria are speed and accuracy of the finite element computation. These in turn impose a number of somewhat conflicting criteria on the mesh: element shape (such as bounds on angles), reasonable complexity (not too many elements), and element orientation (such as aligned with fluid flow). Computational geometers usually work on only one or two optimization criteria at a time; although certain triangulation algorithms, such as Delaunay triangulation and quadtree triangulation, optimize several criteria simultaneously. A typical computational geometry problem disallowing Steiner points is one mentioned above: triangulate a point set while minimizing the maximum angle [64] (Section 2.2.2). A typical problem allowing Steiner points is: triangulate an $n$-vertex polygonal domain using no angles larger than $90°$ and only a polynomial (in $n$) number of Steiner points [19] (Section 2.3.2).

In the optimal triangulation algorithms we present, watch for a recurring theme: local versus global optimization. For many problems that disallow Steiner points, local optimizations can lead to global solutions (Sections 2.2.1 and 2.2.2). For problems that allow Steiner points, a typical approximation algorithm performs some initial global steps, such as defining a grid on the input, followed by some local optimizations (Sections 2.3.1, 2.3.2, and 2.3.4).

### *1.3. Organization*

We organized this chapter at the topmost level into sections on two- and three-dimensional problems, with Section 2.5 serving as a bridge. The major sections are then subdivided by the type of output desired: triangulation without any optimization criteria, optimal triangulation without Steiner points, and finally optimal triangulation with Steiner points. To keep ourselves honest, we include sections surveying the heuristics devised by practitioners.

1. Introduction
   - 1.1. Background
   - 1.2. Formulating the problems
   - 1.3. Organization
2. Two-dimensional Triangulations
   - 2.1. Triangulation without optimization
   - 2.2. Optimal triangulation
     Delaunay triangulation, flip, edge insertion, dynamic programming.
   - 2.3. Steiner triangulation
     No small or large angles, maxmin height, min weight, conforming DT.
   - 2.4. Heuristically generated meshes
     Mesh improvement, quadtrees, polygon decomposition, advancing front.
   - 2.5. Two-and-a-half-dimensional problems
     Interpolation, surfaces for three-dimensional models.
3. Three-dimensional Triangulations
   - 3.1. Tetrahedralization without optimization
   - 3.2. Optimal tetrahedralization

## 2. Two-dimensional Triangulations

In this section, we consider triangular meshes in two dimensions. We distinguish four types of input domains, all of which can be viewed as polygonal regions of the plane, because their boundaries consist of straight (perhaps degenerate) line segments. The task is to partition the domain into triangles, that meet edge to edge, and that may be required to satisfy some other optimality properties. For all four types of domains, a single parameter $n$, the number of vertices, suffices to measure the input complexity.

- **Simple polygon.** The domain is a polygonal region of the plane, and its boundary forms a simple, polygonal, closed curve. The triangulation must use the edges of the boundary as edges in the triangulation. In a *Steiner triangulation* problem, extra vertices may be added to the interior or on the polygon; hence, the edges of the boundary may be subdivided to form several collinear edges in the triangulation.

- **Polygon with holes.** This differs from the previous case in that the boundary may form several disjoint polygonal Jordan curves. These curves surround *holes* within the polygon.

- **Point set.** The input is a set of points in the plane. Without Steiner points, the vertices of the triangulation are exactly the input points, and the boundary of the triangulation is the convex hull. With Steiner points, the vertices of the triangulation are a superset of the input points, and the boundary of the triangulation is a convex region that may be larger than the convex hull.

- **Planar straight line graph (PSLG).** The input is a set of vertices and *non-crossing* (that is, intersecting only at endpoints) line segments in the plane, which must be used as edges (or for Steiner problems, unions of edges) in the triangulation. This most general input occurs in practice for "multiple domains", that is, domains that include boundaries between different materials. One could imagine the segments as holes in a polygon, and for non-Steiner triangulation this is a correct view of the problem. For Steiner triangulations, however, the Steiner points must be identical on both sides of a segment, which would not necessarily happen if we treated each segment as a degenerate hole.

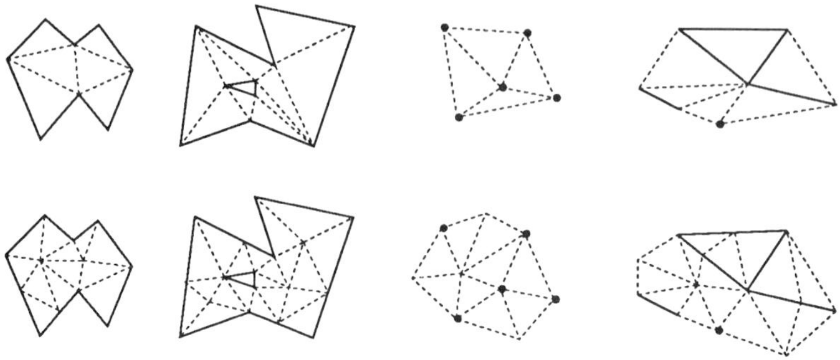

Figure 1. Triangulations with (below) and without Steiner points (above) of a polygon, a polygon with holes, a point set, and a PSLG. Solid lines show input; dotted show triangulation.

### 2.1. *Triangulation without Optimization*

In this section we review triangulation without Steiner points and without optimality criteria. The most basic question about polygon triangulation asks whether a triangulation always exists. A *diagonal* of a simple polygon is a line segment between two vertices, that lies inside the polygon and does not intersect the polygon's boundary except at its endpoints.

**Lemma 1.** *Every polygon with more than three sides has a diagonal.*

**Proof:** Let $b$ be the vertex with minimum $x$-coordinate and $ab$ and $bc$ be its two incident edges. If $ac$ is not cut by the polygon, then $ac$ is a diagonal. Otherwise there must be at least one polygon vertex inside triangle $abc$, as in Figure 2. Let $d$ be the vertex inside $abc$ furthest from the line through $a$ and $c$. Now segment $bd$ cannot be cut by the polygon, since any edge intersecting $bd$ must have one endpoint further from line $ac$. ∎

Once we have found a single diagonal, we can split the polygon in two, and recursively triangulate each part. The proof of the lemma implies a linear-time algorithm for finding a diagonal, so a triangulation can be found in time $O(n^2)$.

Now the following question arises: how quickly can we compute a triangulation? This problem attracted more than a decade of intensive research. It is not difficult to improve the running time to $O(n \log n)$ [77, 138]. For many geometric problems there are matching $\Omega(n \log n)$ lower bounds, but none was known in this case. There were also many faster special case algorithms [38, 77, 86]. After Tarjan and Van Wyk [165] broke through to $O(n \log \log n)$, Chazelle [36] ended the quest by announcing a linear-time triangulation algorithm for simple polygons.

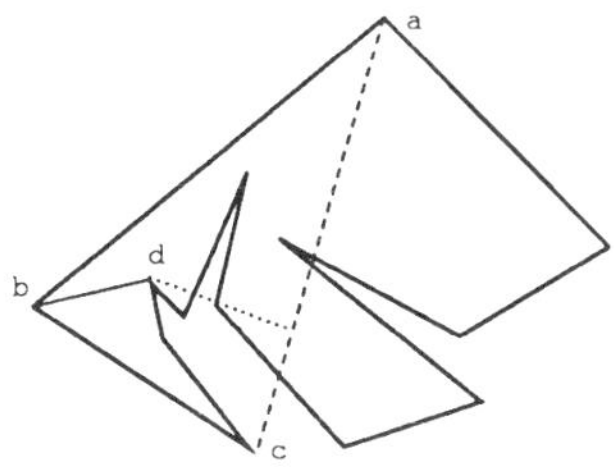


Figure 2. Finding a single diagonal.

We now discuss the other types of input domains. It turns out that any point set can be triangulated in time $O(n \log n)$, and conversely a point set triangulation algorithm can sort real numbers, so there is a matching lower bound in the algebraic decision tree model of computation (see [138]).

Lemma 1 applies to any polygon with holes, and hence to each face in a PSLG. (A *face* is a connected component of the plane minus the PSLG.) Hence every PSLG can be triangulated. It is easy to show by induction that every triangulation of a given PSLG has the same (linear) number of triangles, $n - 2$ in the case of a simple polygon. Triangulating a PSLG can be accomplished in time $O(n \log n)$. Using Chazelle's algorithm, the time for polygons with holes can be improved to $O(n \log h)$, where $h$ is the number of holes.

In contrast to these positive results, determining whether a straight line graph (with crossing edges) contains a triangulation is NP-complete [113]. Also NP-complete is the problem of determining whether a given collection of triangles includes a triangulation of the triangles' vertex set. (These results tend to rule out greedy algorithms for optimal triangulation.)

### 2.2. *Optimal Triangulation*

We have seen that quite efficient algorithms exist for constructing a triangular mesh that covers a given domain. There may, however, be exponentially many triangulations with widely varying appearance. We now turn to the harder task of finding a triangulation that optimizes some measure of quality.

Since all non-Steiner triangulations of a two-dimensional input have the same number of triangles, reasonable quality measures depend upon the *shape* of triangles. Typical measures examine the angles, edge lengths, height, and area of a triangle. The measure of a triangulation is then taken to be the sum, maximum, or minimum of the measure over all triangles.

A number of quality measures find motivation in finite element methods. The numerical condition of matrices in a finite element computation is related to the minimum angle in the triangulation. The error of a finite element approximation is also related to the minimum angle [75, 164], and even more closely related to the maximum angle [3, 90]. The minimum height of a triangle relates to the quality of a curved-surface approximation [80], and to a key step in a new three-dimensional mesh generation algorithm [129]. Several other optimization criteria find application in interpolation [56, 141].

For a given element shape, approximation error grows with element size, which can be measured simply by maximum edge length [164] or by more complicated metrics. For some applications—such as modeling diffusion in a nonisotropic medium—size is appropriately measured by the area of the "min-containment circle", after the domain has been transformed by an appropriate affine map [45]. (Sizes of elements are typically used to weight finite element residuals in a-posteriori error estimates, in order to find regions of the mesh in need of refinement [4, 30].)

### *2.2.1. Delaunay Triangulation and the Flip Algorithm*

A well-known construction, called the *Delaunay triangulation*, simultaneously optimizes several of the quality measures mentioned above: maxmin angle, minmax circumcircle, and minmax min-containment circle.

The Delaunay triangulation (DT) of a point set is the planar dual of the famous *Voronoi diagram*. The Voronoi diagram is a partition of the plane into polygonal cells, one for each input point, so that the cell for input point $a$ consists of the region of the plane closer to $a$ than to any other input point. So long as no four points lie on a common circle, then each vertex of the Voronoi diagram has degree three, and the DT, which has a bounded face for each Voronoi vertex and vice versa, will indeed be a triangulation. If four or more points do lie on a common circle, then these points will be the vertices of a larger face, that may then be triangulated to give a triangulation containing the DT. For more information on Delaunay triangulations and Voronoi diagrams, see the surveys by Fortune [72] and Aurenhammer [2].

There is a nice relationship between Delaunay triangulation and three-dimensional convex hulls [27, 57]. Lift each point of the input to a *paraboloid* in three-space by mapping the point with coordinates $(x, y)$ to the point $(x, y, x^2+y^2)$. The convex hull of the lifted points can be divided into *lower* and *upper* parts; a face belongs to the lower convex hull if it is supported by a plane that separates the point set from $(0, 0, -\infty)$. It can be shown that the DT of the input points is the projection of the lower convex hull onto the $xy$-plane, as depicted in Figure 3.

Finally, a direct characterization: if $a$ and $b$ are input points, the DT contains the edge $\{a, b\}$ if and only if there is a circle through $a$ and $b$ that intersects no other input points and contains no input points in its interior. Moreover, each circumscribing circle (*circumcircle*) of a DT triangle contains no input points in its interior.

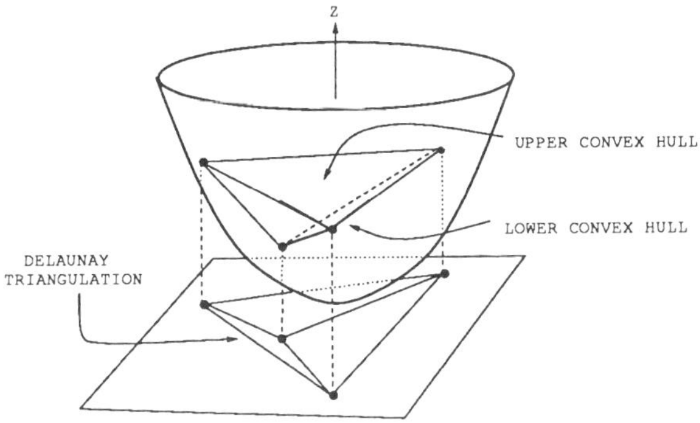


Figure 3. The lifting transformation maps the DT to the lower convex hull.

Many $O(n \log n)$-time algorithms are known for computing the DT of a point set, the first being Shamos and Hoey's divide-and-conquer algorithm [155]. The relation between Delaunay triangulation and convex hulls allows the use of any $O(n \log n)$-time three-dimensional convex hull algorithm, such as that of Preparata and Hong [137]. Fortune developed an elegant sweepline algorithm [71, 72], and Guibas, Knuth, and Sharir devised a simple, randomized incremental algorithm [82]. Maus [122] gave an algorithm for random points with expected linear running time; see also [21, 55, 125] for expected-case analyses.

The typical domain for mesh generation, however, is not a point set, but a polygonal region. In some cases the DT of a nonconvex region's vertices contains a triangulation of the region, but in general DT edges cross the region's boundary, creating an invalid mesh. One approach to this problem adds more vertices to the boundary, so that the boundary edges are covered by the DT of the augmented point set. We consider this *conforming Delaunay triangulation* problem in Section 2.3.5. The alternative approach generalizes the definition of the DT in order to force certain edges (the boundary) into the triangulation. Such a triangulation is known as a *constrained Delaunay triangulation* (CDT) [41, 44, 46, 103, 105, 154].

For generality, we allow the domain to be a planar straight line graph. The CDT of a polygonal region may be obtained by treating the polygon as a PSLG, and then removing all CDT edges exterior to the polygon. We first define the notion of *visibility*. Given a PSLG, we say that point $a$ is visible to point $b$ if line segment $ab$ does not cross any edge of the graph ($ab$ may intersect an edge without crossing it). Point $a$ is visible to line segment $bc$ if it is visible to some point on $bc$.

**Definition 1.** *The* **constrained Delaunay triangulation (CDT)** *contains the edge $\{a, b\}$ between two input vertices, if and only if $a$ is visible to $b$, and some circle through $a$ and $b$ contains no input point $c$, visible to segment $ab$, in its interior.*

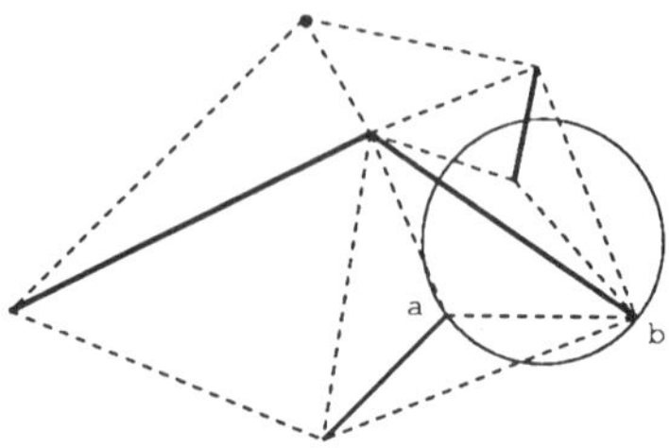


Figure 4. The CDT of a planar straight line graph.

For point sets, this definition reduces to the direct characterization of the DT given above. Definition 1 also implies that the circumcircle of a triangle $abc$ in the CDT cannot contain an input point—other than $a$, $b$, or $c$—that is visible from the interior of $abc$. We now justify Definition 1 by showing that—like the DT—the CDT actually gives a triangulation for inputs in general position. We call the new edges, defined by the empty circle condition above, *Delaunay edges*.

**Lemma 2.** *If no four input vertices lie on a common circle, the CDT of a PSLG will be a triangulation.*

**Proof:** By assumption input edges intersect only at endpoints, and since endpoints of a Delaunay edge must be mutually visible, no Delaunay edge can cross an input edge. Suppose two Delaunay edges cross. But then all four endpoints are visible from the crossing point, and it is easy to show that two circles satisfying the definition cannot exist. Therefore the CDT is a PSLG.

We now show that the CDT contains no face with more than three sides. Suppose that $ab$ and $ac$ are edges in the CDT, but that $bc$ is not. Then the circumcircle of triangle $abc$ must contain some other input point visible to $bc$.

Suppose there is an input point inside triangle $abc$. Such a point is visible to $a$ unless blocked from view by an edge with an endpoint "closer" to $a$, meaning closer when projected onto a line perpendicular to $bc$. Thus the closest input point $d$ to $a$ (again measuring distance from $bc$) must be visible to $a$. Pass a circle through $a$ and $d$, parallel to $bc$ at $d$. Any input points contained in this circle cannot be visible to segment $ad$ because they are blocked by $ab$ and $ac$; hence $ad$ is a Delaunay edge.

Suppose instead that there is an input point on the other side of $bc$ from $a$, visible to $bc$ and lying inside circle $abc$. Imagine a shrinking circle, that starts as circle $abc$, and then shrinks in a way that keeps it tangent to circle $abc$ at $a$. Ultimately, only a single input point $d$ visible to $bc$ will lie in the shrunken circle, and so $d$ must be visible to $a$. There can be no other input points in the circle that are visible to $ad$, so $ad$ is a Delaunay edge.

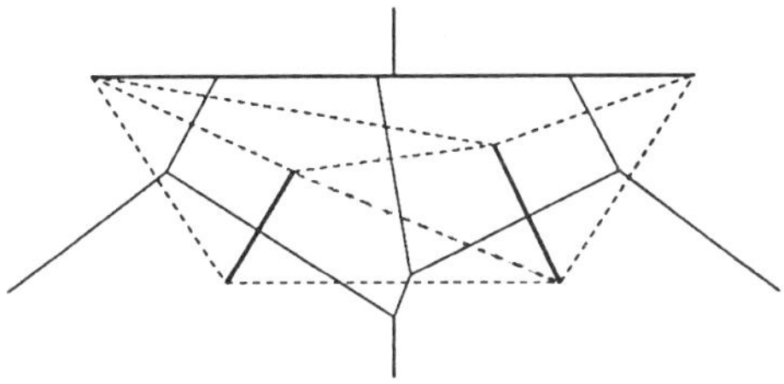

Figure 5. The CDT (dashed) is not the exact dual of the bounded Voronoi diagram.

By these two cases, we have shown that $bac$ cannot be an untriangulated corner. This proves that every connected component is a triangulation. We now show by contradiction that the CDT is connected. Let $a$ and $b$ be nearest-neighbor input points from two different connected components. The diameter circle of $ab$ cannot contain any other input points, so $ab$ must be a Delaunay edge, giving a contradiction. ∎

If the input is not in general position, some faces of the CDT may have more than three vertices, all on a common circle. To avoid this situation, the input can be slightly perturbed before computing the triangulation [60]. Alternatively, the triangulation may be completed after the computation. Mount and Saalfeld [131] showed how to complete a special-position DT or CDT, maintaining some of the triangulation's optimality properties. Dillencourt et al. [50] tie these two approaches together: they show that almost all completions of a Delaunay triangulation can be realized as the DT of a small perturbation of the input.

Just as the DT is dual to the Voronoi diagram, the CDT is related to the *bounded Voronoi diagram* [111, 154, 170], a division of the plane into cells, one for each input vertex, so that the cell for vertex $a$ consists of the region of the plane for which $a$ is the nearest visible input point. Contrary to several claims in the literature, the relation is not exact duality: the CDT may have an edge between a pair of vertices with bounded Voronoi cells that do not meet, but would meet if not cut off by an input edge (Figure 5).

The CDT can be computed directly from Definition 1 by directly testing each candidate edge, but such an algorithm is very inefficient. We now describe the *flip algorithm*, which computes the CDT of a PSLG using a simple local optimization technique; it is derived from a similar algorithm for the DT [102]. The worst-case running time of this algorithm, $O(n^2)$, is not optimal, but its ease of programming makes it quite practical for medium-sized input. Moreover, the algorithm is useful for proving a number of optimality properties.

The flip algorithm starts with any triangulation. For an edge $e$, not an input edge and not on the convex hull, we denote by $Q_e$ the quadrilateral formed by the two triangles on each side of $e$. We say that $Q_e$ is *reversed* if $e$ is not in the

CDT of the four outside edges of the quadrilateral. Equivalently, $Q_e$ is reversed if $e$ forms a smaller minimum angle with the outside edges than the other diagonal does. If $Q_e$ is reversed, trivially the triangulation cannot be the CDT, because edge $e$ violates the circle condition. The converse is also true, giving an example of a local optimization leading to global optimality.

**Lemma 3.** *If no quadrilaterals are reversed, the triangulation is the CDT.*

**Proof:** The idea behind this proof can be traced back to Delaunay [48]. We show that the circumcircle of each triangle $abc$ contains no vertex $d$. Therefore each edge is a correct Delaunay edge. Suppose such a point $d$ does exist for some triangle $abc$. The proof of Lemma 2 shows that we may choose $d$ visible to $a$.

For a circle $C$ with center at coordinates $(x, y)$ and radius $r$, define the *power* of point $s = (x', y')$ to $C$, $p_C(s)$, to be $(x' - x)^2 + (y' - y)^2 - r^2$. Then $p_C(s)$ is positive, negative, or zero exactly when $s$ is outside, inside, or on the boundary of $C$.

Now consider the sequence of triangles $t_1$, $t_2$, ..., $t_k$ crossing line segment $ad$, where $t_1 = abc$ and $t_k$ includes $d$ as a vertex. Construct a corresponding sequence of circumcircles $C_1$, $C_2$, ..., $C_k$. By assumption, each pair $t_i t_{i+1}$ forms a non-reversed quadrilateral. From this, it can be shown that for each $i$, $p_{C_i}(d) > p_{C_{i+1}}(d)$, and therefore that $p_{C_1}(d) > p_{C_k}(d) = 0$. But this contradicts the assumption that $d$ lies inside circle $C_1$. ∎

The flip algorithm maintains a queue of edges whose quadrilaterals might be reversed. In the initial triangulation, the quadrilateral of any non-input edge might be reversed, so the queue initially contains all such edges. Then, we repeatedly remove the first edge $e$ from the queue. If $Q_e$ is not reversed, we simply continue to the next edge. But if $Q_e$ is reversed, we remove $e$ from the triangulation, replacing it with the other diagonal of $Q_e$. This *flip* might change the status of some of the four outside edges of the quadrilateral, so we add the changed ones to the queue if not there already. When the queue is empty, we stop.

**Lemma 4.** *The flip algorithm terminates after $O(n^2)$ flips.*

**Proof:** We use the lifting relation between DTs and convex hulls. Under the mapping that takes $(x, y)$ to $(x, y, x^2 + y^2)$, the DT of the input vertices lifts to the lower convex hull, and—due to the input edges—the CDT lifts to a surface above the lower convex hull. An arbitrary triangulation including the input edges lifts to a surface above the CDT surface. Each flip of a reversed quadrilateral corresponds to gluing a tetrahedron over a reflex edge in this surface. That tetrahedron is never removed, so the reflex edge never returns. After $O(n^2)$ flips, all non-CDT edges will be eliminated. ∎

Coupled with an $O(n^2)$-time algorithm for constructing an initial triangulation (see Section 2.1), we obtain an $O(n^2)$-time algorithm for constructing the CDT. The flip algorithm also gives a number of useful optimality properties of the CDT. The *min-containment circle* of a triangle $t$ is the smallest circle containing $t$.

**Theorem 1.** *Of all triangulations of a PSLG, the CDT (1) minimizes the largest circumcircle; (2) minimizes the largest min-containment circle [45, 139]; and (3) maximizes the minimum angle in the triangulation [102].*

**Proof:** Each of these properties is improved by flipping a reversed quadrilateral. The optimal triangulation cannot be improved, so it has no reversed quadrilaterals, and hence must be the CDT. ∎

A stronger result than (3) holds: the CDT lexicographically maximizes the list of angles, from smallest to largest [103, 105]. An optimal interpolation property of the CDT appears in Section 2.5; there the input vertices include elevations. (The DT of a point set has some further properties. The DT contains the minimum spanning tree; the distance between vertices in the DT is $O(1)$ times their distance in the plane [40, 52, 98]; and, from any viewpoint, DT triangles are acyclically ordered by distance [47, 58].)

For very complex domains, an $O(n^2)$-time algorithm for computing the CDT may be too slow. Chew [41] and others [154, 170] have improved this bound to $O(n \log n)$.

**Theorem 2 (Chew [41]).** *The CDT can be constructed in time $O(n \log n)$.*

**Proof Sketch:** Chew gives a divide-and-conquer algorithm. Line segment endpoints are sorted by $x$-coordinates, and split by a vertical line into two subsets, each with at most $\lceil n/2 \rceil$ endpoints. CDTs are computed recursively for each subset and then merged.

In order to achieve the $O(n \log n)$ bound, the work at each level of the computation tree should be only $O(n)$. Hence the work in a single strip, bounded by two vertical lines, must be proportional to the number of endpoints in the strip, rather than to the number of line segments cutting the strip. To achieve this, the algorithm further divides a strip into regions bounded by segments that cut all the way across it. The algorithm computes the CDT only for those regions containing at least one endpoint.

The major remaining difficulty is merging two adjacent strips. This problem can be reduced to merging adjacent regions. As in some DT algorithms [155], the merger is performed by sweeping a circle along the boundary between the two regions. The work is proportional to the number of old edges removed and new ones added; since the sizes of the old and new triangulations are linear, so is the time bound. ∎

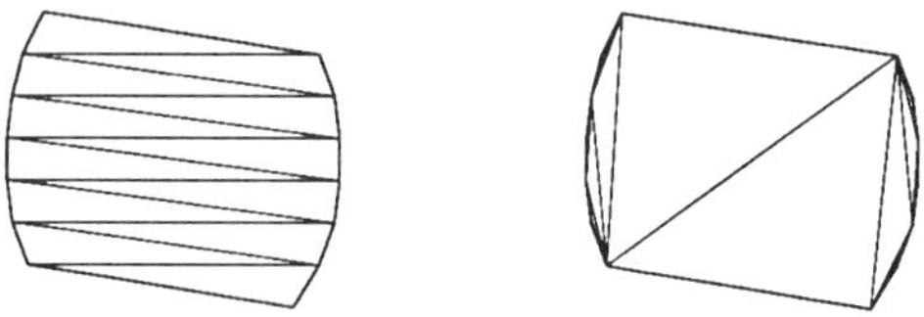

Figure 6. A local optimum for total edge length may be $\Omega(n)$ times the global optimum.

Both of the CDT algorithms just described work for an arbitrary PSLG. There is an $\Omega(n \log n)$ lower bound for constructing any triangulation of a point set or a PSLG, based on a reduction from sorting, but this bound does not apply to polygon triangulation. So the following question arises.

**Open Problem 1.** *Can the CDT of an arbitrary simple polygon be constructed in time $o(n \log n)$, or is there a lower bound of $\Omega(n \log n)$ for this problem?*

For some special cases, the answer is known. Aggarwal et al. [1] showed that, if the input is a convex polygon, the DT (which is also the CDT) can be found in linear time. The algorithm is based on the lifting transformation, and it also solves the problem for certain other types of point sets. In particular, it can update the DT after the removal of a vertex or edge. Djidjev and Lingas [51] extended the linear-time algorithm to compute the DT (which is also the CDT) of a *monotone histogram*. A monotone histogram is a polygon in which the vertex order around the polygon simultaneously sorts all but one of the vertices by both $x$- and $y$-coordinates. A convex polygon is thus the union of four monotone histograms.

### *2.2.2. Edge Insertion*

The flip algorithm described above provides our first example of a local improvement algorithm. As we showed, the flip algorithm in fact produces a global optimum: the resulting triangulation maximizes the minimum angle, and optimizes several other criteria as well. The success of the flip algorithm for Delaunay triangulation has led to the use of edge flipping (with the appropriate definitions of reversed quadrilateral) for finding triangulations that approximately optimize other criteria, such as vertex degree [74], maximum angle [83], total edge length [172], or the ratio of the areas of incircle and triangle [13]. Edge flipping to improve these criteria, however, will not usually compute a global optimum.

The problem is that the algorithm can get stuck in a *local optimum*, in which no flip improves the triangulation. A local optimum can be very far from a global optimum; for example, it may have total edge length $\Omega(n)$ times the true optimal length; see Figure 6.

One way to escape local optima is to allow local moves that do not improve the triangulation, as in *simulated annealing* [100, 169]. A different approach is to generalize the local improvement procedure. This reduces the number of local optima, as a triangulation without an edge to flip may still admit the generalized move. This section describes one such generalization, called *edge insertion*, introduced by Edelsbrunner, Tan, and Waupotitsch [64] for the minmax angle problem.

Consider adding a new edge $e$ to some existing triangulation $\mathcal{T}$ of a PSLG. Edge $e$ crosses other edges in $\mathcal{T}$, causing them to be removed. At this point there will be two simple polygons without diagonals, one on each side of $e$. Optimal triangulation problems on simple polygons tend to be tractable; for now we may assume that these polygons are triangulated optimally using dynamic programming, as in Section 2.2.3. *Edge insertion* is the process of adding a candidate edge, incident to a vertex of a worst triangle (and cutting across that triangle); removing the crossed edges; and retriangulating the remaining regions. The added edge is rejected and the triangulation is returned to its previous state if the triangulation gets worse. Below we explain how to eliminate a possible edge on each insertion, so that the process terminates.

Notice that an edge insertion is more general than an edge flip, as a flip inserts a diagonal of a convex quadrilateral and removes the single edge it crosses. In fact, edge insertion is a significant generalization, as there may be $\Omega(n)$ edges that can be inserted to break a worst triangle, but only one flip. Because of the increased number of possibilities, intuitively edge insertion should reach better local optima than edge flipping, but take longer to do so.

Edelsbrunner et al. [64] showed that edge insertion can in fact compute a global optimum that edge flipping cannot: the triangulation minimizing the maximum angle. The correctness of edge insertion for this problem follows from an abstract property of the maximum angle measure, that also holds for several other natural quality measures, as shown in a subsequent paper by Bern, Edelsbrunner, Eppstein, Mitchell, and Tan [18]. Let $f$ be a function measuring the badness of triangles, and assume that the quality of a triangulation $\mathcal{T}$ is defined to be the maximum (that is, worst) value of $f$ over all triangles in $\mathcal{T}$, denoted $f(\mathcal{T})$.

**Definition 2.** *Let $a$, $b$, and $c$ be vertices in some PSLG $G$. We say that $a$ is an* **anchor vertex** *of triangle $abc$, if in any triangulation $\mathcal{T}$ of $G$, with $f(\mathcal{T}) < f(abc)$, there is an edge $ad$ crossing $bc$. In other words, one cannot improve a triangulation containing triangle $abc$ without cutting $abc$ by an edge incident to the anchor $a$.*

For example, let $f(abc)$ be the measure of the largest angle in triangle $abc$. Then if the largest angle is $\angle bac$, $a$ is an anchor vertex, because in any triangulation of quality better than $f(abc)$, there must be an edge subdividing $\angle bac$ and crossing $bc$. A triangle may have more than one anchor vertex, and all vertices in an optimal triangulation are anchors.

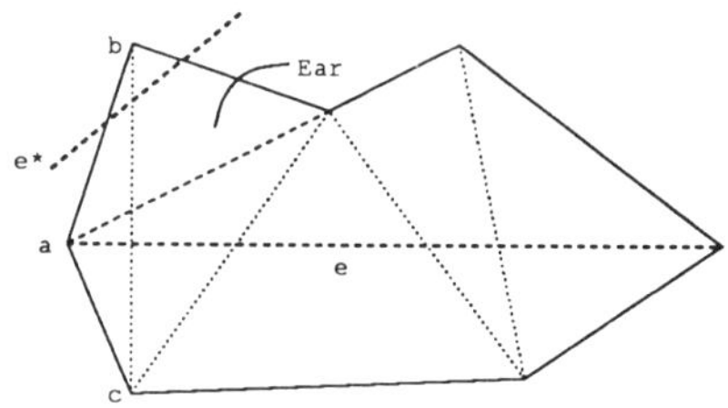


Figure 7. The ear triangle cut by $e^*$ can be no worse than $abc$.

**Definition 3.** *Quality measure $f$ has the* **weak anchor property** *if, for each triangulation $\mathcal{T}$, and each triangle $abc$ in $\mathcal{T}$ with $f(abc) = f(\mathcal{T})$, there is an anchor vertex of $abc$. Similarly, $f$ has the* **strong anchor property** *if all triangles of $\mathcal{T}$ (not just worst triangles) have anchors.*

**Lemma 5 (Bern et al. [18]).** *Let $f$ be a function with the weak anchor property. Then edge insertion finds a triangulation minimizing $f$.*

**Proof Sketch:** Assume $\mathcal{T}$ does not optimize $f$, $abc$ is a worst triangle in $\mathcal{T}$, and $abc$ has anchor vertex $a$. Let $\mathcal{T}^*$ denote a triangulation optimizing $f$; then some edge $e$ of $\mathcal{T}^*$ is incident to $a$ cutting $bc$. We show that inserting $e$ cannot make $\mathcal{T}$ worse, and hence the edge insertion algorithm cannot get stuck. This follows if the simple polygons on each side of $e$ can be triangulated with triangles of quality no worse than $f(abc)$. See Figure 7.

Call one of the simple polygons $P$. If $P$ has more than three sides, some edges of $\mathcal{T}^*$ cut across $P$. It is not hard to see that an edge $e^*$ of $\mathcal{T}^*$ that is "maximally far" from $e$ must cut off a corner of $P$; that is, the side of $e^*$ away from $e$ supports a triangle in the union of $P$ and $\mathcal{T}^*$. We now reduce $P$ by removing the *ear* at that corner (the triangle defined by the endpoints of the sides of $P$ incident to the corner). If this ear is worst in a new (nonoptimal) triangulation, its anchor cannot be the corner vertex, because there is no edge in triangulation $\mathcal{T}^*$ cutting from this vertex through the opposite edge. So its anchor must be one of the two side vertices. But there is no edge in $\mathcal{T}$ cutting from such an anchor through the opposite edge. Hence the ear triangle is either not a worst triangle or it has quality no worse than $f(\mathcal{T})$. Continuing the process of removing ears triangulates $P$. ∎

A successful insertion eliminates the crossed edge of the worst triangle in $\mathcal{T}$ (that is, the weak anchor property implies that the crossed edge need not be in an optimal triangulation); an unsuccessful insertion eliminates the edge inserted. Thus there are $O(n^2)$ insertions. Retriangulating by repeatedly removing ears at least as good as the worst triangle, rather than by dynamic programming, gives time $O(n)$ for each insertion. Thus the total time is $O(n^3)$. If the strong anchor property holds and insertions are performed in a certain order, then all crossed edges—not just the one in the worst triangle—can be eliminated in a successful insertion; then the running time can be improved to $O(n^2 \log n)$ [18, 64].

**Theorem 3 (Edelsbrunner et al. [64], Bern et al. [18]).** *A triangulation minimizing the maximum angle, or maximizing the minimum height, can be computed in time $O(n^2 \log n)$. A triangulation minimizing the maximum distance of a triangle from its circumcenter can be found in time $O(n^3)$.* ∎

The minmax angle triangulation has direct relevance to mesh generation due to the error estimate of Babuška and Aziz [3]. The distance from the circumcenter, or "eccentricity", is a measure of obtuseness, with larger triangles weighted more heavily. In Section 2.5, we show that edge insertion also finds an interpolating surface with minimum slope. All of these criteria are mentioned in [172].

### *2.2.3. Dynamic Programming*

We have seen that a number of optimal triangulation problems admit efficient solutions for PSLGs (including polygons as a special case). We next show that many more such problems can be solved in polynomial time, for simple polygons only, using a dynamic programming approach usually attributed to Klincsek [101].

Let $f$ be a quality measure of triangulations of simple polygons, that is a mapping from triangulations to the real numbers. Let $P$ be a simple polygon. An arbitrary diagonal $\{a, b\}$ splits $P$ into two simple polygons, $P_1$ and $P_2$. Let $\mathcal{T}_1$ and $\mathcal{T}_2$ be triangulations of $P_1$ and $P_2$, and let $\mathcal{T}$ be the triangulation of $P$ that is the union of $\mathcal{T}_1$, $\mathcal{T}_2$, and edge $\{a, b\}$.

**Definition 4.** *We say $f$ is* **decomposable** *if it meets the following conditions: (1) there is a "combining" function $g$ such that $f(\mathcal{T}) = g(f(\mathcal{T}_1), f(\mathcal{T}_2), a, b)$, for all choices of polygon $P$, diagonal $\{a, b\}$, and triangulations $\mathcal{T}_1$ and $\mathcal{T}_2$; (2) $g$ is computable in time $O(1)$ and monotonic in its first two arguments; and (3) if $\mathcal{T}$ is a single triangle, then $f(\mathcal{T})$ is computable in time $O(1)$.*

In other words, the measure of the entire triangulation can be computed quickly from the measures of the two pieces, together with the knowledge of how the pieces are glued together.

**Lemma 6.** *The following are decomposable measures: the minimum (maximum) angle in the triangulation, the minimum (maximum) circumcircle of a triangle, the minimum (maximum) length of an edge in the triangulation, the minimum (maximum) area of a triangle, and the sum of edge lengths in the triangulation.* ∎

Confirming this lemma is straightforward. For example, if $f(\mathcal{T})$ measures the minimum angle, then $g(f_1, f_2, a, b)$ is simply $\min\{f_1, f_2\}$. For the sum of edge lengths, $g(f_1, f_2, a, b) = f_1 + f_2 - |ab|$. This last criterion is especially important, as minimizing it for point sets (the well-known *minimum weight triangulation problem*) seems to be very difficult.

An example of a nondecomposable measure is the maximum degree of a vertex, as the maximum degree in $\mathcal{T}$ does not depend only on the maximum degrees in $\mathcal{T}_1$ and $\mathcal{T}_2$, but also on the degrees at $a$ and $b$. A measure that fails the monotonicity requirement is the difference in areas between the largest and smallest triangles. (Incidentally, both of these measures can be optimized by slightly more complicated dynamic programming.)

**Theorem 4 (Klincsek [101]).** *A triangulation of a simple polygon optimizing any decomposable function can be computed in time $O(n^3)$.*

**Proof:** Assume we are trying to minimize $f(\mathcal{T})$. Number the vertices of polygon $P$ by $v_1, v_2, \ldots, v_n$, in order around the perimeter. If $v_i v_j$ is a diagonal of $P$, we denote by $P(i,j)$ the polygon formed by points $v_i, v_{i+1}, \ldots, v_j$. Let $F(i,j)$ be the minimum value of $f$ on a triangulation of $P(i,j)$. If $v_i v_j$ is not a diagonal, define $F(i,j) = +\infty$. We would like to compute $F(1,n)$.

Note that in any triangulation of $P(i,j)$, $v_i v_j$ must be a side of a triangle, say $v_i v_j v_k$, with $i < k < j$. Using the assumption that $f$ is decomposable, we can compute the measure of the optimal triangulation of $P(i,j)$ by trying all choices for $k$.

$$F(i,j) = \min_{i<k<j} g\,(g(f(v_i v_j v_k), F(i,k), v_i, v_k), F(k,j), v_k, v_j).$$

We compute these values in increasing order of $i$, and for each $i$ in increasing order of $j$; then each value of $F$ will be computed before it is needed. This computes the measure of the triangulation; to compute the triangulation itself we maintain *back pointers* for each pair $(i,j)$ to the $k$ that supplied the minimum.

Each computation of the recurrence takes constant time per possible value of $k$, or $O(n)$ total. Testing whether a pair $(i,j)$ forms a diagonal also takes $O(n)$ time. There are $O(n^2)$ such computations, for a time bound of $O(n^3)$. ∎

In general this is the best time bound known. But the bulk of the work is done for pairs $(i,j)$ that form a diagonal—the other pairs can be quickly ruled out. A sharper time bound is "input-sensitive", depending only on the number of diagonals in the polygon.

**Definition 5.** *The* **visibility graph** *of a polygon $P$ has vertex set consisting of the vertices of $P$, and an edge between vertices $a$ and $b$ if $a$ is visible to $b$ in $P$.*

Let $E$ be the number of edges in the visibility graph. The visibility graph can be computed in time $O(n^2)$ by traversing the boundary of the polygon once per vertex, performing a simple stack algorithm in each traversal [65, 93, 97, 104]. A more complicated algorithm reduces the time to $O(n \log n + E)$ [85].

We can easily improve the dynamic programming algorithm from $O(n^3)$ to $O(En)$, but we can even do a little better. The time bound depends on the number of triangles in the polygon; below we show that few edges imply few triangles.

**Lemma 7.** *A graph with $E$ edges has at most $O(E^{3/2})$ triangles.*

**Proof:** Divide the vertices into two classes: *heavy* vertices with degree at least $\sqrt{E}$, and *light* vertices with smaller degree. If $b$ is light we enumerate the triangles containing $b$ by examining each pair of edges $ab$, $bc$, and testing if $ac$ is also an edge. Each edge belongs to $O(\sqrt{E})$ pairs at its two endpoints, so this produces $O(E^{3/2})$ triangles overall. If $b$ is heavy we enumerate triangles containing $b$ by examining all edges $ac$, and testing if $ab$ and $bc$ are edges. There are $O(\sqrt{E})$ heavy vertices, so again the total is $O(E^{3/2})$ triangles. ∎

**Theorem 5.** *A triangulation of a simple polygon optimizing any decomposable function can be computed in time $O(n^2 + E^{3/2})$, where $E$ is the number of edges in the visibility graph.*

**Proof:** Recall that the dynamic programming algorithm tests diagonals $(i, j)$ in order by $i$, and then for each $i$ in order by $j$. When we start a new value of $i$, we enumerate the triangles containing $v_i$ as in Lemma 7, and store for each $j$ the list of $k$'s that form triangles $v_i v_j v_k$. When we compute $F(i, j)$, we minimize only over the $k$'s stored on this list. ∎

It is curious that the worst case of this algorithm occurs when the polygon is convex (so that the visibility graph contains all possible edges). Typically, convexity makes optimization problems easier. Examples include geometric matching [120] and greedy triangulation (see below) [107, 108].

In the matching problem, one must connect the vertices in pairs by diagonals, minimizing the total edge length. The resulting graph has no crossings, and in this sense resembles the minimum weight triangulation. Matching can be solved by the same dynamic programming techniques as above; however, for convex polygons this can be improved to $O(n \log n)$ [120]. Another problem, construction of optimal binary trees, is also closely related to triangulation of convex polygons [160], and again the $O(n^3)$ dynamic program solves this problem. Yao [175] improved this to $O(n^2)$ using the *quadrangle inequality*, a relation that also holds for diagonal lengths in a convex polygon. In some cases this can be further improved to $O(n \log n)$ [88]. These results suggest that, at least for minimum weight triangulation of convex polygons, $O(n^3)$ is too slow.

**Open Problem 2.** *Can the minimum weight triangulation of a convex polygon be constructed in time $o(n^3)$?*

### 2.2.4. Other Optimal Triangulations

Most of the remaining work on optimal triangulation has used edge length as a quality measure. As mentioned above, edge length can be used as a simple measure of triangle size, which in turn affects finite element approximation error.

Edelsbrunner and Tan [61] considered the triangulation minimizing the maximum edge length. They showed that this triangulation (like the DT) contains the edges of the minimum spanning tree. Therefore it can be found in time $O(n^3)$, by first computing the minimum spanning tree and then triangulating each remaining polygon using dynamic programming. They reduced this time to $O(n^2)$.

The *greedy triangulation* [76, 108] can be found by adding edges one at a time, always choosing the shortest edge that is not already crossed. This triangulation lexicographically minimizes the sorted vector of edge lengths. For arbitrary point sets the greedy triangulation can be computed in time $O(n^2)$ by dynamic maintenance of a CDT [108]. For convex polygons the time bound can be improved to $O(n)$ [108].

Eppstein [66] obtained another min-min triangulation result: the *farthest-point Delaunay triangulation* of a convex polygon minimizes the minimum angle. This triangulation can again be constructed in time $O(n)$ [1]. The farthest-point DT dualizes the farthest-point Voronoi diagram, a data structure for finding the farthest input point from a query point. The farthest-point DT can be defined for arbitrary point sets, but in general it is not a valid triangulation, as it only has edges connecting vertices on the convex hull.

Perhaps the most longstanding open problem in computational geometry is the complexity of the minimum weight triangulation (MWT) for arbitrary point sets [54]. (Recall that the MWT asks for the minimum total edge length.) Indeed, early authors called this the "optimal triangulation" problem. Garey and Johnson [76] included MWT in their list of famous problems neither known to be NP-complete nor known to be solvable in polynomial time. If the MWT problem is generalized slightly, so that the weight of an edge is an arbitrary function unrelated to its length, minimum weight triangulation becomes NP-complete [113]. Therefore, authors have concentrated on *approximating* the MWT.

Any triangulation achieves total edge length $O(n)$ times the minimum [99]. The DT, once claimed to be the MWT, can be as long as $\Omega(n)$ times the optimum [99, 119]. The same is true of the triangulation computed by edge-flipping for minimum length. It remains open how well edge insertion approximates the MWT, but it does not provide an exact solution. The greedy triangulation can be as bad as $\Omega(\sqrt{n})$ [106, 119]. For convex polygons, however, the greedy triangulation is an $O(1)$ approximation [107, 108]. The simple *ring heuristic*, that repeatedly connects every other vertex, gives a triangulation of length $O(\log n)$ times the boundary length of a convex polygon [136].

Lingas [110] suggested the following approach to MWT: start by adding the edges of the convex hull and the minimum spanning tree, and then compute the optimal triangulation within each polygonal region. Building on this approach, Plaisted and Hong [136] gave what is currently the best MWT approximation. Instead of starting with the minimum spanning tree, they partition the convex hull into convex polygons. Then the optimal triangulation, greedy triangulation, or ring heuristic can be used to triangulate these polygons, achieving an $O(\log n)$ approximation. The Plaisted-Hong algorithm has recently been implemented with a running time of $O(n^2 \log n)$ [162].

**Open Problem 3.** *Is it possible to find the MWT (or an $O(1)$ approximation to the MWT) of a point set in polynomial time? Or is this problem NP-complete?*

### *2.3. Steiner Triangulation*

In this section we discuss triangulations using Steiner points. As shown in Figure 1, a Steiner triangulation of a point set may add points outside the convex hull of the input. In a Steiner triangulation of a polygonal region or a PSLG, edges may be subdivided, but they must be *covered*, that is, each original edge must be a union of triangulation edges. Input vertices must be covered by triangle vertices; the remaining triangle vertices are the *Steiner points*.

Any two-dimensional input can be triangulated without Steiner points, so Steiner triangulation only makes sense in the context of some optimality criterion. We consider the following criteria: maxmin angle, minmax angle, maxmin height, and minimum total edge length. Finally, we discuss the conforming Delaunay triangulation problem.

We must exercise a little care in formulating Steiner versions of optimal triangulation problems. For both theoretical and practical reasons, we must concern ourselves with the number of Steiner points. Without any bound on the number of Steiner points, there may be no optimal triangulation—for example, the minimum angle in a triangulation of a point set can be brought arbitrarily close to 60°. And in practice, the number of Steiner points in a mesh directly affects the time to solve a finite element computation.

One might specify the desired number of Steiner points, and then find the optimal triangulation with that many points. Alternatively, one might specify the desired quality measure, and then minimize the number of Steiner points. Either of these formulations, however, results in seemingly intractable problems.

For these reasons, we turn to approximation algorithms. The algorithms we describe achieve quality that is within a constant of the best possible, while using a modest number of Steiner points. In some cases, the algorithms actually use a number of Steiner points that is within a constant factor of the minimum needed for the quality achieved. Allowing approximate optimality offers the additional benefit that we can sometimes simultaneously guarantee bounds on several different measures in the same triangulation.

### *2.3.1. No Small Angles*

The first problem we consider is maximizing the minimum angle, solved in the non-Steiner case by Delaunay triangulation. In the Steiner version of this problem, we demand no small angles: every angle must be greater than some fixed bound.

The smallest angle of a triangle is related to two other quality measures: aspect ratio and height. If the smallest angle in a triangle is $\theta$, the aspect ratio is between $1/\sin\theta$ and $2/\sin\theta$.

**Definition 6.** *The* **height** *of a triangle is the minimum distance from a vertex to a side. The* **aspect ratio** *is the ratio of the length of the longest side to the height.*

If there is any lower bound on the angles, the complexity of a triangulation of a polygonal domain may be nonpolynomial; indeed, inputs of constant complexity may need an unbounded number of triangles. Consider a rectangle with short side length one and long side length $A$. Then any bounded-aspect-ratio triangle contained in the rectangle has area $O(1)$; hence $\Omega(A)$ triangles are necessary to triangulate the rectangle with no small angles.

This lower bound holds in a different form for point set input. Consider a bounded-aspect-ratio Steiner triangulation of the vertices of the same rectangle. There must be an edge incident to the upper left corner of the rectangle that has length at most one. Now imagine walking from this vertex to the upper right corner along edges of the triangulation. Each successive edge can be only a constant multiple longer than the previous edge, and the first edge must have length $O(1)$. Therefore there must be $\Omega(\log A)$ edges in the walk.

Baker, Grosse, and Rafferty [10] first solved the problem of computing no-small-angle triangulations of polygonal regions and PSLGs. Their algorithm uses triangles with angles between 13° and 90°, thereby also solving the nonobtuse triangulation problem described in the next section. They place a uniform square mesh over the input, fine enough that input vertices are several squares apart in the mesh. Using a number of special cases, they show how input edges may be incorporated into the mesh, by retriangulating the squares within a small distance of the edges. They also provide a similar set of special cases to handle the triangulation near each vertex. Melissaratos and Souvaine [124] extend these techniques and give an analysis of the number of Steiner points used by the algorithm.

**Definition 7.** *The* **input feature size** *of a point set is the minimum distance between input points. The input feature size of a PSLG is the minimum distance between a vertex and an edge not incident to that vertex. For a polygon (with*

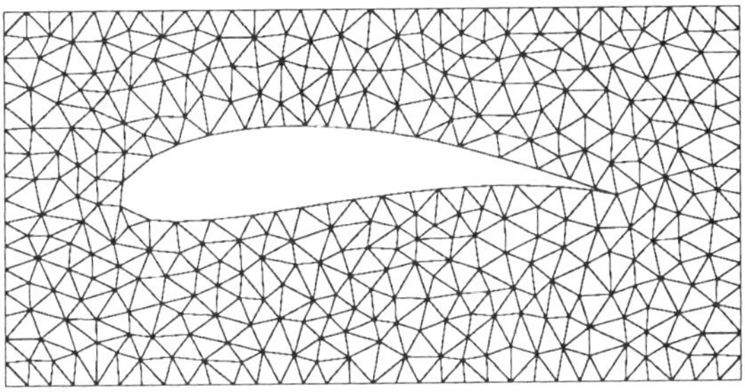

Figure 8. A uniform mesh computed by Chew's algorithm (Chew).

*holes), the input feature size is analogous, only distance is measured by a shortest path interior to the polygon.*

Chew [42] found a quite different approach to no-small-angle triangulation of PSLGs. He assumes that all boundary edges have lengths between $s$ and $\sqrt{3}s$, where $s$ is the input feature size defined above. This condition can be enforced by subdividing edges. Chew starts with the constrained Delaunay triangulation of the input, since it maximizes the minimum angle. Then, while there is a triangle with circumcircle of radius greater than $s$, he adds a Steiner point at the center of the circle, and recomputes the CDT.

**Theorem 6 (Chew [42]).** *The algorithm above computes a Steiner triangulation in which all angles are between* $30°$ *and* $120°$.

**Proof Sketch:** No point is ever added closer than $s$ to another point. Hence the procedure terminates, as only a finite number of points can fit into the polygonal region. At the end of this procedure, all Delaunay circles have radius at most $s$, and all edge lengths are between $s$ and $2s$. No triangle with edge lengths in this range can be more acute than $30°$ or more obtuse than $120°$, without having a circumcircle radius larger than $s$. ∎

Chew also gave a streamlined modification of this algorithm. First place the input in a square grid with side length $s$. Successively add Steiner points to grid squares that have no vertices within distance $s$ already. Each update involves only points and edges within $O(1)$ adjoining grid squares, and hence can be performed in time $O(1)$. Assuming that not too much time is wasted on exterior grid squares, the entire construction takes time $O(n+k)$, where $k$ is the number of Steiner points. Now $k$ may be quite large, but it is within a constant factor of the complexity of any triangulation in which all triangles have side length $O(s)$. In other words, if we require uniform triangle size, the output complexity approximates the optimum. See Figure 8 for an example.

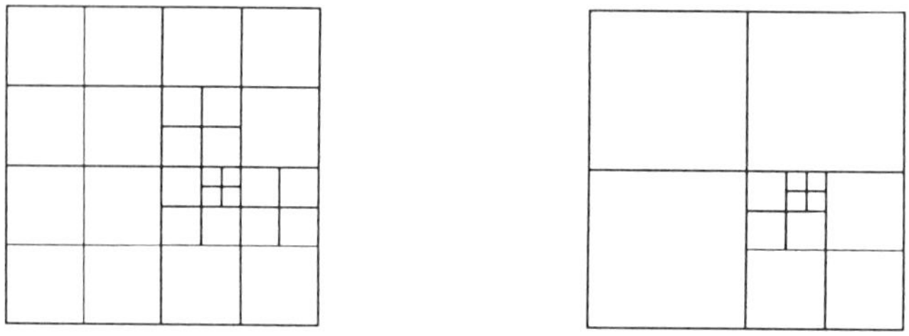

Figure 9. Balanced and unbalanced quadtrees.

In many practical situations, however, widely varying triangle sizes offer greater accuracy and efficiency. For instance, in airflow simulations, we would like small triangles in turbulent regions for accuracy, and large—or at least long—triangles in smooth regions for efficiency.

Bern, Eppstein, and Gilbert [20] gave the first theorems on approximately-optimal-complexity meshes with varying triangle sizes. Their "provably good" algorithms use *quadtrees*, which had been previously suggested by Baker et al. [10] and used in heuristics by Yerry and Shephard [176].

**Definition 8.** *A* **quadtree** *[146, 147] is a recursive partition of a region of the plane into axis-aligned squares. One square, the* **root**, *covers the entire region. A square can be divided into four* **child** *squares, by splitting it with horizontal and vertical line segments through its center. The collection of squares then forms a tree, with smaller squares at lower levels of the tree.*

Bern et al. also maintain a *balance condition* in their quadtrees. Call a quadtree square a *leaf square* if is not subdivided into children. The balance condition requires that for each leaf square $B$, the squares sharing a portion of a side with $B$ must be at most twice the size (side length) of $B$. Equivalently, each side of a leaf square is subdivided into at most two parts by neighboring squares. See Figure 9.

Assume that the input is a simply a point set. The triangulation algorithm builds a quadtree covering the set of input points. The quadtree squares are divided into smaller squares, until each input point is *well separated* from the other points: it must be in the center of a five by five grid of squares, all the same size, none containing another input point. Next the algorithm *warps* the quadtree to conform to the input. Each input point chooses its nearest quadtree vertex. No vertex can be chosen twice, because of the separation of input points. A chosen vertex is removed from the quadtree, and its incident edges are reconnected to the input point.

Finally, the quadtree is triangulated. Because of the balance condition, an unwarped (hence square) region in the quadtree has at most one subdivision point per side; it is easy to confirm that such a region can be triangulated into

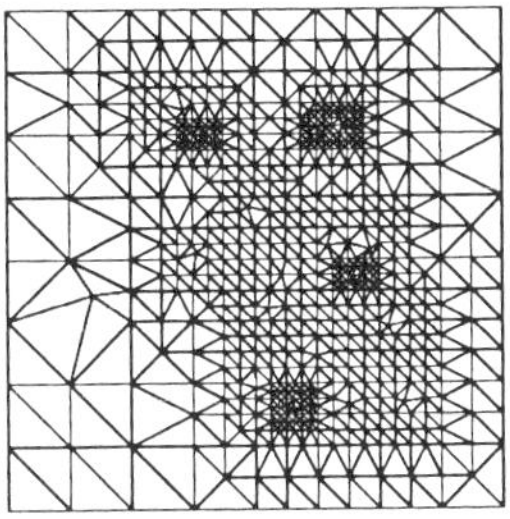

Figure 10. A quadtree triangulation of a point set.

$O(1)$ triangles with no angles smaller than $\arctan(.5) \approx 26.5°$. A warped region is a quadrilateral, with no subdivisions. Three quadrilateral vertices are corners of a square, and the fourth must be fairly near a corner. Such a quadrilateral can be triangulated with no angles smaller than 20° by adding a single diagonal. See Figure 10. A similar, more complicated, algorithm (along the lines of Baker et al. [10]) gives a triangulation of a polygon with holes with no *new* angles smaller than $\arctan(1/3) \approx 18.4°$. (A sharp input angle cannot be erased.) The real innovation, however, is the proof that the number of Steiner points is small, based upon the following lemma.

**Lemma 8.** *The number of squares produced by the algorithm above is $O(n \log A)$, where $A$ is the maximum aspect ratio of a triangle in the DT of the input point set.*

**Proof Sketch:** The only nonlinear behavior occurs when a quadtree square $B$ is split into children, and no input point becomes well separated by this split. In this case we can find a Delaunay triangle $t$ connecting the input points in $B$ to the rest of the DT, and charge the splits of $B$ and its descendants to $t$. The number of splits charged to a single triangle is proportional to the logarithm of its aspect ratio. ∎

**Theorem 7 (Bern et al. [20]).** *The quadtree algorithm above uses $O(k)$ triangles, where $k$ is the minimum number of triangles in a triangulation of the input with no angle smaller than 20°.*

**Proof Sketch:** Let $S$ be the set of input points along with the Steiner points used in an optimal triangulation with no angle smaller than 20°. The DT of $S$ has no angle smaller than 20°, and hence no aspect ratio larger than 6. Then by the lemma above, the quadtree triangulation of $S$ uses $O(k)$ Steiner points. Adding Steiner points to the input set only increases the complexity of the quadtree, so the triangulation of the original point set also has no more than $O(k)$ Steiner points. ∎

This theorem holds no matter what smaller constant replaces 20°, but the constant hidden in the big-$O$ notation increases as the angle bound decreases.

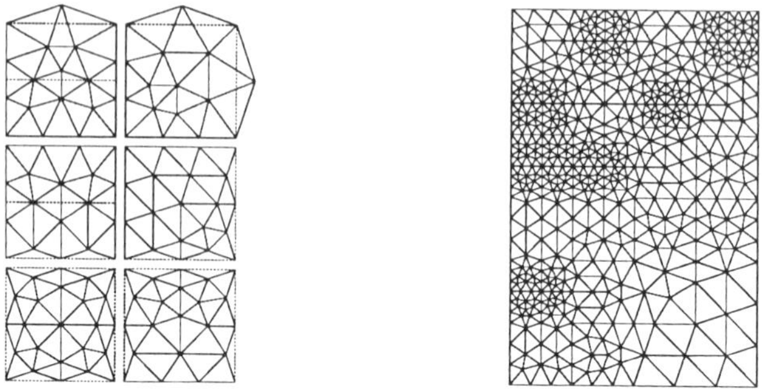

Figure 11. (a) Tiles for all acute triangulation. (b) Example triangulation.

By strengthening the balance condition in the quadtrees, and replacing quadtree squares by tiles with projections and indentations (Figure 11), Bern et al. can guarantee that all angles measure between 36° and 80°. Perhaps one can achieve 51° and 72°, but any further improvement would force the triangulation to be topologically equivalent to a mesh of equilateral triangles, which seems to require a much larger number of Steiner points.

A similar theorem holds for the quadtree algorithm for polygon input. The analog of Lemma 8 guarantees $O(nA)$ quadtree squares, where $A$ is now the maximum aspect ratio of the CDT. Melissaratos and Souvaine [124] combine the algorithms of Baker et al. and Bern et al. to guarantee no obtuse angles as well.

### *2.3.2. No Large Angles*

Many practitioners have suggested largest angle as an important quality measure for both mesh generation [3, 14, 83, 90] and surface interpolation [80, 81]. A bound on the smallest angle implies a bound on the largest angle, but as we have seen, the elimination of small angles requires a triangulation with complexity dependent not only on the number of input vertices $n$, but also on the geometry of the input. In this section, we allow small angles and achieve polynomial bounds on the number of triangles in a no-large-angle triangulation.

If the largest angle in a triangle is $90° - \epsilon$, then the smallest must be at least $\epsilon/2$. Thus the strongest bound on the largest angle, that does not imply a bound on the smallest angle, is 90°; in other words, the triangulation must be *nonobtuse*. Nonobtuse triangulation claims a number of motivations. A nonobtuse mesh guarantees some desirable numerical properties related to diagonal dominance [10]. A second motivation involves planar duality. Each (closed) triangle in a triangulation contains its circumcenter exactly when all angles measure at most 90°. The planar dual of such a triangulation can be formed by simply adding perpendicular bisectors of edges (see [22] for more on constructing planar duals). Practitioners use dual meshes in the "finite volume" method in which each mesh vertex has an associated control volume; a perpendicular planar dual defines especially convenient control volumes, simplifying the calculation of flow or forces across element boundaries. Third, nonobtuse triangulation has application to computational learning theory [145]. Fourth, a nonobtuse triangulation simultaneously guarantees some optimality properties, as shown by the following lemma.

**Lemma 9.** *Any triangulation in which no triangle is obtuse must be the Delaunay triangulation of its vertices. A nonobtuse triangulation also minimizes the maximum angle, and maximizes the minimum height.*

**Proof Sketch:** By Lemma 3, if a triangulation is not the DT then some two adjacent triangles form a reversed quadrilateral. But no quadrilateral in which both triangles are nonobtuse can be reversed. Nonobtuse triangulations are unique, up to the choice of diagonal for points forming a rectangle. Any other triangulation would have to have an obtuse angle, and so could not minimize the maximum angle. The maxmin height triangulation is discussed below. An examination of cases shows that, if the triangle with minimum height is nonobtuse, then in any other triangulation there is an equal or smaller height triangle involving one of the same three vertices. ∎

As noted above, Baker et al. [10] give an algorithm for nonobtuse triangulation of polygons. Their algorithm also avoids small angles, and hence must have complexity that depends on geometry, as well as on $n$. Is it possible to eliminate dependence on geometry? For point sets, a nonobtuse triangulation with $O(n^2)$ triangles is trivial to construct, by simply passing horizontal and vertical lines through each point. The following theorem improves this bound.

**Theorem 8 (Bern et al. [20]).** *For any set of $n$ points, there is a nonobtuse Steiner triangulation of complexity $O(n)$.*

**Proof Sketch:** The algorithm takes as its starting point the quadtree algorithm of Bern et al. [20] described above. The only nonlinear behavior of this algorithm occurs when a square is split without separating any input points. If this happens repeatedly, some tightly spaced cluster of points must be escaping separation by the quadtree sides.

In this algorithm, such a cluster is triangulated recursively inside a small square. A rather complicated construction, using only nonobtuse triangles, then attaches this square to a large quadtree square containing it. Thus we shortcut many levels of the quadtree. ∎

For polygons, the problem requires some new ideas. Bern and Eppstein [19] devised an algorithm for nonobtuse triangulation of polygons, with complexity bounded by $O(n^2)$. We now describe the algorithm.

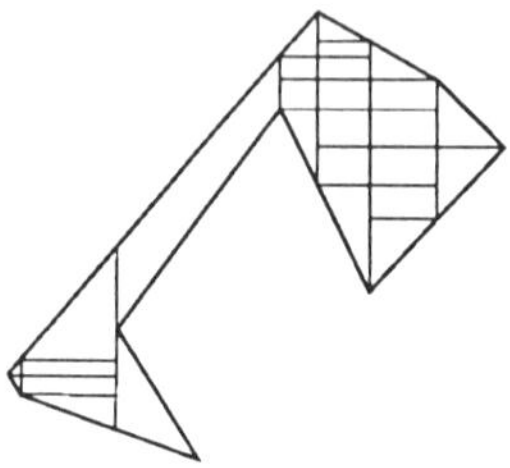

Figure 12. A polygon cut for nonobtuse triangulation.

The first step dices the polygon into rectangles, with nonrectangular portions left over at the boundaries. Start by passing a vertical line through each input vertex, stopping the line at the polyon boundary. Add a Steiner point at all intersections of input edges and vertical lines. This step divides the polygons into *slabs*—quadrilaterals with two vertical sides, possibly having subdivision points on the vertical sides. Next, draw a horizontal line segment through each vertex (input or Steiner), extending the line segment to the last possible vertical segment. In other words, each endpoint of a horizontal segment should lie either on a vertical segment, or on the vertex inducing the horizontal, and each horizontal segment should be as long as possible with this property. A polygon divided as above by horizontal and vertical lines is shown in Figure 12.

At this point, the polygon is divided into regions of four types: (1) rectangles with unsubdivided sides; (2) right triangles with hypotenuse on the boundary of the polygon and vertical leg possibly subdivided; (3) obtuse triangles with two sides on the boundary of the polygon, and one leg vertical and possibly subdivided; (4) slabs with two sides on the boundary of the polygon, and two possibly-subdivided vertical sides, that cannot be simultaneously crossed by a horizontal line. The slabs can be divided by a diagonal into two obtuse triangles. The nonobtuse triangulation problem has now been reduced to triangulating right and obtuse triangles, that have subdivision points on one vertical side, while introducing new subdivisions only on the triangles' longest sides, which lie along the polygon boundary.

**Lemma 10.** *A right or obtuse triangle with $m$ subdivision points on its two shorter sides can be triangulated into $O(m^2)$ nonobtuse triangles, without adding any new Steiner points to its shorter sides.*

**Proof Sketch:** The basic strategy is to replace the shorter sides by sides that form sharper angles with the long side. The region between the old and new sides is triangulated to reduce the number of subdivision points by one.

There are a number of different reduction maneuvers. Two of them are shown in Figure 13. In the first example, the apex $c$ of the triangle (that is, the vertex with the obtuse angle) has been "merged" with the first subdivision point

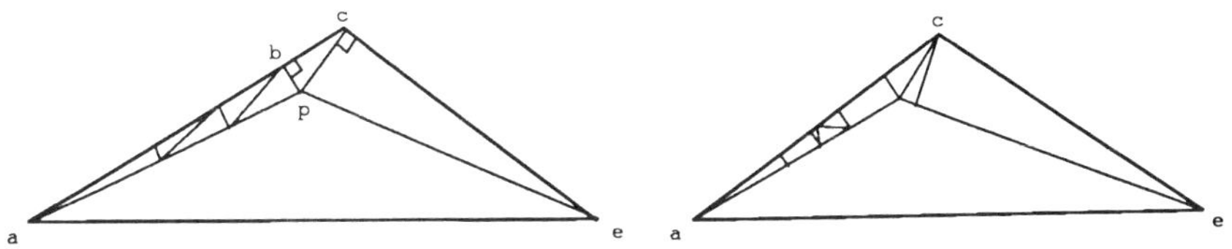


Figure 13. Reduction steps for nonobtuse triangulation of an obtuse triangle.

$b$ by drawing perpendiculars to $ce$ and $ab$. This maneuver succeeds if the meeting point $p$ of the two perpendiculars is sufficiently close to $c$, that is, if $\angle cap$ is small enough. The maneuver fails if two subdivision points along $ac$ are too close together, but in this case, the two close points can themselves be merged, as shown in the second example. A set of reduction steps covers all cases. ∎

**Theorem 9 (Bern and Eppstein [19]).** *An $n$-vertex polygon with holes can be triangulated with $O(n^2)$ nonobtuse triangles.* ∎

We now turn to more complicated domains. Nonobtuse triangulations of PSLGs have application to mesh generation for multiple domains, such as a domain composed of more than one material. Bern and Eppstein extend the technique described above to solve a special case of this problem; they show how to compute an $O(n^4)$-complexity nonobtuse refinement of a given triangulation of a simple polygon. Their method, however, fails to solve a more difficult special case: nonobtuse triangulation of both the interior and exterior (up to the convex hull) of a polygon. This problem arises in computational learning theory [145]. Due to Lemma 9, a solution to this problem would also solve the conforming Delaunay triangulation problem described below. The hard part of the problem is getting "exterior" and "interior" Steiner points to match at the polygon boundary. Nonpolynomial algorithms [10, 124], and an $\Omega(n^2)$ lower bound are known [145].

**Open Problem 4.** *Is there an algorithm for polynomial-complexity nonobtuse triangulation of both the interior and exterior of a polygon?*

Bern, Dobkin, and Eppstein [17] have the following very recent results for no-large-angle triangulation with $o(n^2)$ triangles.

**Theorem 10 (Bern et al. [17]).** *An $n$-vertex convex polygon can be triangulated with $O(n^{1.85})$ nonobtuse triangles.*

**Theorem 11 (Bern et al. [17]).** *Any polygon with $n$ vertices can be triangulated with $O(n \log n)$ triangles, with all angles measuring at most $150°$.*

It seems likely that a combination of the techniques used in these two theorems will yield triangulations with maximum angle $90° + \epsilon$ (for any $\epsilon$), using $O(n^{1.85})$ triangles. We close this section with the following questions.

**Open Problem 5.** *How many nonobtuse triangles does it take to triangulate an arbitrary $n$-vertex polygon? What is the best angle bound (that is, closest to $90°$) achievable with $O(n)$, or $O(n \log n)$, triangles?*

### *2.3.3. Maxmin Height*

We now consider maximizing the minimum height of a triangle. This problem arises in Mitchell and Vavasis's three-dimensional mesh generation algorithm [129].

As we saw above, a minimum-height triangulation without Steiner points can be computed using the edge-insertion paradigm [18]. It is not immediately obvious that Steiner points can improve the height. Consider a regular $n$-gon, however, with sides of unit length. Any triangulation without Steiner points must contain an ear, which has height $\sin(\pi/n)$, but a Steiner point at the center allows all triangles to have nearly unit height. This example is due to Scott Mitchell (personal communication).

For point set input, the linear-complexity nonobtuse triangulation algorithm described above already solves the maxmin height problem. Notice that the minimum distance $s$ between two points, that is, the input feature size defined in Section 2.3.1, gives an upper bound on the minimum height achievable. The nonobtuse triangulation algorithm defines quadtree squares of side length $\Omega(s)$, and each triangle has height proportional to the smallest quadtree square it touches.

For polygons, the input feature size $s$ again gives an upper bound on the minimum height achievable. The no-small-angle quadtree algorithm produces a triangulation with minimum height $\Omega(s)$, but with nonpolynomial complexity. The nonobtuse triangulation algorithm described above does not help, as it can produce triangles with small height.

**Theorem 12 (Bern et al. [17]).** *A polygon with holes can be triangulated into $O(n)$ triangles of height $\Omega(s)$.*

**Proof Sketch:** We give a very rough sketch. The first step starts by cutting off acute corners of the polygon with isosceles triangles; these triangles are triangulated at the very end without adding any new Steiner points. Then the remaining polygon $P$ is cut with vertical and horizontal lines. Imagine the plane divided into an infinite square grid with spacing $s/3$. Now erase all of the grid, except portions of lines bounding squares that either contain—or are adjacent to a square that contains—an input vertex. See Figure 14.

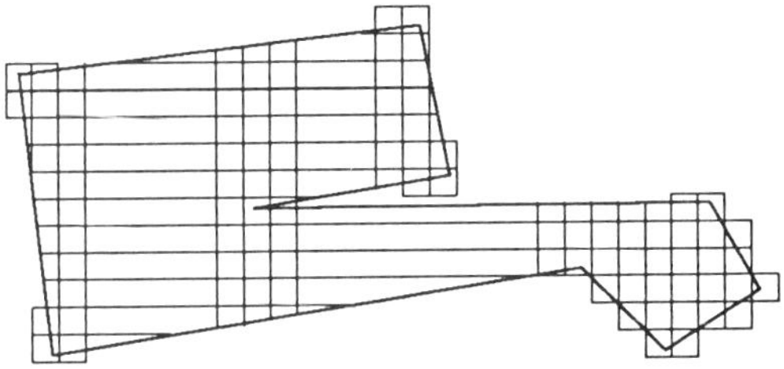

Figure 14. A polygon cut for maxmin-height triangulation.

The second step carefully warps some segments of the horizontal and vertical cutting lines to conform to the boundary of $P$. A case analysis shows that all resulting faces can be triangulated with height $\Omega(s)$. At this point, there are $O(n)$ triangles of height $\Omega(s)$ along the boundary of $P$, and $O(n^2)$ rectangles of height $\Omega(s)$ interior to $P$.

The third step merges interior rectangles into rectilinear polygons to reduce the complexity. Each rectilinear polygon $Q$ will have input feature size $\Omega(s)$ and satisfy a certain matching condition: a horizontal or vertical line extended from a vertex of the polygon across the interior first intersects $Q$ at another vertex.

A sweep algorithm now triangulates such a rectilinear polygon with linear complexity. First draw in all interior vertical line segments between vertices of $Q$. Then sweep from left to right, adding Steiner points to vertical segments. The idea is to project only every other vertex from the current vertical line onto the next vertical line. Horizontal edges are added between corresponding points. When the left-to-right sweep has ended, perform a similar sweep from right to left. After both sweeps, each newly-created face interior to $Q$ has at most one subdivision point on each vertical side; hence, it can be triangulated into $O(1)$ triangles of height $\Omega(s)$. The total number of Steiner points added in the sweeps is $O(n)$, as it is the sum of a geometric series. ■

This result can be combined with the no-large-angle triangulation of Theorem 11 to produce a triangulation with $O(n \log n)$ Steiner points, in which there are no large angles and each triangle has height $\Omega(s)$.

### *2.3.4. Minimum Weight*

As in maxmin height triangulation, it is not obvious that adding Steiner points can reduce the total edge length of a triangulation of a point set. But it is not hard to make examples (say, $n-1$ points on an arc that bows towards one distant point) to show that the minimum weight Steiner triangulation (MWST) can have total edge length $\Omega(n)$ times smaller than the minimum weight triangulation (MWT).

Lingas [110] suggested approximating the MWT of a point set by first choosing all edges from a minimum spanning tree. This partitions the plane into polygons, that can be triangulated optimally using dynamic programming. Lingas did not consider using Steiner points, and his algorithm does not yield a good approximation to the MWST. Clarkson [43] extended Lingas's approach, and gave the first nontrivial MWST approximation. (It can be shown that any triangulation approximates the MWST within a factor of $O(n)$.) Clarkson showed that any polygon has a Steiner triangulation with total edge length $O(\log n)$ times the polygon's perimeter. (The same result can be seen from the quadtree triangulation described below.) Combining this result with Lingas's use of the minimum spanning tree produces a Steiner triangulation of a point set with length $O(\log n)$ times that of the MWST.

Clarkson's result was improved by Eppstein [67], who gave a constant-factor approximation to the MWST. Eppstein showed that the quadtree triangulation algorithm of Bern et al. (Section 2.3.1), which was designed for angle bounds, actually gives such an approximation.

Eppstein first proves that, if $\mathcal{T}$ is any triangulation of the input point set, and $B$ is a quadtree-algorithm square with side length $\ell$, then there is an edge in $\mathcal{T}$ with length $\Omega(\ell)$, that has an endpoint within distance $O(\ell)$ of $B$. Therefore we can charge the length of the triangles associated with $B$ to an edge in $\mathcal{T}$. Each edge is charged at most proportionally to its own length, and hence the quadtree triangulation approximates the MWT. Then, as in the optimality proof of the no-small-angle quadtree triangulation, Eppstein notes that adding Steiner points only increases the total length of the quadtree triangulation. So the quadtree triangulation of the input points has length less than the quadtree triangulation of the input along with an optimal set of Steiner points, which in turn approximates the MWST. This argument works both for the no-small-angle quadtree triangulation (Section 2.3.1), and for the linear-size nonobtuse quadtree triangulation (Section 2.3.2).

The samc techniques can be used to construct an approximate MWST of convex polygons. One simply triangulates the vertices, cuts off the region of the triangulation outside the polygon, and adds diagonals to retriangulate cut polygons. This reduction requires convexity; otherwise the length of the point set triangulation may not approximate the MWST.

**Open Problem 6.** *Is there an efficient algorithm for approximating the MWST of an arbitrary nonconvex polygon?*

Using his quadtree characterization of the MWST length, Eppstein also proves some further properties of the MWST of a point set.

- The MWST has total edge length $O(\log n)$ times that of the minimum spanning tree [43]. For any $n$, there exist point sets for which the MWST has edge length $\Omega(\log n)$ times the minimum spanning tree length.

- For any $n$, there exist point sets for which the MWT has total length $\Omega(n)$ times the MWST length. As noted above, this is tight.
- For any $n$, there exist convex polygons for which the MWT has total length $\Omega(\log n)$ times that of the MWST. Again, this is tight, as both lengths are within $O(\log n)$ of the polygon's perimeter.
- Any convex polygon has a Steiner triangulation in which all Steiner points lie on the polygon boundary, with total length $O(1)$ times that of the MWST.

The last item suggests the question of whether Steiner points in the interior of a polygon really help. For nonconvex polygons, they do; otherwise the weight may be $\Omega(n/\log n)$ times the MWST length. But for convex polygons this question remains open. If we can assume that all Steiner points lie on the boundary, then it seems likely that a dynamic programming algorithm can compute an optimal MWST.

### *2.3.5. Conforming Delaunay Triangulations*

In this section, the input is a PSLG. We consider the problem of finding a set of Steiner points inducing a Delaunay triangulation that is a Steiner triangulation of the input. In other words, each input edge must be a union of edges in the DT of the input vertices and the Steiner points. We call this the *conforming Delaunay triangulation* problem. As noted above, this problem is related to nonobtuse triangulation of the interior and exterior of a polygon. The $\Omega(n^2)$ lower bound for interior-exterior nonobtuse triangulation can be adapted to the conforming DT problem.

We first sketch two algorithms for solving the conforming DT problem that use a number of Steiner points dependent upon the geometry of the input, as well as on $n$. The first algorithm is a modification of one due to Saalfeld [144] (also related to previous algorithms with different complexity analyses [25, 149]). Start with a triangulation of the input, and let $\epsilon$ be the minimum height of any triangle. On each edge, add two Steiner points, at distances $\epsilon/3$ from the endpoints. The diameter circles for the two segments of length $\epsilon/3$ on each edge will be empty, and hence these outer segments are now edges in the DT. Let $\delta$ be the minimum distance between any pair of points in the now augmented PSLG. No interior segment of any edge is within distance $\delta$ of any other edge. Hence if we partition these segments into subsegments of length less than $\delta$, each subsegment will have an empty diameter circle, and the result will be a Delaunay triangulation that covers the input (Figure 15).

Nackman and Srinivasan [132] describe an alternative for polygons with holes. In the previous method, every piece of an input edge can be covered by an empty diameter circle. Actually it suffices to cover each edge by a set of circles not

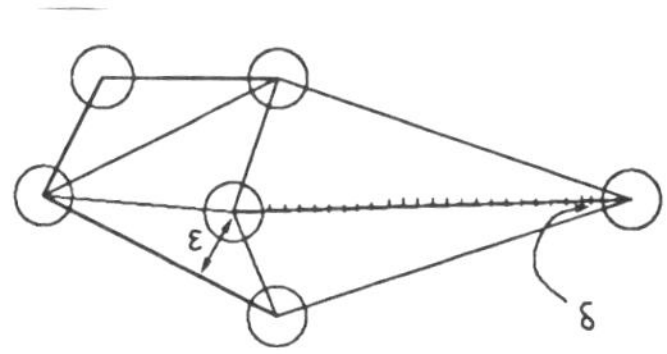


Figure 15. Example of Saalfeld's conforming DT algorithm.

touching any other edge, as in Figure 16. After placing Steiner points at the intersections of circles and edges, every piece of an edge will support a circle containing no other points. The assumption that the input is a polygon implies the existence of a finite set of suitable circles.

Nackman and Srinivasan in fact use the minimum possible number of suitable circles. But there is no guarantee that the algorithm uses a number of Steiner points that is as small as possible—indeed, Nackman and Srinivasan give an example of constant complexity for which their method requires a nonconstant number of Steiner points.

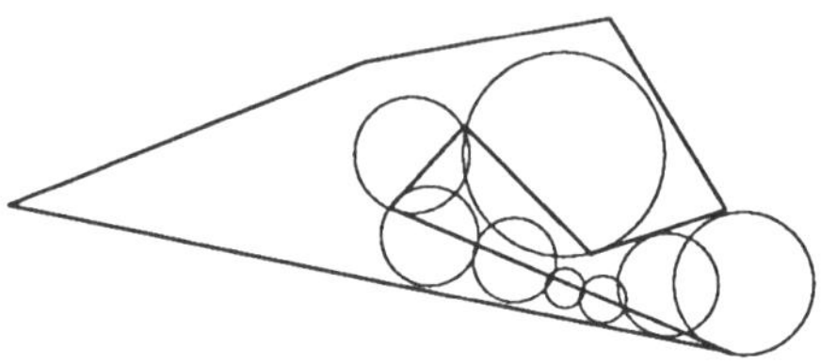

Figure 16. Covering each edge by circles not touching any other edges.

Edelsbrunner [59] described the first algorithm with a bound dependent only on $n$. His technique processes one line segment of an input PSLG at a time. A segment is processed by covering it with a set of circles, none of which contains any of the current vertices (PSLG vertices or previously added Steiner points). New Steiner points are then added at each intersection of an edge and a circle. No Steiner points will be added within any circle, so each portion of an edge contained in a circle (shown dashed in Figure 17) is now discarded. After the last input edge is processed, each portion of an edge is contained in an empty circle, and hence can be part of the DT if we break ties formed by sets of cocircular points in an appropriate way. At each covering step, the number of circles needed is at most proportional to the number of Steiner points already present, and each circle adds at most two new Steiner points per input edge. Hence the total number of Steiner points added is $n^{O(n)}$, which—if not a small number—is at least a function of $n$.

Mehlhorn, Sharir, and Welzl improved this bound to $n^{O(\log n)}$, by interleaving the circle-covering step with a step that recursively splits the set of PSLG

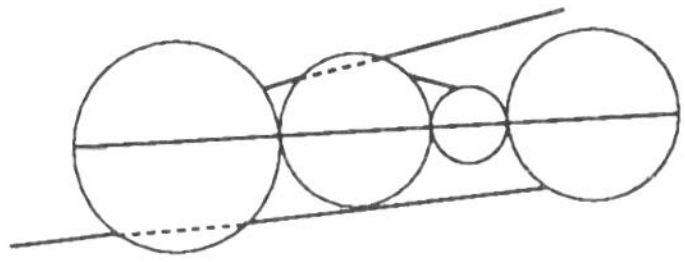

Figure 17. A step in Edelsbrunner's conforming DT algorithm.

endpoints into smaller subsets (Edelsbrunner, personal communication). In a recent breakthrough, Edelsbrunner and Tan [62] found a construction with $O(n^3)$ Steiner points, the first polynomial-size solution to the conforming Delaunay triangulation problem.

### *2.4. Heuristically Generated Meshes*

In this section we describe two-dimensional mesh generation in practice. We do not attempt a thorough literature survey, rather we give informal descriptions of a few mesh generation approaches, chosen to illustrate some of the main ideas in the field. There are already a number of articles that survey and classify the literature on automatic mesh generation [87, 148, 157, 163, 166, 173], although we are not aware of a recent survey of two-dimensional mesh generation with an extensive bibliography. Overall, two-dimensional mesh generation seems fairly mature, and a number of different approaches give good results. As we mentioned in the introduction, we restrict our discussion to generation of unstructured triangular meshes; for structured meshes, such as quadrilateral meshes given by conformal mapping techniques, see [32, 167, 168].

#### *2.4.1. Mesh Improvement Techniques*

Before describing heuristic mesh generation methods, we mention some mesh *improvement* techniques, which can be used as post-processing steps after any of the heuristics.

The first technique, dating to the 1960s [174], is called *Laplacian smoothing* because its repositioning formula can be derived from a finite difference approximation of Laplace's equation [84]. In Laplacian smoothing, a vertex $v$ in the interior of the mesh is moved to the centroid (center of mass) of the polygon formed by its neighbors. The vertex should not be moved if the centroid lies outside the polygon. This repositioning usually improves the size and shape of the triangles around $v$, but it is not guaranteed to do so. A variation weights neighbor vertices by the areas of their surrounding elements. Laplacian smoothing is typically applied successively to each interior node of the mesh, for several (four or five) rounds. See Figure 18.

A second method, due to Frey and Field [74], flips edges to regularize

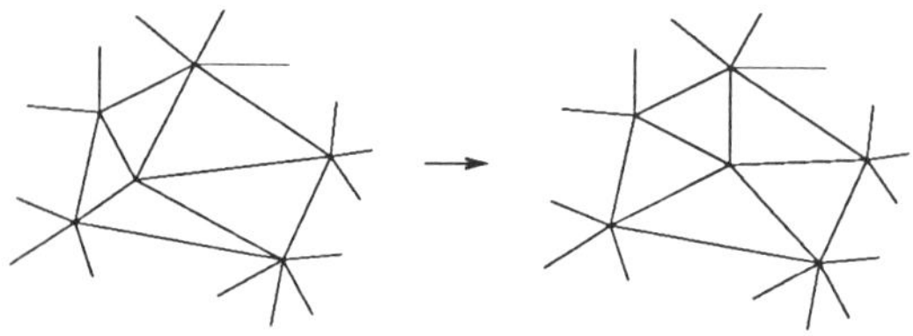

Figure 18. Moving a vertex to its neighbors' center of mass.

degrees. A quadrilateral diagonal is flipped if the sum of its endpoints' degrees exceeds the opposing diagonal's sum by more than two. Laplacian smoothing follows this "mesh relaxation" step.

Finally, we may view the flip algorithm for producing a constrained Delaunay triangulation (Section 2.2.1) as a mesh improvement technique.

### *2.4.2. Quadtrees*

Long before the theoretically good mesh generation methods described in the last section [20], quadtrees had been used in heuristic methods. Yerry and Shephard [176] construct a quadtree representation of a polygonal or curved domain by recursively splitting squares intersected by the boundary of the domain, until squares reach some minimum size bound. Splits may also result from an upper bound on size (that may vary over the domain) or from a balance condition: no quadtree square is adjacent to one more than twice its size. After the quadtree is constructed, each square containing a portion of the boundary is replaced with a shape chosen from a fixed set of patterns. Triangulating each face then yields a mesh that approximates the domain.

A more advanced version of the algorithm uses warping and trimming methods to produce a mesh that does not change the shape of the input [5]. The theoretical paper by Bern et al. [20], contributed warping rules guaranteed not to produce small angles, as well as an analysis of the number of triangles in the mesh. Figure 19 shows an example mesh computed by Mitchell [128], using a modification of the algorithm in [20]. This example demonstrates careful selection of quadtree square sizes around holes and "almost holes".

Quadtrees are the most convenient way to produce *graded* meshes, that is, meshes with small elements near complicated parts of the boundary that grade up to larger elements elsewhere. Another advantage is that the quadtree itself may be computed entirely in integer arithemetic, so that floating-point operations are carried out only within small squares containing simple parts of the domain boundary.

Quadtree methods, however, have been criticized for occasionally producing

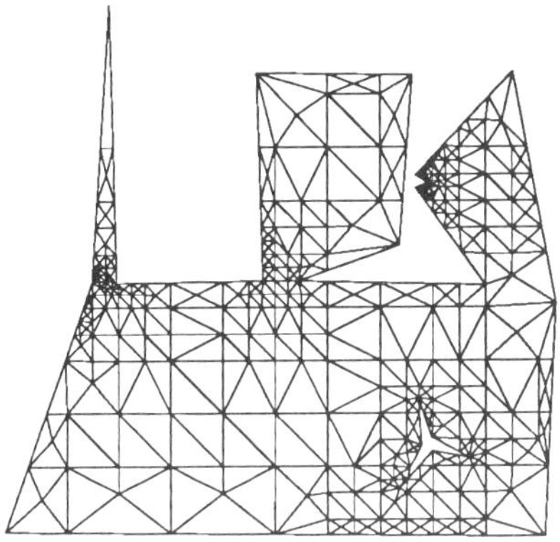

Figure 19. A mesh derived from a quadtree (S. Mitchell).

poorly-shaped boundary elements, and for introducing artificial preferred directions (namely parallel to the $x$- or $y$-axes) [163]. The problem of poorly-shaped elements can be solved by the warping methods of Bern et al. [20]. The second problem may be more inherent, although non-square quadtree tiles, as used in [20, 130], or some sort of randomization procedure, coupled with Laplacian smoothing, may sufficiently break up the directionality.

### *2.4.3. Polygon Decomposition*

The polygon decomposition approach to mesh generation also initially divides the domain (most generally a PSLG) into simple regions. This approach, however, attempts to find intrinsic dividing lines, rather than dividing lines from a rectilinear grid.

Joe and Simpson [96, 91] first divide the domain into convex polygons by cutting along lines extending from reflex vertices (that is, vertices at which the interior angle measures more than 180°). The resulting convex polygons are further subdivided into convex polygons with boundary edge lengths that do not vary too much. Cutting lines are chosen heuristically, attempting to avoid small angles. Finally each convex polygon is triangulated using triangles of approximately equal size, taking care to match Steiner points at the cutting lines. See Figure 20 for an example.

In Joe and Simpson's method, two input parameters control the mesh: a target number of triangles (typically exceeded by a small amount), and a "smoothing parameter" that controls the allowable variation in size between two neighboring polygons. These parameters are combined with something dependent on *local feature size* to yield a "mesh distribution function", that gives a target triangle size at

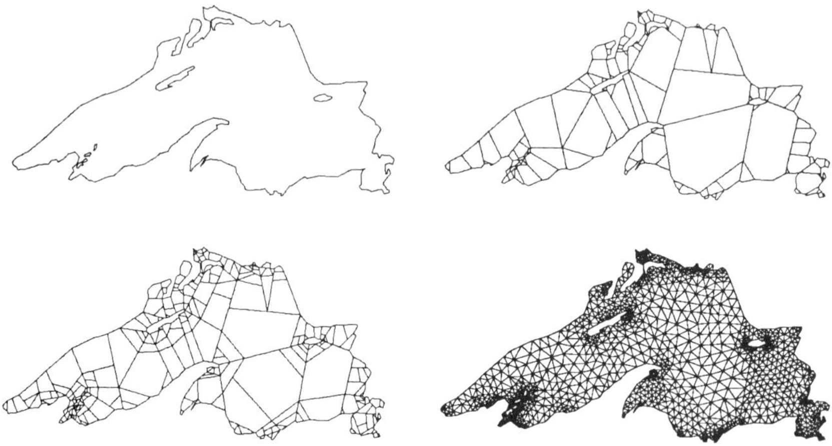

Figure 20. Lake Superior (Joe) (a) input, (b) decomposition, (c) refinement, (d) mesh.

each point within the domain.

The concept of local feature size recurs in most mesh generation approaches. We may define the local feature size at point $a$ to be the size of the quadtree box that contains $a$, as produced by the polygon version of the algorithm of Bern et al. [20], described in Section 2.3.2. This sets the local feature size at a vertex $v$ of the polygonal boundary $P$ to be proportional to the minimum distance (within the domain) to an edge of $P$ not incident to $v$. Local feature size then varies fairly smoothly between vertices. Most of the mesh generators described in this section define their own versions of local feature size, but the definition just given is sufficient for understanding. Many of the mesh generators also allow the user to control the local feature size in some way, perhaps through input parameters. This extra control is important in applications in which the solution to the finite element computation is expected to show features smaller than the features of the domain.

Srinivasan et al. [163] recently developed an interesting polygon decomposition mesh generator, using the *symmetric axis transform*. The symmetric axis is the set of all centers of disks contained in $P$ that contact $P$ at two or more points; it consists of straight lines and parabolic arcs.

Figure 21 shows the operation of this mesh generator on a multiple domain. The input is shown in (a), with different materials shown by different shades. The first step computes the Voronoi diagram of the edges and vertices of the domain; this contains each face's symmetric axis. Parabolic arcs in each symmetric axis are then replaced by one or two straight edges (chords of the arc), as shown in (b). Each vertex on a symmetric axis is then joined to two or more points on the domain boundary by the touching radii of the disk centered at the vertex, resulting

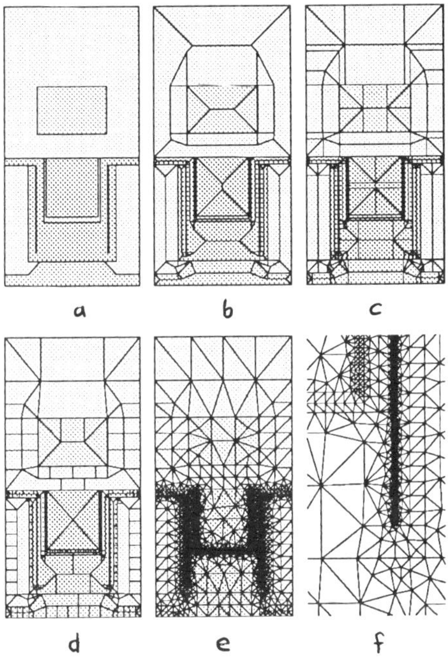


Figure 21. The steps in meshing a multiple domain (Srinivasan et al.).

in a PSLG in which each face is a triangle or trapezoid, as shown in (c). A "sliver processing" step then removes or breaks up faces with bad aspect ratio (see (d)). Boundary edges and touching radii are then used to extract local feature size values. These values induce a node spacing at each vertex in the PSLG, and interpolation then gives a node spacing function over the entire domain. The node spacing function guides an iterative process that adds and deletes more Steiner points. Final triangulation is accomplished by the constrained Delaunay triangulation (shown in (e) and in a zoom in (f)).

### *2.4.4. Advancing Front*

The *advancing front* approach to mesh generation [109, 114] is especially well-suited to fluid dynamics problems. In this approach, the domain's boundary $P$ is first subdivided appropriately, and then Steiner points are placed in successive layers around each connected component of $P$. This yields triangles oriented with the flow field. Figure 22 shows a mesh computed by Barth and Jespersen [16]; this mesh is the Delaunay triangulation of vertices placed by the advancing front method.

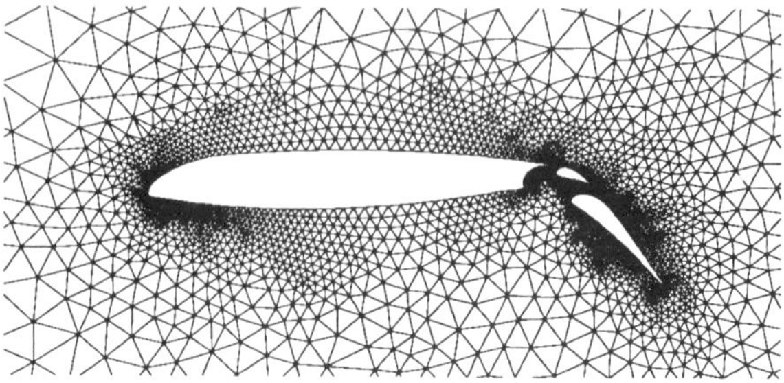

Figure 22. A Delaunay triangulation of points placed in layers (Barth and Jespersen).

Lo [116] has developed a mesh generator that places Steiner points along predefined contour lines, that need not follow the domain boundary. This generalization allows the output of an initial finite element computation to control the generation of the next mesh.

Mavriplis [123] has computed meshes for high Reynolds number flows using an idea related to contour lines. His method identifies "stretching" lines and places the first Steiner points along these curves. Local, structured meshes generate interior Steiner points. The method then computes Delaunay triangulations in locally transformed regions in order to generate long, thin—but not overly obtuse—triangles oriented with the flow. Such a triangulation is an especially efficient mesh for laminar flows. (See [45] for another use of the DT in a transformed space.)

### *2.4.5. Mesh Refinement*

The problem of refining a given mesh occurs in practice, for example, when an initial finite element computation reveals a region that requires greater resolution. A number of researchers have taken mesh refinement as the central step in mesh generation itself [12, 73, 142]. For example, Chew's algorithm [41], discussed in the last section, refines the constrained Delaunay triangulation by adding centers of circumcircles.

An earlier, somewhat related, heuristic method is due to Frey [73]. In Frey's method, Steiner points are initially added to the boundary according to a "spacing function" that approximates local feature size. After this step, for most practical inputs, a Delaunay triangulation of the input vertices and the Steiner points includes the domain boundary. (Using the recent results of [132, 144], this step can be made exact; see Section 2.3.5 above.) Next interior Steiner points are added. Satisfactory results were obtained by the following method: (1) find a triangle $t$ containing its circumcenter; (2) generate a prospective Steiner point $a$ partway between the incenter and circumcenter of $t$; (3) if $a$ is not too close to the

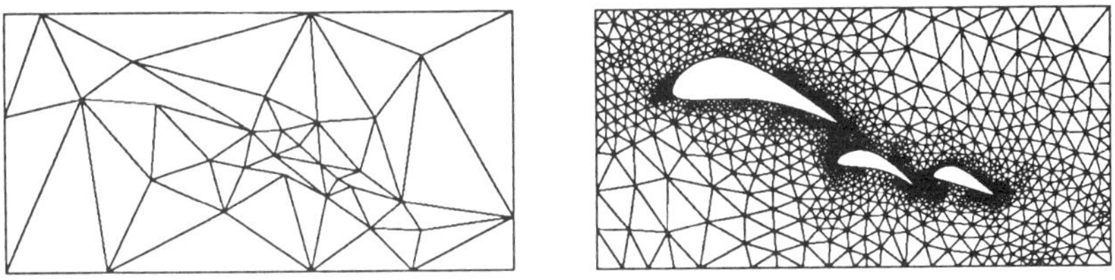

Figure 23. Input and output of a refinement-based generator (Shaw).

vertices of $t$ (where close is defined by the local feature size), then add $a$ and rebuild the Delaunay triangulation.

Shaw [156] has recently developed a simple, refinement-based mesh generator. The user inputs a very rough, triangular, initial mesh, with the correct number of boundary components in roughly correct locations. The user assigns a local feature size to each node of the initial mesh.

Triangles larger than their vertices' smallest feature size are then split into four similar copies of themselves by adding Steiner points at the midpoints of sides. A midpoint is assigned the average feature size of the endpoints. (Splitting triangles into four similar copies was previously used by Bank [11, 12].) Neighboring triangles now have subdivision points; this situation is corrected by triangulating these faces with diagonals. Correction of subdivision points is followed by a few cycles of Delaunay flipping and weighted Laplacian smoothing, and the refinement cycle repeats. (Field [69] also interleaves Laplacian smoothing and Delaunay flipping.) As the boundary triangles refine, they are parametrically "pulled" to the correct geometry (which may include spline curves), thereby computing a valid boundary element mesh at the same time as the rest of the mesh. See Figure 23 for an example of an input mesh and the resulting refined mesh; the three nontriangular faces in the input mesh correspond to the holes in the domain. Final aspect ratios are quite insensitive to the quality of the input mesh.

### *2.5. Two-and-a-half-dimensional Problems*

A 2.5-dimensional problem asks for a triangulated surface embedded in three dimensions. We first discuss interpolating surfaces for point set data with elevations, and then triangulated surfaces for three-dimensional models.

#### *2.5.1. Interpolation of Bivariate Functions*

The input is a set of points $S$ in the plane, along with a real-valued *elevation* $f(a)$ at each point $a \in S$. Any two-dimensional triangulation $\mathcal{T}$ of the input points induces a piecewise-linear function $f_{\mathcal{T}}$ defined on the region $R$ bounded by the convex hull of

$S$. For each point $d$, $f_{\mathcal{T}}(d)$ is the weighted average of the elevations at the vertices, $a$, $b$, and $c$, of the triangle $abc$ in $\mathcal{T}$ that contains $d$. Writing $d$ as $c_1a + c_2b + c_3c$, with $c_1 + c_2 + c_3 = 1$ and $c_1, c_2, c_3 \geq 0$, we have $f_{\mathcal{T}}(d) = c_1 f(a) + c_2 f(b) + c_3 f(c)$. We say that $f_{\mathcal{T}}$ *interpolates* $S$.

The question arises: which triangulations are good for interpolation? This question has been discussed in the literature [13, 46, 53, 102, 152]. Rippa [140] recently proved a surprising result. Regardless of the input elevations, the Delaunay triangulation gives an interpolating surface, or *elevated triangulation*, optimal in a certain least energy sense.

**Theorem 13 (Rippa [140]).** *Let $f_{\mathcal{T}}$ be a piecewise-linear function interpolating $S$ induced by a triangulation $\mathcal{T}$. The piecewise-linear function $f_{\mathcal{DT}}$ induced by the Delaunay triangulation satisfies*

$$\int\!\!\int_R (\nabla f_{\mathcal{DT}})^2 \leq \int\!\!\int_R (\nabla f_{\mathcal{T}})^2.$$

We now explain the integrals above in terms more familiar to computational geometers. Let $abc$ be a triangle of triangulation $\mathcal{T}$. Let the plane passing through $f(a)$, $f(b)$, and $f(c)$ have the equation $z = Ax + By + C$ in Cartesian coordinates. Then over $abc$, the gradient squared $(\nabla f_{\mathcal{T}})^2$ is simply the constant $A^2 + B^2$, and the contribution of $abc$ is its area times this constant. Rippa's theorem states that the sum of these contributions over all triangles is minimized by the Delaunay triangulation. The proof of Rippa's theorem is an intricate calculation showing that the flip procedure cannot increase the integral. Hence the CDT is also an optimal interpolating surface. (De Floriani et al. [46] had previously proposed the use of the CDT for this purpose.)

Rippa and Schiff [141] show that the minimization above corresponds to the energy functional associated with the nonhomogeneous Laplace equation. We may think of the DT as giving the stretched membrane (a "drumhead") with least potential energy, among all elevated-triangulation membranes. Rippa and Schiff also discuss other energy functionals, and use the flip algorithm as a heuristic for their minimization. See [13, 56] for heuristic solutions to other interpolation problems.

Bern et al. [18] recently considered the problem of finding the *minimum slope* interpolating surface for input points with elevations (an optimization criterion mentioned in [172]). The slope of an elevated triangle is the slope in the direction of steepest descent, and the slope of an elevated triangulation is the maximum slope of any of its elevated triangles. Bern et al. showed that this problem can be solved in time $O(n^3)$ using the edge-insertion paradigm, discussed in Section 2.2.2. The following lemma shows that the "weak anchor property" holds, thus establishing the applicability of edge insertion.

**Lemma 11.** *Assume $abc$ is a maximum-slope triangle in elevated triangulation $\mathcal{T}$. Assume the line of steepest descent on $abc$ passes through $a$ (either ascending or*

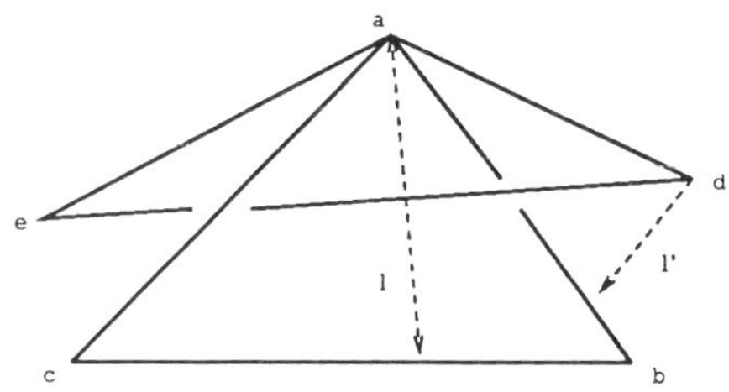


Figure 24. The weak anchor property holds for minmax slope.

*descending from a). Assume elevated triangulation $\mathcal{T}'$ has smaller slope than $\mathcal{T}$. Then there is an edge of $\mathcal{T}'$ incident to a that crosses bc (in the projection onto the plane).*

**Proof:** Assume without loss of generality that the line of steepest descent $\ell$ descends from $a$ to $bc$. If the lemma is false, then $\mathcal{T}'$ must contain an elevated triangle $ade$, with $de$ intersecting both $ab$ and $ac$ in the projection onto the plane containing the input. (Edge $de$ does not necessarily cross both $ab$ and $ac$, so $e$ could be identical to $c$.) A vertical plane $V_\ell$ through $\ell$ must cut $de$ above $\ell$, since $\mathcal{T}'$ has smaller slope than $\mathcal{T}$. So at least one of $d$ and $e$, say $d$, must lie strictly above the plane containing $abc$. See Figure 24.

Now consider the elevated triangle $adb$ (which is not necessarily a triangle of $\mathcal{T}$ or $\mathcal{T}'$). Because $d$ lies above the plane containing $abc$, the slope of $adb$ must be greater than the slope of $abc$. (Here notice that $V_\ell$ intersects the plane containing $adb$ in a line with steeper slope than $\ell$.) Let $\ell'$ be the line of steepest descent on $adb$. If $\ell'$ connects $d$ with $ab$, we consider the triangles of $\mathcal{T}$ that intersect $V_{\ell'}$, a vertical plane through $\ell'$. These triangles intersect $V_{\ell'}$ in a polygonal path from $d$ down to $ab$; at least one edge of this path must have slope at least that of $\ell'$, a contradiction to the assumption that $abc$ is a maximum slope triangle in $\mathcal{T}$. Contradictions also follow in the other two cases: when $\ell'$ connects $b$ with $ad$ or $a$ with $bd$. ∎

This lemma also holds for constrained triangulations, giving an $O(n^3)$ algorithm for finding a least-slope interpolating surface for polygonal inputs with holes and elevations. There are a host of open questions on optimal interpolation; we list three. Many interpolation problems, including the second one listed below, also make sense when Steiner points are allowed.

**Open Problem 7.** *For point set data with elevations, can a triangulation maximizing the minimum angle on an elevated triangle be computed in polynomial time?*

**Open Problem 8.** *For point set data with elevations, can a triangulation mini-*

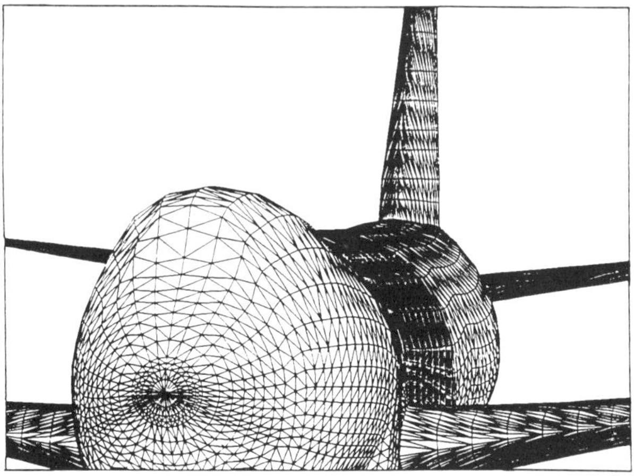

Figure 25. Surface triangulation of a Boeing 747-200 (Baker).

*mizing (or approximately minimizing) the total surface area be computed in polynomial time?*

**Open Problem 9.** *For point set data with elevations, can a triangulation with least-sharp sharpest dihedral angle be computed in polynomial time? (See [56].)*

### *2.5.2. Surfaces for three-dimensional models*

Here the input is a "solid model" and the output is a triangulated surface. This step often precedes three-dimensional mesh generation, especially in advancing-front mesh generators.

Solid models take a number of rather varied forms, which complicates even the definition of problems. Solids may be approximately defined by point sets, either all interior to the solid, or labeled "interior" and "exterior". Constructive solid geometry (CSG) defines a solid as the intersection and union of primitive solids, such as half-spaces. Inputs in medical and aerospace applications often take the form of regularly spaced planar cross-sections [9, 24]. Inputs in computer graphics applications may be defined by spline patches [15], or by level sets of functions of three variables ("implicit surfaces" [23]).

Solids defined by point sets in three dimensions occur in scientific visualization, learning theory, and computer graphics. One may want to represent such an input with an enclosing or interpolating polyhedron. The convex hull (the intersection of all half-spaces containing the point set) is an enclosing polyhedron, but it does not usually give a good representation of the "shape" of the point set.

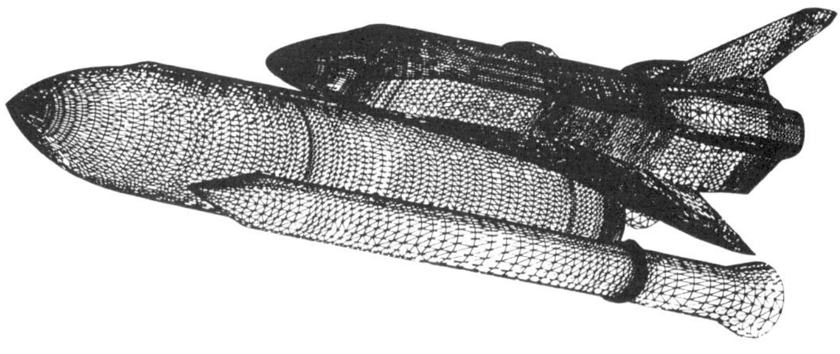

Figure 26. Surface triangulation of the U.S. Space Shuttle (Baker).

An $\alpha$-hull generalizes the convex hull by replacing half-spaces with balls or complements of balls, with radius $1/\alpha$ [57]. This generalization allows more faithful shape representation, but may have complexity $\Omega(n^2)$. Another approach to shape representation, about which little is known, is to find an enclosing or interpolating polyhedron (without Steiner points) that is optimal for some criterion. The following problem is an example.

**Open Problem 10.** *For points in three dimensions, can an approximate min-surface-area enclosing polyhedron be found in polynomial time?*

Triangulating the surface of a solid given as a polyhedron reduces to triangulating a PSLG. Lindholm [109] uses an advancing-front generator within each face.

Boissonat [24] and Baker [9] consider the problem of giving a triangulated surface interpolating a number of polygonal, planar cross-sections. Both authors devise heuristics using the DT of a three-dimensional point set. Baker's goal is to compute a surface triangulation of an aircraft; he places points interior to each polygonal cross section, roughly one point for each of the polygon's vertices. He then computes the three-dimensional DT of all input vertices and interior vertices. Each tetrahedron interior to the aircraft has a Steiner vertex, so that the union of the tetrahedra defined by four input vertices defines a surface triangulation of the aircraft. See Figures 25 and 26. For this algorithm to succeed, input vertices must be closely spaced relative to the thickness and separation of aircraft parts.

Chew (personal communication) has extended his two-dimensional mesh generator (Section 2.3.1) to meshing curved surfaces defined by a single patch, by generalizing planar Delaunay triangulation. As before, the resulting mesh has all angles greater than 30°, but the mesh's density is no longer uniform, rather it is controlled locally by deviation from the curved surface. See Figure 27.

Bloomenthal [23] uses octrees to polygonize implicit surfaces; these surfaces can then be triangulated if desired. Octrees have been applied to polygonizing spline surfaces as well. Overall, computing provably good triangulated surfaces for solid models appears to be an open area.

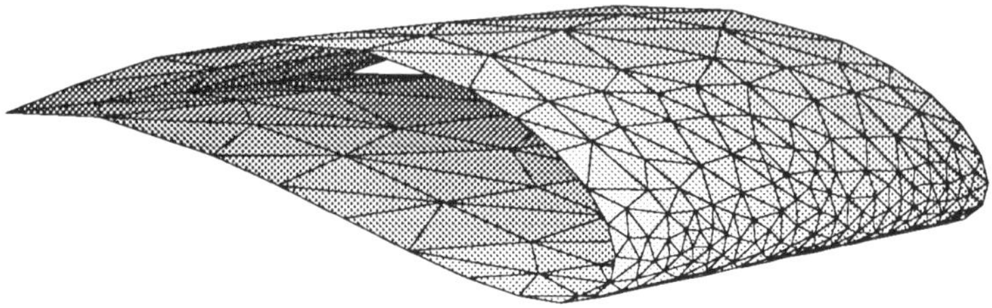

Figure 27. Surface triangulation of a wing (Chew).

## 3. Three-dimensional Triangulations

Triangulation in three dimensions is called *tetrahedralization* (or sometimes *tetrahedrization*). A tetrahedralization is a partition of the input domain, point set or polyhedron, into a collection of tetrahedra, that meet only at shared faces (vertices, edges, or triangles). Tetrahedralization turns out to be significantly more complicated than triangulation.

As in two dimensions, $n$ represents the number of vertices of the input domain, and we distinguish several different types of domains.

- **Simple polyhedron.** A simple polyhedron is topologically equivalent to a sphere; it does not meet itself in a handle, or touch itself at a point or an edge. The boundary of such a polyhedron forms a connected planar graph. In triangulations without Steiner points, each tetrahedron's vertices must be vertices of the polyhedron.
- **Nonsimple polyhedron.** A nonsimple polyhedron may be multiply connected, topologically equivalent to a torus or a higher-genus surface. It may also have holes, meaning that its boundary is not connected.
- **Point set.** As in two dimensions, a triangulation of a point set fills the convex hull. If Steiner points are allowed, then the boundary of the triangulation may be a larger convex polytope.

### *3.1. Tetrahedralization without Optimization*

In this section, we concentrate on existence and construction of tetrahedralizations, without concern for optimality. Existence and construction are already interesting, since many two-dimensional triangulation properties break down in three dimensions.

The first surprise is that different triangulations of the very same input may contain different numbers of tetrahedra. For example, choose $n$ points $v_i = (i, i^2, i^3)$

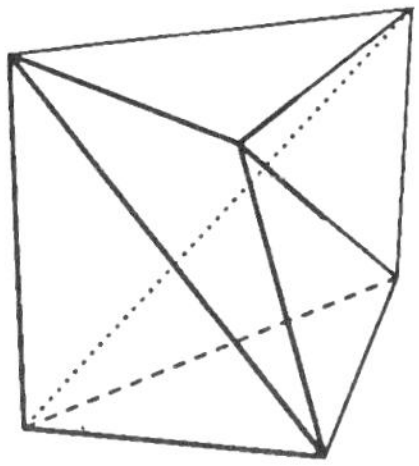

Figure 28. An untetrahedralizable polyhedron.

on the *moment curve*. It is not hard to show that their convex hull can be triangulated with the $\binom{n-2}{2}$ tetrahedra of the form $v_i v_{i+1} v_j v_{j+1}$. (In fact this is the Delaunay triangulation of these points.) A generalization of Euler's formula shows that any tetrahedralization of an $n$-vertex polyhedron has at most this many tetrahedra [59]. If we choose the tetrahedralization carefully, however, we can achieve linear, rather than quadratic, complexity for this same input. In fact, any strictly convex polyhedron can be tetrahedralized with at most $2n - 7$ tetrahedra: choose a vertex $v$, triangulate each face of the polyhedron that is not adjacent to $v$, and then connect $v$ to each triangle. This bound is within a factor of two of optimal, as any tetrahedralization of a simple polyhedron has at least $n - 3$ tetrahedra.

Edelsbrunner, Preparata, and West [63] show how to construct a linear-complexity tetrahedralization of point sets. After tetrahedralizing the convex hull with only linear complexity as above, interior points are added one at a time. When a point is added, the tetrahedron containing it is replaced by four smaller tetrahedra.

When we try to extend these results to nonconvex polyhedra, we meet a second surprise: not all polyhedra are tetrahedralizable. The following counterexample is due to Schönhardt [150]. Start with a triangular prism, and twist one triangle relative to the other so that each rectangular face of the prism folds into two triangles with a reflex edge between them (Figure 28). Any set of four vertices must include a pair that face each other across such a reflex edge. So the polyhedron contains no tetrahedron, and tetrahedralization is impossible.

Schönhardt's polyhedron can be tetrahedralized if we add one Steiner point. This leads to the question of how many Steiner points may be required for tetrahedralization. Chazelle [35] found a simple polyhedron in which $\Omega(n^2)$ Steiner points are needed even to partition the polyhedron into convex regions. Clearly, this is also a lower bound for tetrahedralization.

Chazelle's polyhedron (Figure 29) can be viewed as a cube, from which numerous thin wedges have been removed. Wedges parallel to the $y$-axis are removed from the top face of the cube, and wedges parallel to the $x$-axis are removed from the bottom face. The reflex edges at the tips of the wedges form two sets of lines, that almost meet at the center of the polyhedron, near a ruled surface in the form

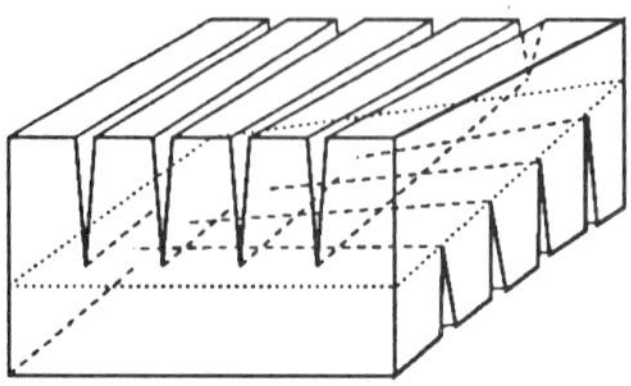

Figure 29. Chazelle's lower bound example.

of a hyperbolic sheet. Viewed from above, the lines partition this sheet into $\Omega(n^2)$ small squares. Noncorresponding points on any two squares are not visible to each other, and hence must be in different convex regions. Therefore, $\Omega(n^2)$ Steiner points are needed. As we now show, this bound is tight.

**Theorem 14.** *Any polyhedron can be triangulated with $O(n^2)$ Steiner points and $O(n^2)$ tetrahedra.*

**Proof:** Extend a vertical "wall" from each edge of the polyhedron boundary, up and down from that edge until it reaches some other part of the boundary. These walls divide the polygon into generalized cylinders. Triangulating the top and bottom faces of the cylinders partitions the polygon into $O(n^2)$ triangular prisms. Each vertical prism side is crossed at most once by a polyhedron edge, so the prisms are polyhedra with at most twelve vertices. Triangulate the faces of these polyhedra, making sure that tetrahedra from different prisms will meet face to face, and then triangulate each prism with at most 20 tetrahedra incident to a single interior Steiner point. (We need the Steiner point as the edges crossing the vertical faces make the prisms not strictly convex.) ∎

Figure 30. Extending a vertical wall from a reflex edge.

As we have seen, convex polyhedra can be triangulated with only $O(n)$ tetrahedra, while Chazelle's polyhedron requires $\Omega(n^2)$. This suggests the possibility that an appropriate measure of nonconvexity would interpolate these bounds and achieve $o(n^2)$ tetrahedra for "slightly nonconvex" polyhedra. One natural measure

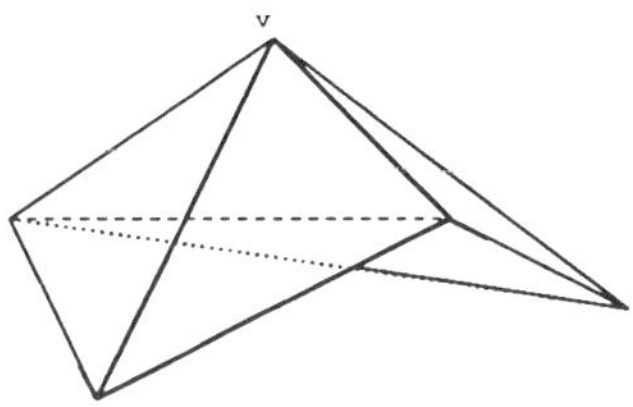


Figure 31. The cap of vertex $v$.

is $r$, the number of reflex edges. Chazelle and Palios [39] developed a triangulation algorithm sensitive to this measure.

Let $N(v)$ be the set of neighbors of vertex $v$, and define the *cap* of $v$ to be the star-shaped polyhedron formed by removing the convex hull of $N(v)$ from the convex hull of $N(v) \cup \{v\}$. See Figure 31. Since the boundary of the polyhedron forms a planar graph, one can always find a vertex $v$ with a cap with at most five other vertices. One would like to remove such a cap from the polyhedron, replacing it with at most three tetrahedra, and continue until the polyhedron is triangulated. This would be analogous to triangulating a polygon in the plane by removing a single ear triangle at a time. Since not all polyhedra are tetrahedralizable, this approach does not work—the difficulty is that the rest of the polyhedron might penetrate into the cap, so that it could not be removed without causing the polyhedron to intersect itself.

**Lemma 12.** *Let $S$ be a set of vertices of a simple polyhedron with triangular faces, such that no vertex of $S$ is adjacent to another vertex of $S$ or to a reflex edge. Then at most $2r$ caps of vertices in $S$ are penetrated by other portions of the polyhedron.*

**Proof:** Suppose the cap of vertex $v$ is penetrated. The faces touching $v$ are boundaries of the polyhedron, so the penetration must occur through the remaining faces of the cap. Moreover, since no edge can completely cross the cap without crossing a face adjacent to $v$, there must be a polyhedron vertex interior to the cap.

Project all interior vertices onto a line extending from $v$ through the cap, and let $w$ be the vertex with projection closest to $v$. Then $w$ cannot be in any other cap penetrating the cap of $v$, and $w$ cannot be chosen as closest for some cap not penetrating the cap of $v$. Line segment $vw$ must be entirely contained in the polyhedron; otherwise it would cross a face, one vertex of which would be closer to $v$. And finally $w$ must be an endpoint of a reflex edge, because otherwise (as segment $vw$ is inside the polyhedron) there must be a face incident to $w$ with a vertex closer to $v$. So we can charge each penetrated cap to an endpoint of a reflex edge. ∎

**Theorem 15 (Chazelle and Palios [39]).** *Any simple polyhedron with $n$ vertices and $r$ reflex edges can be partitioned into $O(n+r^2)$ tetrahedra using $O(n+r^2)$ Steiner points.*

**Proof:** Start by triangulating the faces of the polyhedron. As in any planar graph, we can find $\Omega(n)$ nonadjacent vertices, all with degree at most six. By Lemma 12, unless $n$ is $O(r)$, one of these vertices has a low-degree cap that is not penetrated. Remove this cap, leaving a smaller polyhedron. The cap itself can be split into $O(1)$ tetrahedra. After we remove all but $O(r)$ vertices, the remaining polyhedron can be triangulated by the vertical wall method. ∎

The resulting partition is not yet a triangulation, because the vertical walls subdivide the faces of the reduced polyhedron, without matching the caps removed from those faces. If $m$ wall edges occur on the faces of the cap of vertex $v$, then $v$'s cap can be triangulated with $O(m)$ tetrahedra with apex $v$, and this subdivision does not propagate into other caps. The complexity of the triangulation may grow if an edge of the reduced polyhedron is shared by many removed caps. This complication can be handled by surrounding each such edge with a narrow prism-shaped polyhedron before doing the vertical wall construction. Now vertical walls subdivide the faces of the prism, rather than the edge itself; the prism can be triangulated with a single Steiner point. We omit the details of handling the tips of the prisms where they meet at vertices of the reduced polyhedron. The final product is a triangulation with $O(n + r^2)$ tetrahedra, that with careful implementation can be constructed in time $O(nr + r^2 \log r)$ [39].

This result reveals that nearly convex polyhedra require few Steiner points. The question arises: can we find an efficient algorithm that uses the minimum number of Steiner points? Ruppert and Seidel [143] gave a negative answer to this question. They showed that testing whether Steiner points are necessary to triangulate a given polyhedron is NP-complete (see [76]), even for star-shaped polyhedra (which can trivially be triangulated with a single Steiner point). They similarly prove that, for any $k$, it is NP-hard to test whether $k$ Steiner points suffice. The following open question asks for an approximation algorithm.

**Open Problem 11.** *Is there an efficient algorithm for triangulating any $n$-vertex polyhedron into $O(m)$ tetrahedra, where $m$ is the minimum possible number?*

### *3.2. Optimal Tetrahedralization*

In this section, we consider three-dimensional optimal triangulation without Steiner points. Since Steiner points are required simply to tetrahedralize nonconvex polyhedra, this section treats only point sets (and as a special case, convex polyhedra). Even here very little is known, and this section has more open problems than results.

Since a single input has tetrahedralizations of different complexity, a natural optimization question is the following. A more general open question asks for a minimum-complexity tetrahedralization of a point set.

**Open Problem 12.** *Is there a polynomial-time algorithm for triangulating an arbitrary convex polyhedron with the minimum number of tetrahedra?*

The Delaunay triangulation (DT) in three dimensions contains each tetrahedron with vertices from the input point set, whose circumsphere contains no other input points on its surface or in its interior. Assuming general position, no five points lie on a single sphere, so this defines a triangulation. The complexity of the DT may be as high as $\Omega(n^2)$, as shown by the moment curve example (Section 3.1). There does not seem to be a reasonable definition of constrained DT in three dimensions.

The *lifting* transformation defined in Section 2.2.1 generalizes to three (and higher) dimensions. We map an input point with Cartesian coordinates $(x, y, z)$ to the point $(x, y, z, x^2 + y^2 + z^2)$. The image points all lie on a paraboloid in four dimensions; the projection of the lower convex hull back onto the $xyz$-hyperplane gives the DT. Coupled with an algorithm for computing four-dimensional convex hulls [153], this gives a worst-case optimal, quadratic-time algorithm to compute the DT.

There are also direct algorithms. Bowyer [26] and Watson [171, 68] gave incremental algorithms that are quite popular in practice. Watson's algorithm inserts points in sorted order by one coordinate, testing all old circumspheres that intersect the current sweep plane. Bowyer includes evidence that his algorithm runs in time $O(n^{4/3})$ for a random point set. Dwyer [55] gives a linear-expected-time algorithm for random points in the unit ball.

Joe [92] and Rajan [139] generalize the flip algorithm for DT construction. In three dimensions, flips involve sets of five points, forming a tetrahedral bipyramid. Such a figure can be tetrahedralized in two ways: either as a pair of tetrahedra separated by a face, or as three tetrahedra surrounding an interior diagonal. Thus flips trade two tetrahedra for three, or vice versa. See Figure 32. Starting from an arbitrary tetrahedralization, however, the flip algorithm can get stuck in a local optimum and fail to produce the DT [92].

Joe [94] showed that, if we start with the DT of some point set, and add a single point (dividing the tetrahedron containing it into four, or if the new point is outside the convex hull, adding tetrahedra connecting it to the triangles it can see), then flipping from the resulting triangulation never gets stuck. All tetrahedra involved in flips are neighbors of the new vertex, so in some sense this flipping procedure becomes two- rather than three-dimensional. This result gives another $O(n^2)$-time algorithm for computing the DT: add points one by one (say, in sorted order by $x$-coordinate) and, after each addition, flip until the DT is reached. Rajan [139] described a similar procedure for incrementally adding points and flipping

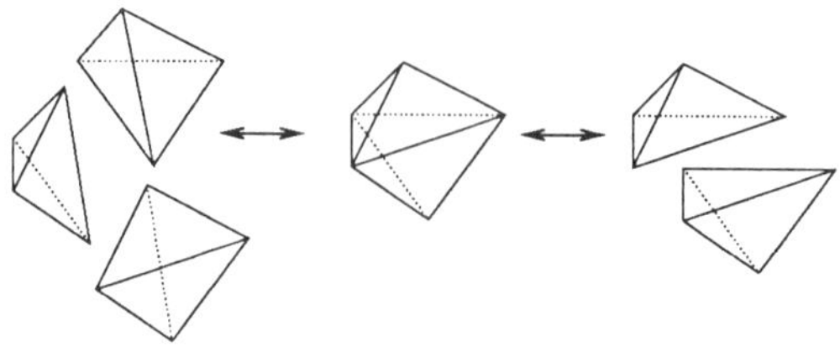

Figure 32. Two ways of tetrahedralizing five points.

tetrahedra to find the DT. His procedure flips tetrahedra in the order corresponding to the changes in the convex hull of the lifted points as the new point is moved vertically down onto the paraboloid. Thus, Rajan's algorithm generalizes to higher-dimensional DT construction. Very recently, Edelsbrunner and Shah extended Joe's algorithm to higher dimensions as well (personal communication).

Though the worst-case time bound for computing the three-dimensional DT must be $\Omega(n^2)$, an "output-sensitive" algorithm runs much more quickly on simple input instances. Techniques recently developed by Matoušek [121] for half-space range queries give an algorithm with running time $O(n^{4/3+\epsilon} + k \log n)$, where $k$ is the complexity of the DT. See Fortune's survey [72] for more details on three-dimensional DT algorithms, including some important implementation issues.

Because the DT possesses so many optimality properties in two dimensions, geometers long suspected that it should optimize something in three dimensions. Recently, Rajan [139] discovered the first such result. (His result actually holds in all dimensions.) The *min-containment sphere* of a simplex $t$ is the smallest sphere containing $t$. If $t$ contains its circumcenter, then the min-containment sphere is identical to the circumsphere. Otherwise, the min-containment sphere circumscribes a lower-dimensional face of $t$. For example, in two dimensions, the min-containment sphere is either the circumcircle or the diameter circle of the longest edge. Rajan proved the following, which generalizes a result of D'Azevedo and Simpson for the planar case [45].

**Theorem 16 (Rajan [139]).** *The Delaunay triangulation is the triangulation that minimizes the maximum radius of a min-containment sphere.*

**Proof Sketch:** Lift the points to the paraboloid $(x, y, z, x^2+y^2+z^2)$. Any sphere corresponds to a hyperplane cutting this paraboloid. Let $\mathcal{T}$ be any triangulation, and for any tetrahedron $t$ in $\mathcal{T}$, define $H(t)$ to be the half-space above the hyperplane corresponding to the circumsphere of $t$. If $S$ is the min-containment sphere of $t$, the radius of $S$ corresponds to the vertical distance between the lifted center of $S$ and $H(t)$. Now form a polytope $P(\mathcal{T})$ as the intersection of all such halfspaces. (This can be thought of as a *power diagram* in the original space.) Then the largest min-containment sphere corresponds to the largest distance between this polytope

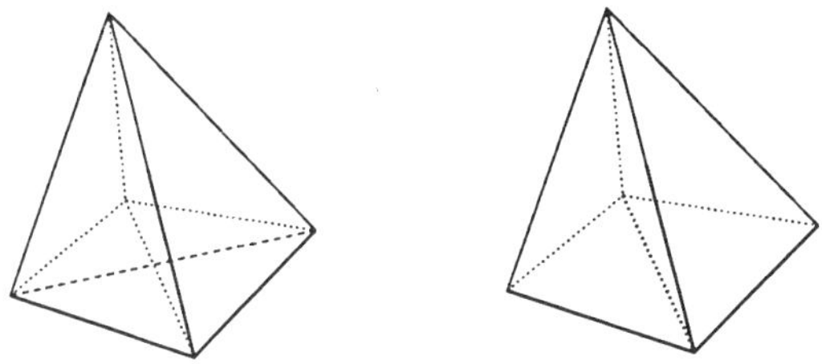

Figure 33. The DT (left) contains a very flat "sliver" tetrahedron.

and the portion of the paraboloid to which the convex hull of the input can be lifted. The DT is the convex hull of the lifted points, so it is lower than any other possible polytope $P(\mathcal{T})$, and hence minimizes this distance. ∎

Dey (personal communication) observed that for cospherical points, all triangulations have the same maximum min-containment radius, so any completion of the Delaunay triangulation solves the degenerate case. Figure 33 gives a counterexample to the plausible conjecture that the DT also minimizes the radius of the largest circumsphere. We close with two open problems.

**Open Problem 13.** *Are any other quality measures optimized by the DT?*

**Open Problem 14.** *What is the appropriate three-dimensional generalization of the connection between maximum angle and error [3, 81]?*

### *3.3. Steiner Tetrahedralization*

We have already touched on the subject of Steiner tetrahedralization in Section 3.1, because tetrahedralizing a nonconvex polyhedron may require Steiner points. We now discuss problems in which Steiner points are used to improve the quality of the solution.

#### *3.3.1. Reducing Delaunay Triangulations*

The Delaunay triangulation of a set of $n$ points in three dimensions may have $\Omega(n^2)$ tetrahedra, though a "typical" point set has only $O(n)$ [55]. This raises the question of whether Steiner points can be used to reduce the complexity of the DT.

Chazelle et al. [37] answered this question affirmatively by showing that for any point set, there exists a set of $O(n^{1/2}\log^3 n)$ points, such that the Delaunay triangulation of the union of the two point sets has $O(n^{3/2}\log^3 n)$ tetrahedra. Their method repeatedly finds a point that lies inside a large number of Delaunay circumspheres; the addition of such a point removes all the corresponding tetrahedra and replaces them by $O(n)$ new tetrahedra.

The success of this method follows from a combinatorial lemma of independent interest that holds in arbitrary fixed dimension $d$. If there are $m$ spheres, each passing through a pair of points, then some point of space (not necessarily one of the $n$ input points) is interior to $\Omega(m^2/(n^2 \log^{2d}(n^2/m)))$ spheres. Results stronger by log factors hold for diameter spheres and rectangular boxes.

Bern, Eppstein, and Gilbert [20] showed how to use more Steiner points, and reduce the complexity of a Delaunay triangulation to $O(n)$. Their technique is completely different, and it works for any fixed dimension. The algorithm first computes a balanced octree (in general, a $2^d$-ary tree) such that each point is alone in a cube surrounded by empty cubes the same size. The closest cube vertex to each input point is then replaced by that input point.

To reduce the size of the tree, the algorithm identifies long chains of cubes, say more than $2^d$ levels, in which each cube has only one nonempty child cube. Next it removes the middle-sized cubes of these chains, leaving small cubes floating inside large cubes; then the algorithm surrounds each small cube with a constant number of layers of cubes its own size. This guarantees that every $d$-sphere that contains input points from both inside and outside a small floating cube, also contains at least one vertex from these layers. Now every point is incident on $O(1)$ maximal empty spheres, so using the vertices of cubes as Steiner points gives a linear-size Delaunay triangulation. As just explained, the running time of this algorithm depends on the size of the initial tree, but long chains can be identified without actually computing them to give a time bound of $O(n \log n)$.

### *3.3.2. Provably Good Mesh Generation*

The most commonly used definition of the *aspect ratio* of a simplex is the ratio of the radii of the circumscribed sphere to the inscribed sphere [68]. In three dimensions, there are a number of different types of simplices with poor aspect ratios [8, 49].

- **Needle.** A needle is a tetrahedron in which the longest edge to shortest edge ratio is very large.
- **Cap.** A cap is a tetrahedron in which the circumsphere's radius is much larger than the longest edge.
- **Sliver.** A sliver is a tetrahedron formed by four nearly coplanar points, fairly evenly spaced around a great circle of the circumsphere.

It is possible to distinguish subtypes, for example, a cap may or may not also be a needle, but the types defined above are sufficient for discussion.

Bern et al. [20] use octrees (or in general $2^d$-ary trees) to triangulate point set input, using only a constant times the minimum number of bounded-aspect-ratio tetrahedra. Dey, Bajaj, and Sugihara [49] generalize Chew's mesh generation algorithm to three dimensions. For point sets or convex polyhedra with point holes, this method avoids all bad tetrahedra except slivers.

Mitchell and Vavasis [129] have recently generalized the "provably good" polygon mesh generation of Bern et al. [20] to three dimensions, avoiding all types of bad tetrahedra. The generalization is not straightforward, primarily because vertices of polyhedra may have arbitrary degree.

Their algorithm first computes a balanced octree that safely separates—that is, by some constant number of cubes—faces of the input polyhedron $D$. The octree is refined in three phases: first, vertices are separated from nonincident edges and faces; second, boxes away from vertices are split to separate edges from other edges and facets; and finally, boxes away from both vertices and edges are split to separate facets from other facets. Cubes are duplicated in Riemann sheets for faces close together in space, but far apart by geodesic distance.

Next the boxes around vertices are merged in order to approximately center each vertex in its box. The intersection of the boundary of $D$ and the surface of a box must be triangulated with triangles of height a constant fraction of the maximum possible (as in Section 2.3.3). A complicated set of warping rules conforms the octree to the edges and facets of $D$. Finally, warped boxes are triangulated by adding Steiner points near their centers, with tetrahedra radiating from these points.

Let $A$ be the minimum aspect ratio of a Steiner triangulation of domain $D$. (Here the aspect ratio of a triangulation is the maximum aspect ratio of its tetrahedra.) The algorithm just sketched gives a triangulation with aspect ratio at most $cA$ (where $c$ is a constant), that uses at most a constant factor times the minimum number of tetrahedra needed to achieve $cA$. The proof of this theorem also requires a new idea beyond [20]. Because no analog of the CDT is known in three dimensions, Mitchell and Vavasis must compare their tetrahedralization to an optimal tetrahedralization, showing that at each point in $D$ the tetrahedron chosen by their algorithm is no more than a constant times smaller than the largest possible tetrahedron at that point.

**Theorem 17 (Mitchell and Vavasis [129]).** *There is an algorithm, based on octrees, that computes an approximate optimal-aspect-ratio tetrahedralization of an arbitrary polyhedral domain, using no more than a constant times the optimal number of tetrahedra.*

The theorem has special importance because the edge skeleton of a bounded-aspect-ratio tetrahedralization has a "separator" of complexity $O(n^{2/3})$ [126, 127]. (A separator is a set of vertices whose removal disconnects the graph into two pieces of roughly equal size.) Nested dissection then saves a factor of $O(n)$ in the asymptotic time to solve the linear equations that arise in the finite element method [112].

Finally, we mention an open question about Steiner tetrahedralization. There are at least two three-dimensional analogs of nonobtuse triangulation: bounding dihedral angles by 90°, and requiring tetrahedra to contain their circumcenters. As in the planar case, a nonobtuse mesh possesses some desirable numerical properties. However, no algorithms are known for nonobtuse tetrahedralization of polyhedra.

### *3.4. Heuristically Generated Three-Dimensional Meshes*

Most of the techniques for generating and improving two-dimensional meshes can be generalized to three dimensions, though not without some difficulties. See [7, 117] for surveys discussing both structured and unstructured meshes for fluid flow problems. Overall, three-dimensional unstructured mesh generation is still in its early stages of development.

#### *3.4.1. Mesh improvement*

Laplacian smoothing (see Section 2.4.1) generalizes to three dimensions [34, 177], but the improvement it offers may not be as significant as in two dimensions.

There are local tranformations that trade two tetrahedra for three, and three tetrahedra for two, as discussed above (Section 3.2.1). These transformations give the analog of the flip procedure in two dimensions, and may be used to improve a triangulation according to some criterion, such as the Delaunay empty circumsphere condition. In three dimensions, however, the flip procedure may get stuck in a local optimum that is not a global optimum [92]. Joe [94] and Rajan [139] have shown that, for the Delaunay criterion, special starting triangulations always lead to a global optimum. Joe has also used the flip procedure to locally maximize the minimum solid angle [95]. Starting from the Delaunay triangulation, he significantly improved aspect ratios, while also slightly decreasing the number of tetrahedra.

#### *3.4.2. Octrees*

Yerry and Shephard generalized their quadtree algorithm to a three-dimensional algorithm using balanced octrees [177]. They kept the number of "patterns" for boundary cubes manageable by assuming that each cube was cut by at most three facets of the input polyhedron. In the final steps, patterns are warped to approximate the actual boundary of the input, and Laplacian smoothing is applied. Further progress on their octree algorithm is reported by Shephard et al. [158]. Perucchio et al. [135, 148] have also advanced the octree approach with a different way of handling boundary cubes.

Buratynski [28] uses rectangular octrees, that is, noncubical boxes, and a hierarchical set of warping rules. Boxes are first warped to input corners, then input edges, and finally input faces. The rules are somewhat simplified by the fact that the octree is initially refined so that input edges intersect boxes of only one size. (Hence, this method does not come with the theoretical size guarantee of Mitchell and Vavasis's method [129].) Empirically, Buratynski's method seems to give no tetrahedra with bad aspect ratio.

Field and Smith [70] and Moore and Warren [130] suggest the use of a tetrahedral octree, built by recursively cutting a "bcc tetrahedron" (for body-centered cubic) into eight copies of itself. Tetrahedra spawn fewer boundary patterns than hexahedra.

### *3.4.3. Polyhedron Decomposition*

Cavendish, Field, and Frey [34] developed one of the first three-dimensional mesh generators. Their approach cuts the polyhedron into polygonal cross-sections, adds randomly chosen Steiner points with average spacing determined by a measure of local feature size [33], and then computes the Delaunay triangulation. (See [21, 125] for analyses of DTs of random points.) If Steiner points are sufficiently closely spaced on the boundaries of the polygonal cross-sections, the DT will be conforming. A final improvement step merges tetrahedra or moves Steiner points to remove slivers (very flat tetrahedra). Joe [95] is currently working on a generalization of his two-dimensional convex-decomposition generator [96].

### *3.4.4. Advancing Front*

Löhner [117, 118], Baker [6, 7], Jameson et al. [89], Peraire et al. [134], and others have generated tetrahedral meshes for entire aircraft. Typically, something akin to the advancing front method places Steiner points in layers around the aircraft.

Jameson et al. [89] use a number of overlapping structured grids to place Steiner points, and then produce an unstructured tetrahedral grid with the Delaunay triangulation.

Baker's method [6] starts with a surface triangulation, as described in Section 2.5 above. Steiner points are then added exterior to the aircraft. Regular lattices of Steiner points surround the aircraft, with the density of Steiner points decreasing away from the surface. A shell of closer-in Steiner points is created by adding a few points along a normal to each surface point. A Delaunay triangulation is computed incrementally, but with violations allowed where the DT would pierce the surface. A final improvement step removes slivers. Empirically, the only remaining tetrahedra of poor aspect ratio are needles near junctures of aircraft parts, such as where the wing joins the fuselage.

## 4. Conclusions

We have described work in computational geometry motivated by finite element mesh generation. This material spans a spectrum from purely theoretical results (for example, Chazelle's linear-time triangulation algorithm), through a middle ground (our own work on Steiner triangulations), to practical heuristics devised by numerical analysts.

We believe that worthwhile research is spread throughout this spectrum. We have attempted to gather together these scattered results, and hope this compilation proves useful to both theorists and practitioners.

## Acknowledgments

We would like to thank Tim Baker, Paul Chew, Mike Dillencourt, David Dobkin, Herbert Edelsbrunner, David Field, John Gilbert, David Goldberg, Barry Joe, Scott Mitchell, Lee Nackman, John Shaw, Shang-Hua Teng, Steve Vavasis, and Frances Yao for various helpful suggestions.

## References

[1] A. Aggarwal, L.J. Guibas, J. Saxe, and P.W. Shor. A linear time algorithm for computing the Voronoi diagram of a convex polygon. *Disc. and Comp. Geometry* 4 (1989) 591–604.

[2] F. Aurenhammer. Voronoi diagrams—a survey of a fundamental geometric data structure. *ACM Computing Surveys* 23 (1991) 345–405.

[3] I. Babuška and A. Aziz. On the angle condition in the finite element method, *SIAM J. Numer. Analysis* 13 (1976) 214–227.

[4] I. Babuška and W.C. Rheinboldt. A-posteriori error estimates for the finite element method. *Int. J. Numer. Meth. Eng.* 12 (1978) 1597–1615.

[5] P.L. Baehmann, S.L. Wittchen, M.S. Shepard, K.R. Grice, and M.A. Yerry. Robust geometrically-based automatic two-dimensional generation, *Int. J. Numer. Meth. Eng.* 24 (1987) 1043–1078.

[6] T.J. Baker. Automatic mesh generation for complex three-dimensional regions using a constrained Delaunay triangulation. *Eng. with Computers* 5 (1989) 161–175.

[7] T.J. Baker. Developments and trends in three-dimensional mesh generation. *Appl. Numer. Math.* 5 (1989) 275–304.

[8] T.J. Baker. Element quality in tetrahedral meshes. *7th Int. Conf. on Finite Element Models in Flow Problems*, Huntsville, Alabama, 1989.

[9] T.J. Baker. Unstructured meshes and surface fidelity for complex shapes. In *Proc. 10th AIAA Comp. Fluid Dynamics Conf.*, Hawaii, 1991.

[10] B.S. Baker, E. Grosse, and C.S. Rafferty. Nonobtuse triangulation of polygons. *Disc. and Comp. Geom.* 3 (1988) 147–168.

[11] R.E. Bank. *PLTMG User's Guide.* SIAM, 1990.

[12] R.E. Bank, A.H. Sherman, and A. Weiser. Refinement algorithms and data structures for regular local mesh refinement. In R. Stepleman et al., *Scientific Computing.* IMACS/North-Holland, 1983, 3–17.

[13] R.E. Barnhill. Representation and approximation of surfaces. *Math. Software III,* J.R. Rice, ed., Academic Press, 1977, 69–120.

[14] R.E. Barnhill. Computer aided surface representation and design, *Surfaces in Computer Aided Geometric Design*, R. Barnhill and W. Boehm, eds., North-Holland, Amsterdam, 1983, 1–24.

[15] R.H. Bartels, J.C. Beatty, and B.A. Barsky. *An Introduction to Splines for Use in Computer Graphics and Modeling*, Morgan Kaufmann, 1987.

[16] T.J. Barth and D.C. Jespersen. The design and application of upwind schemes on unstructured meshes. In *Proc. AIAA 27th Aerospace Sciences Meeting*, Reno, 1989.

[17] M. Bern, D. Dobkin, and D. Eppstein. Triangulating polygons without large angles. Submitted for publication, 1991.

[18] M. Bern, H. Edelsbrunner, D. Eppstein, S. Mitchell, and T.S. Tan. Edge-insertion for optimal triangulations. To appear in *Proc. Latin American Theoretical Informatics*, 1992.

[19] M. Bern and D. Eppstein. Polynomial-size nonobtuse triangulation of polygons. In *Proc. 7th ACM Symp. Comp. Geometry* (1991) 342–350.

[20] M. Bern, D. Eppstein, and J.R. Gilbert. Provably good mesh generation. In *Proc. 31st IEEE Symp. Foundations of Computer Science* (1990) 231–241. To appear in *J. Comp. System Science.*

[21] M. Bern, D. Eppstein, and F. Yao. The expected extremes in a Delaunay triangulation. *Int. J. Comp. Geometry and Applications* 1 (1991) 79–92.

[22] M. Bern and J.R. Gilbert. Drawing the planar dual. Submitted for publication, 1991.

[23] J. Bloomenthal. Polygonization of implicit surfaces. *CAGD* 5 (1988) 341–355.

[24] J.D. Boissonat. Shape reconstruction from planar cross sections. *Comp. Vision, Graphics, and Image Processing* 44 (1988) 1–29.

[25] J.D. Boissonat, O.D. Faugeras, and E. Le Bras-Mehlman. Representing stereo data with the Delaunay triangulation. Tech. Rep. 788, INRIA, France, 1988.

[26] A. Bowyer. Computing Dirichlet tessellations. *Computer J.* 24 (1981) 162–166.

[27] K.Q. Brown. Voronoi diagrams from convex hulls. *Inform. Process. Lett.* 9 (1979) 223–228.

[28] E.K. Buratynski. A fully automatic three-dimensional mesh generator for complex geometries. *Int. J. Numer. Meth. Eng.* 30 (1990) 931–952.

[29] A. Bykat. Design of a recursive, shape controlling mesh generator. *Int. J. Numer. Meth. Eng.* 19 (1983) 1375–1390.

[30] G.F. Carey and J.T. Oden. *Finite Elements: Computational Aspects.* Prentice-Hall, 1984.

[31] G.F. Carey, M. Sharma, and K.C. Wang. A class of data structures for 2-d and 3-d adaptive mesh refinement. *Int. J. Numer. Meth. Eng.* 26 (1988) 2607–2622.

[32] J.E. Castillo, ed. *Mathematical Aspects of Numerical Grid Generation. SIAM*, 1991.

[33] J.C. Cavendish. Automatic triangulation of arbitrary planar domains for the finite element method. *Int. J. Numer. Meth. Eng.* 8 (1974) 679–696.

[34] J.C. Cavendish, D.A. Field, and W.H. Frey. An approach to automatic three-dimensional finite element mesh generation. *Int. J. Numer. Meth. Eng.* 21 (1985) 329–347.

[35] B. Chazelle. Convex partitions of polyhedra: a lower bound and worst-case optimal algorithm. *SIAM J. Comput.* 13 (1984) 488–507.

[36] B. Chazelle. Triangulating a simple polygon in linear time. *Disc. and Comp. Geometry* 6 (1991), 485–524.

[37] B. Chazelle, H. Edelsbrunner, L.J. Guibas, J.E. Hershberger, R. Seidel, and M. Sharir. Selecting multiply covered points and reducing the size of Delaunay triangulations. In *Proc. 6th ACM Symp. Comp. Geometry* (1990) 116–127.

[38] B. Chazelle and J. Incerpi. Triangulation and shape complexity. *ACM Trans. on Graphics* 3 (1984) 135–152.

[39] B. Chazelle and L. Palios. Triangulating a nonconvex polytope. *Disc. and Comp. Geometry* 5 (1990), 505–526.

[40] L.P. Chew. There are planar graphs almost as good as the complete graph. *J. Comp. System Science* 39 (1989) 205–219.

[41] L.P. Chew. Constrained Delaunay triangulations. *Algorithmica* 4 (1989) 97–108.

[42] L.P. Chew. Guaranteed-quality triangular meshes. Tech. Rep. TR-89-983, Cornell University, 1989.

[43] K. Clarkson. Approximation algorithms for planar traveling salesman tours and minimum-length triangulations. In *Proc. 2nd ACM-SIAM Symp. Disc. Algorithms* (1991) 17–23.

[44] A.K. Cline and R.J. Renka. A constrained two-dimensional triangulation and the solution of closest node problems in the presence of barriers. Tech. Rep., U. of Texas at Austin, 1985.

[45] E.F. D'Azevedo and R.B. Simpson. On optimal interpolation triangle incidences. *SIAM J. Sci. Stat. Comput.* 10 (1989) 1063–1075.

[46] L. De Floriani, B. Falcidieno, and C. Pienovi. Delaunay-based representation of surfaces defined over arbitrarily shaped domains. *Computer Vision, Graphics, and Image Processing* 32 (1985) 127–140.

[47] L. De Floriani, B. Falcidieno, C. Pienovi, and G. Nagy. On sorting triangles in a Delaunay tesselation. Tech. Rep., Inst. Matematica Applicata, Cons. Nazionale delle Richerche, Genova, Italy, 1988.

[48] B. Delaunay. Sur la sphère vide. *Izvestia Akademia Nauk SSSR, VII Seria, Otdelenie Matematicheskii i Estestvennyka Nauk* 7 (1934) 793–800.

[49] T. Dey, C. Bajaj, and K. Sugihara. On good triangulations in three dimensions. In *Proc. ACM Symp. on Solid Modeling Foundations and CAD/CAM Applications*, 1991.

[50] M.B. Dillencourt, I. Rivin, and W.D. Smith. Combinatorial structure of Delaunay triangulations of the plane. Manuscript, 1991.

[51] H. Djidjev and A. Lingas. On computing the Voronoi diagram for restricted planar figures. 2nd Worksh. Algorithms and Data Structures. Springer-Verlag LNCS 519 (1991) 54–64.

[52] D. Dobkin, S. Friedman, and K. Supowit. Delaunay graphs are almost as good as complete graphs. *Disc. and Comp. Geometry* 5 (1990) 399–407.

[53] D. Dobkin, S. Levy, W. Thurston, and A. Wilks. Contour tracing by piecewise linear approximations. *ACM Trans. on Graphics* 9 (1990) 389–423.

[54] R.D. Düppe and H.J. Gottschalk. Automatische Interpolation von Isolinien bei willkürlichen Stützpunkten. *Allgemeine Vermessungsnachrichten* 77 (1970) 423–426.

[55] R.A. Dwyer. Higher-dimensional Voronoi diagrams in linear expected time. *Proc. 5th Annual ACM Symp. on Comp. Geometry* (1989) 326–333.

[56] N. Dyn, D. Levin, and S. Rippa. Data dependent triangulations for piecewise linear interpolation. *IMA J. Numer. Analysis* 10 (1990) 137–154.

[57] H. Edelsbrunner. *Algorithms in Combinatorial Geometry.* Springer-Verlag, 1987.

[58] H. Edelsbrunner. An acyclicity theorem for cell complexes in $d$ dimensions. *Combinatorica* 18 (1990) 251–260.

[59] H. Edelsbrunner. Triangulations. Course notes for CS 497, U. Illinois, Spring 1991.

[60] H. Edelsbrunner and E.P. Mücke. Simulation of simplicity, a technique to cope with degenerate cases in geometric computations. *ACM Trans. Graphics* 9 (1990) 66–104.

[61] H. Edelsbrunner and T.S. Tan. A quadratic time algorithm for the minmax length triangulation. In *Proc. 32nd IEEE Symp. Foundations of Comp. Science* (1991) 414–423.

[62] H. Edelsbrunner and T.S. Tan. A cubic bound for conform Delaunay triangulations. Manuscript, 1992.

[63] H. Edelsbrunner, F.P. Preparata, and D.B. West. Tetrahedrizing point sets in three dimensions. *J. Symbolic Comp.* 10 (1990) 335-347.

[64] H. Edelsbrunner, T.S. Tan, and R. Waupotitsch. A polynomial time algorithm for the minmax angle triangulation. In *Proc. 5th ACM Symp. Comp. Geometry* (1990) 44–52. To appear in *SIAM J. Sci. Stat. Comp.*.

[65] H. ElGindy and D. Avis. A linear algorithm for computing the visibility polygon from a point. *J. Algorithms* 2 (1981) 186–197.

[66] D. Eppstein. The farthest point Delaunay triangulation minimizes angles. Report ICS-90-45, Univ. California, Irvine (1990). To appear in *Comp. Geometry Theory and Applications*.

[67] D. Eppstein. Approximating the minimum weight triangulation. To appear in *Proc. 3rd ACM-SIAM Symp. Disc. Algorithms* (1992).

[68] D.A. Field. Implementing Watson's algorithm in three dimensions. In *Proc. 2nd ACM Symp. Comp. Geometry* (1986) 246–259.

[69] D.A. Field. Laplacian smoothing and Delaunay triangulations. *Comm. in Applied Numer. Analysis* 4 (1988) 709–712.

[70] D.A. Field and W.D. Smith. Graded tetrahedral finite element meshes. *Int. J. Numer. Meth. Eng.* 31 (1991) 413–425.

[71] S. Fortune. A sweepline algorithm for Voronoi diagrams. *Algorithmica* 2 (1987) 153–174.

[72] S. Fortune. Voronoi diagrams and Delaunay triangulations. In *Computing in Euclidean Geometry*, F.K. Hwang and D.Z. Du, eds., World Scientific, 1992.

[73] W.H. Frey. Selective refinement: a new strategy for automatic node placement in graded triangular meshes. *Int. J. Numer. Meth. Eng.* 24 (1987) 2183–2200.

[74] W.H. Frey and D.A. Field. Mesh relaxation: a new technique for improving triangulations. *Int. J. Numer. Meth. Eng.* 31 (1991) 1121–1133.

[75] I. Fried. Condition of finite element matrices generated from nonuniform meshes. *AIAA J.* 10 (1972) 219–221.

[76] M.R. Garey and D.S. Johnson. *Computers and Intractability: A Guide to the Theory of NP-Completeness*. W.H. Freeman, 1979.

[77] M.R. Garey, D.S. Johnson, F.P. Preparata, and R.E. Tarjan. Triangulating a simple polygon. *Inform. Process. Lett.* 7 (1978) 175–179.

[78] P.L. George, F. Hecht, and E. Saltel. Constraint of the boundary and automatic mesh generation. In *Proc. 2nd Int. Conf. on Numer. Grid Generation in Comp. Fluid Mechanics* (1988).

[79] P.D. Gilbert. New results in planar triangulations. Report R–850, Univ. Illinois Coordinated Science Lab (1979).

[80] C. Gold, T. Charters, and J. Ramsden. Automated contour mapping using triangular element data structures and an interpolant over each irregular triangular domain, In *Proc. Siggraph*, 1977, 170–175.

[81] J. Gregory. Error bounds for linear interpolation on triangles, *The Mathematics of Finite Elements and Application II*, J. R. Whiteman, ed., Academic Press, London, 1975, 163–170.

[82] L.J. Guibas, D.E. Knuth, and M. Sharir. Randomized incremental construction of Delaunay and Voronoi diagrams. In *Proc. 17th Int. Colloq.— Automata, Languages, and Programming*. Springer-Verlag LNCS 443 (1990) 414–431.

[83] D. Hansford. The neutral case for the min-max triangulation. *CAGD* 7 (1990) 431–438.

[84] L.R. Hermann. Laplacian-isoparametric grid generation scheme. *J. of the Eng. Mechanics Div. of the American Soc. of Civil Engineers* 102 (October 1976) 749–756.

[85] J. Hershberger. Finding the visibility graph of a polygon in time proportional to its size. *Algorithmica* 4 (1989) 141–155.

[86] S. Hertel and K. Mehlhorn. Fast triangulation of simple polygons. *4th Conf. Foundations of Computation Theory* . Springer-Verlag LNCS 158 (1983) 207–218.

[87] K. Ho-Le. Finite element mesh generation methods: a review and classification. *Computer-Aided Design* 20 (1988) 27–38.

[88] T.C. Hu and A.C. Tucker. Optimal computer search trees and variable length alphabetic codes. *SIAM J. Applied Math.* 21 (1971) 514–532.

[89] A. Jameson, T.J. Baker, and N.P. Weatherill. Calculation of inviscid transonic flow over a complete aircraft. In *Proc. AIAA 24th Aerospace Sciences Meeting*, Reno, 1986.

[90] P. Jamet. Estimations d'erreur pour des elements finis droits presque degeneres. Tech. Report CRM-447, Centre d'Etudes de Limeil.

[91] B. Joe, Delaunay triangular meshes in convex polygons. *SIAM J. Sci. Stat. Comput.* 7 (1986) 514–539.

[92] B. Joe. Three-dimensional triangulations from local transformations. *SIAM J. Sci. Stat. Comput.* 10 (1989) 718–741.

[93] B. Joe. On the correctness of a linear-time visibility polygon algorithm. *Intern. J. Computer Math.* 32 (1990) 155-172.

[94] B. Joe. Construction of three-dimensional Delaunay triangulations using local transformations. *CAGD* 8 (1991) 123–142.

[95] B. Joe. Delaunay versus max-min solid angle triangulations for three-dimensional mesh generation. *Int. J. Numer. Meth. Eng.* 31 (1991) 987–997.

[96] B. Joe and R.B. Simpson. Triangular meshes for regions of complicated shape. *Int. J. Numer. Meth. Eng.* 23 (1986) 751–778.

[97] B. Joe and R.B. Simpson. Corrections to Lee's visibility polygon algorithm. *BIT* 27 (1987) 458–473.

[98] J.M. Keil and C.A. Gutwin. The Delaunay triangulation closely approximates the complete Euclidean graph. *1st Workshop on Algorithms and Data Structures*. Springer-Verlag LNCS 382 (1989) 47–56.

[99] D.G. Kirkpatrick. A note on Delaunay and optimal triangulations. *Inform. Process. Lett.* 10 (1980) 127–128.

[100] S. Kirkpatrick, C.D. Gelatt, and M.P. Vecchi. Optimization by simulated annealing. *Science* 220 (1983) 671–680.

[101] G.T. Klincsek. Minimal triangulations of polygonal domains. *Ann. Disc. Math.* 9 (1980) 121–123.

[102] C.L. Lawson. Software for $C^1$ surface interpolation. in J. Rice, ed., *Mathematical Software III*, Academic Press (1977) 161–194.

[103] D.T. Lee. Proximity and reachability in the plane. Tech. Rep. R–831, Coordinated Science Laboratory, Univ. Illinois, 1978.

[104] D.T. Lee. Visibility of a simple polygon. *Comput. Vision, Graphics, and Image Proc.* 22 (1983) 207–221.

[105] D.T. Lee and A. Lin. Generalized Delaunay triangulation for planar graphs. *Disc. and Comp. Geometry* 1 (1986) 201–217.

[106] C. Levcopoulos. An $\Omega(\sqrt{n})$ lower bound for non-optimality of the greedy triangulation. *Inform. Process. Lett.* 25 (1987) 247–251.

[107] C. Levcopoulos and A. Lingas. On approximation behavior of the greedy triangulation for convex polygons. *Algorithmica* 2 (1987) 175–193.

[108] C. Levcopoulos and A. Lingas. Fast algorithms for greedy triangulation. 2nd Scand. Worksh. Algorithm Theory, Springer-Verlag LNCS 447 (1990) 238–250.

[109] D.A. Lindholm. Automatic triangular mesh generation on surfaces of polyhedra. *IEEE Trans. Magnetics* MAG-19 (1983) 2539–2542.

[110] A. Lingas. Advances in minimum weight triangulation. Ph.D. thesis, Linköping Univ., 1983.

[111] A. Lingas. Voronoi diagrams with barriers and their applications. *Inform. Process. Lett.* 32 (1989) 191–198.

[112] R.J. Lipton, D.J. Rose, and R.E. Tarjan. Generalized nested dissection. *SIAM J. Numer. Analysis* 16 (1979) 346–358.

[113] E.L. Lloyd. On triangulations of a set of points in the plane. In *Proc. 18th IEEE Symp. Found. Comp. Sci.* (1977) 228–240.

[114] S.H. Lo. A new mesh generation scheme for arbitrary planar domains. *Int. J. Numer. Meth. Eng.* 21 (1985) 1403–1426.

[115] S.H. Lo. Delaunay triangulation of nonconvex planar domains. *Int. J. Numer. Meth. Eng.* 28 (1989) 2695–2707.

[116] S.H. Lo. Automatic mesh generation and adaptation by using contours. *Int. J. Numer. Meth. Eng.* 31 (1991) 689–707.

[117] R. Löhner. Finite elements in CFD: what lies ahead. *Int. J. Numer. Meth. Eng.* 24 (1987) 1741–1756.

[118] R. Löhner. Generation of three-dimensional unstructured grids by the advancing-front method. In *Proc. AIAA 26th Aerospace Sciences Meeting*, Reno, 1988.

[119] G.K. Manacher and A.L. Zobrist. Neither the greedy nor the Delaunay triangulation approximates the optimum. *Inform. Process. Lett.* 9 (1979) 31–34.

[120] O. Marcotte and S. Suri. Fast matching algorithms for points on a polygon. *SIAM J. Comput.* 20 (1991) 405–422.

[121] J. Matoušek. Reporting points in halfspaces. In *Proc. 32nd IEEE Symp. Foundations of Comp. Science* (1991) 207–215.

[122] A. Maus. Delaunay triangulation and the convex hull of $n$ points in expected linear time. *BIT* 24 (1984) 151–163.

[123] D.J. Mavriplis. Unstructured and adaptive mesh generation for high Reynolds number viscous flows. ICASE Report 91-25, NASA Langley Research Center, 1991.

[124] E. Melissaratos and D. Souvaine. Coping with inconsistencies: a new approach to produce quality triangulations of polygonal domains with holes for the the finite element method. Tech. Report, Dept. of Comp. Science, Rutgers University, 1991.

[125] R.E. Miles. On the homogeneous planar Poisson point process. *Mathematical Biosciences* 6 (1970) 85–127.

[126] G.L. Miller and W. Thurston. Separators in two and three dimensions. In *Proc. 22nd ACM Symp. Theory of Computing* (1990) 300–309.

[127] G.L. Miller, S.-H. Teng, and S.A. Vavasis. A unified geometric approach to graph separators. In *Proc. 32nd IEEE Symp. on Foundations of Comp. Science* (1991) 538–547.

[128] S. Mitchell. Center for Applied Mathematics, Cornell University, 1991.

[129] S. Mitchell and S. Vavasis. Quality mesh generation in three dimensions. Manuscript, Center for Applied Mathematics, Cornell University, 1991.

[130] D. Moore and J. Warren. Adaptive mesh generation I: packing space. Tech. Report TR 90-106, Dept. of Computer Science, Rice University, 1990.

[131] D.M. Mount and A. Saalfeld. Globally-equiangular triangulations of co-circular points in $O(n \log n)$ time. In *Proc. 4th ACM Symp. Comp. Geom.* (1988) 143–152.

[132] L.R. Nackman and V. Srinivasan. Point placement for Delaunay triangulation of polygonal domains. In *Proc. 3rd Canadian Conf. Comp. Geometry* (1991) 37-40.

[133] C.H. Papadimitriou, A.A. Schäffer, and M. Yannakakis. Simple local search problems that are hard to solve. *SIAM J. Comput.* 20 (1991) 56–87.

[134] J. Peraire, J. Peiro, L. Formaggia, K. Morgan, and O.C. Zienkiewicz. Finite element Euler computations in three dimensions. In *Proc. AIAA 26th Aerospace Sciences Meeting*, Reno, 1988.

[135] R. Perucchio, M. Saxena, and A. Kela. Automatic mesh generation from solid models based on recursive spatial decomposition. *Int. J. Numer. Meth. Eng.* 28 (1989) 2469–2502.

[136] D.A. Plaisted and J. Hong. A heuristic triangulation algorithm. *J. Algorithms* 8 (1987) 405–437.

[137] F.P. Preparata and S.J. Hong. Convex hulls of finite sets of points in two and three dimensions. *Commun. Assoc. Comput. Mach.* 20 (1977) 87–93.

[138] F.P. Preparata and M.I. Shamos. *Computational Geometry: An Introduction*, Springer-Verlag, 1985.

[139] V.T. Rajan. Optimality of the Delaunay triangulation in $R^d$. In *Proc. 7th ACM Symp. Comp. Geometry* (1991) 357–363.

[140] S. Rippa. Minimal roughness property of the Delaunay triangulation. *CAGD* 7 (1990) 489–497.

[141] S. Rippa and B. Schiff. Minimum energy triangulations for elliptic problems. *Comp. Meth. in Applied Mech. and Eng.* 84 (1990) 257–274.

[142] M.C. Rivara. Algorithms for refining triangular grids suitable for adaptive and multigrid techniques. *Int. J. Numer. Meth. Eng.* 20 (1984) 745–756.

[143] J. Ruppert and R. Seidel. On the difficulty of tetrahedralizing 3-dimensional nonconvex polyhedra. In *Proc. 5th ACM Symp. Comp. Geometry* (1989) 380–393. To appear in *Disc. and Comp. Geometry.*

[144] A. Saalfeld. Delaunay edge refinements. In *Proc. 3rd Canadian Conf. Comp. Geometry* (1991) 33-36.

[145] S. Salzberg, A. Delcher, D. Heath, and S. Kasif. Learning with a helpful teacher. In *Proc. 12th Int. Joint Conf. Artificial Intelligence* (1991).

[146] H. Samet. The quadtree and related hierarchical data structures. *Computing Surveys* 16 (1984) 188-260.

[147] H. Samet. *The Design and Analysis of Spatial Data Structures.* Addison-Wesley, 1990.

[148] N. Sapidis and R. Perucchio. Advanced techniques for automatic finite element meshing from solid models. *Computer-Aided Design* 21 (1989) 248–253.

[149] N. Sapidis and R. Perucchio. Delaunay triangulation of arbitrarily shaped planar domains. To appear in *CAGD.*

[150] E. Schönhardt. Über die Zerlegung von Dreieckspolyedern in Tetraeder. *Math. Annalen* 98 (1928) 309–312.

[151] W.J. Schroeder and M.S. Shephard. Geometry-based fully automatic mesh generation and the Delaunay triangulation. *Int. J. Numer. Meth. Eng.* 26 (1988) 2503–2515.

[152] L.L. Schumaker. Triangulation methods. In *Topics in Multivariate Approximation*, C.K. Chui, L.L. Schumaker, and F.I. Utreras, eds., Academic Press, 1987, 219–232.

[153] R. Seidel. A convex hull algorithm optimal for point sets in even dimensions. Report 81-14, Dept. of Computer Science, U. British Columbia, 1981.

[154] R. Seidel. Constrained Delaunay triangulations and Voronoi diagrams with obstacles. In H.S. Poingratz and W. Schinnerl, eds., *1978-1988 Ten Years IIG* (1988) 178–191.

[155] M.I. Shamos and D. Hoey. Closest-point problems. In *Proc. 16th IEEE Symp. Foundations of Comp. Science* (1975) 151–162.

[156] J.G. Shaw. Xerox Design Research Institute, 400 Engineering and Theory Center, Cornell University, 1991.

[157] M.S. Shephard. Approaches to the automatic generation and control of finite element meshes. *Appl. Mech. Rev.* 41 (1988) 169–185.

[158] M.S. Shephard, F. Guerinoni, J.E. Flaherty, R.A. Ludwig, and P.L. Baehmann. Finite octree mesh generation for automated adaptive three-dimensional flow analysis. In *Proc. of 2nd Int. Conf. on Numer. Grid Generation in Computational Fluid Mechanics* (1988) 709–718.

[159] R. Sibson. Locally equiangular triangulations. *Computer J.* 21 (1978) 243–245.

[160] D. Sleator, R.E. Tarjan, and W. Thurston. Rotation distance, triangulations, and hyperbolic geometry. *J. Amer. Math. Soc.* 1 (1988) 647–682.

[161] W.D. Smith. *Studies in Discrete and Computational Geometry.* PhD thesis, Princeton University, 1988.

[162] W.D. Smith. Implementing the Plaisted-Hong min-length plane triangulation heuristic. Manuscript cited by [43], 1989.

[163] V. Srinivasan, L.R. Nackman, J-M. Tang, and S.N. Meshkat. Automatic mesh generation using the symmetric axis tranformation of polygonal domains. Tech. Rept. RC 16132, Comp. Science, IBM Research Division, Yorktown Heights, NY, 1990.

[164] G. Strang and G.J. Fix. *An Analysis of the Finite Element Method.* Prentice-Hall, 1973.

[165] R.E. Tarjan and C.J. Van Wyk. An $O(n \log\log n)$-time algorithm for triangulating a simple polygon. *SIAM J. Comput.* 17 (1988) 143–178.

[166] W.C. Thacker. A brief review of techniques for generating irregular computational grids. *Int. J. Numer. Meth. Eng.* 15 (1980) 1335–1341.

[167] J.F. Thompson, ed. *Numerical Grid Generation.* North-Holland, 1982.

[168] J.F. Thompson, Z.U.A. Warsi, C.W. Mastin. *Numerical Grid Generation: Foundations and Applications.* North-Holland, 1985.

[169] P.J.M. van Laarhoven and E.H.L. Aarts. *Simulated Annealing: Theory and Practice.* Kluwer Academic Publishers, Dordrecht, the Netherlands, 1987.

[170] C. Wang and L. Schubert. An optimal algorithm for constructing the Delaunay triangulation of a set of line segments. In *Proc. 3rd ACM Symp. Comp. Geometry* (1987) 223–232.

[171] D.F. Watson. Computing the $n$-dimensional Delaunay tessellation with application to Voronoi polytopes. *Computer J.* 24 (1981) 167–171.

[172] D.F. Watson and G.M. Philip. Systematic triangulations. *Computer Vision, Graphics, and Image Processing* 26 (1984) 217–223.

[173] R.D. Williams. Adaptive parallel meshes with complex geometry. Tech. Report CRPC-91-2, Center for Research on Parallel Computation, California Inst. of Technology, 1991.

[174] A.M. Winslow. An irregular triangle mesh generator. Report UCXRL-7880, National Technical Information Service, Springfield, VA, 1964.

[175] F.F. Yao. Speed-up in dynamic programming. *SIAM J. Algebraic and Disc. Methods* 3 (1982) 532–540.

[176] M.A. Yerry and M.S. Shephard. A modified quadtree approach to finite element mesh generation. *IEEE Computer Graphics and Applications* 3 (January/February 1983) 39–46.

[177] M.A. Yerry and M.S. Shephard. Automatic three-dimensional mesh generation by the modified-octree technique. *Int. J. Numer. Meth. Eng.* 20 (1984) 1965–1990.

*Computing in Euclidean Geometry*, pp. 91–115
Eds. D.-Z. Du and F. K. Hwang


# MACHINE PROOFS OF GEOMETRY THEOREMS

Shang-Ching Chou and Mahesh Rathi
*Department of Computer Science, The Wichita State University*
*Wichita, KS 67208*

ABSTRACT

In the past decade highly successful algebraic methods for machine proofs of geometry theorems have been developed. The first step in these methods is to assign (variable) coordinates to key points, and then translate the hypotheses and conclusion of a geometric proposition into (multivariate) polynomial equations and inequalities. To date the most practically successful algebraic techniques have been Wu's method and the Gröbner basis (GB) method. This survey mainly concentrates on Wu's (or GB) method and related issues by using understandable examples. Also the relation of algebraic methods and the traditional method for proving geometry theorems is discussed.

## 1. Introduction

One of the most important and intellectually challenging tasks in mathematical work is proving theorems. This makes the field of automated theorem proving a very interesting and highly sophisticated research area. The main aim of researchers in this field is finding better methods and the design and implementation of computer software capable of proving theorems based on these methods. The field of automated theorem proving is emerging from the ivory tower of academic research into real world applications. At the same time, it is also asserting a definite place in many university curricula

In this paper, we are particularly interested in the field of *geometry* theorem proving. In addition to being an exciting branch of mathematics, geometry is also a part of everyday life. The field of geometry theorem proving is not only an appealing and attractive research area but also very crucial to the development of the field of *geometric reasoning*. Geometric reasoning, in turn, is of great relevance to such interesting and sophisticated applications as *computer-aided design, robotics,* etc.

We can divide the proofs of geometric theorems into two categories: traditional (or Euclidean or *synthetic*) proofs and algebraic (or analytic) proofs. Both of these approaches are discussed in detail in later sections. We illustrate these approaches with examples and point out their advantages and shortcomings.

The earliest successful work in the area of automated geometry theorem proving was done by H. Gelernter and his collaborators. Their method was based on the Euclidean traditional proof method. A. Tarski, on the other hand, gave the famous Tarski decision procedure (carried out in real closed fields [42]), for what he called elementary geometry, based on an algebraic method. Subsequent improvements were made by A. Seidenberg [40], G. Collins [21] and others along this line. Then, in 1977, another algebraic method was introduced by Chinese mathematician Wu Wen-Tsün [45]. Wu's method has been successfully used to prove many geometry theorems whose traditional proofs need enormous amounts of human intelligence [11].

In section 2, we trace the history of automated theorem proving, with particular emphasis on geometry theorem proving. Sections 3 and 4 describe the traditional and algebraic approaches, respectively, for proving geometry theorems and illustrate these approaches with examples. The most attention, by far, is devoted to section 5 where we describe Wu's method and its variants. Detailed examples are provided along with the proofs generated by a geometry theorem prover. In section 6, we briefly discuss the relationship between geometry and algebra and illustrate the inclusion relationship among the models of the various geometries discussed. And, finally, section 7 contains some concluding remarks.

## 2. History of Automated Theorem Proving

In this section, we have tried to trace the history of automated theorem proving in general and geometry theorem proving in particular, adhering to [24]. We *know* that, because of space limitations, we have omitted several significant developments and many important contributors to the field and we apologize, in advance, to all.

The word "geometry" has its root in the Greek word *geometrein* (geo = earth, metrein = to measure). In about 600 B.C., Thales of Miletus brought geometry to Greece from a journey to Egypt and applied to it the deductive procedures of Greek philosophy, thus founding deductive geometry. One of the main achievements of ancient Greek mathematics as a whole was the development of deductive geometry. Thales insisted on establishing geometric statements by deductive reasoning rather than by trial and error and, thus, he can be called the "father of geometry theorem proving."

In about 530 B.C., Pythagoras and his disciples established the first systematic treatment of plane geometry. They proved the famous *Pythagorean theorem* and made many other significant contributions.

In about 300 B.C., Euclid founded the classic axiomatic geometry. His main contribution was the famous 13 volume work *Elements*. It represents the mathematical knowledge of his time, ingeniously arranged in a deductive manner, starting from a few definitions and axioms. With more than one thousand editions it has dominated geometry teaching for more than two thousand years and, obviously, has had a great impact on the development of mathematics.

A major breakthrough occurred in 1637 when Deścartes founded analytic geometry. He discovered that algebraic methods could be used to develop classical

Greek geometry. He said (in [27]),

> ... it is possible to construct all the problems of ordinary geometry by doing no more than the little covered in the four figures[1] that I have explained. This is one thing which I believe the ancients did not notice, for otherwise they would not have put so much labor into writing so many books in which the very sequence of the propositions showed that they did not have a sure method ...

What had seemed in Euclid to be the result of cunning and mathematical ingenuity was now revealed as being accessible to relatively mechanical treatment. Deścartes observed that a point in a plane can be completely determined if its signed distances from two fixed perpendicular lines drawn in the plane are given, thus giving rise to the coordinate system.

Leibniz, in the late $17^{th}$ century, attempted to do for all deductive reasoning what Deścartes had done for geometry. He created spectacular prospects in proposing two projects which later turned out to be overly ambitious. These projects were *calculus ratiocanator* (a calculus of reason) and *lingua characteristica* (a universal language). Even though his technical contributions were minimal, Leibniz was able to take an enormous step in conceiving of logic itself as a deductive science whose principles can be formally derived from axioms.

In 1899, Hilbert laid the foundation of modern axiomatic geometry when he wrote a book *Grundlagen der Geometrie* which was later expanded through several editions and the final edition was translated into English [29]. Hilbert formulated fifteen axioms of the plane geometry. Following Euclid's approach, he used the five primitive terms *point, line, on, between*, and *congruent*. Hilbert's book is definitely a landmark in the history of mathematics in general and geometry in particular. In 1928, together with Ackermann, he provided a formal axiomatization for the predicate calculus.

In 1954, after digital computers became available, Martin Davis wrote a program [25] to prove first order sentences about linear equalities and inequalities over the domain of non-negative integers. It was called the *Presburger prover* since it was a direct implementation of the classical Presburger decision procedure for additive arithmetic. The complexity of this decision procedure is worse than exponential. It is not surprising, therefore, that the Presburger prover could only prove very simple facts. Its best result was to prove that the sum of two even numbers is even.

In 1957, Newell, Shaw, and Simon wrote a program [36] that could prove theorems in the propositional calculus using the axiomatization in *Principia Mathematica*. This program attempted to simulate, mechanically, the deduction processes used by humans in proving propositional logic theorems and was able to prove 38 theorems from *Principia Mathematica*.

[1] These "four figures" showed how to reduce arithmetic operations to geometric constructions.

Hao Wang explicitly sought to counter the impression Newell, Shaw, and Simon imparted about the efficiency of proof procedures. His work on mechanizing mathematical thought was published in 1960 [44]. He developed three programs, one for propositional calculus, one for a decidable part of the predicate calculus, and the third for all of predicate calculus. He proved many theorems and demonstrated that one does not necessarily have to use human-oriented techniques. Wang received the first *milestone* prize for research in automatic theorem proving for his "fundamental contributions to the field".

A major contribution was made by Martin Davis and Hilary Putnam when they introduced the D-P procedure [23]. They introduced skolem functions and proposed the now familiar conjunctive normal form clausal arrangement. Prawitz refined the D-P procedure and showed in [38] that only "relevant" clauses need to be generated. Robinson, in 1965, introduced his method of resolution [39] which revolutionized the subject of automatic theorem proving. The elegant simplicity of resolution led many researchers to try and find "refinements" to control the explosive clause growth. These refinements include *unit preference*, *set-of-support*, *hyperresolution*, *model elimination*, *paramodulation*, etc. and are all discussed in detail in [34]. Perhaps the most successful group to implement a resolution based system is the one at Argonne National Laboratory led by Larry Wos. They produced a very powerful proof system named *Otter* [37]. Larry Wos and Steve Winker received the first "Current Research" prize for their research in automatic theorem proving.

## 3. Traditional Proofs

Traditional Euclidean proofs employ the synthetic approach. In other words, the traditional approach uses properties of points, lines, triangles, etc. It employs the definitions and properties of these key geometric concepts to prove theorems in plane geometry using diagrams and geometric construction.

In the 1950s and the early 1960s, when there was considerable excitement about theorem proving in AI, Gelernter and his associates [28] used the traditional approach to initiate an effort in mechanical theorem proving in geometry. They were reportedly inspired by an observation, made by Minsky at a 1956 conference, that the diagram that traditionally accompanies plane geometry problems was a simple model for the theorem that could greatly prune the proof search. The diagram proved to be a powerful heuristic [35].

However, successes using the traditional approach have been very limited so far. The reason for this is very simple: unlike commonly believed, a typical traditional proof is full of gaps. As pointed out by Adler in his book on modern geometry [1], "one of the main defects in the traditional Euclidean proof is its almost complete disregard of such notions as the *two sides of a line* and the *interior of an angle*. Without clarification of these ideas, absurd consequences result."

The following two examples illustrate the points made above:

**Example 1:** Let ABCD be a parallelogram (i.e., AB||CD, BC||AD), E be the intersection of the diagonals AC and BD. Show AE ≡ CE (Figure 1).

The traditional proof of this theorem is first to prove △ACB ≅ △CAD (hence AB ≡ CD), then to prove △AEB ≅ △CED (hence AE ≡ CE). In proving the congruence of these triangles, we repeatedly use the fact that ∠CAB ≡ ∠ACD. This fact is quite evident because the two angles are the alternative angles with respect to parallels AB and CD. However, here we have implicitely assumed the *trivial fact* that point D and B are on either sides of line AC. The last fact is harder to prove than the original statement. (Please try it!)

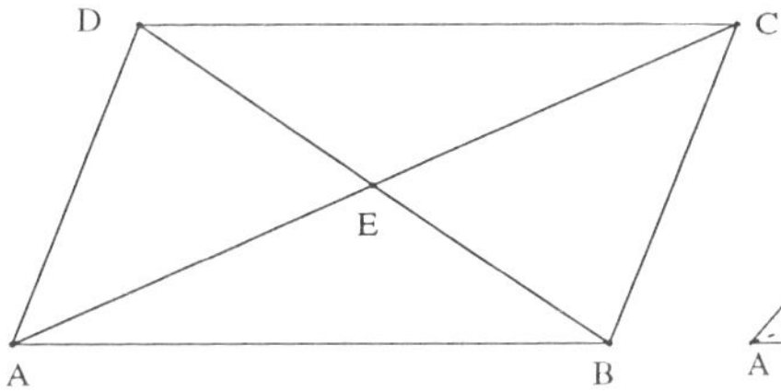


Figure 1

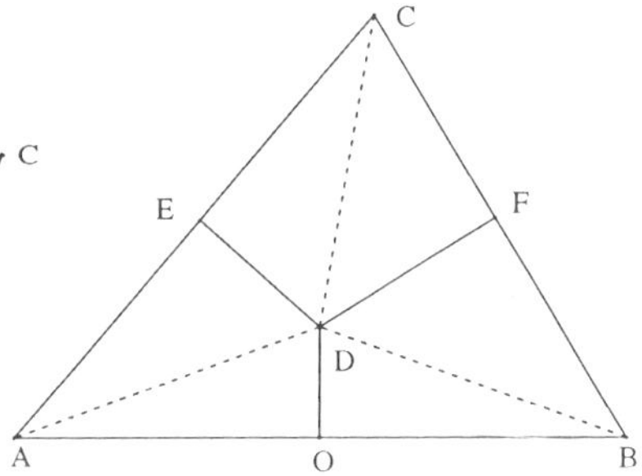


Figure 2

This extremely simple example reveals difficulty in implementing a powerful and sound geometry theorem prover based on traditional proofs. One can, of course, develop an interactive prover so that the user can input some trivial facts such as the one in the previous paragraph. These facts can be stored in a data base by programs. But, then one would face a much more severe problem of the consistency of such proofs.

**Example 2:** Every triangle is isosceles. Let ABC be a triangle as shown in Figure 2. We want to prove CA ≡ CB.

*Proof.* Let D be the intersection of the perpendicular bisector of AB and the internal bisector of angle ACB. Let DE⊥AC and DF⊥CB. It is easy to see that △CDE≅ △CDF and △ADO≅ △BDO which then makes it trivial to show that △ADE≅ △BDF. Hence CE+EA=CF+FB, i.e., CA ≡ CB. .QED.

This is a very well known paradox of last century. The problem, of course, lies in the (false) assumption that point D is in the *interior* of the triangle ABC!

## 4. Algebraic Proofs

Algebraic methods of proving geometry theorems are based on translating geometry statements into first-order formulae using the operations of an ordered (or unordered) field (e.g., $+, *, <, =$, etc.) with variables ranging over that field. Simply stated, in the algebraic approach, one transforms the geometric problem into a problem about some set of polynomials. Hypotheses and conjectures (conclusions) in a geometry problem are represented as polynomial equations or inequalities. Variables or indeterminates in these polynomials are coordinates of points.

In comparison to synthetic approaches to geometric theorem proving, the experience with algebraic methods has been quite encouraging. Many plane geometry theorems which humans find difficult to prove can be easily proven automatically by provers based on the algebraic methods. In the beginning, the algebraic approach was not widely used. This was perhaps because the algebraic manipulation is conceptually quite different from the approach humans take in solving geometric problems. It could also be due to the enormously high complexity of early decision procedures. For example, the famous Tarski decision procedure, due to its very high complexity, could only prove very simple and trivial theorems of elementary geometry. Subsequent improvements were made by Seidenberg [40], Collins [21], and others. In particular, Collins and his students used the cylindrical algebraic decomposition method [22] to substantially improve Tarski's decision procedure.

It was really a surprise when Chinese mathematiciam Wu Wen-Tsün introduced a new algebraic method in 1977 [45]. Wu's method has been found to be extremely successful in proving many nontrivial geometry theorems whose traditional proofs require enormous amounts of human intelligence [11].

We now illustrate how the parallelogram theorem of Example 1 can be formulated algebraically and also illustrate the proof technique.

**Example 1a**[2] : Let ABCD be a parallelogram (i.e., AB||CD, BC||AD), E be the intersection of the diagonals AC and BD. Show AE $\equiv$ CE (Figure 1).

Without loss of any generality, we can assume A to be the origin (0,0) and the line AB to be along the x-axis. Taking $u_1, u_2, u_3$ to be any three arbitrary real numbers, let us fix points B and C as well. Let the coordinates of B be ($u_1$, 0) and C = ($u_2$, $u_3$). We can solve the line equations for AD and CD to obtain the coordinates of point D, and the line equations for BD and AC to obtain the coordinates of point E. Let D = ($x_2$, $x_1$) and E = ($x_4$, $x_3$). Then we want to prove $g = 2u_2x_4 + 2u_3x_3 - u_3^2 - u_2^2 = 0$, i.e., AE $\equiv$ CE. So, after we determine the values of $x_i$'s, we can substitute these in the conclusion polynomial g to see whether the result is zero.

Thus, we have two line equations for $x_1$ and $x_2$:

$$h_1 = u_1x_1 - u_1u_3 = 0 \qquad \textit{AB is parallel to DC.}$$

[2]Example 1a is just a restatement of Example 1. We use this scheme for other examples as well.

$h_2 = u_3x_2 - (u_2 - u_1)x_1 = 0$ *DA is parallel to CB.*

Also, we have two line equations for $x_3$ and $x_4$:

$h_3 = x_1x_4 - (x_2 - u_1)x_3 - u_1x_1 = 0$ *E is on BD.*

$h_4 = u_3x_4 - u_2x_3 = 0$ *E is on AC.*

Solving the first two equations, we have $x_1 = u_3, x_2 = u_2 - u_1$. Solving the last two equations, using the solutions for $x_1$ and $x_2$, we have $x_3 = u_3/2, x_4 = u_2/2$. Now, substituting the solutions in g, we have $g = 0$. Thus, we prove the theorem.

The above approach is essentially what P.J. Davis used to prove Pappus' theorem on the computer [26]. Actually, a general mechanical proof method for affine geometry was already given by D. Hilbert in his classic book [29].

One defect in the above solution is that it does not specify the *degenerate* conditions under which the conclusion of the theorem might be invalid, or does not emphasize explicitly the fact that we only prove geometry statements to be *generically* or *generally* true (for the definitions and further discussion of these terms we refer the reader to [11]). For example, in order to have the above *normal* solutions for $x_1, \ldots, x_4$, we have to assume (at least) that $u_1 \neq 0$ and $u_3 \neq 0$, i.e., the *nondegenerate condition* that A, B, and C are not collinear.

## 5. Wu's Method and Its Variants

As we said before, variants of the Tarski decision method, in spite of the subsequent improvements especially by A. Seidenberg, still seemed far away from mechanically proving difficult geometry theorems in practice during the 70s. Thus Wu's work in 1977 [45] and subsequent developments [49, 5] were really a surprise to the community of theorem proving. It was Wu who first realized the importance of nondegenerate conditions in mechanical geometry theorem proving and introduced a method which can deal with such conditions mechanically.

The success of Wu's method stimulated researchers to apply the Gröbner basis (GB) method to the same class of geometry theorems which can be addressed by Wu's method. Around the end of 1985, three groups were successful in this direction: Chou–Schelter [20], Kapur [30], and Kutzler–Stifter [33].

The GB method, as an algebraic method, has long been known before Wu's work. But, as applied to geometry theorem proving, the GB approach is actually a kind of a variant of Wu's approach.

The method introduced by Wu (and its variants) addresses statements in elementary geometries for which, in their algebraic forms[3], the hypotheses and the conclusion can be expressed by *polynomial equations*[4]. For such a geometry statement, after adopting an appropriate coordinate system, the hypotheses can be

[3] For the relationship between axiomatic geometries and number systems, the reader is strongly urged to read [29].

[4] Dealing with inequality is generally beyond the method.

expressed as:

$$
\begin{aligned}
h_1(u_1, \ldots, u_d, x_1, \ldots, x_t) &= 0 \\
h_2(u_1, \ldots, u_d, x_1, \ldots, x_t) &= 0 \\
&\vdots \\
h_n(u_1, \ldots, u_d, x_1, \ldots, x_t) &= 0,
\end{aligned}
\tag{1}
$$

and the conclusion as $g(u_1, \ldots, u_d, x_1, \ldots, x_t) = 0$, where $h_1, \ldots, h_n$ and $g$ are polynomials with rational coefficients. Variables $u_1, \ldots, u_d$ are parameters or independent variables and variables $x_1, \ldots, x_t$ are algebraically dependent on $u$'s under *normal conditions*, being restricted by (1). The selection of the parameters $u$'s comes from the assumption that the figure in the geometry statement should be in a general position, an assumption that is often implicit in geometry statements. The use of two different kinds of variables goes back to the Hilbert mechanization theorem for affine geometry (see theorem 62 in [29]).

Now let us use three examples to illustrate how Wu's method works. The proofs of these three examples were actually produced by the prover, developed by the first author [5], including the selection of coordinates.

The implicit assumption that the parallelogram is in general position means that any three points among the four points A, B, C, and D can be arbitrarily chosen. Thus $u_1, u_2$, and $u_3$ can be chosen as parameters. We have four equations $h_1 = 0, \ldots, h_4 = 0$ for the hypotheses and one equation $g = 0$ for the conclusion (see Section 4).

Now, one might think that the algebraic formulation of the problem would be

$$(h_1 = 0 \wedge h_2 = 0 \wedge h_3 = 0 \wedge h_4 = 0) \Rightarrow g = 0. \tag{2}$$

However, (2) is not a valid statement. The reason is that the important nondegenerate conditions were overlooked. Wu's method can produce a sufficient number of nondegenerate conditions to make (2) valid.

The next step is to triangulate $h_1, h_2, h_3, h_4$ so that each polynomial introduces only one new (dependent) variable $x_i$. Thus, $h_1, h_2, h_3, h_4$ are not in triangular form because $h_3$ introduces two new variables, $x_3$ and $x_4$, at the same time. We can use a simple elimination procedure to obtain a triangular form: let $f_1 = h_1, f_2 = h_2, f_3 = prem^5\ (h_4, h_3, x_4), f_4 = h_4$. Then we have a triangular form:

$$
\begin{aligned}
f_1 &= u_1 x_1 - u_1 u_3 = 0 \\
f_2 &= u_3 x_2 - (u_2 - u_1) x_1 = 0 \\
f_3 &= (u_3 x_2 - u_2 x_1 - u_1 u_3) x_3 + u_1 u_3 x_1 = 0
\end{aligned}
$$

[5]Here $prem(h_4, h_3, x_4)$ denotes the pseudo remainder obtained by dividing $h_4$ by $h_3$ in variable $x_4$. For the basic algebraic operations underlying Wu's method, see [11].

$$f_4 = u_3x_4 - u_2x_3 = 0.$$

Now, we do successive pseudo divisions:

$$R_4 = prem(g, f_4, x_4) = (2u_3^2 + 2u_2^2)x_3 - u_3^3 - u_2^2u_3$$

$$R_3 = prem(R_4, f_3, x_3)$$
$$= (-u_3^4 - u_2^2u_3^2)x_2 + ((u_2 - 2u_1)u_3^3 + (u_2^3 - 2u_1u_2^2)u_3)x_1 + u_1u_3^4 + u_1u_2^2u_3^2$$

$$R_2 = prem(R_3, f_2, x_2) = (-u_1u_3^4 - u_1u_2^2u_3^2)x_1 + u_1u_3^5 + u_1u_2^2u_3^3$$

$$R_1 = prem(R_2, f_1, x_1) = 0.$$

Because the *final remainder* $R_1$ is zero, the theorem follows from appropriate subsidiary conditions. To see this, let us recall the simple and important *remainder formula* for successive pseudo divisions of $g$ with respect to a triangular form $f_1, \ldots, f_r$:

$$I_1^{s_1} \ldots I_r^{s_r} g = Q_1 f_1 + \ldots + Q_r f_r + R_1 \quad (3)$$

where $I_k$ are the leading coefficients of $f_k$ in $x_k$.

Since $R_1 = 0$, we have that $g = 0$ under $I_k \neq 0\,(k = 1, \ldots, r)$. The subsidiary conditions $I_k \neq 0$ are usually connected with nondegeneracy and are also called nondegenerate conditions. For this particular case, (2) is valid under

$$I_1 = u_1 \neq 0$$

$$I_2 = u_3 \neq 0$$

$$I_3 = u_3x_2 - u_2x_1 - u_1u_3 \neq 0$$

$$I_4 = u_3 \neq 0$$

The conditions $u_1 \neq 0$ and $u_3 \neq 0$ mean that A, B, and C are not collinear. The condition $u_3x_2 - u_2x_1 - u_1u_3 \neq 0$ means that line AC and line BD should have a normal intersection. By a more sophisticated study, these nondegenerate conditions (in geometric form) can be generated mechanically by machines. For more details, see [17].

Before going to the next example, let us summarize this simple use of Wu's method:

*Step 1*: Assign coordinates to the points involved, then translate the geometric hypotheses into a set (conjunction) of polynomial equations $h_1 = 0, \ldots, h_r = 0$; also the conclusion is translated into a polynomial equation $c = 0$. For geometric statements of constructive type[6], this can be done totally automatically.

*Step 2*: Transform the set of hypothesis polynomials into a triangular form (an ascending chain) $ASC = f_1, \ldots, f_p$ (usually $r = p$). For geometric statements of constructive type, since new points are introduced one by one, the new dependent variables are introduced at most two by two, and this transformation is then easy to do locally (involving at most two equations at a time). In general, we have a

[6] A geometric statement of constructive type means that new points are introduced one at a time by geometric constructions such as taking a point on the intersection of a line with another line or a circle, etc. For details, see [11].

complete triangular algorithm presented, and referred to as Ritt's principle, by Wu [46, 11].

*Step 3*: Divide the conclusion polynomial $c$ successively with respect to the triangular form $ASC = f_1, \ldots, f_p$ to obtain the final remainder $R_1 = prem(c; ASC)$. If $R_1 = 0$, then the statement is confirmed under subsidiary conditions $I_k \neq 0$, where $I_k$ is the leading coefficient of $f_k$.

*Step 4*: Analyze the subsidiary conditions $I_k \neq 0$ which are usually connected with nondegeneracy. Again, for geometric statements of constructive type, this analysis can be done automatically: nondegenerate conditions can be generated mechanically in their geometric form [6, 17].

The experienced reader might notice that for the simple parallelogram example, we do not have to use this general schema. We can just use the method in section 4, i.e., solve the dependent variables $x_1, ..., x_4$ successively in terms of the $u$'s, then substitute the solutions into the conclusion polynomial $c$ to see whether it is identical to zero. This is exactly what Hilbert's mechanical proof method [29] does for constructive affine geometric statements. However, Hilbert did not mention nondegenerate conditions which are important for a geometry statement to be valid; or, put another way, what Hilbert considered (a geometric statement) to be true is what we consider to be generally (or generically) true. Moreover, if we cannot solve dependent variables explicitly, then we have to use the above general schema. The following theorem is such an example.

**Example 3:** (Simson's Theorem). Let $D$ be a point on the circumscribed circle $(O)$ of triangle $ABC$. Perpendiculars are drawn from $D$ to the three sides $BC$, $CA$ and $AB$. Let $E$, $F$ and $G$ be the three feet. Show that $E$, $F$ and $G$ are collinear (Fig 3).

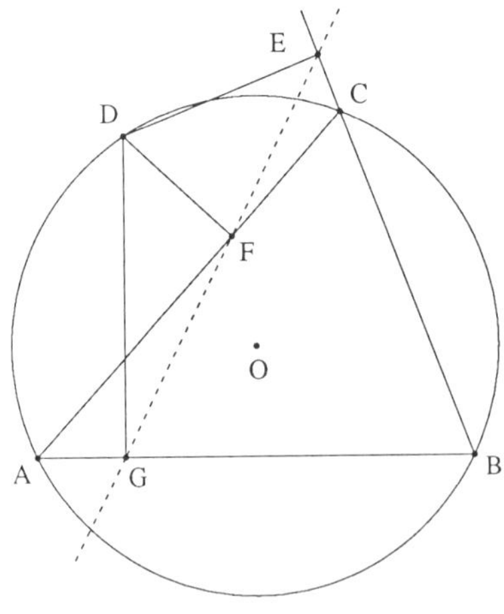


Figure 3

*Step 1*: Converting the statement into its algebraic form, we let $A = (0,0)$, $B = (u_1, 0)$, $C = (u_3, u_2)$, $O = (x_1, x_2)$, $D = (x_3, u_4)$, $E = (x_5, x_4)$, $F = (x_7, x_6)$, and $G = (x_9, x_8)$. We then have the following nine hypothesis equations:

$$
\begin{aligned}
h_1 &= 2u_1x_1 - u_1^2 = 0 && OA = OB \\
h_2 &= 2u_2x_2 + 2u_3x_1 - u_3^2 - u_2^2 = 0 && OA = OC \\
h_3 &= x_3^2 - 2x_1x_3 + u_4^2 - 2x_2u_4 = 0 && DO = OA \\
h_4 &= u_2x_5 - (u_3 - u_1)x_4 - u_1u_2 = 0 && E \text{ is on } BC \\
h_5 &= (u_3 - u_1)x_5 + u_2x_4 - (u_3 - u_1)x_3 - u_2u_4 = 0 && ED \perp BC \\
h_6 &= u_2x_7 - u_3x_6 = 0 && F \text{ is on } AC \\
h_7 &= u_3x_7 + u_2x_6 - u_3x_3 - u_2u_4 = 0 && FD \perp AC \\
h_8 &= u_1x_8 = 0 && G \text{ is on } AB \\
h_9 &= u_1x_9 - u_1x_3 = 0 && GD \perp AB.
\end{aligned}
$$

The conclusion that $E$, $F$, and $G$ are collinear, can be expressed by the equation $c = (x_6 - x_4)x_9 - (x_7 - x_5)x_8 + x_4x_7 - x_5x_6 = 0$.

*Step 2*: Using the pseudo remainder (*prem*) computation, we can triangulize $h_1, ..., h_9$ by letting $f_4 = prem(h_5, h_4, x_5)$, $f_6 = prem(h_7, h_6, x_7)$, $f_i = h_i$ for $i \neq 4, 6$. We thus have a triangular form $ASC = f_1, ..., f_9$:

$$
\begin{aligned}
f_1 &= 2u_1x_1 - u_1^2 \\
f_2 &= 2u_2x_2 + 2u_3x_1 - u_3^2 - u_2^2 \\
f_3 &= x_3^2 - 2x_1x_3 + u_4^2 - 2x_2u_4 \\
f_4 &= (u_3^2 - 2u_1u_3 + u_2^2 + u_1^2)x_4 - (u_2u_3 - u_1u_2)x_3 - u_2^2u_4 + u_1u_2u_3 - u_1^2u_2 \\
f_5 &= u_2x_5 - (u_3 - u_1)x_4 - u_1u_2 \\
f_6 &= (u_3^2 + u_2^2)x_6 - u_2u_3x_3 - u_2^2u_4 \\
f_7 &= u_2x_7 - u_3x_6 \\
f_8 &= u_1x_8 \\
f_9 &= u_1x_9 - u_1x_3.
\end{aligned}
$$

*Step 3*: Now, we use successive pseudo division to compute the final remainder $R_1 = prem(c; ASC)$. That is,

$$
\begin{aligned}
R_9 &= prem(R_{10}, f_9, x_9) = -(u_1x_7 - u_1x_5)x_8 + u_1x_4x_7 - (u_1x_5 - u_1x_3)x_6 - u_1x_3x_4 \\
R_8 &= prem(R_9, f_8, x_8) = u_1^2x_4x_7 - (u_1^2x_5 - u_1^2x_3)x_6 - u_1^2x_3x_4 \\
R_7 &= prem(R_8, f_7, x_7) = -(u_1^2u_3x_5 + u_1^2u_2x_4 - u_1^2u_3x_3)x_6 + u_1^2u_2u_4x_4 \\
R_6 &= prem(R_7, f_6, x_6) \qquad \text{(6 terms)}^7 \\
R_5 &= prem(R_6, f_5, x_5) \qquad \text{(6 terms)} \\
R_4 &= prem(R_5, f_4, x_4) \qquad \text{(11 terms)} \\
R_3 &= prem(R_4, f_3, x_3) \qquad \text{(11 terms)} \\
R_2 &= prem(R_3, f_2, x_2) \qquad \text{(8 terms)}
\end{aligned}
$$

[7] We omit the rest of the remainder polynomials, the largest of which has only 11 terms. Step 3 can be done in a fraction of one second on a Mac Plus. In view of the fact that the traditional proofs of Simson's theorem are nontrivial and subject to many gaps, the simple mechanical proof by the computer here is really a surprise.

$$R_1 = prem(R_2, f_1, x_1) = 0.$$

Since the final remainder $R_1$ is 0, by the remainder formula (3), $c = 0$ follows from $h_i = 0$ and subsidiary conditions $I_i \neq 0$, where the $I_i$ are the leading coefficients the of $f_i$:

$$I_1 = 2u_1$$
$$I_2 = 2u_2$$
$$I_3 = 1$$
$$I_4 = u_3^2 - 2u_1u_3 + u_2^2 + u_1^2$$
$$I_5 = u_3 - u_1$$
$$I_6 = u_3^2 + u_2^2$$
$$I_7 = u_3$$
$$I_8 = I_9 = u_1.$$

*Step 4*: Finally, we analuze the subsidiary conditions ($I_k \neq 0$). Here, the non-zeroness of $I_1$, $I_2$, $I_8$, and $I_9$ mean that $A$, $B$, and $C$ are not collinear. $I_7 \neq 0$ and $I_5 \neq 0$ are not necessary by a more careful analysis based on the work in [6, 17]. The role of conditions $I_4 \neq 0$ and $I_6 \neq 0$ is very subtle, and they are *necessary* when using Wu's method or the GB method (see Example 4 on the next page and see [17] for a further discussion). We would emphasize that all this analysis can be done mechanically [6, 17]. A simple version of the prover for the Macintosh Plus is available [18], which can do all of the above steps totally automatically. For example, the input of Simson's theorem to the prover is:

```
'((point-order A B C O D E F G)
  (collinear A B G)
  (collinear B C E)
  (collinear A C F)
  (perpendicular E D B C)
  (perpendicular F D A C)
  (perpendicular G D A B)
  (eqdistance A O B O)
  (eqdistance A O C O)
  (eqdistance A O D O)
  (collinear E F G))            ;The last line is the conclusion
```

*Remark*: This way of proving theorems is not the one adopted by most provers based on the logic approach. We start with a set of hypotheses which do not necessarily imply the conclusion and end up with the confirmation of the conclusion by adding some subsidiary conditions. Would it be possible that the original geometric statement is substantially weakened or changed? To answer this question we need to address the formulation problem – in what sense do we prove geometry theorems? Generally, for a geometric statement, the equality part of the hypotheses (i.e., $H = \{h_1 = 0, ..., h_r = 0\}$) is easy to identify. If some of them are missing and the statement is not confirmed (or disproved), it is the user's mistake. However, the inequation part of the hypotheses, which is usually connected with nondegeneracy, is not so simple to identify. For a given geometric statement, a nondegenerate condition that is obvious to one person might not be obvious to a second, and a third person might even refuse to accept the condition as relevant or appropriate. In most of the recent research in geometry theorem proving, two different, but related, formulations for geometry statements have been considered.

**Formulation (Approach) F1.** Introducing parameters and the notion of "generally true" for a geometry statement. The present techniques can prove a statement to be generally true, at the same time giving nondegenerate conditions (usually in algebraic form) automatically. See [11] for details.

**Formulation (Approach) F2.** Giving nondegenerate conditions in geometric form manually (or mechanically) at the beginning as an integral part of the hypotheses. Then the prover only needs to answer whether the conclusion follows from the hypotheses *without adding* any other conditions.

The method originally developed by Wu is for Formulation F1. One of the advantages of Formulation F1 is that nondegenerate conditions can be taken care of without explicitly specifying them. Formulation F1 gives a clear insight into the nature of a geometric statement: if a geometry statement is proved to be generally false, then it cannot be valid no matter how many additional nondegenerate conditions are added so long as the hypotheses remain consistent.

**Example 4:** (The Butterfly Theorem). $A$, $B$, $C$ and $D$ are four points on circle $(O)$. $E$ is the intersection of $AC$ and $BD$. Through $E$ draw a line perpendicular to $OE$, meeting $AD$ at $F$ and $BC$ at $G$. Show $FE \equiv GE$ (Figure 4).

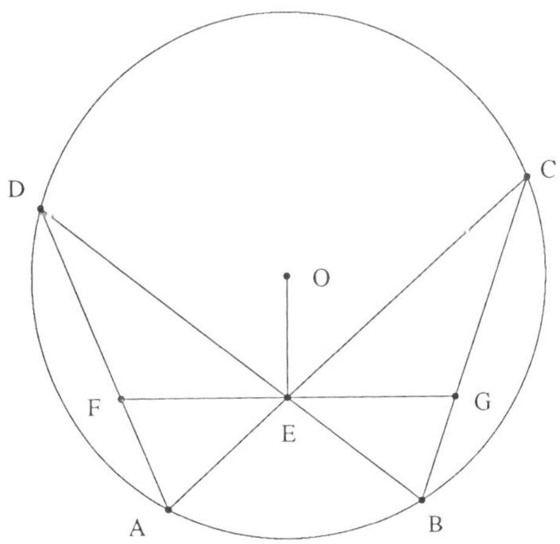


Figure 4

Let $E = (0,0)$, $O = (u_1, 0)$, $A = (u_2, u_3)$, $B = (x_1, u_4)$, $C = (x_3, x_2)$, $D = (x_5, x_4)$,

$F = (0, x_6)$, and $G = (0, x_7)$. Then the hypothesis equations are:

| | |
|---|---|
| $h_1 = -x_1^2 + 2u_1x_1 - u_4^2 + u_3^2 + u_2^2 - 2u_1u_2 = 0$ | $OA \equiv OB$ |
| $h_2 = -x_3^2 + 2u_1x_3 - x_2^2 + u_3^2 + u_2^2 - 2u_1u_2 = 0$ | $OA \equiv OC$ |
| $h_3 = -u_3x_3 + u_2x_2 = 0$ | Points $C$, $A$ and $E$ are collinear |
| $h_4 = -x_5^2 + 2u_1x_5 - x_4^2 + u_3^2 + u_2^2 - 2u_1u_2 = 0$ | $OA \equiv OD$ |
| $h_5 = -u_4x_5 + x_1x_4 = 0$ | Points $D$, $B$ and $E$ are collinear |
| $h_6 = (-x_5 + u_2)x_6 + u_3x_5 - u_2x_4 = 0$ | Points $F$, $A$ and $D$ are collinear |
| $h_7 = (-x_3 + x_1)x_7 + u_4x_3 - x_1x_2 = 0$ | Points $G$, $B$ and $C$ are collinear. |

and the conclusion is $g = x_7 + x_6 = 0$.

We can let $f_1 = h_1$, $f_2 = prem(h_2, h_3, x_3)$, $f_3 = h_3$, $f_4 = prem(h_4, h_5, x_5)$, $h_5 = f_5$, $f_6 = h_6$, $f_7 = h_7$ to obtain a triangular form:

$$\begin{aligned}
f_1 &= -x_1^2 + 2u_1x_1 - u_4^2 + u_3^2 + u_2^2 - 2u_1u_2 = 0 \\
f_2 &= (-u_3^2 - u_2^2)x_2^2 + 2u_1u_2u_3x_2 + u_3^4 + (u_2^2 - 2u_1u_2)u_3^2 = 0 \\
f_3 &= -u_3x_3 + u_2x_2 = 0 \\
f_4 &= (-x_1^2 - u_4^2)x_4^2 + 2u_1u_4x_1x_4 + (u_3^2 + u_2^2 - 2u_1u_2)u_4^2 = 0 \\
f_5 &= -u_4x_5 + x_1x_4 = 0 \\
f_6 &= (-x_5 + u_2)x_6 + u_3x_5 - u_2x_4 = 0 \\
f_7 &= (-x_3 + x_1)x_7 + u_4x_3 - x_1x_2 = 0.
\end{aligned}$$

If we do successive pseudo divisions of $g$ with respect to the above triangular form, the final remainder $R_1$ is *not* zero. The reason is that the above triangular form is *reducible* (see [11]). This reducibility comes from a special kind of degeneracy (or *ambiguity*) when using algebraic equations to encode certain geometric conditions. Equations $h_2 = 0$ and $h_3 = 0$ specify point $C$: $C$ is on circle $(O)$ and on line $AE$. However, there are two points satisfying the equations: one is $C$, which we really want; the other point is $A$. Because $A$ has been constructed before $C$, reducibility arises. We might use the complete method of Wu to deal with reducibility. However, for such special reducibility we can use the following elementary method, which is more satisfactory in geometry.

Since $f_2 = 0$ and $f_3 = 0$ have two solutions (one is for $A = (u_2, u_3)$ and the other is for $C = (x_3, x_2)$), $f_2 = 0$ is satisfied when $x_2$ is substituted by $u_3$. In other words, $x_2 - u_3$ is a factor of $f_2$: $f_2 = (x_2 - u_3)((-u_3^2 - u_2^2)x_2 - u_3^3 + (-u_2^2 + 2u_1u_2)u_3)$.

Thus we can use division to obtain $f_2' = (-u_3^2 - u_2^2)x_2 - u_3^3 + (-u_2^2 + 2u_1u_2)u_3$ and replace $f_2$ by $f_2'$. In the same way, we have:

$$f_4 = (x_4 - u_4)f_4' + r$$

where $f_4' = (-x_1^2 - u_4^2)x_4 - u_4x_1^2 + 2u_1u_4x_1 - u_4^3$, and $r = -u_4^2x_1^2 + 2u_1u_4^2x_1 - u_4^4 + (u_3^2 + u_2^2 - 2u_1u_2)u_4^2$.

We have $prem(r, f_1, x_1) = 0$. Hence, under the conditions $f_i = 0$ and $I_i \neq 0$ $(i = 1, 2, 3)$: $f_4 = (x_4 - u_4)f_4'$. We can replace $f_4$ by $f_4'$ to obtain the nondegenerate triangular form $f_1, f_2', f_3, f_4', f_5, f_6, f_7$.

Now we can do the successive divisions of $g$ with respect to the new triangular form. This time, *final remainder* $R_1 = 0$. Thus, the theorem follows from the following subsidiary conditions: $I_2 = u_3^2 + u_2^2 \neq 0$, $I_3 = u_3 \neq 0$, $I_4 = x_1^2 + u_4^2 \neq 0$, $I_5 = u_4 \neq 0$, $I_6 = x_5 - u_2 \neq 0$, $I_7 = x_3 - x_1 \neq 0$. In addition, the following nondegenerate conditions, which come from reducibility and have been detected by our prover, should be added as well: $D \neq B$ (i.e., $x_2 - u_3 \neq 0$); $C \neq A$ (i.e., $x_4 - u_4 \neq 0$).

With programs based on further developments of this simple use of Wu's method [47, 49, 5, 8, 11], hundreds of geometry theorems have been proved, including Morley's trisector theorem (see Figure 5), a conjecture of V. Thèbault (see Figure 6), and several possibly new theorems.

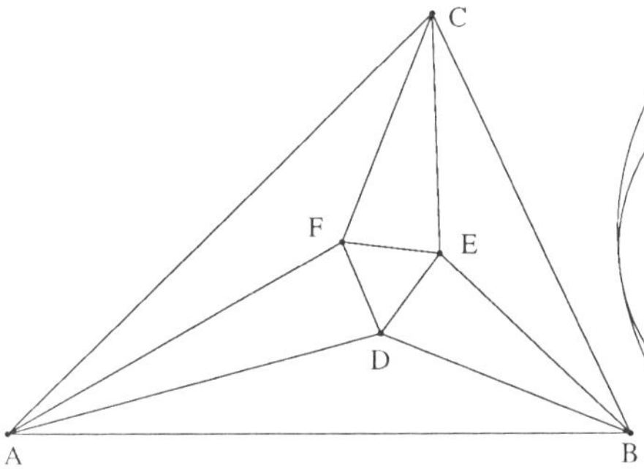


Figure 5

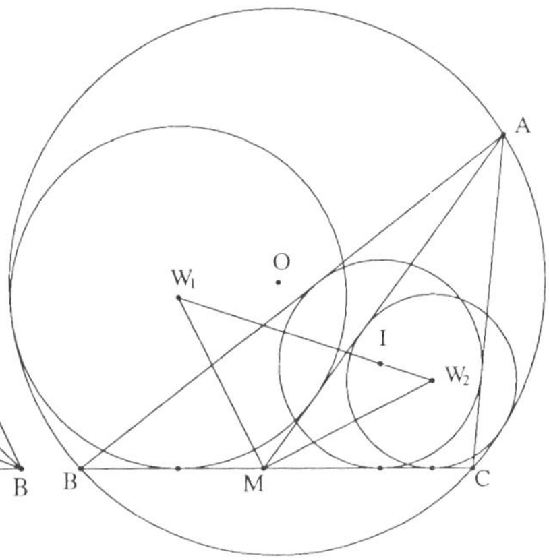


Figure 6

*Morley's Theorem.* The three points of intersection of the adjacent trisectors of the angles of any triangle form an equilateral triangle (Figure 5). (In fact, there are 27 triangles thus formed; among them 18 are equilateral. Our prover proves this most general form).

*Thèbault's Conjecture* (proposed in 1938 and proved in 1983). Through the vertex $A$ of a triangle $ABC$ a straight line $AM$ is drawn cutting the side $BC$ in $M$. Let $O$ and $I$ be the centers of the circumscribed circle $(O)$ and the inscribed circle $(I)$ of $ABC$. The circle $(W_1)$ and $(W_2)$ with centers $W_1$ and $W_2$ are each tangent to $(O)$ and the first is tangent also to two sides of angle $AMB$, while the second is tangent to the two sides of angle $AMC$. Prove that the straight line joining $W_1$ and $W_2$ passes through $I$ (Figure 6).

## *5.1. Gröbner Basis Method*

Now we use the same three examples to illustrate how the Gröbner basis method can be used to prove geometry theorems. As we know, one of the important applications of the Gröbner basis method is to decide ideal membership for polynomial ideals.

**Example 1b:** Example 1 of Parallelogram.

Let $u$, $x$, $h_1, ..., h_4$ and $g$ be the same as in Example 1. Thus the question can be asked whether $g \in \text{Ideal}(h_1, h_2, h_3, h_4)$, or, more precisely, whether $g \in \text{Radical}(h_1, h_2, h_3, h_4)$. However, neither of the above assertions is true if we consider $h_1$, $h_2$, $h_3$, $h_4$ as polynomials in $Q[u_1, u_2, u_3, x_1, x_2, x_3, x_4]$. The reason is the same as we explained earlier in this section: nondegenerate conditions are missing. Nondegenerate conditions can be produced if we work on polynomial ring $Q(u)[x]$ instead of $Q[u, x]$, where $Q(u)$ is the quotient field of polynomial ring $Q[u]$.

Thus we can ask whether $g \in \text{Radical}(h_1, ..., h_4)$ in $Q(u)[x]$. By Hilbert's Nullstellensatz, this is equivalent to showing whether $\text{Radical}(h_1, \ldots, h_4, gz - 1) = (1)$ (the unit ideal) in $Q(u)[x]$. We can check this by computing the Gröbner basis of $\{h_1, \ldots, h_4, gz - 1)$[8] to see whether it is $\{1\}$. This time, our GB program confirmed the theorem of parallelogram. The subsidiary (nondegenerate) conditions were implicitly used in the computation: the denominators in the computation should be nonzero. It is easy to collect these denominators and to remove the repeated factors. For this example, we need "$u_3 \neq 0$, $u_1 \neq 0$" in the computation.

As a first approximation, we can check whether $g \in \text{Ideal } (h_1, \ldots, h_4)$ in $Q(u)[x]$. Thus, we can compute the Gröbner basis of $h_1, .., h_4$ in the pure lexicographic ordering $x_1 < x_2 < x_3 < x_4$; then to check whether it reduces the conclusion polynomial $g$ to zero. $\text{GB}(h_1, ..., h_4) = \{x_1 - u_3,\ x_2 - u_2 + u_1,\ x_3 - u_3/2,\ x_4 - u_2/2\}$, which reduces the conclusion polynomial $g = 2u_2x_4 + 2u_3x_3 - u_3^2 - u_2^2$ to zero. Thus, the statement is confirmed. Again, we need subsidiary conditions, $u_3 \neq 0$ and $u_1 \neq 0$, in the computation.

For all the theorems we found in practice, this first approximation is enough. It is faster and more informative than using radical ideals.

**Example 3a:** Example 3 (Simson's Theorem). $\text{GB}(h_1, ..., h_7)$ (in $Q(u)[x]$) =

$$x_1 - (u_3^2 + u_2^2 - u_1u_2)/2u_3$$
$$x_2 - u_1/2$$
$$x_3^2 - (u_1u_3x_3 - u_3u_4^2 + (u_3^2 + u_2^2 - u_1u_2)u_4)/u_3$$
$$x_4 - ((u_2 - u_1)u_3x_3 + u_3^2u_4 + (-u_1u_2 + u_1^2)u_3)/(u_3^2 + u_2^2 - 2u_1u_2 + u_1^2)$$
$$x_5 - ((u_2^2 - 2u_1u_2 + u_1^2)x_3 + (u_2 - u_1)u_3u_4 + u_1u_3^2)/(u_3^2 + u_2^2 - 2u_1u_2 + u_1^2)$$
$$x_6 - (u_2u_3x_3 + u_3^2u_4)/(u_3^2 + u_2^2)$$
$$x_7 - (u_2^2x_3 + u_2u_3u_4)/(u_3^2 + u_2^2).$$

The Gröbner basis reduces the conclusion polynomial $g$ to zero. Thus, we prove Simson's theorem to be generically true. The subsidiary conditions needed for the computation are: $u_2 \neq 0$, $u_1 \neq 0$, $u_3 \neq 0$, $u_3^2 + u_2^2 - 2u_1u_2 + u_1^2 \neq 0$, $u_3^2 + u_2^2 \neq 0$.

[8]When we talk about a Gröbner basis of an ideal, we always mean the unique reduced Gröbner basis with respect to certain compatible ordering among monomials and denote it by $\text{GB}(h_1, ..., h_n)$, where $h_1, ..., h_n$ are generators of the ideal.

**Example 4a:** The Butterfly theorem: Example 4.

We choose slightly different coordinates: $E = (0,0)$, $O = (u_1, 0)$, $A = (u_2, u_3)$, $B = (u_4, x_1)$, $C = (x_2, x_3)$, $D = (x_4, x_5)$, $F = (0, x_6)$, and $G = (0, x_7)$. Then, the hypotheses and the conclusion are:

$$\begin{aligned}
h_1 &= x_1^2 + u_4^2 - 2u_1u_4 - u_3^2 - u_2^2 + 2u_1u_2 = 0\\
h_2 &= x_3^2 + x_2^2 - 2u_1x_2 - u_3^2 - u_2^2 + 2u_1u_2 = 0\\
h_3 &= u_2x_3 - u_3x_2 = 0\\
h_4 &= x_5^2 + x_4^2 - 2u_1x_4 - u_3^2 - u_2^2 + 2u_1u_2 = 0\\
h_5 &= u_4x_5 - x_1x_4 = 0\\
h_6 &= (x_4 - u_2)x_6 + u_2x_5 - u_3x_4 = 0\\
h_7 &= (x_2 - u_4)x_7 + u_4x_3 - x_1x_2 = 0\\
g &= x_7 + x_6 = 0
\end{aligned}$$

However, neither $\mathrm{GB}(h_1, ...h_7, zg-1) = \{1\}$, nor $\mathrm{GB}(h_1, ..., h_7)$ reduces $g$ to zero. The reason is the same as we explained in example 4: in specifying points $C$ and $B$, we have ambiguities. These ambiguities can be easily detected by the program and, in the case of using the Gröbner basis method, can be removed by adding two new equations $s_1 = (x_2 - u_2)v_1 - 1 = 0$ $(A \neq C)$, $s_2 = (x_4 - u_4)v_2 - 1 = 0$ $(B \neq D)$. Thus $\mathrm{GB}(s_1, s_2, h_1, ..., h_7) =$

$$\begin{aligned}
&v_1 + (2u_1u_4 + u_3^2 + u_2^2 - 2u_1u_2)/(2u_1u_4^2 + (2u_3^2 + 2u_2^2 - 4u_1u_2)u_4)\\
&v_2 + (u_3^2 + u_2^2)/(2u_2u_3^2 + 2u_2^3 - 2u_1u_2^2)\\
&x_1^2 + u_4^2 - 2u_1u_4 - u_3^2 - u_2^2 + 2u_1u_2\\
&x_2 + (u_2u_3^2 + u_2^3 - 2u_1u_2^2)/(u_3^2 + u_2^2)\\
&x_3 + (u_3^3 + (u_2^2 - 2u_1u_2)u_3)/(u_3^2 + u_2^2)\\
&x_4 + ((u_3^2 + u_2^2 - 2u_1u_2)u_4)/(2u_1u_4 + u_3^2 + u_2^2 - 2u_1u_2)\\
&x_5 + ((u_3^2 + u_2^2 - 2u_1u_2)x_1)/(2u_1u_4 + u_3^2 + u_2^2 - 2u_1u_2)\\
&x_6 + ((u_2u_3^2 + u_2^3 - 2u_1u_2^2)x_1 + (-u_3^3 + (-u_2^2 + 2u_1u_2)u_3)u_4)/\\
&\qquad ((u_3^2 + u_2^2)u_4 + u_2u_3^2 + u_2^3 - 2u_1u_2^2)\\
&x_7 - ((u_2u_3^2 + u_2^3 - 2u_1u_2^2)x_1 + (-u_3^3 + (-u_2^2 + 2u_1u_2)u_3)u_4)/\\
&\qquad ((u_3^2 + u_2^2)u_4 + u_2u_3^2 + u_2^3 - 2u_1u_2^2).
\end{aligned}$$

The Gröbner basis reduces $g$ to zero. Thus, we have proved the Butterfly theorem. In the computation we needed the following subsidiary conditions: $(u_3^2+u_2^2)u_4+u_2u_3^2+u_2^3-2u_1u_2^2 \neq 0$, $u_4-u_2 \neq 0$, $2u_1u_4+u_3^2+u_2^2-2u_1u_2 \neq 0$, $u_4 \neq 0$, $u_1u_4+u_3^2+u_2^2-2u_1u_2 \neq 0$, $u_3^2 + u_2^2 \neq 0$, $u_2 \neq 0$, $u_3^2 + u_2^2 - u_1u_2 \neq 0$.

Note that the computation of the Gröbner basis is very sensitive to the variable ordering. If we choose the same coordinates that are in example 4, then the computation would take much longer. For the present choice of coordinates, the computation took about 24 seconds on a Symbolics 3600.

Alternatively, we can add (expected) nondegenerate conditions to the statement as an integral part of the hypothesis, then prove the statement is true or not without adding any other conditions (formulation $F_2$). This has been done for both Wu's method and the GB method [30, 32, 13, 31].

For the example of parallelogram (Example 1), we can add $d_1 \neq 0$ (where $d_1 = u_1u_3$) as the only nondegenerate condition and prove the statement

$$(h_1 = 0 \wedge \cdots \wedge h_4 = 0 \wedge d_1 \neq 0) \Rightarrow g = 0$$

to be valid. One of the approaches based on the GB method [30, 20] is conceptually very simple (but not practically efficient). It is based on Hilbert's Nullstellensatz: we only need to check whether the set $S$ of polynomials $h_1, ..., h_4, d_1z_1 - 1, gz - 1$ (here $z$ and $z_1$ are two new variables) generates the unit ideal. This can be done by computing a GB of $S$ to see whether it is $\{1\}$.

*An additional, important remark*: if Wu's method or the GB method proves a statement to be true (or generally true), then the statement is valid for all fields, including real closed fields. However, if the methods disprove a statement, then it is not valid in algebraically closed fields, i.e., it is not a theorem in the theory of unordered metric geometry (see section 6). But, it is still possible that the statement might be valid in, say, real closed fields. However, most statements from Euclidean geometry are **R**–generic, for which, Wu's method can also be used to disprove them in real closed fields. See [11] for details.

## 6. Algebraic v/s Traditional Proofs

In order to discuss this issue, it is important to understand the relationship between algebra and geometry. There is a variety of geometries. For example, proving or disproving geometry statements in affine geometry and Euclidean geometry is different since the field associated with Euclidean geometry is the field of real numbers, whereas almost any field can be the field associated with affine geometry; in particular, it can be an algebraically closed field.

### *6.1. Axiomatic Geometries and Number Systems*

We shall discuss, briefly, the relationship between geometry and algebra in order to show that the Cartesian product of the field of real numbers is not the *only* realistic (plane) geometry. There are two approaches to defining geometry.

**(A) The Algebraic Approach.** Starting from some number system $E$ (usually a field), we can define geometric objects and relations between those objects in the Cartesian product $E^n$ (or $E^n/E^*$ in projective geometry). In modern geometry, especially in algebraic geometry, this approach indisputably prevails. If we take this approach, then there are only a few differences between algebra and geometry; geometry can be regarded as a part of algebra.

However, the second approach is more attractive from the point of view of axiomatic geometry and the point of view of traditional proofs of geometry theorems in Euclidean style.

**(B) The Axiomatic Geometry Approach.** By this approach we mean the one that was approximately used by Hilbert in his classic book [29]. We shall restrict ourselves to first order logic. Depending on the context, the term "a geometry" has two meanings. When we mention ***the theory of a geometry*** we mean ***a first-order theory*** defined in [41], i.e., a formal system in first-order logic. Here "a geometry" means a class of geometry, e.g., affine geometry. On the other hand, when ordinary geometry textbooks mention "a geometry", it usually means *a model* of a certain theory of geometry. Thus Euclidean geometry (i.e., $R^2$) is an affine geometry; i.e., Euclidean geometry is a model of the theory of affine geometry. Of course, this model has *much richer structures* than an ordinary affine geometry.

A very attractive feature of this approach is that algebra can be introduced in an elegant way (see [29]). For each model of a theory of geometry, we can prove the existence of a number system (usually a field) inherent to that model. This field is called the *field associated E* with that model. That model can then be represented as the Cartesian product $E^2$ of its associated field.

### 6.1.1. Affine Geometry

The only basic relation in the theory of affine geometry is "*incidence*", which satisfies three groups of axioms: *axioms of incidence*, *axiom of parallels*, and *Pappus' axiom*. Each model of the theory can be represented as $E^2$, where $E$ is a field with certain restriction on its characteristic.

### 6.1.2. Metric Geometry

Metric geometry is a class of affine geometry in which there are two new basic relations "*perpendicularity*" and "*Congruence*" [48]. These new relations satisfy two groups of new axioms (*axioms of perpendicularity* and *axioms of congruence*, see [48], or [6]). These two groups of axioms, together with axioms of affine geometry, make up *the theory of metric geometry*. Each model of the theory can be represented as $E^2$, where $E$ is a Hilbert field of characteristic zero (a *Hilbert field* is a field in which the sum of the squares of two elements has a square root).

### 6.1.3. Hilbert Geometry

The theory of Hilbert geometry consists of the first four groups of axioms in [29]. As we know, each model in this theory can be represented as $E^2$, where $E$ is an *ordered* Hilbert field.

Hilbert geometry is *ordered* metric geometry. We can define an angle as two rays from one point. In general (unordered) metric geometry, we cannot define rays. However, we can define an angle as a pair of two lines which have a common point. Many theorems involving angle congruence in Euclidean geometry are still valid in unordered geometry as our prover and the prover of Wu verified. Feuerbach's theorem, Morley's trisector theorem, V. Thèbault theorem are among such theorems.

### 6.1.4. Tarski Geometry

The theory of Tarski geometry was given fully in [43]. Each model of the theory of Tarski geometry can be represented as $R^2$, where $R$ is a real closed field [43]. Due to a fundamental contribution by Tarski [42], the theory of Tarski geometry is decidable. Note that the decision procedure is carried out in number systems (real closed fields), i.e., at the model level. This geometry is what Tarski called elementary geometry. As we know, there are Non-Archimedean Tarski geometries. The meaning of the term "elementary geometry" depends on who uses it. Following Wu, we prefer elementary geometry to include all geometries in which no differentiation is involved. Elementary geometry in this sense includes not only all geometries discussed so far, but also those "advanced" geometries, such as Bolyai-Lobachevskian geometry and Riemannian geometry. The inclusion relationship among the models of the four geometries discussed above can be illustrated by the following diagram.

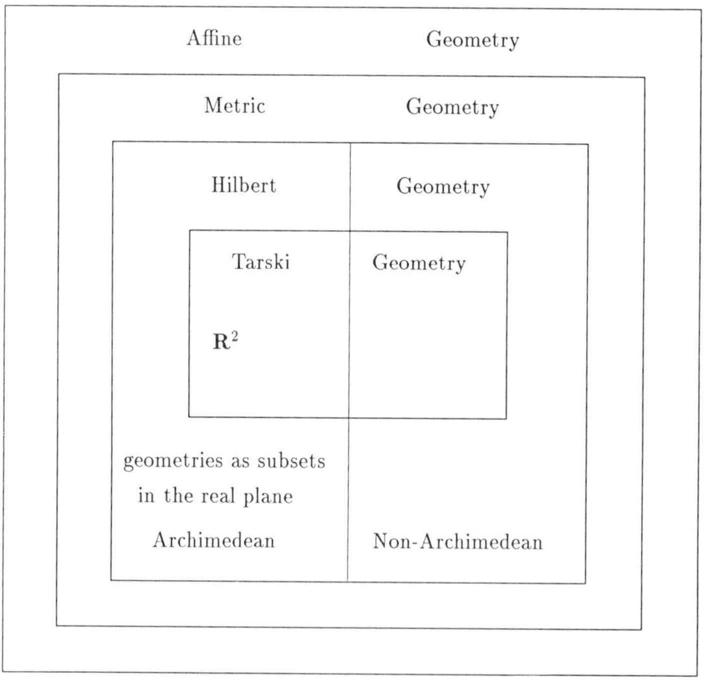

The above four geometries are compatible with Euclidean geometry in the sense that Euclidean geometry is a model of the theories of the four geometries. Among geometries incompatible with Euclidean geometry, we mention projective geometry, Bolyai-Lobachevskian geometry, Riemannian geometry, etc. For those geometries, algebra and fields can be also introduced.

### *6.2. Algebraic Proofs v/s Formal Proofs*

From the above discussion it is clear that an algebraic proof of a geometric statement can be considered a proof of the statement in models of the theory of related geometry. If the proof succeeds, the statement is valid in *all* models of the theory; thus it is a theorem of the theory, i.e., there is a formal deduction of the statement from the theory (this is an application of the Gödel completeness theorem for first order logic).

Since it resembles a traditional proof in Euclidean style, a formal proof could be expected to be very long and cumbersome. On the other hand, though, an algebraic proof of a statement only proves the existence of a formal proof of the statement in the theory of an underlying geometry.

The interesting question is whether the algebraic proof itself can be considered a formal proof of the statement. For simple theories of geometry, such as the theory of affine geometry, Hilbert's argument (see Theorem 62 in [29]; also see pages 38–39 of [11]) almost convinces us that this is the case. However, in general, the answer to this question is unclear to us.

## 7. Concluding Remarks

Highly successful algebraic methods for mechanical geometry theorem proving have been developed since Wu's work. Hundreds of hard theorems from Euclidean and non-Euclidean geometries have been totally automatically proven by computer programs based on these methods. However, we should not be over-optimistic. There is still a lot of work to be done.

First, present techniques still cannot, in practice, prove relatively hard geometry theorems involving inequalities. In this direction, G. Collins has made an important contribution along the Tarski line [21]. His cylindrical algebra decomposition (CAD) algorithm is currently the best of the Tarski type algorithms. This method was implemented by D. Arnon, and several difficult problems in algebra and/or geometry were solved by Arnon's program [2, 3]. More recently, another major practical improvement of Collins' method was made by Collins and H. Hong [22]. The persistent effort by Collins and his collaborators for 17 years convinces us to believe that, with further developments, the CAD algorithm can possibly solve a substantial number of difficult and practical problems involving inequalities.

A theorem involving inequalities, generally, also involves equalities. Since Wu's method and the GB method work so well for equality problems, we expect a combination of Wu's (or GB) method with Collins' method could solve problems not in the scope of Wu's method, but which cannot be solved by Collins' method alone within the available time and space. The work in [4] was in this direction, and a number of hard problems were solved with some human interaction. More recently, a more concrete combination of Wu's method and Collins' method was proposed [16].

Second, only people connected with Wu's group have been actively doing research in proving theorems in differential geometries [19, 49]. Some topics in the local theory of surfaces are still in the frontiers of the current research in mathematics. We hope that, with further efforts, some interesting results can be obtained with the aid of geometry theorem provers.

Third, all algebraic methods mentioned in this survey can only deal with a fixed number of geometric objects. We can prove properties of, say, a 100-point geometric configuration (at least in theory if not in practice). But we cannot use these methods to prove properties of an $n$ point configuration where $n$ is a variable. One can easily find many such examples which are totally beyond the scopes of algebraic methods discussed in this paper.

## 8. References

1. C. F. Adler, *Modern Geometry,* McGraw–Hill Book Company Inc., 1958.
2. D. Arnon and M. Mignotte, "On Mechanical Quantifier Elimination for Elementary Algebra and Geometry", *J. Symbolic Computation,* **5** (1988), 237-259.
3. D. Arnon, "Geometric Reasoning with Logic and Algebra", *Artificial Intelligence Journal,* V. **37** (1988), pp. 37-60.
4. D. Arnon, S.C. Chou and X.S. Gao, "On the Mechanical Proof of Geometry Theorems Involving Inequalities", 1988, to appear in the "*Issues in Robotics and Nonlinear Geometry*" Volume of *Advances in Computing Research.*
5. S.C. Chou, "Proving Elementary Geometry Theorems Using Wu's Algorithm" in Automated Theorem Proving: After 25 years, Ed. By W.W. Bledsoe and D. Loveland, AMS Contemporary Mathematics Series V. **29** 1984, 243-286.
6. S.C. Chou, "Proving and Discovering Theorems in Elementary Geometries Using Wu's Method", PhD Thesis, Department of Mathematics, University of Texas, Austin (1985).
7. S.C. Chou, "A Method for Mechanical Derivation of Formulas in Elementary Geometry", *Journal of Automated Reasoning,* **3**(1987), 291–299.
8. S.C. Chou, " Proving Geometry Theorems Using Wu's Method: A Collection of Geometry Theorems Proved Mechanically", Technical Report 50, Institute for Computing Science, University of Texas at Austin, July 1986.

9. S.C. Chou, "GEO-Prover – A Geometry Theorem Prover Developed at UT" in *Proceedings of CADE-8*, Oxford, 1986.
10. S.C. Chou, "An Introduction to Wu's Method for Mechanical Theorem Proving in Geometry", *Journal of Automated Reasoning*, **4** (1988), 237–267.
11. S.C. Chou, *Mechanical Geometry Theorem Proving,* D. Reidel Publishing Company, Dordrecht, Netherlands, 1988.
12. S.C. Chou, W. F. Schelter and J. G. Yang, "An Algorithm for Constructing Gröbner Bases from Characteristic Sets and its Application to Geometry", *Algorithmica*, (1990) 5: 147–154.
13. S. C. Chou and X. S. Gao, "Ritt-Wu's Decomposition Algorithm and Geometry Theorem Proving" in *Proceedings of CADE–10*, *Lecture Notes in AI*, Vol.449, 207–220, 1990.
14. S. C. Chou and X. S. Gao, "Methods for Mechanical Geometry Formula Deriving" in *1990 Proceedings of International Symposium on Symbolic and Algebraic Computation*, ACM Press, 265–270.
15. S.C. Chou and X.S. Gao, "A Collection of 120 Computer Solved Geometry Problems in Mechanical Formula Derivation", TR-89-22, Computer Sciences Department, The University of Texas at Austin, August 1989.
16. S.C. Chou, X.S. Gao and Nicholas McPhee, "A Combination of Ritt–Wu's Method and Collins' Method", TR-89-28, Computer Sciences Department, The University of Texas at Austin, October, 1989.
17. S.C. Chou and X.S. Gao, "A Class of Geometry Statements of Constructive Type and Geometry Theorem Proving", TR-89-37, Computer Sciences Department, The University of Texas at Austin, November, 1989.
18. S.C. Chou, "A Geometry Theorem Prover for Macintoshes", TR-91-8, Computer Sciences Department, The University of Texas at Austin, March, 1991.
19. S.C. Chou and X.S. Gao, "Mechanical Theorem Proving in Differential Geoemtry II: Bertrand Curves", TR-90-9, CS Department, The University of Texas at Austin, 1990.
20. S.C. Chou and W.F. Schelter, "Proving Geometry Theorems with Rewrite Rules", *Journal of Automated Reasoning*, V. 2(4) 1986, 253–273.
21. G.E. Collins, "Quantifier Elimination for Real Closed Fields by Cylindrical Algebraic Decomposition", *Lecture Notes In Computer Science*, V. 33 1975, Springer-Verlag, Berlin, 134-183.
22. G.E. Collins and H. Hong, "Partial CAD Construction in Quantifier Elimination", 1989, to appear in *Journal of Symbolic Computation.*
23. M. Davis and H. Putnam, "A computing procedure for quantification theory", *Journal of the ACM*, 1960, 201–215.
24. M. Davis, "The Prehistory and Early History of Automated Deduction" in *Automation of Reasoning*, V. 1, pages 1–28, Edited by J. Siekmann and G.

Wrightson, Springer-Verlag, 1983.

25. M. Davis, "A computer program for Presburger's procedure" in *Summaries of Talks Presented at the Summer Institute for Symbolic Logic*, 1957. Second edition, published by Institute for Defense Analysis, 1960.
26. P. J. Davis and E. Cerutti, "FORMAC meets Pappus", *The American Mathematical Monthly*, V. 76, 1969, 895–905.
27. R. Descartes, "Geometry" (translated by D. Smith and M. Latham), Dover, 1954. Original publication as appendix to *Discours de la Methode*, Jan Maire, Leyden, 1637.
28. H. Gelernter, J.R. Hanson, and D.W. Loveland, "Empirical explorations of the geometry-theorem proving machine", *Proc. West. Joint Computer Conf.*, 143-147, 1960.
29. D. Hilbert, *Foundations of Geometry*, Open Court Publishing Company, La Salla, Illinois, 1971.
30. D. Kapur, "Geometry Theorem Proving Using Hilbert's Nullstellensatz", *Proceedings of the 1986 Symposium on Symbolic and Algebraic Computation*, 202-208.
31. D. Kapur and Hoi K. Wan, "Refutational Proofs of Geometry Theorems via Characteristic Set Computation", *Proceedings of International Symposium on Symbolic and Algebraic Computation*, ACM Press, 1990, 277–284.
32. H.P. Ko, "Geometry Theorem Proving by Decomposition of Quasi-Algebraic Sets: An Application of the Ritt-Wu Principle", *Artificial Intelligence*, Vol. 37, pp95-122 (1988).
33. B. Kutzler and S. Stifter, "Automated Geometry Theorem Proving Using Buchberger's Algorithm", *Proceedings of the 1986 Symposium on Symbolic and Algebraic Computation*, 209-214.
34. D. Loveland, *Automated Theorem Proving: a logical basis*, North Holland, Amsterdam, 1978.
35. D. Loveland, "Automated Theorem Proving: a quarter-century review", *Contemporary Mathematics*, V. 29, 1984, 1–45.
36. A. Newell, J. Shaw, and H. Simon, "Empirical explorations of the logic theory machine: a case study in heuristics", *Proceedings of Western Joint Computer Conference*, 1956, 218–239.
37. W. McCune, *Otter 2.0 User's Guide*, Argonne National Laboratory REport ANL-90/9, 1990.
38. D. Prawitz, "An improved proof procedure", *Theoria*, 1960, 102–139.
39. J. Robinson, "A machine-oriented logic based on the resolution principle", *Journal of the ACM*, 1965, 227–234.
40. A. Seidenberg, "A New Decision Method for Elementary Algebra", *Annals of Math.*, V. 60, 1954, 365–371.

41. J. R. Shoenfield, *Mathematical Logic*, Addison-Wesley Publishing Company, (1967).
42. A. Tarski, *A Decision Method for Elementary Algebra and Geometry*, Report R-109, second revised ed. Santa Monica, CA: The Rand Corporation, 1951.
43. A. Tarski, "What is Elementary Geometry", *Proc. Intl. Symp. on the Axiomatic Method, with special reference to geometry and physics*, University of California, Berkeley, Dec. 1957 - Jan. 1958, *Studies in Logic and the Foundations of Mathematics*, (L. Henkin, P. Suppes, A. Tarski eds.), North Holland, Amsterdam, 1959, pp. 16-29.
44. H. Wang, "Toward mechanical mathematics", *IBM Journal*, V. 4, 1960, 2–22.
45. Wu Wen-tsün, "On the Decision Problem and the Mechanization of Theorem Proving in Elementary Geometry", *Scientia Sinica* V. 21, 1978, 157-179.
46. Wu Wen-tsün, "Basic Principles of Mechanical Theorem Proving in Geometries", *J. of Sys. Sci. and Math. Sci.* **4(3)**, 1984, 207-235, republished in *Journal of Automated Reasoning* V. 2(4), 1986, 221-252.
47. Wu Wen-tsün, "Some Recent Advances in Mechanical Theorem-Proving of Geometries" in *Automated Theorem Proving: After 25 years*, American Mathematical Society, Contemporary Mathematics V. 29, 1984, 235-242.
48. Wu Wen-tsün, *Basic Principles of Mechanical Theorem Proving in Geometries*, (in Chinese) Peking 1984.
49. Wu Wen-tsün, "Mechanical Theorem Proving in Elementary Geometry and Differential Geometry", *Proc. 1980 Beijing Symposium on Differential Geometry and Differential Equations*, V. 2, 1982, 125-138, Science Press.

*Computing in Euclidean Geometry*, pp. 117–162
Eds. D.-Z. Du and F. K. Hwang


# RANDOMIZED GEOMETRIC ALGORITHMS

KENNETH L. CLARKSON
*AT&T Bell Laboratories, 600 Mountain Avenue*
*Murray Hill, NJ 07974 USA*

## 1 Introduction

Randomization has found many applications in computational and combinatorial geometry in recent years. For a variety of geometric problems, the resulting algorithms are simpler than those previously known, or asymptotically faster, or both. Randomization helps in building geometric subdivisions, and in building data structures to quickly answer questions about such subdivisions. Randomization gives a general way to "divide and conquer" geometric problems, and can be used in the setting of parallel as well as serial computation. There are several facts of combinatorial geometry that have simple probabilistic proofs. This paper will explore some of these ideas and applications. We will also briefly discuss connections with the theory of "learning" geometric concepts.

It is important to note that the main focus here is *not* on random input, such as a collection of points distributed uniformly and independently in a region. Combinatorial and algorithmic questions concerning such point sets, or problems in *stochastic geometry*,[68] have an enormous literature. Here we will mainly look at algorithms that use a source of random numbers, and consider their performance for arbitrary input. We generally look at *Las Vegas* algorithms, whose output is guaranteed to be correct, and not *Monte Carlo* algorithms, whose output may be incorrect. (Monte Carlo algorithms generally return an estimate of some quantity, with a performance guarantee that the estimate is accurate within some range with high probability.)

It is worth mentioning at the outset that there is nothing magical about randomization: several algorithms described here have deterministic competitors, or can be "derandomized" to yield deterministic versions. (See the works of Matoušek, P. Agarwal, and Chazelle.[18,14,49,50,2]) While the resulting algorithms are more complicated, they often satisfy the same asymptotic performance guarantees as the randomized algorithms, without the need for random bits. We hope the reader will see, however, how useful a "random" point of view can be.

While the emphasis here is on theorems and proofs, several of the algorithms discussed have been implemented. (For example, the algorithms of §4,§7.1,§4.4 have been implemented, and the resulting programs seem to be quite acceptable in complexity and efficiency. Fortune's survey gives some rough results for one randomized incremental algorithm.[39]) We also emphasize general techniques, and not the specifics of applications; the closing remarks of §8 give some references for these.

### 1.1 Outline

This paper will consider two main ways to apply randomization: a randomized incremental approach to building geometric structures, and a randomized divide-and-conquer technique for geometric computations. The divide-and-conquer scheme is to use a random subset of the input to create small subproblems, solve them, and then merge the solutions to obtain a solution for the original problem. If the input comprises $n$ sites (points, for example), and the random subset has size $r$, the subproblems will have size about $n/r$. The number of subproblems will be bounded by some function $m(r) > r$.

When divide-and-conquer is applied to a computational problem, ideally the problem is partitioned into equally-sized subproblems, so that no subproblem has size larger than $n/r$ and the subproblem sizes add up to the input size. Much of this survey discusses how closely this ideal situation can be approached in a geometric setting.

Section 2 discusses the use of randomized divide-and-conquer for answering closest-point queries. It shows how to obtain $O(r)$ subproblems of size $O(\log r)n/r$;[1] in many instances, this is within $O(\log r)$ of the divide-and-conquer ideal: the total of the subproblem sizes is $O(n \log r)$. The section also discusses the Vapnik-Chervonenkis (VC) dimension, a combinatorial property closely linked to range query algorithms, PAC-learnability of geometric concepts, and statistics.

Section 3 provides sharper bounds for the analysis of algorithms using the probabilistic approach of §2; we see that on average, the subproblems have size $O(n/r)$, within a constant factor of the ideal. As an application, we describe a nearly optimal algorithm for finding the trapezoidal diagram of a simple polygon; this is a subdivision of the interior of the polygon into simple regions. Finally, the section mentions the generalization of the approach to finding the trapezoidal subdivision of the plane induced by lines, or by line segments that cross.

Section 4 describes an incremental approach to building geometric structures, as applied to convex hulls: maintaining the hull of a subset as more and more points are included in that subset. The section also discusses a simple approach to planar point location.

Section 5 gives some results of combinatorial geometry. One result is an upper bound on the number of incidences between a set of points and a set of lines in the plane. The relevance of this result to the *Hough transform* technique of computer vision is noted, as are Monte Carlo algorithms for that technique. The section also gives results on $(\leq k)$-*regions*, bounds that are helpful in the following section.

Section 6 gives still more careful results on probabilistic divide-and-conquer; here we see that the average of the *squares* of the subproblem sizes is $O(n/r)^2$, so that subproblems are small quite uniformly. The section also gives a construction of *cuttings*, giving subproblems that are *all* of size $O(n/r)$.

Section 7 describes two applications of *reweighting* techniques: linear programming in small dimension, and a data structure for triangular range queries.

[1] Remember that $O(\log r)n/r$, $O((\log r)n/r)$, and $O(1)(\log r)n/r$ all denote the same set of functions. This survey will use the variations of this notation that seem clearest.

Section 8 gives some references for applications.

Appendix A summarizes some simple facts of probability and geometry, and some useful inequalities; these are applied many times throughout. Appendix B gives a list of the common features of various geometric problems considered in this paper. These features allow similar arguments to be applied in proving expected bounds for algorithms.

## 2 Probabilistic divide-and-conquer

### *2.1 Closest-point queries*

Just how can randomization help in geometric computations? Let's look at *closest-point problems.* Let $S$ be a set of $n$ points in the plane; we'll call the points of $S$ *sites* for convenience. One closest-point problem:

> Construct a data structure for $S$ so that given a query point $p$, a site closest to $p$ in Euclidean distance can be found quickly.

Of course, there is the following obvious approach, which we'll dignify as **Algorithm O**; its "data structure" is just a list of sites.

- compute the distance to $p$ from each site;
- return a site for which the computed distance is smallest.

This algorithm is simple, and often the best approach, but can be too slow if $S$ is large and we need to answer many queries.

To beat the obvious approach, we will use a subset $R$ of $S$ to limit the possibilities for the closest site, following a simple observation: if we know the site $q$ of $R$ nearest to a query point $p$, then we know that the closest site in $S$ to $p$ must be at least as close as $q$. In terms of the "locus" approach to geometric problems,[34] we have the open disk $C_R(p)$, which is bounded by a circle with center $p$ and passing through $q$. Either $q$ itself is a closest site, or a closest site to $p$ is in the (open) disk $C_R(p)$.

Now for a region $T$ in the plane, let $C_R(T)$ be the union of the disks $C_R(s)$ for all $s \in T$. If the query point $p$ falls in $T$, a closest site to $p$ must be in $C_R(T)$, or in $R$.

Suppose $\Delta$ is a subdivision of the plane into simple regions. Suppose that for $\Delta$ and $R \subset S$, we have lists of the "candidate" set of sites $S \cap C_R(T)$ for all regions $T \in \Delta$. (We ignore here the details of this preprocessing.) We are ready to answer queries, as follows:

- for a query point $p$, find the region $T \in \Delta$ containing $p$;
- apply Algorithm **O** to $p$ and the candidate sites $S \cap C_R(T)$;
- apply Algorithm **O** to $p$ and sites of $R$;

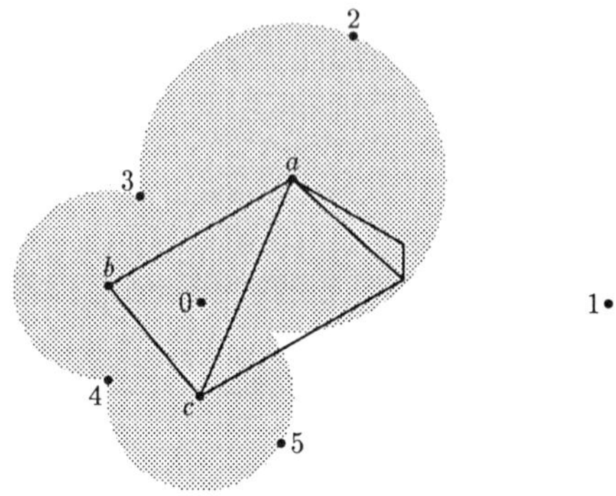


Figure 1: A triangle $T$ with vertices $a$, $b$ and $c$, and $C_R(T)$.

- pick the closer of the sites in $R$ and $S \cap C_R(T)$.

If the number of regions in $\Delta$, the number of candidate sites for $p$, and the number of sites in $R$ are all small relative to the size of $S$, we have answered the query faster than Algorithm **O**.

How can we choose $R$ and $\Delta$ to get a speedup, for any query point? Conveniently, there is a subdivision $\Delta(\mathcal{V}(R))$ into triangular regions with the property that for $T \in \Delta(\mathcal{V}(R))$, the region $C_R(T)$ is the union of the open disks $C_R(v)$, $v$ a vertex of $T$. Moreover, these three disks are *Delaunay* disks of $R$: they contain no sites of $R$, and each has a bounding circle that passes through three sites of $R$. (Here we ignore unbounded regions, but similar conditions apply to them as well. For those interested, the subdivision $\Delta(\mathcal{V}(R))$ is a triangulation of the Voronoi diagram of $R$: see Fortune's survey.[39]) Figure 1 gives an example with $R$ comprising the six numbered sites. The figure shows three triangular regions of $\Delta(\mathcal{V}(R))$, including a region $T$ with vertices $a$, $b$, and $c$. The shaded area is $C_R(T)$. Note that $C_R(T) = C_R(a) \cup C_R(b) \cup C_R(c)$.

If $R$ is a subset of size $r$, then $\Delta(\mathcal{V}(R))$ will have $O(r)$ regions; a value for $r$ is discussed below.

The randomization at last comes in: we choose $R$ to be a random subset of $S$, where all subsets of size $r$ are equally likely to be chosen. With this choice of $R$, we're very likely to get candidate sets that are all small.

Why is this statement plausible? Suppose some arbitrary open disk $D$ contains a large proportion of the sites. If $R$ is a large random subset, we are very likely to have some site of $R$ in $D$, and so $D$ is unlikely to be a Delaunay disk of $R$. Roughly, the fact that Delaunay disks of $R$ contain no sites of $R$ is evidence that they contain few sites of $S$.

Here is a more precise statement.

**Lemma 2.1** *Let $S$ be a set of points in the plane, and $R$ a random subset of $S$ of size $r$. With probability at least $1 - O(1/n^2)$, every candidate set $S \cap C_R(T)$*

*for $T \in \Delta(\mathcal{V}(R))$ contains $O(\log n)n/r$ sites. More precisely, if $r > 5$ then with probability at least $1 - e^{-K/2+3\ln r}$, the random subset $R$ is* good*: every Delaunay disk of $R$ contains no more than $Kn/r$ sites.*

To apply this lemma to obtain a good subset $R$, choose $r$ large enough that $8\ln r < r$ and $K = 8\ln r$. Then choose $R$ at random and check if every Delaunay disk of $R$ contains no more than $Kn/r$ sites. If $R$ is not good, we try again. Each attempt fails with probability less than $1/r$, so in $k$ attempts it fails with probability only $1/r^k$.

If we take $r = \sqrt{n}$, queries can be answered in $O(\sqrt{n}\log n)$ time, since the number of regions in $\Delta(\mathcal{V}(R))$ is $O(r)$. We could also make $r$ a sufficiently large constant, and recursively apply this scheme to the candidate sets of $\Delta(\mathcal{V}(R))$ to obtain a search tree, with a search time that is $O(\log n)$.

*Proof.* [of Lemma 2.1] Two observations make the proof, as discussed below:

- the number of disks that are potentially Delaunay disks of a subset is not too large (only polynomially large);
- the probability that a given disk that contains more than $Kn/r$ sites will be a Delaunay disk of a random subset is exponentially small.

More specifically, let $\mathcal{F}_S$ be the set of open disks bounded by circles inscribed on three sites. (A circle is *inscribed* on three points if it passes through them. There can be only one inscribed circle if the points are distinct.) There are at most $\binom{n}{3} = n!/3!(n-3)! = n(n-1)(n-2)/6$ such disks. Note that the Delaunay disks of any $R \subset S$ are contained in $\mathcal{F}_S$, since Delaunay disks are inscribed on three sites of $R$.

Suppose some open disk $D \in \mathcal{F}_S$ contains $j$ sites. What is the probability that $D$ will be a Delaunay disk of $R$? There are $\binom{n}{r} = n!/r!(n-r)!$ choices for $R$, but $D$ will be a Delaunay disk if and only if the three sites on bounding circle of $D$ are chosen, and if none of the $j$ sites in $D$ are chosen. That is, there are only $\binom{n-j-3}{r-3}$ ways to pick $R$ so that $D$ is a Delaunay disk of $R$, and so the probability is

$$\binom{n-j-3}{r-3} \Big/ \binom{n}{r}. \tag{1}$$

Using some standard manipulations that we omit (but which are very similar to (9) of §6), probability (1) is no more than $(1-j/(n-3))^{r-3}\binom{r}{3} \Big/ \binom{n}{3}$; since $1+x < e^x$ for all $x$, this is no more than

$$e^{-j(r-3)/(n-3)} \binom{r}{3} \Big/ \binom{n}{3}. \tag{2}$$

(It is instructive to consider the corresponding probability for a sample (subset) "taken with replacement." Here each of the $r$ sample sites $\hat{R} = \{x_1 \ldots, x_r\}$ is chosen independently at random, with each site in $S$ having equal probability to be $x_i$ for each $i$. The probability is about $r/n$ that a given site is chosen to be in the sample $\hat{R}$, so the probability that the three sites defining a given disk $D \in \mathcal{F}_S$ are chosen is about $(r/n)^3$. The probability that no sites in $D$ are chosen is $(1-j/n)^r$, since the chance that no site in $D$ is chosen as $x_1$ is $1-j/n$, and the same must hold true for each of the independently chosen $x_i$.)

Now to put the two observations together: we want to bound the probability that the random subset $R$ is not good. This is the event that some disk $D \in \mathcal{F}_S$ containing at least $j \geq Kn/r$ sites is a Delaunay disk of $R$. The probability of the union of a set of events is no more than the sum of the probabilities of those events. From (2), the probability of one of the events is less than $e^{-j(r-3)/(n-3)}\binom{r}{3}/\binom{n}{3}$.

The number of events is no more than the number of disks in $\mathcal{F}_S$ containing more than $Kn/r$ sites, which is no more than $\binom{n}{3}$. Thus the probability that $R$ is not good is no more than

$$\binom{n}{3} e^{-j(r-3)/(n-3)}\binom{r}{3}\Big/\binom{n}{3} = e^{-j(r-3)/(n-3)}\binom{r}{3} < e^{-K/2+3\ln r},$$

for $j > Kn/r$. □

This approach to closest-point problems is due to Clarkson.[24]

The proof of Lemma 2.1 didn't depend on anything particular to closest-point queries, or really on anything geometric. For example, we can generalize this argument to the more general $\epsilon$-transversal problem for range spaces. A *range space* $(U, X)$ is an underlying set $U$ and a collection $X$ of subsets of $U$. The members of $X$ are the ranges; sometimes we refer only to $X$ and leave $U$ implicit. An *$\epsilon$-transversal* is a subset $R \subset U$ that has nonempty intersection with every range $x \in X$ with $\epsilon|X| > |U|$. (Here $|X|$ and $|U|$ denote the sizes of these sets.) For given $\epsilon$, how small is an $\epsilon$-transversal of $X$? Conversely, given $r$, for how small an $\epsilon$ is there an $\epsilon$-transversal of size $r$? The proof above suggests the following lemma, whose proof is omitted.

**Lemma 2.2** *If $(U, X)$ is a range space and $K > 0$, then with probability at least $1 - e^{-K/2+\ln|X|}$ a random subset of $U$ of size $r$ is a $K/r$-transversal.*

This lemma implies the first statement of Lemma 2.1: the underlying set is $S$ and the ranges are the subsets of the form $S \cap D$ for all $D \in \mathcal{F}_S$. A $K/r$-transversal $R$ has the property that no Delaunay disk of $R$ contains more than $Kn/r$ sites.

An $\epsilon$-transversal is sometimes called a *cover* of the ranges it meets, or a *puncture set*. The lemma, and related bounds, yield several combinatorial results proved using the "probabilistic method," as noted by Spencer.[76,38] Even in the one-dimensional case, Lemma 2.1 has interesting implications; Reischuk used a special case to analyze a parallel algorithm for sorting.[67] The use of a random subset as a cover is also discussed in the context of network problems.[6] Another application, to estimating the Hough transform, is mentioned in §5.1.

Delaunay disks have the useful property that every disk containing no sites of $R$ is contained in the union of three Delaunay disks of $R$. (Strictly speaking, this holds only if we include as "Delaunay disks" the empty halfplanes of $R$, as defined in §4.1.) Thus if $R$ is good in the sense of Lemma 2.1, *any* disk containing no sites of $R$ contains few sites of $S$. An $\epsilon$-transversal $R$ is thus a $3\epsilon$-transversal of the range space with underlying set $S$ and with ranges of the form $S \cap D$, where $D$ is *any* open disk. (More commonly $\epsilon$-transversals are called $\epsilon$-nets in papers on geometric algorithms, but this sometimes makes for confusion with the earlier idea of $\epsilon$-nets in metric spaces, and so we follow Blumer *et al.*[8] in using a different term.)

Next we'll consider $\epsilon$-transversals in greater generality, and see a general condition that implies that random subsets are $\epsilon$-transversals with high probability.

### 2.2 The Vapnik-Chervonenkis dimension

The *Vapnik-Chervonenkis dimension*[79,80] is a property of range spaces; it has a fundamental relation to some questions about algorithms, learning, and statistics. In this subsection, we'll briefly note some of these relations.

Call a range space $(U, X)$ *polynomial* of degree $k$ if $|X| \leq |U|^k$. Lemma 2.2 immediately implies that if $(U, X)$ is polynomial of degree $k$, then a random subset $R \subset U$ of size $r$ is a $K/r$-net with probability $1 - e^{-K/2+k\ln|U|}$, and so $(U, X)$ has $3(\ln|U|)k/r$-transversals of size $r$.

If $S$ is a set of $n$ sites in the plane, the range space $(S, \mathcal{F}_S)$ of §2.1 is polynomial of degree 3. Again, this follows from the fact that any disk in $\mathcal{F}_S$ is "defined by" three sites: any such disk is bounded by a circle inscribed on three sites of $S$. Thus Lemma 2.1 can be generalized to range spaces where the ranges are defined by some small number $b$ of elements of $U$. We'll say that such range spaces have *specification complexity* $b$; they are polynomial of degree $b$ and so have small $\epsilon$-transversals. Appendix B gives several more examples.

There are more general conditions, however, for range spaces to have small $\epsilon$-transversals. If $(U, X)$ is a range space, for $U' \subset U$ define the *subspace induced by* $U'$ as the range space $(U', X')$ with

$$X' = \{x \cap U' \mid x \in X.\}$$

If $U$ is infinite, and there is some constant $k'$ so that every subspace induced by $U' \subset U$ with finite $|U'| > k'$ is polynomial of degree $k$, say that $(U, X)$ has *polynomial discrimination* of degree $k$. That is, the ranges in $X$ don't break up any subset of $U$ into too many pieces. It follows from the paragraph before last that if $(U, X)$ has polynomial discrimination of degree $k$, then any subspace has small $\epsilon$-transversals.

The range spaces with polynomial discrimination are precisely those with bounded *Vapnik-Chervonenkis dimension*, or VC dimension for short. The VC dimension of $(U, X)$ is the maximum size of a subset $U'' \subset U$ such that the subspace induced by $U''$ is *complete*: its ranges comprise all subsets of $U''$. If no such maximum exists, the VC dimension is unbounded, or infinite.

Many classes of geometric regions have bounded VC dimension: disks in the plane, balls in $E^d$, halfspaces in $E^d$, regions bounded by polynomial curves of bounded degree, axis-oriented rectangular regions, and many others. (See Blumer *et al.*[8] for references.)

We quote without proof the following theorem, which gives explicitly and sharply the relation between polynomial discrimination and VC dimension. (This result is due to Sauer and to Perles and Shelah;[69,75] see also Bollabás's text,[11] where the term *trace* is used for what we call the VC dimension.)

**Theorem 2.3** *Let $(U', X')$ be range space with $|U'|$ finite. If $|X'| > \sum_{0 \leq i < k} \binom{|U'|}{i}$, then for some $U''$ of size at least $k$, the subspace of $(U', X')$ induced by $U''$ is complete.*

If $(U, X)$ has bounded VC dimension $d$, then so does any subspace $(U', X')$, and from the theorem, $|X'| < |U'|^d$ if $|U'|$ is finite, and so $(U, X)$ has polynomial discrimination of degree $d$. Thus if a range space has bounded VC dimension, all its subspaces have small $\epsilon$-transversals.

### *2.2.1 Range queries*

A key algorithmic problem associated with range spaces is that of processing *range queries*:

> Given a set $S \subset U$ of size $n$, build a data structure so that for any range $A \in X$, the set $A \cap S$ can be found quickly.

Here the query range $A$ is given by some compact description: a halfspace might be given as the equation of its bounding plane. Sometimes the answer can be reported as the union of some predefined subsets, so that the reporting cost is independent of the answer size.

If $X$ has unbounded VC dimension, then any subset of $S$ might be an answer to a query; the time required to look up one answer out of the $2^n$ possible is $\Omega(n)$ by any reasonable cost measure, so there is no way to answer a query faster than by looking at each element of $S$ and checking for containment in $A$.

On the other hand, if $X$ has bounded VC dimension, then as Haussler and Welzl showed in pioneering work, queries can be answered more quickly than $\Omega(n)$, even if only $O(n)$ storage is allowed.[42] More recently, Welzl has shown that an answer can be expressed as the union of $O(n^{1-1/d})$ subsets of $S$, though an algorithm to find those subsets quickly for a given query is unknown in general.[82,19] Section 7.2 discusses an approach to range queries that applies in a geometric setting.

### *2.2.2 PAC-learnability of geometric concepts*

The VC dimension is intimately related to the general problem of *learning* in range spaces:

> Suppose $(U, X)$ is a range space, and a random subset $R = \{x_1, \ldots, x_r\}$ is chosen from $U$. Given $R \cap A$ for some unknown range $A \in X$, find a "good approximation" $A(R)$ to $A$: the hypothesis $A(R)$ attempts to account for the "concept" $A$.

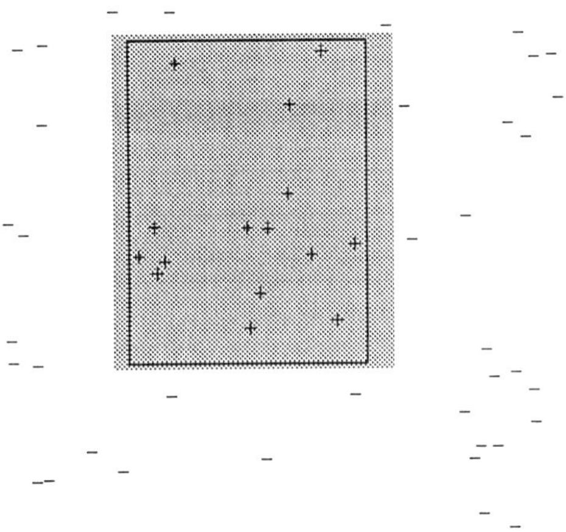

Figure 2: Sample points, the concept range $A$, and hypothesis $A(R)$.

In Figure 2, for example, the underlying set $U$ is the set of points in a square, and the ranges are rectangular regions bounded by horizontal and vertical line segments. The range $A$ is shaded, and the subset $R$ is chosen uniformly at random from square $U$. The points in $A \cap R$ are shown with +'s, the points in $R \setminus A$ are shown with −'s, and the hypothesis $A(R)$, with its boundary rectangle shown, agrees with $A$ on $R$. (Since there are uncountably many points in the square, we are not really examining a random subset of a finite set, and the proper setting is a probability distribution on $U$; to keep comfortably finite, imagine that $U$ is the set of grid points $\{(i/m, j/m) \mid 1 \leq i, j \leq m\}$ for some very large $m$.)

One measure of the badness of $A(R)$ is the probability of misclassification of a random element of $U$. the probability that random $x_{r+1}$ is in the symmetric difference $A(R) \oplus A$. A range space $(U, X)$ is said to be PAC-learnable only if there is always a range $A(R)$ so that with high probability, this misclassification probability is small. (PAC stands for "Probably Approximately Correct.")

A range space has bounded VC dimension if and only if it is PAC-learnable.[8] If the VC dimension $d$ is bounded, a satisfactory hypothesis $A(R)$ is any range agreeing with the data, so that $A(R) \cap R = A \cap R$. If $(U, X)$ has bounded VC dimension, then $(U, X')$ does, where $X' = \{x \oplus y \mid x, y \in X\}$. If $r$ is large enough, then with high probability $R$ is an $\epsilon$-transversal for $X'$, with $\epsilon = O(\log r)d/r$. Hence the probability that random $x_{r+1}$ is in $A(R) \oplus A$ is less than $\epsilon$, since otherwise $R \cap (A(R) \oplus A)$ is not empty.

### 2.2.3 Conclusions on the VC dimension

In this discussion of the Vapnik-Chervonenkis dimension, we have seen that the property of bounded VC dimension is necessary and sufficient for a range space to have polynomial discrimination, $\epsilon$-transversals, efficient range query structures, and PAC-learnability.

The bounds on the size of an $\epsilon$-transversal given above can be sharpened;[42,8] the best bound known is that a range space of VC dimension $d$ has an $\epsilon$-transversal of size $(d/\epsilon)\log(1/\epsilon)(1+o(1))$ as $1/\epsilon \to \infty$.[47]

Pach and Wöginger have shown that there are range spaces of dimension $d$ with $\epsilon$-transversals of size $\Omega((1/\epsilon)\log(1/\epsilon))$.[62] Thus the $\log(1/\epsilon)$ factor is required in the $\epsilon$-transversal size bounds for general range spaces. In contrast, point sets in $E^3$ have $\epsilon$-transversals with respect to halfspaces of size $O(1/\epsilon)$, as do points in the plane with respect to disks, as can be shown using arguments similar to those of §6.2.[51] Moreover, the results of §3 and §6 show that the logarithmic factor is not needed for divide-and-conquer applications with range spaces of bounded specification complexity. Thus while the combinatorial condition of bounded VC dimension is elegant and fundamental, there remain geometric problems for which it is insufficient to yield the best results.

## 3 Sharper expected bounds for divide-and-conquer

### 3.1 Trapezoidal diagrams of simple polygons

We have seen one example of a divide-and-conquer approach to a geometric problem using a random subset: we get some $m(r)$ subproblems of size $n_1, \ldots, n_{m(r)}$, for some function $m(r)$ of the subset size $r$. In §2.1, the $m(r) = O(r)$ subproblems were all small when $R$ was a good subset, with $n_i = O(\log r)n/r$ for all $i$, where $n_i$ is the number of sites in the $i$th Delaunay disk of $R$. For some purposes, a uniform bound on the subproblem sizes is critical, but for others it is enough that the subproblems are small on average, so that the total $\sum_i n_i$ of the sizes of all the subproblems is small. This section shows that sharper expected bounds apply for this total than are implied by the uniform bounds alone. Specifically, while the high-probability bounds imply that the total of the subproblem sizes $\sum_{1\leq i\leq m(r)} n_i$ is no more than $m(r)O(\log r)n/r$, we can get expected bounds on this sum like $m(r)O(1)n/r$.

With the logarithmic factor removed, optimal or near-optimal algorithms can be obtained for a variety of problems. The problem of computing *trapezoidal diagrams* of simple polygons is one example. Let $S$ be a set of line segments that do not cross. The trapezoidal diagram of $S$ is the planar subdivision $\mathcal{T}(S)$ with regions bounded by the segments, and also by all *visibility* segments. Figure 3 shows a trapezoidal diagram of five noncrossing segments. A visibility segment $v$ is a vertical segment whose endpoints meet $S$, but otherwise does not meet $S$, and with one endpoint of $v$ also an endpoint of a segment of $S$. We can think of the trapezoidal diagram as a collection of simple *cells*, or regions, that are often trapezoids. These cells also have adjacency relations that must be represented in an implementation: for

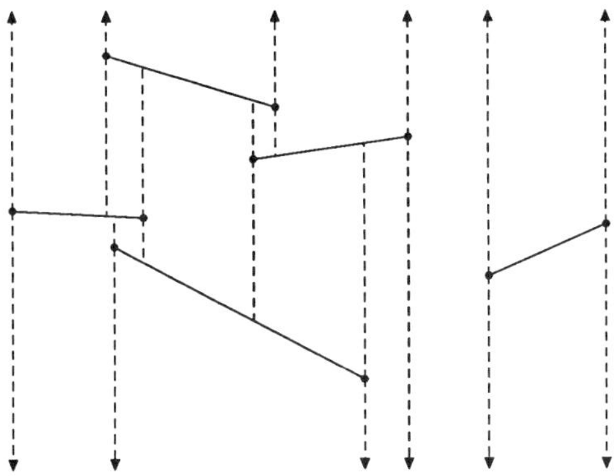

Figure 3: A set of five segments, and its trapezoidal diagram.

now, two cells will be considered adjacent only if their boundaries share a common visibility segment. A representation of a trapezoidal diagram should include this visibility information. Note that each cell has at most four neighbors (under some non-degeneracy assumptions.)

The boundary of a simple polygon is a closed polygonal chain: a collection $S$ of segments meeting at their endpoints, and only there. Many algorithmic problems involving simple polygons were long known to require $O(n)$ time for a polygon with $n$ sides, given the trapezoidal diagram of the polygon. Algorithms were also long known that require $O(n \log n)$ time to find the trapezoidal diagram of a set of $n$ noncrossing segments, and in particular a simple polygon. An outstanding open question was whether the diagram could itself be found in $O(n)$ time, given the polygon as a list of connected segments. Recently, Chazelle found such a linear-time algorithm.[15] While Chazelle's algorithm is optimal in the worst case, here we'll look at a slightly superlinear randomized algorithm that is much simpler than Chazelle's. (The algorithm is due to Clarkson, Cole, and Tarjan[26]; Seidel has given a similar algorithm.[72] The first algorithm with comparable performance is due to Clarkson, Tarjan and Van Wyk.[30]) The idea is to use randomized divide-and-conquer, with the general framework as follows:

- Compute $\mathcal{T}(R)$ for random $R \subset S$ of size $r$;
- insert the segments of $S$ into $\mathcal{T}(R)$: that is, find the cells that meet each segment, and so the set of segments $S_T$ that meet each cell $T \in \mathcal{T}(R)$;
- for each cell $T \in \mathcal{T}(R)$, compute the subdiagram $T \cap \mathcal{T}(S_T)$;
- merge appropriate cells of the subdiagrams to build cells of $\mathcal{T}(S)$.

Figure 4 shows a simple polygon and the trapezoidal diagram of a subset $R$ comprising two segments, shown in bold.

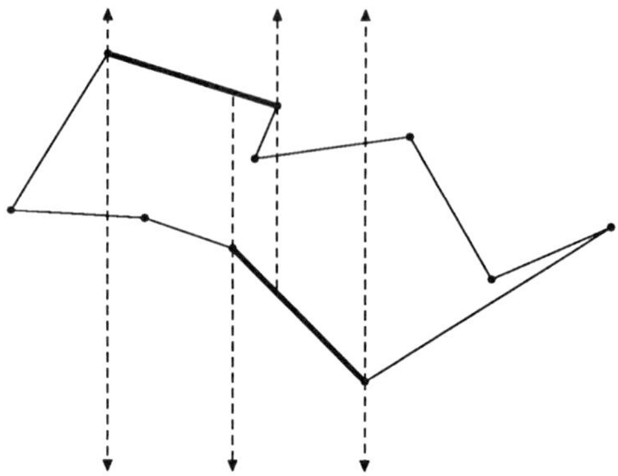

Figure 4: Using $\mathcal{T}(R)$ for divide-and-conquer.

Why does this approach produce the desired output? A visibility segment found in a subdiagram is a visibility segment of $\mathcal{T}(S)$, and no other visibility segments need be found, since such segments do not cross the segments of $R$. On the other hand, visibility segments of $\mathcal{T}(R)$ are not necessarily visibility segments of $\mathcal{T}(S)$, so we need to merge together some cells of subdiagrams in the last step. If $T$ and $T'$ are cells of $\mathcal{T}(R)$ that share a visibility segment $v$, the cells of the subdiagrams of $T$ and $T'$ to merge are those that meet $v$. That is, there are (sorted) lists of cells along $v$ in $T$ and in $T'$ that must be matched up and merged.

The next lemma gives several bounds for the sizes of the sets $S_T$. The general idea here, analogous to that for Lemma 2.1, is that a cell $T \in \mathcal{T}(R)$ meets no segments of $R$, and this is good evidence that it meets few segments of $S$. (We regard $T$ as open, and we are not interested in segments that meet its boundary. Appendix B may help to fill out the details of the analogy here.)

**Lemma 3.1** *Given a set $S$ of $n$ line segments, let $R \subset S$ be a random subset of $S$ of size $r$, with all subsets of size $r$ equally likely. Then:*

(i) *With probability $1 - 1/n$, $\max_{T\in\mathcal{T}(S)} |S_T| = O(\log n)n/r$.*

(ii) *The expected value of $\sum_{T\in\mathcal{T}(S)} |S_T|$ is $O(n)$.*

(iii) *The expected value of $\sum_{T\in\mathcal{T}(S)} |S_T| \log |S_T|$ is $O(n)\log(n/r)$.*

*Proof.* The tail estimate (i) can be proven with the same kind of argument as for Lemma 2.1: for a set of segments $S$, let $\mathcal{F}_S$ denote the set of regions that are cells of some diagram $\mathcal{T}(A)$, where $A \subset S$ has four or fewer segments. For any $R \subset S$, the cells of $\mathcal{T}(R)$ are in $\mathcal{F}_S$, which has size $O(n^4)$. For each region $T \in \mathcal{F}_S$ there is a set of segments that meet it: this set would be $S_T$ if $T \in \mathcal{T}(R)$. If any segment in this set is in $R$, then $T$ in not in $\mathcal{T}(R)$. Thus we want a $K/r$-transversal $R$ of the range space $(S, X)$, where $X = \{S_T \mid T \in \mathcal{F}_S\}$, and the number of ranges in $X$ is $O(n^4)$.

Lemma 2.2 applies, and this gives us the tail estimate (i). Part (ii) is given in §3.2, and part (iii) is proven at the beginning of §6. □

Such results for general probabilistic divide-and-conquer were first proven by Clarkson and Shor.[29] See also the work of Reif and Sen.[65]

As a first step in applying this approach, suppose $r = |R| = n/\log n$. Computing $\mathcal{T}(R)$ then requires $O(n)$ time, using an algorithm needing $O(r \log r) = O(n)$ time. Each subdiagram can be computed in $O(|S_T| \log |S_T|)$ time, which from Lemma 3.1(iii) requires $O(n \log\log n)$ expected time overall. The insertion step, finding the sets $S_T$, can be done by walking along the bounding segments $S$, and simultaneously through $\mathcal{T}(R)$, recording intersections of segments and cells as the walk proceeds. It is not hard to show that the time required by the walk is proportional to the number of such intersections recorded, which by Lemma 3.1(ii) is $O(n)$. Finally, the merge step requires $O(|S_T|)$ for each $T$, since it amounts to merging two sorted lists; thus it requires $O(n)$ expected time overall. So with $r = n/\log n$, we have an algorithm requiring $O(n \log\log n)$ expected time to compute the trapezoidal diagram. We used the connectivity information among the segments of $S$ to divide the problem into many small subproblems, and got leverage to reduce a factor of $\log n$ to $\log\log n$.

To get an algorithm that is even faster (at least asymptotically), we build a sequence of diagrams of random subsets

$$S^1 \subset S^2 \subset \cdots \subset S^{\log^* n} = S,$$

where $\log^* n$ is the smallest number $k$ so that $\log^{(k)} n < 1$, and $\log^{(k)} n = \log\log^{(k-1)} n$ is the $k$th iterate of the logarithm function. Each subset $S^i$ has size

$$r_i \approx n/\log^{(i)} n;$$

to find such a nested sequence of random subsets, imagine the segments of $S$ to be randomly permuted, and take $S^i$ as the first $r_i$ segments in the permuted order. (For more on this, see Appendix A.)

We use $\mathcal{T}(S^{i-1})$ to compute $\mathcal{T}(S^i)$, in the same way that $\mathcal{T}(R)$ was used above to find $\mathcal{T}(S)$. The only difference in this computation is that in the insertion phase, we walk $S$ to find the intersections with cells of $\mathcal{T}(S^{i-1})$, and then throw away most of the intersection information we get that way: we retain such information only for segments of $S^i$. The time required for the walk is still expected $O(n)$, however.

Since $S^i$ is a bit smaller than $S$, so too are the sets $S^i_T = S_T \cap S^i$ for $T \in \mathcal{T}(S^{i-1})$; as a result, the computation of the subdiagrams is a bit faster. Since $S^{i-1}$ is a random subset of $S^i$, Lemma 3.1 can be applied, with $S^i$ taking on the role of $S$, and $S^{i-1}$ the role of $R$. The result is that computing all the subdiagrams requires $O(n)$ total expected time.

Thus each phase of computing the diagram of a subset $S^i$ requires expected $O(n)$ time. The total expected time required by the algorithm is $O(n \log^* n)$.

## *3.2 Proving the expected bounds*

For now we'll only prove Lemma 3.1(ii).

*Proof.* [of Lemma 3.1(ii)] We want to show that a typical segment of $S \setminus R$ meets $O(1)$ cells. Since there are $n-r$ segments in $S \setminus R$, this implies the bound. To find such a typical segment, take a segment $s$ at random from $S \setminus R$. Now the cells of $\mathcal{T}(R)$ that $s$ meets are those in $\mathcal{T}(R) \setminus \mathcal{T}(R')$, where $R' = R \cup \{s\}$ and we view the diagrams as sets of cells. That is, we are interested in counting the number of cells that $s$ "destroys" when added to $R$.

There is a helpful preservation of randomness here: since $R$ is a random subset of $S$ and $s$ is a random element of $S \setminus R$, we know that $R'$ is a random subset of $S$ and $s$ is a random element of $R'$. So we can think of these random choices taking place "backwards": first pick random $R' \subset S$, and then pick random $s \in R'$.

For finite sets $A$ and $B$,

$$|A| + |B \setminus A| = A \cup B = |B| + |A \setminus B|,$$

so

$$|\mathcal{T}(R) \setminus \mathcal{T}(R')| = |\mathcal{T}(R)| - |\mathcal{T}(R')| + |\mathcal{T}(R') \setminus \mathcal{T}(R)|. \tag{3}$$

For a given $R' \subset S$, consider the expected values of these quantities for random $s \in R'$: by the linearity of expectation, we have

$$E|\mathcal{T}(R) \setminus \mathcal{T}(R')| = E|\mathcal{T}(R)| - |\mathcal{T}(R')| + E|\mathcal{T}(R') \setminus \mathcal{T}(R)|.$$

Here the expectation is with respect to the choice of $s$, which does not affect $R'$.

Consider the last quantity, $E|\mathcal{T}(R') \setminus \mathcal{T}(R)|$, the expected number of cells of $\mathcal{T}(R')$ deleted when random $s$ is removed from $R'$. A cell $T$ is removed from $\mathcal{T}(R')$ if $s$ is one of the segments defining it; that is, $s$ contains an upper or lower boundary segment of $T$, or an endpoint of $s$ gives a visibility segment bounding $T$. At most four segments define $T$, so the probability that $s$ is one of them is at most $4/(r+1)$. We have

$$E|\mathcal{T}(R') \setminus \mathcal{T}(R)| \leq 4|\mathcal{T}(R')|/(r+1).$$

Now we take expectations with respect to the random choice of $R'$, to obtain

$$|E\mathcal{T}(R) \setminus \mathcal{T}(R')| \leq E|\mathcal{T}(R)| - E|\mathcal{T}(R')| + 4E|\mathcal{T}(R')|/(r+1).$$

(Again, with the random choice of $R'$ and $s \in R'$, the subset $R$ is a random subset of $S$.) We have reduced Lemma 3.1(ii) to the problem of accurately estimating the expected complexity of the trapezoidal diagram of a random subset of a given size. It is enough to consider sets of segments that don't meet at all, even at endpoints: if we perturb segments a tiny bit so that they no longer meet at endpoints, there is a new cell/segment intersection for every old one. (We are only looking at intersections of segments with *open* regions.) When $S$ is a set of line segments with no two intersecting, the number of cells in $\mathcal{T}(R)$ for $R$ any subset of $S$ of size $r$ is $3r+1$, as the industrious reader can show. This gives us

$$\begin{aligned} |E\mathcal{T}(R) \setminus \mathcal{T}(R')| &\leq (3r+1) - (3(r+1)+1) + 4(3(r+1)+1)/(r+1) \\ &= 9 + 12/(r+1). \end{aligned}$$

So the expected number of cells met by a random member of $S \setminus R$ is $9 + 12/(r+1)$, and the expected total number of cell/segment intersections is at most $n - r$ times this. □

This analysis is based on that of Clarkson, Mehlhorn, and Seidel.[28] Seidel first showed the usefulness of the "backwards" style of analysis.[73]

By similar arguments applied to the visibility segments (rather than cells), it can be shown that the expected number of cell/segment intersections is (exactly) $5(n-r)$, and this holds for boundary segments of a simple polygon as well.

### *3.3 General segments and lines*

While we have assumed that input line segments do not cross, trapezoidal diagrams can also be defined for those that do. Simply include visibility segments extending from intersection points, as well as from endpoints. The result is a subdivision into simple "trapezoidal" regions, as before, with similar results for divide-and-conquer using random subsets. If a set $S$ of $n$ segments has $I$ pairs of intersecting segments, then the number of cells is $O(I+n)$, using the planarity of the graph of edges of the diagram and Euler's relation.

It is useful here to notice that the expected size of the trapezoidal diagram of a random subset is a bit smaller than you might think: a random subset $R$ of size $r$ has expected $I\binom{r}{2}/\binom{n}{2}$ intersecting pairs. Briefly, a given intersecting pair is in $\binom{n-2}{r-2}$ subsets, and so has $\binom{n-2}{r-2}/\binom{n}{r} = \binom{r}{2}/\binom{n}{2}$ probability of appearing in $R$. The expected number appearing in $R$ is the sum of the probabilities that a given one appears and so is $I\binom{r}{2}/\binom{n}{2}$.

This fact implies, for example, that trapezoidal diagrams can be computed in $O(I+n\log n)$ expected time, as shown by Mulmuley and by Clarkson and Shor.[55,58,29] These algorithms use randomized incremental approaches akin to those discussed in the next section. (Chazelle and Edelsbrunner gave a deterministic algorithm with the same complexity.[16]) The existence of an algorithm with complexity linearly dependent on $I$ was a longstanding open question.

Similarly, we can define trapezoidal diagrams for *arrangements* of lines, which are the subdivision of the plane induced by a set of lines. This subdivision is a collection of convex polygonal regions,and we carry over the terminology of vertices and edges in the obvious way. The visibility segments have the effect of partitioning the polygonal cells into trapezoids or triangles.

We can apply divide-and-conquer to build a search tree for locating points in a trapezoidal diagram by building the trapezoidal diagram of a random subset of a size $r$ which is constant but "sufficiently large"; each cell will correspond to a child node of the root of the search tree. Now build the search tree recursively for the lines (or segments) meeting each cell. Each cell meets $O(\log n)n/r$ lines with high probability. (The proof is almost identical to that of Lemma 3.1(i).) Hence this construction will succeed, with high probability, in producing a search tree whose leaves correspond to cells contained in at most one cell of the trapezoidal diagram of the whole set. The space required is $O(n^{2+\epsilon})$ for trapezoidal diagrams of lines, where

$\epsilon$ is a constant whose value decreases in $r$.

There are better ways to locate points in a trapezoidal diagram (see for example §4.4); however, this scheme can be generalized to higher dimensions. Several applications apply this kind of divide-and-conquer to higher dimensional problems.

## 4 Randomized incremental algorithms

While the divide-and-conquer technique was the main topic earlier, here we'll look at the different approach of building a geometric structure incrementally, making changes as needed.

For example, if we have the trapezoidal diagram $\mathcal{T}(S)$ of some set of segments $S$, it would be convenient given some new segment $s$ to make only the necessary changes in $\mathcal{T}(S)$ to get $\mathcal{T}(S \cup \{s\})$. Similarly we might want to maintain the Voronoi diagram of a set of points, or its convex hull.

For some geometric structures, such as trapezoidal diagrams and planar Voronoi diagrams, the change in structure when including one new site is $\Omega(n)$ in the worst case, matching the complexity of the whole diagram up to a constant factor. Thus these update problems are not very interesting in a worst-case setting.

However, suppose that the updates are randomly chosen from some set; that is, we maintain $\mathcal{T}(R)$ for random $R \subset S$, and then update by inserting into $R$ a random element of $S \setminus R$, or deleting a random element from $R$. These update operations preserve the randomness of $R$. Here (as we'll show) the insertions can be done fast enough that we have optimal algorithms for computing $\mathcal{T}(S)$.

If we are given the set $S$ and want to insert all of its members, yielding $\mathcal{T}(S)$, we can ensure that these randomness assumptions are satisfied, since we simply repeatedly insert random elements of $S$. Put another way, in this *off-line* setting, we can pick a random permutation of $S$ and insert into $R$ based on that permutation.

The randomness assumptions are also satisfied if $S$ is a collection of *random* points that are distributed independently and identically. Here the randomness is in the data, and we can compute $\mathcal{T}(R)$ on-line, without knowing $S$ beforehand.

### *4.1 Convex hulls*

This subsection gives a randomized incremental algorithm for maintaining convex hulls of point sets.

Before discussing the algorithm, a few words about convex hulls. (For more discussion and some motivation for computing them, see Fortune's survey.[39]) The convex hull $P = \mathcal{C}(S)$ of a set of sites (points) $S$ is the "smallest" convex set that contains $S$: it is a subset of any convex set containing $S$. A set of points is convex if any line segment with endpoints in the set is contained in the set. In particular, halfspaces are convex, and $\mathcal{C}(S)$ is the intersection of all halfspaces containing $S$. We'll assume the sites (the points of $S$) are in "general position," so that no $d+1$ are coplanar when $S \subset E^d$. With sites in three-dimensional space, so $d = 3$, the boundary of $\mathcal{C}(S)$ is a collection of triangles; each triangle, called a *facet*, has an associated *supporting* plane containing it; the *supporting halfspace* of a facet is bounded by its

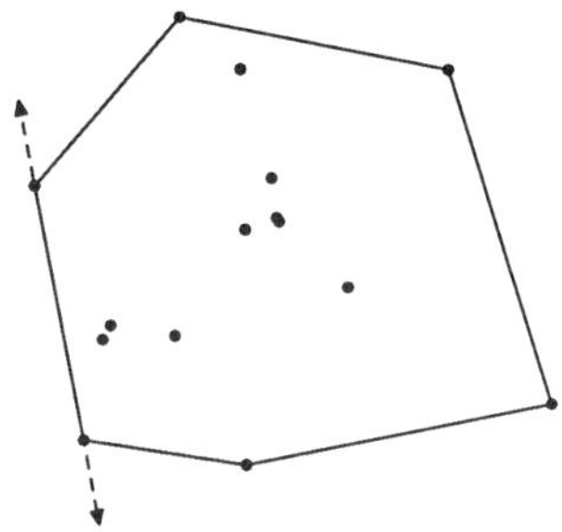

Figure 5: A set of sites in the plane and its hull.

supporting plane and contains $\mathcal{C}(S)$. In fact, $\mathcal{C}(S)$ is the intersection of the supporting halfspaces of its facets. The vertices of each facet are sites.

We'll call the open complement of a supporting halfspace of a facet an *empty halfspace* of $S$. Note that each empty halfspace of $S \subset E^3$ is "specified" by three sites contained in its bounding plane, the vertices of the corresponding facet. As the name should suggest, each empty halfspace contains no sites of $S$.

As in previous sections for Delaunay balls and cells in trapezoidal diagrams, useful bounds hold for the number of sites (of $S$) in the empty halfspaces of a random subset $R$ of $S$. For example, on average such an empty halfspace contains $O(n/r)$ sites, for $|R| = r$, and with high probability every empty halfspace contains $O(\log r)n/r$ sites.

While we'll sometimes confuse $\mathcal{C}(R)$ with its associated set of empty halfspaces, a computer representation of $\mathcal{C}(R)$ should include the adjacency relations between facets: each facet will know the facets with which it shares all but one vertex.

Figure 5 shows a set of sites in the plane ($d = 2$), the boundary facets (edges) of its convex hull, and the line that bounds both a supporting halfspace (halfplane) and its corresponding empty halfspace.

Now consider the problem of maintaining the convex hull in three dimensions under insertions. Given a new site $s \in S \setminus R$, how is $\mathcal{C}(R')$ different from $R$, where $R' = R \cup \{s\}$? Some empty halfspaces of $R$ are not empty halfspaces of $R'$, since they contain $s$, and so some facets of $\mathcal{C}(R)$ are not facets of $\mathcal{C}(R')$. These facets are said to be *visible* to $s$ (and $s$ is visible to them), since line segments between $s$ and such a facet are not "blocked" by $\mathcal{C}(R)$. On the other hand, $R'$ may have some facets with $s$ as a vertex, and so some empty halfspaces not in $\mathcal{C}(R)$.

Figure 6 shows a planar hull and a new site $s$ that results in the deletion of several visible edges and the creation of two new ones.

There are two problems to be solved to obtain $\mathcal{C}(R')$: the *search* problem, to find the facets of $\mathcal{C}(R)$ visible to $s$, and the *update* problem, to delete visible facets and determine the new facets with $s$ as a vertex (*incident* to $s$). Note if $s$ is inside the current hull, there will not be *any* facets visible to $s$, and no updating at all need be done.

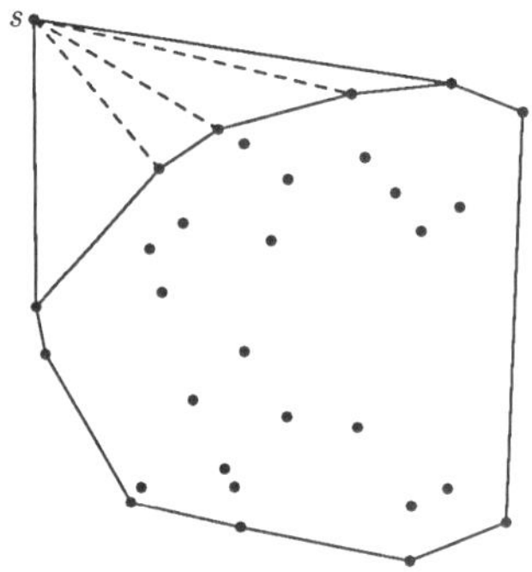


Figure 6: The effect of a new site $s$.

#### 4.1.1 Updates

While searching is the more expensive of the two operations, let's look at updating first, in the setting $d = 3$. As noted, in the worst case updating can take $\Omega(n)$ time: a good fraction of the facets of $\mathcal{C}(R)$ might be visible to $s$, and need to be deleted. However, when $R$ is a random subset of $S$, then as in the proof of Lemma 3.1(ii), when $d = 3$ on average a point $s \in S \setminus R$ is visible to a constant number of facets of $\mathcal{C}(R)$.

The number of new facets incident to $s$ is also expected to be constant. Moreover, each new facet incident to $s$ has an edge (not containing $s$) that is contained in a facet visible to $s$, and also in another facet not visible to $s$; such *horizon* edges give all and only the new facets. Since each visible facet is adjacent to at most three facets that aren't visible, the time needed to create the new facets is proportional to the number of old facets.

It will be convenient for updating and searching to maintain a triangulation $\mathcal{H}$ of the convex hull: a collection of tetrahedra whose vertices are sites, whose union is the hull, and for which the intersection of two tetrahedra in the collection is either empty or a common face of the two. (Here a face is a vertex, edge, or triangular facet.)

The triangulation $\mathcal{H}$ will be maintained for insertions by including, for each visible facet, the tetrahedron with that facet, and also a vertex that is the inserted site. (If $s$ is in the hull, no updating of the triangulation need be done.) Call the visible facet the *base* facet of the tetrahedron, and the inserted site its *peak* vertex. The facets of $\mathcal{C}(R') \setminus \mathcal{C}(R)$, incident to $s$, will be among the facets of the new tetrahedra.

We'll maintain adjacencies among these tetrahedra, so that each tetrahedron knows the four others with which it shares a facet. We can represent adjacencies for current facets by including tetrahedra with current "base" facets and with a "dummy" peak vertex that may later be replaced by a site visible to the base facet.

When fewer than $d+1$ sites have been inserted, we simply maintain $R$. When $R$ has $d+1$ sites, we initialize $\mathcal{H}$ with an *origin* tetrahedron whose vertices are all sites, and with $d+1$ tetrahedra adjacent to the origin tetrahedron, which all have the dummy site as a peak vertex. It will be convenient to have the convention that the origin tetrahedron contains the coordinate origin, and has a peak vertex and base facet, though that facet was never a hull facet. Unlike every other base facet, the base facet of the origin tetrahedron will be (by convention) visible to sites on the *same* side of the facet as the origin.

### *4.1.2 Search*

As with many known ways to search for the visible facets, here we'll look at a way to use visibility of old, deleted facets to find current visible ones. That is, after a site visible to a facet has been inserted, that facet is marked "deleted," but lives on as a base facet of a tetrahedron of $\mathcal{H}$, to help with searches for other sites.

How can deleted facets help with searches? First we must extend the notion of visibility to deleted facets: call a facet visible to a site, and vice versa, if that relation held when the facet was created. In other words, a facet is visible to a site if the site is in the halfspace that was the empty halfspace for the facet when it was created.

With this extended idea of visibility, our search procedure will find all visible facets, deleted as well as current: to visit a tetrahedron, check if its base facet is visible, and if so, visit tetrahedra adjacent to it. Start with the origin tetrahedron and the neighbor with which it shares a base facet.

Figure 7 shows the triangulation $\mathcal{H}$ for a set of 400 sites uniformly distributed in a disk. Not all the sites are shown, and the number of vertices of the triangulation is much smaller than 400, since most sites were in the hull when inserted. Base facets are shown as solid segments, and many have small triangles pointing toward their associated peak vertex. The remaining segments are dashed. (Note that many edges are base facets of one triangle they bound, though not the other.) The shaded triangles are some of those visited when searching for the site $s$.

Why does this procedure work? One argument is as follows: draw a line segment from a point in the origin tetrahedron to the new site. The base facet of every tetrahedron that meets that segment is visible to the new site, and the tetrahedra along the segment form a series of adjacent pairs, so the outermost tetrahedron, with a visible and current base facet, would be visited by the search procedure. Given that one current visible facet, the remaining such facets can be found using the tetrahedron adjacency information, since (as can be shown) the visible current facets form a connected set under the adjacency relation. (More exactly, in the undirected graph whose nodes are tetrahedra and with edges between adjacent tetrahedra, the visible current facets form a connected subgraph.)

## *4.2 Analysis of insertions*

The time for inserting a new site is dominated, up to a constant factor, by the search time. This is proportional to the number of facets, deleted and current, visible to the new site. The space required by the triangulation is proportional to

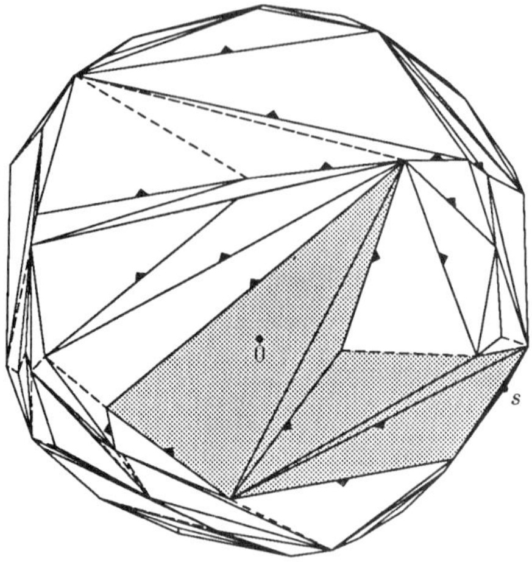


Figure 7: Some triangles visited in searching for a facet seen by site $s$.

the total number of facets, deleted and current. We'll analyze the expected sizes of these quantities when the sites are inserted in the order $x_1, x_2, \ldots, x_n$, and this is a random permutation of $S$. This implies $x_i$ is a random element of $S \setminus R_{i-1}$, inserted into $R_{i-1} = \{x_1, \ldots, x_{i-1}\}$, which is a random subset of $S$. We'll let $\mathcal{H}_i$ denote the triangulation $\mathcal{H}$ after $i$ insertions.

The analysis is not much different when the dimension $d$ is not equal to 3, so we'll put it in terms of $d$. The analysis uses the fact that facets have $d$ vertices: in the terms of §2.2, they have a "specification complexity" of $d$. In $d$ dimensions, the convex hull of $m + 1 \leq d + 1$ points is called an *$m$-simplex*. (Generally a $d$-simplex is called a simplex, omitting the $d$.) A 1-simplex is a line segment, a 2-simplex is a triangle, and a 3-simplex is a tetrahedron. With the general position assumption, the facets are $(d-1)$-simplices, and the triangulation comprises $d$-simplices.

The analysis will express the expected time and space requirements in terms of $f_j$, the expected number of facets of $\mathcal{C}(R_j)$. When $d = 3$ we have $f_j < 2r - 4$, and in general $f_j = O(j^{\lfloor d/2 \rfloor})$.[52,12] We'll say that $f_d \equiv 2$ for convenience, and $f_i \equiv 0$ for $i < d$.

**Theorem 4.1** *Let $p_r$ be the expected size of the triangulation $\mathcal{H}_r$ of $\mathcal{C}(R_r)$. Then $p_r = \sum_{j \leq r} df_j/j$.*

*Proof.* After $d + 1$ points have been added, the triangulation $\mathcal{H}_{d+1}$ has $d + 2$ simplices, as discussed above. For $j \geq d + 2$, the base facets of $\mathcal{H}_j$ that are not base facets of $\mathcal{H}_{j-1}$ are those incident to $x_j$ and facets of the hull of $R_j$. Since each facet is incident to $d$ vertices, the expected number of vertex-facet incidences is $df_j$, since $R_j$ is a random subset of $S$. Since $x_j$ is a random element of $R_j$, it is incident to an expected $df_j/j$ facets. □

In $d = 3$, $p_r = O(r)$, and in general the triangulation size is within a constant factor of the output size for "well-behaved" values $f_j$. As an example of a point set that is *not* well-behaved in this sense, pick a set with $f_j = \Omega(j)$, and then include the vertices of a tetrahedron containing the set. While the output has only four facets, the triangulation will have $\Omega(r)$ tetrahedra. On the other hand, for many natural probability distributions random data will have $f_j = O(j)$, or even $f_j = \log^{O(1)} j$, and the triangulations will be about the "right size."

**Theorem 4.2** *The expected number of base facets of $\mathcal{H}_{r-1}$ that see $x_r$ is*

$$-df_r/r + \sum_{j\leq r} d(d-1)f_j/j(j-1).$$

When $d = 3$ the search time is expected $O(1)\sum_{1\leq j\leq r} 1/j = O(\log r)$. This means the expected total time to build $\mathcal{C}(S)$ is $O(n \log n)$, which is optimal for arbitrary point sets.

The proof shifts the analysis from estimating the number of facets visible to $x_r$, to one of estimating the number that would be been incident to $x_r$ (with $x_r$ as a vertex) if it had been inserted first. The latter is easy to estimate.

*Proof.* Let $\mathcal{H} = \mathcal{H}_{r-1} = \mathcal{H}(x_1, \ldots, x_{r-1})$ and $\mathcal{H}' = \mathcal{H}(x_r, x_1, \ldots, x_{r-1})$, that is, in $\mathcal{H}'$ we "pretend" that $x_r$ was put in first. (Note that each simplex of a triangulation has exactly one corresponding base facet, so we'll confuse $\mathcal{H}$ and $\mathcal{H}'$ with the corresponding sets of base facets.)

Let $X$ be the number of base facets of $\mathcal{H}_{r-1}$ that see $x_r$. Now $X = |\mathcal{H} \setminus \mathcal{H}'|$ since $\mathcal{H} \setminus \mathcal{H}'$ is the set of base facets in $\mathcal{H}$ which see $x_r$, and hence are not in $\mathcal{H}'$. We have

$$X = |\mathcal{H} \setminus \mathcal{H}'| = |\mathcal{H}| - |\mathcal{H}'| + |\mathcal{H}' \setminus \mathcal{H}|.$$

(Compare with (3) of §3.2.) By the linearity of expectation, and using Theorem 4.1,

$$EX = E|\mathcal{H} \setminus \mathcal{H}'| = p_{r-1} - p_r + E|\mathcal{H}' \setminus \mathcal{H}|, \tag{4}$$

averaging over the random choices of $x_1$ through $x_r$, and noting that $(x_r, x_1, \ldots, x_{r-1})$ is a random permutation of $R_r$.

It remains to estimate $E|\mathcal{H}' \setminus \mathcal{H}|$. What base facets are in $\mathcal{H}' \setminus \mathcal{H}$? This is a set of base facets incident to $x_r$: the collection of facets of the hull with $x_r$ as a vertex, over the insertion sequence $x_r, x_1, \ldots, x_{r-1}$. To count these base facets, we count the expected number that appear when $x_j$ is inserted, for $j \leq r$. That is, letting $R'_j = R_j \cup \{x_r\}$, for each base facet $F$ of $\mathcal{H}'$, either $F$ is a facet of$\mathcal{C}(R'_{d-1})$, or there is exactly one $j \geq d$ such that $F$ is a facet of $\mathcal{C}(R'_j)$ and $x_j$ is incident to $F$. A base facet in $\mathcal{H}' \setminus \mathcal{H}$ is also incident to $x_r$, and so for given $j$ the number of base facets we count is the expected number of facets of $\mathcal{C}(R'_j)$ incident to $x_r$ and $x_j$. Since each facet of $\mathcal{C}(R'_j)$ is incident to $\binom{d}{2}$ ordered pairs of vertices, the expected number of facets incident to a random pair of vertices in $R'_j$ like $x_r$ and $x_j$ is $\binom{d}{2}E|\mathcal{C}(R'_j)|/\binom{j+1}{2}$. Hence,

$$E|\mathcal{H}' \setminus \mathcal{H}| = f_d + \sum_{d\leq j<r} \binom{d}{2} E|\mathcal{C}(R'_j)|/\binom{j+1}{2}.$$

Putting this expression for $E|\mathcal{H}' \setminus \mathcal{H}|$ with (4),

$$EX = p_{r-1} - p_r + f_d + \sum_{d<j\leq r-1} d(d-1)E|\mathcal{C}(R'_j)|/(j+1)j$$

Since $R'_j$ is a random subset of $S$, $E|\mathcal{C}(R'_j)| = f_{j+1}$, and a bit of rearranging gives the result. □

### 4.2.1 References

The above algorithm and analysis follows Clarkson, Mehlhorn, and Seidel,[28] who also discussed tail estimates for the space required by the triangulation. Boissannatt and Teillaud gave an on-line algorithm for Delaunay triangulations;[10] Clarkson and Shor gave a similar off-line algorithm and analyzed its performance.[29] Independently Mulmuley described and analyzed an off-line randomized incremental algorithm for trapezoidal diagrams.[55] Subsequently Boissannat and others, and Guibas and others, gave on-line algorithms and analyzed them,[9,40] and recently several papers have given on-line algorithms that allow deletion from $R$, with fast expected times if the deleted point is a random element of $R$.[71,61,59,60,28,32] Klein has described a (greatly) generalized planar Voronoi diagram;[46] it can be computed in optimal expected time with a randomized incremental algorithm.[46,54]

## 4.3 Other search schemes

### 4.3.1 Off-line algorithms

The approach of building a structure to help search is not the only way to find visible facets if the algorithm is off-line and looks at all of $S$ at once. For example, we might maintain the visibility relation between all facets of $\mathcal{C}(R)$ and all sites in $S \setminus R$. With this information, the search problem is trivial. Notice that the expected total number of visible facet-site pairs can be estimated: the algorithm at any moment has the information that might be used with $R$ for divide-and-conquer. The algorithm must maintain the visibility relation as facets in the hull are deleted and created. The work to eliminate a facet is proportional to the number of sites that see it; that is, we pay $O(1)$ for every facet that a site sees before the site is inserted. In other words, the total work for building $\mathcal{C}(S)$ by this algorithm is the same, up to a constant, as the total work for building $\mathcal{C}(S)$ by the on-line algorithm.

The relation between these on-line and off-line versions is the same as between quicksort and binary search trees: indeed, sorting can be reduced to a special case of trapezoidal diagram construction and planar convex hull construction; the off-line algorithm is a clumsy version of quicksort, and the on-line algorithm is a clumsy version of building a binary search tree.

The convex hull algorithm only needs one visible current facet to find the rest quickly; thus an off-line algorithm need only maintain for each site $s \in S \setminus R$ one visible facet $v_s$ of the current hull. This gives an approach that is space-efficient in the worst case, unlike either the algorithm based on triangulations or the above off-line algorithms. There is an updating problem for this information: when $v_s$ is deleted because some site $t$ is inserted into $R$ that also sees $v_s$, some facet seen by $s$ and not by $t$ must be found. This can be done by walking among the facets of the hull, starting at $v_s$, until one seen by $s$ and not $t$ is found. (If no such facet exists, $s$ is inside the hull, and it can be disregarded.)

These two off-line algorithms were given by Clarkson and Shor.[29]

### *4.3.2 Avoiding search*

Several algorithms use a randomized incremental approach, but obtain improved performance by avoiding the work that a general search procedure requires: the visibility information is obtained by other, faster means. (This may be true in a practical as well as theoretical setting: searching may be faster in practice with bucketing or quadtrees.)

For example, Seidel notes that the problem of finding a facet of $\mathcal{C}(R)$ visible to a given point can be reduced to small-dimensional linear programming, for which several $O(r)$ algorithms have been proposed.[73] (See §7.1.) This approach is faster for worst-case sites in $d > 3$ dimensions than the triangulation-searching method given here, and needs little space. (On the other hand, since for many distributions of random input we have $f_j = O(j)$, the expected search time of the above scheme is $O(\log r)$. On the other, other hand, Seidel's scheme is a bit simpler.)

In one of the earliest randomized incremental algorithms with provably good performance, Chew gave an algorithm for the Voronoi diagram of points that are vertices of a convex polygon.[20] That is, the points are in *convex position*, and their order along the polygon is given as input. Chew shows that with a small amount of auxiliary processing, the searches can be done in $O(1)$ time, giving a randomized incremental algorithm requiring $O(n)$ expected time.

Here is a sketch of his ideas, assuming some knowledge of Voronoi diagrams: the analog of site-facet visibility for Voronoi diagrams is site-Delaunay disk intersection. It turns out that two sites that are successive vertices of a polygon are Delaunay neighbors, and so have two incident Delaunay disks in common. This holds also for successive sites in a subset $R$, and so when a site is inserted between two successive vertices, the search need only examine the two disks incident to those vertices. Thus search is trivial if the neighbors of the site in $R$ are known. This information can be obtained by running the insertions backward, maintaining polygon adjacencies as sites are deleted. A similar trick is helpful for maintaining convex hulls in higher dimension under deletions, when the sites are in convex position.[28]

The algorithm of §3.1 for trapezoidal diagrams of simple polygons can be viewed as using the connectivity information about segments to avoid search. Devillier has extended these observations to speed the computation of medial axis transforms and Delaunay triangulation for sites with a known minimum spanning tree.[31]

### 4.4 Trapezoidal diagrams and planar point location

Many of the above schemes have analogs for randomized incremental construction of trapezoidal diagrams; here the analog of facet-site visibility is cell-segment intersection. (Appendix B discusses this analogy in more detail.) An algorithm might maintain all such intersections; alternatively, for space efficiency, only the location of one endpoint in each uninserted segment (in $S \setminus R$) could be maintained. The remaining intersections could then be found by walking through the current diagram when inserting a segment.

An analog of the use of a triangulation for on-line maintenance is to keep old trapezoids around and use them for search. A cell that meets a new segment $s$ is contained in the union of two new cells, each with $s$ containing their upper or lower boundaries. Searching is aided by providing "influence" pointers from the deleted cell to each of the two new ones. If a point (say, a segment endpoint) is in that cell, then it is in one of the new ones as well. Searching entails tracing through the influence pointers from an initial cell to the current one containing the point. Searching is a means of finding the *location* of the point in the current diagram.

The time required is $O(1)$ per cell visited in this search process. As for convex hull searching, the expected search time for randomly chosen endpoints of segments in $S \setminus R$ is $O(\log r)$; this bound also holds for the expected number of cells, deleted and current, that meet a random uninserted segment.

What if we seek the location of an *arbitrary* point? Previous analysis techniques in this paper tell us little about this case. (At least, they say nothing directly, and don't provide the best bound.) However, bounds on the search time do hold for any given point: as discussed below, with probability $1 - 1/n^A$, for any fixed $A$, the search time is no more than $B \log n$ for some constant $B$. Here $B$ is increasing in $A$. We'd like a bound that holds for *all* points, however. Fortunately there are only $n^{O(1)}$ distinct classes of points: two points are equivalent here if the same sequence of cells is visited for them while searching. Thus the probability that search time exceeds $B \log n$ for *any* point is $n^{O(1)}/n^A$. By making $A$ and $B$ sufficiently large constants, we have $O(\log n)$ search time for all points, with high probability.

Thus the search data structure for randomized incremental construction of trapezoidal diagrams solves the *planar point location* problem, with high probability: given a subdivision of the plane by noncrossing line segments, build a data structure so that given a point, its location in the subdivision can be found quickly. Moreover, the data structure can be maintained under random insertions and deletions with rapid execution of these operations on average.[28,61,71]

Given a triangulation of the plane, with its adjacency information, the data structure can be found in $O(n)$ expected time: by starting with the triangulation and maintaining the trapezoidal diagram of its segments under random deletions, we avoid searching but obtain the cells and pointers that an incremental algorithm would produce with random insertions; the cells to delete when deleting a segment can be found in constant time per cell, since we have the cell adjacencies, and finding the extensions of the visibility edges incident to a deleted segment is equivalent to merging sorted lists. Therefore the $O(n)$ cells in the data structure require $O(1)$ time per cell.

**Theorem 4.3** *For any positive $A$, there is a constant $B$ so that for any given point, the number of cells visited while searching through influence pointers is less than $B \log n$ with probability $1 - n^A$.*

*Proof.* [Sketch] Consider the construction of the data structure by deleting the segments of the triangulation in random order. The cell containing a given point changes only when one of the segments defining it is deleted. This occurs with probability at most $4/r$ when $r$ segments remain. Thus the number of cells visited is the sum of $n$ random variables, with variable $y_r = 1$ with probability $4/r$, and $y_r = 0$ otherwise. The expected value of this sum is $4H_n = \ln n + O(1)$, where $H_n = \sum_{1 \le r \le n} 1/r$. By arguments as for the Chebyshev bounds for binomial variates, the probability that the sum exceeds $c$ times its mean is no more than $e^{H_n(1+c\ln(c/e))}$. These observations are due to Seidel.[72] (See also Reif and Sen's work.[65] Clarkson, Mehlhorn, and Seidel also give a proof.[28] □

## 5 Some combinatorial bounds

In this section, we'll look at some questions of *combinatorial geometry*, which are counting problems regarding geometric objects. These problems can be attacked with some of the tools we have discussed; conversely, the answers to these problems can help us analyze some geometric algorithms.

### *5.1 Incidences*

Given a set $S$ of $m$ points and a set $L$ of $n$ lines, how many incidences can there be between them? How large is

$$I(S, L) \equiv \{(p, l) \mid p \in S, l \in L, p \in l\},$$

the set of incident pairs? Let's look at some extreme cases. Suppose $S$ comprises a single point. Then $|I(S, L)| \le n$, and can be that large, if all the lines meet at that point. Suppose $S$ contains at least $n^2$ points. Each line of $L$ can contain no more than $n - 1$ points of $S$ that are also incident to other lines of $L$; therefore $|I(S, L)| \le m + n(n-1)$. In the former case, the single point of $S$ was incident to $n$ lines. In the latter, an average point of $S$ is incident to $1 + n(n-1)/m \le 2$ lines. What happens in between? We seek a general bound on the quantity $I(m, n)$, the maximum size of $I(S, L)$ over all $m$-point sets $S$ and $n$-line sets $L$. It is worth noting that $I(m, n) = I(n, m)$, using a duality map, since the dual line of a point of $S$ and the dual point of a line of $L$ are incident just when the point and line are. (For more on duality, see Appendix A, or for example Edelsbrunner's text.[34])

First we'll show that $I(m, n) < m + n\sqrt{m}$. Number the points of $S$ from 1 to $m$. Let $k_i$ denote the number of lines incident to point $i$. We can bound $\sum_i \binom{k_i}{2}$: the quantity $\binom{k_i}{2} = k_i(k_i - 1)/2$ counts the number of *pairs* of lines incident to point

$i$, so the sum counts the number of incidences of points with pairs of lines. Since two distinct lines meet at no more than one point, a given pair contribute only one incidence; thus $\sum_i \binom{k_i}{2}$ is no more than $\binom{n}{2}$, or more loosely,

$$\sum_i (k_i - 1)^2 \leq n^2. \tag{5}$$

We can now invoke the Cauchy-Buniakowskii-Schwartz inequality, which implies

$$m \sum_i (k_i - 1)^2 \geq \left( \sum_i (k_i - 1) \right)^2.$$

(That is, the dot product of the vectors $(k_1 - 1, \ldots, k_m - 1)$ and $(1, \ldots, 1)$ is no more than the product of the Euclidean norms of the two.) Hence from (5),

$$\sum_i (k_i - 1) \leq n\sqrt{m},$$

or $\sum_i k_i \leq m + n\sqrt{m}$.

Using projective duality, as discussed in Appendix A, we also have $I(m, n) = O(n + m\sqrt{n})$.

This kind of bound is due to Canham.[13] The bound fits our extreme cases $m = 1$ and $m = n^2$, but is it best possible? We'll prove the following theorem.

**Theorem 5.1** *The maximum number of incidences between a set of $m$ points and a set of $n$ lines in the plane is $O((mn)^{2/3} + m + n)$.*

*Proof.* To get a better bound, divide-and-conquer the combinatorial problem: choose a random $R \subset L$ of size $r$, and consider the trapezoidal diagram $\mathcal{T}(R)$. A point of $S$ is either incident to a line of $R$, in an open cell of $\mathcal{T}(R)$, or on a visibility segment, incident to no lines of $L$.

Let $S'$ denote the points of $S$ incident to lines of $R$. The number of incidences involving points of $S'$ is no more than $m + 2nr$: such a point is either incident to a line of $R$ and no others, or to a line of $R$ and a line of $L$. Each line of $L$ can contain no more than $r$ points incident to a line of $R$. Hence the number of incidences involving the points of $S'$ is no more than

$$m + 2nr. \tag{6}$$

Each point of $S \setminus S'$ is in some open cell of $\mathcal{T}(R)$. If cell $C \in \mathcal{T}(R)$ contains $m_C$ points of $S$ and meets $n_C$ lines of $L$, then by the above Canham bound, the number of incidences between these points and lines is no more than $n_C + m_C\sqrt{n_C}$. That is, "charge" each point of $S \cap C$ a number of incidences $\sqrt{n_C}$, and the total charges will bound the number of incidences in $C$, up to $n_C$. Put another way, charge each point $i$ of $S$ with $\sqrt{n_i}$, the square root of the number of lines meeting the cell containing that point. The total of the charges $\sum_i \sqrt{n_i}$ will then bound the number of incidences involving points of $S \setminus S'$, up to $\sum_{C \in \mathcal{T}(R)} n_C$. The expected value of the latter sum is $O(n)$, using Theorem 3.1.

The next paragraph shows that the expectation of $\sum_i \sqrt{n_i} = O(m)\sqrt{n/r}$, so the expected number of incidences involving points in open cells of $\mathcal{T}(R)$ is $O(m)\sqrt{n/r} + O(n)$. (Since this expectation holds for a random subset, there must exist a subset for which the bound holds.) With (6), the total number of incidences is no more than

$$O(m)\sqrt{n/r} + O(n) + m + 2nr.$$

Choosing $r = (m^2/n)^{1/3}$ to balance these terms, this quantity is $O((mn)^{2/3} + m + n)$, as claimed.

To complete the proof, it suffices to show that for each point $i$, the expected value of $\sqrt{n_i}$ is $O(\sqrt{n/r})$. Again using the Cauchy-Buniakowskii-Schwartz inequality, one can show that $E\sqrt{n_i} \leq \sqrt{En_i}$, and it is enough to bound $En_i$ by $O(n/r)$. As in the proof of Theorem 3.1, consider the probability that a random line $l \in L \setminus R$ meets the cell $C \in \mathcal{T}(R)$ that contains $p$. This is the probability that the cell $D$ of $\mathcal{T}(R')$ that contains $p$ is *not* $C$. Here $R' = R \cup \{l\}$, and $R'$ is a random subset of $L$. Now $D \in \mathcal{T}(R')$ is not in $\mathcal{T}(R)$ only if $l$ is chosen from among the four (or fewer) lines defining the boundary of $D$. The probability of this is no more than $4/(r+1)$, and so the probability that random $l \in L \setminus R$ meets $C$ is no more than $4/(r+1)$, and the expected number of lines of $L \setminus R$ meeting $C$ is no more than $4(n-r)/(r+1)$, as needed. □

Besides the mathematical interest of these questions, why should anyone care about them? Here is one possible motivation: given a set $S$ of $m$ sites (points), we may want to find patterns in the sites. One sort of pattern is collinearity: many sites fall on some common straight line. For example, suppose we have the image of a black object against a white background. Many schemes are available to check the neighborhood of a point in the image to decide if it is on or near the black/white image boundary. Such schemes are not entirely reliable, though, so some way is needed to find a large collection of boundary points suggesting a boundary line. This kind of problem is closely related to the *Hough transform*, an approach computer vision researchers have used for finding straight boundary lines, or visual structure of similar kinds. Our bounds show that no more than $O(m^2/t^3)$ lines can each be incident to $t$ points in a set of $m$ points; this gives a bound on the size of the output of a structure-finding algorithm.

On the algorithmic side, it is worth noting that several computer vision researchers have suggested *Monte Carlo* algorithms for the Hough transform.[7,44,83] If a collinear subset of the sites has size $t$, then with high probability a random subset of $O(\log m)m/t$ sites will contain at least two sites of that subset, as Lemma 2.2 implies. Thus with high probability, the set of lines containing at least $t$ sites is contained in the set of lines defined by pairs of sites in the random subset. This gives an algorithm for finding lines containing large collinear subsets; the algorithm is only *Monte Carlo*, however: as the above bound shows, no more than a $O(\log m)^2/t$ fraction of the reported lines will contain $t$ sites, and some such lines may not be reported. However, fast randomized algorithms for this problem can be obtained using the techniques of this survey.

The approach of this subsection can be extended to other questions about arrangements, for example, the *many faces* question: what is the maximum possible total number of vertices of a set of $m$ cells in an arrangement of $n$ lines? This quantity $K(m,n)$ obeys a bound asymptotically similar to that of $I(m,n)$.

The discussion here follows the results in a series of papers[27,36,37,35] which also discuss similar bounds for the total complexity of collections of faces in arrangements of curves and surfaces, and related algorithmic questions.

### 5.2 Bounds on ($\leq k$)-regions

A convex polygon in the plane with $n$ vertices has $n$ edges, and hence $n$ empty halfplanes (as defined in §5). That is, there are $n$ lines with the property that each line contains two vertices of the polygon, and the vertices not on the line are contained in a halfplane bounded by the line. A natural generalization of this question: how many distinct open halfplanes have the property that their bounding line contains two vertices, and they contain no more than $k$ vertices, for some $k$? Such halfplanes are "within $k$" of being empty.

We might ask a similar question about Delaunay disks: how many disks can be within $k$ sites of being Delaunay disks of a set of sites in the plane? Such *$k$-disks* are open disks bounded by circles inscribed on three sites, and contain no more than $k$ sites.

As yet another example, consider trapezoidal diagrams of noncrossing segments: while these diagrams have $O(n)$ cells, how many $k$-cells do they have? Such a region is within $k$ segments of being a cell: it is a cell of some diagram $\mathcal{T}(S')$ where $S' \subset S$ and has at least $n-k$ segments. That is, such a region $T$ is a cell of some diagram $\mathcal{T}(A)$ for $A \subset S$ of size no more than four, and no more than $k$ segments of $S$ meet $T$.

In all these cases, we can use the bounds on the number of *0-regions*, which in the example above are, respectively, empty halfplanes, trapezoidal diagram cells, and Delaunay disks. With these bounds we can bound the number of ($\leq k$)-regions, which are "within $k$" of being 0-regions. For these structures we will show a bound of the form $f_{n/k}k^b$; here $f_{n/k}$ is the number of 0-regions of a random subset of size $n/k$, and $f_{n/k} = O(n/k)$ for the above structures. The parameter $b$ is the "specification complexity" of 2, 4, and 3 for the examples of halfplanes, trapezoidal diagram cells, and Delaunay disks, respectively. Thus bounds of the form $O(n)k^{b-1}$, with $b$ as given, hold for the number of ($\leq k$)-regions.

Before looking at proof techniques for this collection of problems, it is worth noting that the ($\leq k$)-halfplanes question is a special case of the many-faces problem for arrangements, mentioned at the end of the last subsection. Consider the ($\leq k$)-halfplanes problem, when stated in terms of the projective dual: we have a set of lines, the dual lines of the vertices, and look at the halfplanes that are bounded by these lines and contain the origin. The intersection of these halfplanes is a polygon that the dual of the original polygon. A line segment between the origin and a vertex of this dual polytope does not cross the dual lines, and we know that there are no more than $n$ vertices of the arrangement of these lines with this property. For how many of the vertices of the arrangement does the corresponding line segment cross no more than $k$ dual lines? Each such vertex corresponds to a $(\leq k)$-region of the original polygon. Thus we are bounding the complexity of a particular family of cells of the arrangement of the dual lines: those whose "crossing distance" from the origin is no more than $k$.

As an example, here is a proof of bounds for $(\leq k)$-halfplanes of point sets. We want to bound the maximum, over all sets $S$ of $n$ sites (points), of the number of $(\leq k)$-halfplanes of the sites. These open halfplanes contain no more than $k$ sites, and are bounded by lines that contain two sites. (We will assume that no three points of the set are collinear.)

The basic idea for proving the bounds is to use the relationship of these half-planes to the 0-halfplanes (empty halfplanes) of subsets. That is, we know that any subset of size $r$ has no more than $r$ associated 0-halfplanes, each halfplane corresponding to an edge of the convex hull of the subset. We will show that a $(\leq k)$-halfplane of $S$ has a relatively high likelihood of being an empty halfplane of a random subset $R$ of size $r = \lfloor n/k \rfloor$. Since there are at most $r$ such empty halfplanes, we have a bound on the number of $(\leq k)$-halfplanes.

More precisely, suppose $R$ is a random subset of $S$ of size $r$, with all subsets of that size equally likely. We can write the expected number of empty halfplanes of $R$ in the following way: let $\mathcal{F}_S$ denote the set of $j$-halfplanes of $S$, over all $j$ (not only the set of $(\leq k)$-halfplanes); that is, $\mathcal{F}_S$ is the set of open halfplanes bounded by a line containing two points of $S$. For $H \in \mathcal{F}_S$, let $I_H$ be a random variable that is 1 when $H$ is an empty halfplane of $R$, and zero otherwise. Then the number of empty halfplanes of $R$ is

$$\sum_{H \in \mathcal{F}_S} I_H,$$

and the expectation of this quantity is the expected number of 0-halfplanes of $R$. By the linearity of expectation, this is

$$\sum_{H \in \mathcal{F}_S} EI_H.$$

The expectation $EI_H$ is the probability that $H$ is an empty halfplane of $R$. Since $R$ has at most $r$ empty halfplanes, we know that

$$\sum_{H \in \mathcal{F}_S} \text{Prob}\{H \in \mathcal{F}_R^0\} \leq r, \tag{7}$$

where $\mathcal{F}_R^0$ denotes the set of empty halfspaces of $R$.

Next we find an expression for the probability that some $j$-halfplane $H$ is an empty halfplane of $R$. This event requires two conditions:

(i) The two sites on the line bounding $H$ must be in $R$;

(ii) The $j$ sites in $H$ must *not* be in $R$.

Two points in $R$ are fixed by condition (i) above, and so from among the $n-j-2$ sites that aren't in $H$ or on its boundary, any $r-2$ can be chosen. That is, the number of subsets of $S$ of size $r$ that satisfy these conditions is $\binom{n-j-2}{r-2}$. Since there are $\binom{n}{r}$ subsets of $S$ with size $r$, the probability that $H$ appears as an empty halfspace of $R$ is

$$\frac{\binom{r}{2}}{\binom{n}{2}} \frac{\binom{n-j-2}{r-2}}{\binom{n-2}{r-2}}. \tag{8}$$

The factor $\binom{r}{2}/\binom{n}{2}$ is the probability that the two sites on the boundary of $H$ appear in $R$. The other factor expresses the probability that the $j$ sites in $H$ are not in $R$, and we can bound it from below:

$$\begin{aligned}\frac{(n-j-2)!(r-2)!(n-r)!}{(n-j-r)!(r-2)!(n-2)!} &= \frac{(n-r)(n-r-1)\cdots(n-r+1-j)}{(n-2)(n-3)\cdots(n-j-1)} \\ &= \left(1-\frac{r-2}{n-2}\right)\left(1-\frac{r-2}{n-3}\right)\cdots\left(1-\frac{r-2}{n-j-1}\right) \\ &\geq \left(1-\frac{r-2}{n-j-1}\right)^j\end{aligned}$$

For $j \leq k$, this is at least $(1-(r-2)/(n-k-1))^k$, which for $r = \lceil n/(k+1) \rceil$ is at least

$$(1-1/(k+1))^k = \frac{1}{(1-1/k)^k} \geq 1/e,$$

using $1+x \leq e^x$. Substituting $1/e$ for the second term in (8), the probability that a $j$-halfplane $H$ is an empty halfplane of $R$ is at least $\binom{r}{2}/e\binom{n}{2}$.

Putting this bound with (7), if $M$ is the number of $(\leq k)$-halfplanes of $S$, then

$$\frac{M\binom{r}{2}}{e\binom{n}{2}} \leq r,$$

or

$$M \leq \frac{en(n-1)}{r-1} \leq \frac{en(n-1)}{n/(k+1)-1} < 3n(k+1)$$

for $k < n/11$.

**Theorem 5.2** *Any set of $n$ points in the plane in general position (no three collinear) has no more than $3n(k+1)$ $(\leq k)$-halfplanes, for $k < n/11$.*

We can prove similar results for $k \geq n/11$. This bound is not tight: the maximum number of $(\leq k)$-halfplanes is $n(k+1)$, and this bound is tight.[4] However, our proof is easily generalized: the only thing here specific to planar convex hulls and halfplanes was the fact that two points specify a halfplane, via its bounding line. In other instances of this general problem, we simply substitute a different number $b$ of specifiers, yielding a different power of $k$ in the bound, and use the appropriate bound for the number of 0-regions.

For example, in higher dimensions, we have ($\leq k$)-halfspaces, where the 0-halfspaces are bounded by planes containing facets of the convex hull of the points. From the bound $O(n^{\lfloor d/2 \rfloor})$ for the number of facets of the convex hull of $n$ points in $d$ dimensions, we obtain $O((n/k)^{\lfloor d/2 \rfloor} k^d) = O(n^{\lfloor d/2 \rfloor} k^{\lceil d/2 \rceil})$ as a bound for the number of ($\leq k$)-halfspaces.

These results were given in excruciating generality by Clarkson and Shor;[29]; subsequently they were sharpened a bit by Mulmuley.[56,57] The lower bound argument given here for (8) uses some arguments of Sharir,[74] who gave several applications of such bounds.

## 6 Still sharper bounds for divide-and-conquer

This section bounds the distribution of subproblem sizes more carefully, using analogs of the bounds for ($\leq k$)-halfplanes. For example, with divide-and-conquer using a random subset, an average subproblem has a *squared* size that is $O(n/r)^2$. These bounds hold because the number of subproblems of size $in/r$ falls off (almost) exponentially in $i$: about $1/e$ of the subproblems are of size less than about $n/r$, $1/e^2$ are of size less than $2n/r$ and greater than $n/r$, and so on. Here is a more precise statement of this claim.

**Lemma 6.1** *Let $S$ be a set of $n$ noncrossing line segments in the plane, and $R$ a random subset of $S$ of size $r$. There is a constant $K$ so that the expected number of $j$-cells of $S$ which satisfy $i = \lfloor j(r-4)/(n-4) \rfloor$ and appear in $\mathcal{T}(R)$ is no more than $Kr(i+1)^3 e^{-i}$.*

*Proof.* Let $\mathcal{T}_i$ be the set of $j$-cells of $S$ with $\lfloor j(r-4)/(n-4) \rfloor = i$. We want bound the expected number of cells of $\mathcal{T}_i$ that appear in $\mathcal{T}(R)$. We'll prove this only for the cells defined by exactly four segments; the proofs for those cells defined by fewer segments are analogous.

The probability that a given $j$-cell $T$ of $S$ appears as a 0-cell of $R$ is

$$\frac{\binom{r}{4}}{\binom{n}{4}} \frac{\binom{n-j-4}{r-4}}{\binom{n-4}{r-4}},$$

similarly to (8); the factor on the left is the probability that the four segments defining $T$ appear in $R$, and is no more than $(r/n)^4$. The factor on the right reflects the probability that the $j$ segments meeting $T$ are *not* in $R$. We bound the latter from above:

$$\begin{aligned} \frac{(n-j-4)!(r-4)!(n-r)!}{(n-j-r)!(r-4)!(n-4)!} &= \frac{(n-j-4)(n-j-5)\cdots(n-j-3-r)}{(n-4)(n-3)\cdots(n-r-3)} \\ &= \left(1-\frac{j}{n-4}\right)\left(1-\frac{j}{n-5}\right)\cdots\left(1-\frac{j}{n-r-3}\right) \\ &\leq \left(1-\frac{j}{n-4}\right)^{r-4} < e^{-j(r-4)/(n-4)} \qquad (9) \end{aligned}$$

Hence the probability that $T \in \mathcal{T}_i$ is a cell in $\mathcal{T}(R)$ is no more than $(r/n)^4 e^{-i+1}$. As in §5.2, the number of $j$-cells with $j \leq (i+1)(n-4)/(r-4)$ is $O(n((i+1)n/r)^3)$,so the expected number of cells of $\mathcal{T}_i$ in $\mathcal{T}(R)$ is

$$(r/n)^4 e^{-i+1} O(n((i+1)n/r)^3) = O(r)(i+1)^3 e^{-i}.$$

□

As an application of this lemma, here is a proof of one of the results needed for the analysis of the algorithm of §3.1.

*Proof.* [of Lemma 3.1(iii).] We need to show that the expected value of $\sum_{T \in \mathcal{T}(S)} |S_T| \log |S_T|$ is $O(n) \log(n/r)$. A $j$-cell with $i = \lfloor jr/n \rfloor$ contributes no more than $((i+1)n/r) \log((i+1)n/r)$ to the sum; the expected number of such cells is no more than $Kr(i+1)^3 e^{-i}$, by the lemma, and so the expected value of the sum is no more than

$$\sum_{i \geq 0} ((i+1)n/r) \log((i+1)n/r) Kri^3 e^{-i} = Kn \sum_{i \geq 0} e^{-i} i^3 (i+1) \log((i+1)n/r),$$

which is $O(n) \log(n/r)$. □

Such a proof was given in generality by Clarkson.[23]

### 6.1 Bounds for the sum of squares

As given below, similar bounds hold for trapezoidal diagrams of lines. The following subsection uses the sharper bounds to show how to make *cuttings*: subdivisions of a collection of lines into $O(r^2)$ trapezoidal (or triangular) cells so that for *every* cell, the number of segments meeting it is $O(n/r)$.

The following lemma has a proof analogous to the previous one.

**Lemma 6.2** *There is a constant $K$ so that, for a set $S$ of $n$ lines in the plane and $R$ a random subset of $S$, the expected number of $j$-cells of $S$ which satisfy $i = \lfloor jr/n \rfloor$ and appear in $\mathcal{T}(R)$ is no more than $Kr^2(i+1)^2 e^{-i}$.*

This lemma readily implies the following theorem, for which we'll also supply another proof.

**Theorem 6.3** *For a set $S$ of $n$ lines in the plane and $R$ a random subset of $S$, the expectation of $\sum_{T \in \mathcal{T}(R)} |S_T|^2$ is $O(n^2)$.*

*Proof.* The previous lemma implies the bound; here is another proof giving a smaller constant factor. Let $\mathcal{F}_S$ be the set of all possible cells in $\mathcal{T}(R)$: the set of cells that are in $\mathcal{T}(A)$, for some $A \subset S$ with $|A| \leq 4$. Let $I_T$ be an indicator function for the event that $T \in \mathcal{F}_S$ is in $\mathcal{T}(R)$. Then

$$\sum_{T \in \mathcal{T}(R)} \binom{|S_T|}{2} = \sum_{T \in \mathcal{F}(S)} I_T \binom{|S_T|}{2},$$

and using the linearity of expectation and the fact that $EI_T = \text{Prob}\{T \in \mathcal{T}(R)\}$,

$$E \sum_{T\in\mathcal{T}(R)} \binom{|S_T|}{2} = \sum_{T\in\mathcal{F}_S} \text{Prob}\{T \in \mathcal{T}(R)\} \binom{|S_T|}{2}.$$

Letting $i_T$ denote the number of segments that define $T$, the probability that $T$ is in $\mathcal{T}(R)$ is

$$\binom{n - |S_T| - i_T}{r - i_T} \Big/ \binom{n}{r},$$

since out of the $\binom{n}{r}$ subsets, a subset must be picked that includes the $i_T \leq 4$ segments defining $T$, and for the remaining $r - i_T$ segments, there are only $n - |S_T| - i_T$ to choose from, since the $|S_T|$ meeting $T$ cannot be picked. Thus the quantity to estimate is

$$\sum_{T\in\mathcal{F}_S} \binom{|S_T|}{2} \binom{n - |S_T| - i_T}{r - i_T} \Big/ \binom{n}{r},$$

which is no more than

$$2\frac{n^2}{r^2} \sum_{T\in\mathcal{F}_S} \binom{|S_T|}{2} \binom{n - |S_T| - i_T}{r - i_T - 2} \Big/ \binom{n}{r},$$

for large $r$ and $n - r$. The trick here is that the summand is the probability that the cell $T$ is a 2-cell of $R$. For $T$ to be a 2-cell of $R$, the segments defining $T$ and two of the segments meeting it must be in $R$; the remaining $r - i_T - 2$ segments of $R$ must be chosen from the segments that neither meet $T$ nor define it. The summand is the probability of this event. The sum is the expected number of 2-cells of $R$. The number of 2-cells of any set of lines of size $r$ is $O(r^2)$, using a proof analogous to that of Theorem 5.2. The sum obeys this bound, and the theorem follows. □

This proof follows that of Clarkson and Shor.[29]

## 6.2 *Uniform bounds and cuttings*

This section shows the existence of optimal *cuttings*: given a set $S$ of $n$ lines and a parameter $r$, a $(1/r)$*-cutting* is a subdivision into simple cells so that every cell meets no more than $n/r$ lines. The methods of §2 imply only that the trapezoidal diagram of a random subset of $S$ of size $O(r \log n)$ is a $(1/r)$-cutting with high probability; the number of cells is $O(r \log n)^2$. The methods of §3 show that, on average, a cell of a trapezoidal diagram of a random subset of $O(r)$ lines meets $n/r$ lines, but still some cells may meet more.

Here we'll show that a $(1/r)$-cutting of size $O(r^2)$ can be built; this is asymptotically optimal. The idea is to start with the trapezoidal diagram of a random subset and recursively "repair" the cells which meet too many lines. By the results of the last subsection, the sizes of the repairs will be small. Such repairs have some resemblance to deletion methods of the "probabilistic method."[77]

**Theorem 6.4** *There is a constant $C$ so that given a set $S$ of $n$ lines and a parameter $m$, there is a $(1/m)$-cutting of $S$ of size no more than $Cm^2$.*

*Proof.* Choose a random subset $R$ of $S$ of size $r = m\alpha\sqrt{C}$ and build its trapezoidal diagram, choosing $\alpha$ small enough that $\mathcal{T}(R)$ has no more than $Cm^2/4$ cells. For every cell $T$ meeting $j > n/m$ lines, recursively subdivide $T$ with a $1/\lceil jm/n \rceil$-cutting. This completes the construction.

The expected size of the resulting cutting is no more than

$$Cm^2/4 + \sum_{i>\alpha\sqrt{C}} (C(i+1)^2)Kr^2(i+1)^2e^{-i+1} = Cm^2/4 + Cr^2 \sum_{i>\alpha\sqrt{C}} K(i+1)^4e^{-i+1},$$

using Lemma 6.2. For sufficiently large $C$, the sum is less than $1/4\alpha\sqrt{C}$, so the expected size of the cutting is no more than $Cm^2/2$. By Markov's inequality, the probability that the cutting will have size greater than $Cm^2$ is no more than $1/2$, so with probability $1/2$ the construction succeeds in obtaining a cutting with no more than $Cm^2$ cells. Repeat until success, requiring two expected attempts. □

In the literature, the cells of a cutting are triangles, so as a last step here, split the cells into triangles.

This construction follows ideas of Chazelle and Friedman.[18] Similar ideas shows that cuttings into simplices in higher dimensions can be found. A more recent construction, and a "derandomized" version, is described by Chazelle,[14] who uses it for a worst-case optimal algorithm for computing convex hulls in higher dimensions.

## 7 Reweighting methods

This section gives two applications of a *reweighting* technique. The idea is to assign weights to input sites, so that the weight reflects the "importance" of the site in some way. By choosing random subsets where the probability of choosing a site is proportional to its weight, we can divide into subproblems with low weight, and hence importance.

In early applications, Welzl applied this approach to building range query data structures, and Littlestone applied it to learning.[82,48]

We'll always use integral weights; in fact, the weights will be powers of 2. The weight of site $s$ will be denoted by $w(s)$, and the total weight of a set of sites $Q$ will be $w(Q) = \sum_{s\in Q} w(s)$. To take a random subset $R$ using the weights, for each site $s$ we conceptually make ordered pairs $(s,1),(s,2)\dots,(s,w(s))$, yielding a resulting (unweighted) set $S'$. Now take a random subset $R'$ of $S'$ of size $r$, and then project: put in $R$ all sites $s$ with $(s,i) \in R'$, for some $i$. With this scheme, weighted versions of the divide-and-conquer theorems can be shown to follow.

### *7.1 Linear programming*

In this section we'll look at an algorithm for the following problem:

Given a set $S$ of $n$ points in $d$ dimensions, and a line $l$, find the lower endpoint of $l \cap \mathcal{C}(S)$, the intersection of $l$ with the convex hull of $S$.

Here "lower" means having the smallest $x_1$ coordinate. Perhaps surprisingly, this problem is linear programming, from the "column space geometry" point of view. (In a review article, Stone and Tovey look at simplex and interior point methods from this perspective.[78])

The algorithm will find the facet $T^*(S)$ of $\mathcal{C}(S)$ that meets $l$ at its lower endpoint. The algorithm is interesting only when $n \gg d$. Here we can view $d$ as a constant, and try to minimize the dependence of the running time on $n$.

The basis of the algorithm is the following simple observation:

> If $R \subset S$, the set $V(R)$ of points of $S \setminus R$ that see $T^*(R)$ must include at least one vertex of $T^*(S)$, unless $T^*(R) = T^*(S)$.

That is, at least one vertex of $T^*(S)$ must be in the empty halfspace of $T^*(R)$, below it. Otherwise, all vertices of $T^*(S)$ would be contained in the supporting halfspace $H$ of $T^*(R)$, above it, and so would their convex hull, which is $T^*(S)$. This would imply that $l \cap T^*(S)$ is above $T^*(R)$, a contradiction.

To use this observation, we give all elements of $S$ weights, that are all one at first. To mark the importance of $V(R)$, or at least one member of it, we double the weights of its elements when $w(V(R)) < 2dw(S)/r$. Here $r = 4d^2$ is the size of $R$. Naturally, we'll make $R$ a random subset of $S$, using the weighted sampling scheme discussed above; this implies that the expected weight $w(V(R))$ is no more than $dw(S)/r$, as shown below.

In short, the main iteration of the algorithm is: take sample $R$, compute $T^*(R)$, and double the weights of elements of $V(R)$ when $w(V(R)) < 2dw(S)/r$. This is iterated until the current set $V(R)$ is empty, and we have $T^*(S)$. This is the entire algorithm, but for the problem of computing $T^*(R)$. Since $r = 4d^2$ and we're assuming $n \gg d$, we can use any algorithm we like, including computing the entire convex hull of $R$. Let $L(n, d)$ denote the time taken by this "brute force" algorithm to solve the problem for $n$ sites in set $S$ in $d$ dimensions. From the upper bound theorem, we have $L(r, d) = L(4d^2, d)$ proportional to $\binom{4d^2}{d} < d^{O(d)}$.

**Theorem 7.1** *The algorithm terminates in expected $O(d \log n)$ iterations, and so has expected running time $O(d^2 n \log n)$.*

*Proof.* We'll show that in the unweighted case the expected size of $V(R)$ is $d(n - r)/(r + 1)$. A similar bound $d(w(S) - r)/(r + 1)$ holds for the weighted case, though that won't be proven here.

The proof is analogous to the last paragraph of the proof of Theorem 5.1. It is enough to show that a random element of $S \setminus R$ sees $T^*(R)$, and so is in $V(R)$, with probability $d/(r + 1)$. If $s \in S \setminus R$ sees $T^*(R)$, then $T^*(R') \neq T^*(R)$, where $R' = R \cup \{s\}$. This occurs only if $s$ is a vertex of $T^*(R')$. Since $s$ is a random element of $R'$, this occurs with probability $d/(r + 1)$, as desired.

How many iterations will the algorithm need to "converge" to $T^*(S)$? First we bound the expected number of iterations, versus the number of "successful" ones when $w(V(R)) < 2dw(S)/r$ and the weights of elements of $V(R)$ are doubled. Since the expected weight $Ew(V(R)) < dw(S)/r$, the probability that $V(R)$ is doubled is at least 1/2 for each iteration by Markov's inequality, and so the expected number of iterations is twice the number of successful ones.

After a successful iteration, the weight $w(S)$ increases by a factor no more than $1 + 2d/r = 1 + 1/2d$, and so is no more than $(1 + 1/2d)^k < e^{k/2d}$ after $k$ successful iterations. On the other hand, by the critical observation above, some vertex of $T^*(S)$ is in $V(R)$, and so has its multiplicity doubled after a successful iteration, so after $k$ successful iterations, the total weight $w(T^*(S))$ is $\sum_{1 \le j \le d} 2^{i_j}$, for some integers $i_j \ge 1$ with $\sum_{1 \le j \le d} i_j = k$. The sum is minimized when all $i_j = k/d$, so $w(T^*(S)) > d2^{k/d}$. Since $w(S)/w(T^*(S)) > 1$, the number $k$ of successful iterations must satisfy $(ne^{k/2d})/(d2^{k/d}) > 1$, which implies $k = O(d \log n)$. □

This algorithm is due to Clarkson;[22] an earlier, recursive, algorithm requiring $O(n)$ expected time can be combined with it, and the approach can be applied to integer programming. Adler and Shamir have shown that these algorithms can be applied to general convex programming.[1] Chazelle has noted that the recursive algorithm can be derandomized, giving a deterministic algorithm requiring $d^{O(d)}n$ time.[14] The first $O(n)$-time algorithms for $d = 3$ are due to Megiddo, and to Dyer.[53,33]

Seidel gave a different randomized algorithm,[73] requiring $O(d!n)$ expected time; recently Sharir and Welzl found a variant of Seidel's algorithm requiring time subexponential in $d$. Their algorithm is a randomized instance of the simplex algorithm.[81] Kalai was the first to find a subexponential simplex algorithm.[43] Problem instances have long been known for which versions of the simplex algorithm require at least $2^d$ operations.[45] These results cast new light on the complexity of the simplex algorithm, and on the possibility that linear programming problems can be solved in "strongly polynomial" time, with $(nd)^{O(1)}$ operations, and with the number of operations independent of the size of the numbers specifying a problem instance. (For more on these issues, see for example Schrijver's textbook.[70])

### *7.2 Triangular partitions*

This subsection gives a data structure for answering halfplane queries: given a set $S$ of $n$ points, the resulting data structure requires $O(n)$ space; given a halfplane $H$, the data structure can be used to find the size of $S \cap H$ in $\sqrt{n}(\log n)^{O(1)}$ time. The data structure uses a *triangular partition* of $S$: a collection $\Pi$ of pairs $(S_i, t_i)$, with $1 \le i \le m$, where $t_i$ is a triangular region and $S_i \subset S \cap t_i$. Also $|S_i| < 2n/m$ and the $S_i$ form a partition of $S$. That is, $\Pi$ splits $S$ into roughly equally sized pieces, and each piece is contained in a triangle. An important property of such a partition is its *crossing number*: the maximum, over all lines, of the number of triangles of $\Pi$ cut by the line. Matoušek, extending ideas of Welzl, proved the following remarkable fact:[50]

**Theorem 7.2** *For any parameter $r \le n$ and set of points $S$, there is a triangular partition $\Pi$ of $S$ of size $m = O(r)$ and with crossing number $O(\sqrt{r})$.*

To apply this to halfplane queries, build a *partition tree* by recording the size of $S_i$ for each $t_i$, and recursively applying the construction to the sets $S_i$. The total size of the resulting tree is clearly $O(n)$. To answer a query $H$, check for each $i$ if

$t_i \subset H$; if so, add $|S_i|$ to the answer. If $t_i$ is crossed by the boundary line of $H$, recursively find $|S_i \cap H|$ and add that quantity to the answer. By choosing $r = \sqrt{n}$, the query time is $\sqrt{n}(\log n)^{O(1)}$.

The proof of the theorem will need two lemmas. The first lemma says that for the purpose of building a triangular partition, we can use a small *test set* of lines to check the crossing number property of a partition.

**Lemma 7.3** *Given a set $S$ of $n$ points in the plane, there is a set $Q$ of lines such that for every triangular partition $\Pi$ of $S$ satisfying $|S_i| \geq n/r$ for all $i$, the following holds: if $\kappa_Q$ is the crossing number of $\Pi$ relative to $Q$, the crossing number of $\Pi$ is no more than $3\kappa_Q + O(\sqrt{r})$.*

*Proof.* [sketch] Construct a cutting (§6.2) of the dual lines of the points of $S$. A suitable set $Q$ is the dual of the vertices of the triangles of that cutting. We omit the proof that $Q$ satisfies the conditions of the lemma. □

The next lemma says that $S$ has a large subset with a triangular partition.

**Lemma 7.4** *Given a set $S$ of $n$ points and parameter $r \leq n$, there is a subset $W$ of $S$ at least half the size of $S$, so that $W$ has a triangular partition $\Pi$ with sets $W_i$ of size $\lfloor n/r \rfloor$ for all $i$, and with crossing number $O(\sqrt{r})$.*

Given this lemma, a triangular partition for the whole set $S$, as promised by Theorem 7.2, is built as follows: apply the lemma to $S^0 = S$ with parameter $r_0 = r$, yielding a triangular partition for subset $W^0$. Then apply the lemma to $S^1 = S \setminus W^0$ with parameter $r_1 = r/2$, and so on $j = O(\log r)$ times, until $S^j$ has fewer than $n/r$ points. A single triangle containing $S^j$, and all the triangles built along the way, completes the construction.

*Proof.* [of Lemma 7.4.] The partition is built triangle-by-triangle. To make the first pair $(S_1, t_1)$, build a $1/t$-cutting of the test set $Q$ for $S$, where $t$ is chosen so that the cutting has no more than $r/2$ triangles; by Theorem 6.4, a value of $t = \Theta(\sqrt{r})$ will suffice. Since $S$ has $n$ points, some triangle of the cutting contains at least $\lfloor n/r \rfloor$ points. Shrink that triangle until it contains exactly that many points; those points and the triangle are $S_1$ and $t_1$ for the cutting.

So far, so good: each line of the test set cuts only one triangle, because there *is* only one triangle. The application of reweighting here is to make sure that the lines cutting that triangle are less likely to cut future triangles. To do this, we double the weight of the lines meeting $t_1$, and repeat the above procedure with $S \setminus S_1$. The fact that, by construction, no more than $O(\sqrt{r})$ lines of $Q$ cut $t_1$, means that we are not greatly altering the conditions of the next cutting for the lines that do not meet $t_1$.

In the general step, we have pairs $(S_1, t_1), \ldots, (S_{i-1}, t_{i-1})$. If the total of the size of these $S_j$'s is more than $n/2$, we're done. Otherwise, we want to find $(S_i, t_i)$.

Each line $l \in Q$ has a weight $w_i(l) = 2^{\kappa_{i-1}(l)}$, where $\kappa_{i-1}(l)$ is the number of triangles $t_1, \ldots, t_{i-1}$ crossed by $l$. Using a variant of the cutting construction for these weighted lines, we get a $(1/t)$-cutting for $Q$ with these weights, again so that the cutting has $r/2$ triangles, and some triangle $t$ contains at least $\lfloor n/r \rfloor$ points of $S \setminus \cup_{1 \leq j < i} S_j$. We shrink that triangle until it contains exactly $\lfloor n/r \rfloor$ points, and call the result $t_i$. The weights of the lines cutting $t_i$ are doubled, and the construction of $(S_i, t_i)$ is complete.

The triangle $t_i$ is cut by lines of weight no more than $w_i(Q)/t$. Therefore doubling makes $w_{i+1}(Q)$ no more than $(1+1/t)w_i(Q)$.

When the construction is done we have $w_{m+1}(Q) = \sum_{l \in Q} 2^{\kappa(l)}$, where $\kappa(l)$ is the crossing number of $l$ for the triangular partition. By construction, $w_{m+1}(Q) \leq (1+1/t)^m w_1(Q) < e^{m/t} r$. Hence for any $l \in Q$, $2^{\kappa(l)} < e^{m/t} r$, or $\kappa(l) < m/t + \lg r$; since $m < 2r$ and $t = \Theta(\sqrt{r})$, we have $\kappa(l) = O(\sqrt{r})$, and by Lemma 7.3, this bound extends to all lines. □

The results here follow Matoušek,[49] who gave a construction that was applicable in higher dimensions, and which he made deterministic.

## 8 Closing Remarks

The reader may now be convinced of the wide-ranging use of randomization in computational geometry. If not, here is a list (not exhaustive, with apologies) of some other application areas: parallel algorithms;[5,66,41,26] geometric minimum spanning trees;[3] ray shooting;[2,64] line orientation queries;[17] levels in arrangements and high-order Voronoi diagrams;[57,21] the planar traveling salesman problem;[25] and binary space partition trees.[63]

## Appendix A. Some probability, geometry, and notation

For $A$ and $B$ finite sets, let the set difference $A \setminus B = \{a \in A \mid a \notin B\}$, and the symmetric difference $A \oplus B = (A \setminus B) \cup (B \setminus A)$. The size of a set $A$ is denoted $|A|$.

**Expectations.** Let $X$ be a random variable taking a finite set $A$ of real values. The expectation of $X$ is $EX \equiv \sum_{x \in A} x \text{Prob}\{X = x\}$. If $X$ and $Y$ have expectations $EX$ and $EY$, then the expectation of the random variable $X + Y$ is $EX + EY$. If $X$ takes on only nonnegative values, then *Markov's inequality* is that $\text{Prob}\{X > kEX\}$, the probability that $X$ is greater than $kEX$, is no more than $1/k$, since

$$k\text{Prob}\{X > kEX\} = k \sum_{x \in A, x > kEX} \text{Prob}\{X = x\} < \sum_{x \in A, x > kEX} x\text{Prob}\{X = x\} < EX.$$

**Random permutations.** We need some simple facts about random permutations. We can label a set $S$ with $n$ elements by the numbers 1 through $n$, and so consider here only permutations of the set $\mathbf{n} = \{1, 2, \ldots, n\}$. If $x = (x_1, x_2, \ldots, x_n)$ is a random permutation of $\mathbf{n}$, and $y = (y_1, y_2, \ldots, y_n)$ is a fixed permutation of $\mathbf{n}$, then the product $xy = (y_{x_1}, y_{x_2}, \ldots, y_{x_n})$ is also a random permutation: exactly one $x$ yields a given product $z = xy$, so the probability of $x$ and of $z$ is $1/n!$.

If $x$ is a random permutation, then the set $R_j(x) = \{x_1, x_2 \ldots, x_j\}$ is a random subset of $\mathbf{n}$. To see this, suppose $E$ and $F$ are two subsets of $\mathbf{n}$ of size $j$. Then there is a permutation $y = (y_1, y_2, \ldots, y_n)$ such that $y_i \in F$ if and only if $i \in E$. Thus a permutation $x$ has $R_j(x) = E$ if and only if $R_j(xy) = F$. Thus all subsets of n of size $j$ are equally likely.

Similar arguments show that $x' = (x_1, x_2 \ldots, x_j)$ is a random permutation of the set $R_j(x)$, and that $x_j$ is a random element of $R_j(x)$ and $\{x_j, x_{j+1}, \ldots, x_n\}$.

**Projective duality.** A duality mapping is a one-to-one correspondence between points and lines, so that, among other things, incidence is preserved: a point $p$ is on line $l$ if and only if its dual point is on the dual line of $p$. One important form of duality is *projective*: let $(w, x, y)$, with $w \geq 0$, be a representation of points in the plane by *homogeneous coordinates*, so that $(w, x, y)$ and $(w', x', y')$ represent the same point in $E^2$ if and only if $w'x = wx'$ and $w'y = wy'$. (No point has the representation $(0, 0, 0)$.) For any lines $l$, there is a triple $a = (a_w, a_x, a_y)$ so that $l$ comprises all points whose homogeneous coordinates $p = (w, x, y)$ satisfy $a \cdot p \equiv a_w w + a_x x + a_y y = 0$. The dual point $\mathcal{D}(l)$ of $l$ is has representation $\mathcal{D}(l) = a = (a_w, a_x, a_y)$, and the dual line $\mathcal{D}(p)$ of a point $p = (w, x, y)$ is the line comprising all points $(w', x', y')$ satisfying $(w, x, y) \cdot (w', x', y') = 0$. Thus point $p$ with homogeneous coordinates $(w, x, y)$ is on $l$ if and only if $\mathcal{D}(l)$ is on $\mathcal{D}(p)$, since both conditions are equivalent to $a \cdot p = 0$. Moreover, the mapping $\mathcal{D}()$ is *orientation preserving*: $p$ is on the same side of $l$ as the origin if and only $\mathcal{D}(l) = (a_w, a_x, a_y)$ is on the same side of $\mathcal{D}(p)$ as the origin, since both conditions are equivalent to $a \cdot p \geq 0$.

## Appendix B

This paper repeatedly uses analogies among different geometric structures; this appendix explicitly gives the roles played by the elements of each structure.

In the following list, $S$ is a set of *sites* and $\mathcal{F}_S$ is a set of regions "defined" by those sites. (Each region in $\mathcal{F}_S$ is defined by $b$ or fewer elements of $S$, where $b$ is the specification complexity of the particular problem, as discussed in §2.2. Hence we have $\mathcal{F}_S$ in one-to-one correspondence with a multiset whose elements are subsets of $S$ of size $b$ or less.)

In each domain, there is a *conflict* relation between the sites and the regions, and for a subset $R$ of $S$ we are generally interested in the set $\mathcal{F}_R^0$ of all regions $F \in \mathcal{F}_S$ such that the sites defining $F$ are in $R$, but the sites conflicting with it are not.

(The reader may now wish to look at the list below for several instances of these general notions.)

Sometimes, as in §5.2, we may be interested in the set of regions $F_R^k \subset \mathcal{F}_S$, the set of all regions $F \in \mathcal{F}_S$ such that the sites defining $F$ are in $R$ and exactly $k$ sites conflicting with it are not. When we use a random subset $R$ for divide-and-conquer, the regions $\mathcal{F}_R^0$ each give a subproblem, and the subproblem size is $k$ for a region $F \in \mathcal{F}_S^k$.

For each of these geometric structures, an associated range space has the set of sites $S$ as the underlying set, and for each region $F \in \mathcal{F}_S$, there is a range $S_F$ comprising all sites conflicting with $F$. If $|S_F| = k$ then $F \in \mathcal{F}_S^k$. Lemma 2.2 can be applied to this range space to show the existence of $\epsilon$-transversals.

We give below for each geometric structure the main sections discussing it, the name and description of the sites $S$, the regions $\mathcal{F}_S$, the conflict relation between sites and regions and its name, and the name of the set $\mathcal{F}_R^0$.

- Voronoi diagrams in the plane, closest point problems; §2.1.

  $S$: Points in the plane $E^2$.

  $\mathcal{F}_S$: The set of open disks bounded by circles inscribed on 3 sites.

  **conflict:** $s \in S$ and $D \in \mathcal{F}_S$ have $s \in D$.

  $\mathcal{F}_R^0$: The Delaunay disks of $R$.

- Trapezoidal diagrams; §3.1,§3.3.

  $S$: Line segments or lines in the plane.

  $\mathcal{F}_S$: Open cells of every diagram $\mathcal{T}(A)$ where $A \subset S$ with $|A| \leq 4$.

  **conflict:** $s \in S$ and $T \in \mathcal{F}_S$ have nonempty intersection (they meet).

  $\mathcal{F}_R^0$: Cells of the trapezoidal diagram of $R$.

- The cell of the trapezoidal diagram containing a given point $p$; §5.1.

  $S$: Lines in the plane.

  $\mathcal{F}_S$: Open cells of every diagram $\mathcal{T}(A)$ where $A \subset S$ with $|A| \leq 4$, such that the cell $T$ has $p \in T$.

  **conflict:** $s \in S$ and $T \in \mathcal{F}_S$ have nonempty intersection (they meet).

  $\mathcal{F}_R^0$: The cell of $\mathcal{T}(R)$ containing $p$.

- Convex hulls; §4.1.

  $S$: Points in $E^d$.

  $\mathcal{F}_S$: Open halfspaces bounded by planes that contain $d$ sites.

  **conflict:** $s \in S$ and $H \in \mathcal{F}_S$ have $s \in H$. The site $s$ sees the facet with empty halfspace $H$.

  $\mathcal{F}_R^0$: Empty halfspaces of $R$.

- Linear programming; §7.1.

  $S$: Points in $E^d$.

  $\mathcal{F}_S$: Open halfspaces bounded by planes that contain $d$ sites, such that the convex hull of the $d$ sites meets $l$.

  **conflict:** $s \in S$ and $H \in \mathcal{F}_S$ have $s \in H$.

  $\mathcal{F}_R^0$: The empty halfspace of the facet $T^*(R)$ that meets $l$. (The lower of the two such facets.)

**Acknowledgements**

The author is very grateful to Jon Bentley, Steve Fortune, John Hobby, and Margaret Wright for their vigorous yet tactful comments. Thanks also to Jon for suggesting Appendix B, and John for MetaPost help.

**References**

1. I. Adler and R. Shamir. A randomization scheme for speeding up algorithms for linear and convex quadratic programming problems with a high constraints-to-variables ratio. Technical Report 21-90, Rutgers Univ., May 1990. To appear in *Math. Programming.*

2. P. K. Agarwal. *Intersection and Decomposition Algorithms for Planar Arrangements.* Cambridge University Press, New York, 1991.

3. P. K. Agarwal, H. Edelsbrunner, O. Schwarzkopf, and E. Welzl. Euclidean minimum spanning trees and bichromatic closest pairs. *Discrete and Computational Geometry*, pages 407–422, 1991.

4. N. Alon and E. Győri. The number of small semispaces of a finite set of points in the plane. *J. Combin. Theory Ser. A*, 41:154–157, 1986.

5. N. Alon and N. Megiddo. Parallel linear programming in fixed dimension almost surely in constant time. In *Proc. 31st IEEE Symp. on Foundations of Computer Science*, pages 574–582, 1990.

6. B. Awerbuch, A. Bar-Noy, N. Linial, and D. Peleg. Compact distributed data structures for adaptive routing. In *Proc. 21st Annual SIGACT Symp.*, pages 479–489, 1989.

7. J. R. Bergen and H. Shvaytser. A probabilistic algorithm for computing Hough transforms. *J. Algorithms*, 4:639–656, 1991.

8. A. Blumer, A. Ehrenfeucht, D. Haussler, and M. Warmuth. Learnability and the Vapnik-Chervonenkis dimension. *Journal of the ACM*, pages 929–965, 1989.

9. J.D. Boissonnat, O. Devillers, R. Schott, M. Teillaud, and M. Yvinec. Applications of random sampling to on-line algorithms in computational geometry. *Discrete and Computational Geometry*, to appear. Available as Technical Report INRIA 1285. Abstract published in IMACS 91, Dublin.

10. J.D. Boissonnat and M. Teillaud. On the randomized construction of the Delaunay tree. *Theoretical Computer Science*, to appear. Available as Technical Report INRIA 1140.

11. B. Bollabás. *Combinatorics: Set Systems, Hypergraphs, Families of vectors, and Combinatorial Probability.* Cambridge University Press, New York, 1986.

12. A. Brøndsted. *An Introduction to Convex Polytopes.* Springer-Verlag, Berlin, 1983.

13. R. J. Canham. A theorem on arrangements of lines in the plane. *Israel J. Math.*, 7:393–397, 1969.

14. B. Chazelle. An optimal convex hull algorithms and new results on cuttings. In *Proc. 32nd IEEE Symp. on Foundations of Computer Science*, pages 29–38, 1991.

15. B. Chazelle. Triangulating a simple polygon in linear time. *Discrete and Computational Geometry*, pages 485–524, 1991.

16. B. Chazelle and H. Edelsbrunner. An optimal algorithm for intersecting line segments in the plane. In *Proc. 29th IEEE Symp. on Foundations of Computer Science*, pages 590–600, 1988.

17. B. Chazelle, H. Edelsbrunner, L. Guibas, and M. Sharir. Lines in space: Combinatorics, algorithms and applications. In *Proc. 21st Annual SIGACT Symp.*, pages 382–393, 1989.

18. B. Chazelle and J. Friedman. A deterministic view of randomization and its use in geometry. *Combinatorica*, 10:229–249, 1990.

19. B. Chazelle and E. Welzl. Quasi-optimal range searching in spaces of finite VC-dimension. *Discrete and Computational Geometry*, 4, 1989.

20. L. P. Chew. Building Voronoi diagrams of convex polygons in linear expected time. Unpublished manuscript.

21. K. L. Clarkson. New applications of random sampling in computational geometry. *Discrete and Computational Geometry*, 2:195–222, 1987.

22. K. L. Clarkson. A Las Vegas algorithm for linear programming when the dimension is small. In *Proc. 29th IEEE Symp. on Foundations of Computer Science*, pages 452–456, 1988. Revised version: Las Vegas algorithms for linear and integer programming when the dimension is small (preprint).

23. K. L. Clarkson. Random sampling in computational geometry, II. *Proc. Fourth ACM Symp. on Comp. Geometry*, pages 1–11, 1988.

24. K. L. Clarkson. A randomized algorithm for closest-point queries. *SIAM Journal on Computing*, 17:830–847, 1988.

25. K. L. Clarkson. Approximation algorithms for planar traveling salesman tours and minimum-length triangulations. In *Proc. 2nd ACM-SIAM Symp. Discrete Algorithms*, pages 17–23, 1991.

26. K. L. Clarkson, R. Cole, and R. E. Tarjan. Randomized parallel algorithms for trapezoidal diagrams. In *Proc. Seventh ACM Symp. on Comp. Geometry*, pages 152–161, 1991.

27. K. L. Clarkson, H. Edelsbrunner, L. Guibas, M. Sharir, and E. Welzl. Combinatorial complexity bounds for arrangements of curves and surfaces. *Discrete and Computational Geometry*, 5:99–160, 1990.

28. K. L. Clarkson, K. Mehlhorn, and R. Seidel. Four results on randomized incremental constructions. In *Proceedings of the Symposium on Theoretical Aspects of Computer Science*, 1992. To appear.

29. K. L. Clarkson and P. W. Shor. Applications of random sampling in computational geometry, II. *Discrete and Computational Geometry*, pages 387–421, 1989.

30. K. L. Clarkson, R. E. Tarjan, and C. J. Van Wyk. A fast Las Vegas algorithm for triangulating a simple polygon. *Proc. Fourth ACM Symp. on Comp. Geometry*, 1988.

31. O. Devillers. Randomization yields simple $O(n \log^\star n)$ algorithms for difficult $\omega(n)$ problems. *International Journal on Computational Geometry and Aplications*, to appear. Full paper available as Technical Report INRIA 1412. Abstract published in the Third Canadian Conference on Computational Geometry 1991 in Vancouver.

32. O. Devillers, S. Meiser, and M. Teillaud. Fully dynamic Delaunay triangulation in logarithmic expected time per operation. In *WADS 91*, volume LNCS 519. Springer Verlag, 1991. Full version available as Technical Report INRIA 1349.

33. M. E. Dyer. Linear time algorithms for two- and three-variable linear programs. *SIAM Journal on Computing*, 13:31–45, 1984.

34. H. Edelsbrunner. *Algorithms in Combinatorial Geometry*. Springer-Verlag, New York, 1987.

35. H. Edelsbrunner, L. Guibas, J. Hershberger, R. Seidel, M. Sharir, J. Snoeyink, and E. Welzl. Implicitly representing arrangements of lines or segments. *Discrete and Computational Geometry*, pages 433–466, 1989.

36. H. Edelsbrunner, L. Guibas, and M. Sharir. The complexity and construction of many faces in arrangements of lines and of segments. *Discrete and Computational Geometry*, pages 161–196, 1990.

37. H. Edelsbrunner, L. Guibas, and M. Sharir. The complexity of many cells in arrangements of planes and related problems. *Discrete and Computational Geometry*, pages 197–216, 1990.

38. P. Erdős and J. Spencer. *Probabilistic Methods in Combinatorics.* Academic Press, New York, 1974.

39. S. J. Fortune. Voronoi diagrams and Delaunay triangulations. *This volume.*

40. L.J. Guibas, D.E. Knuth, and M. Sharir. Randomized incremental construction of Delaunay and Voronoi diagrams. *Proc. of ICALP*, pages 414 – 431, 1990. To appear in *Algorithmica.*

41. T. Hagerup, H. Jung, and E. Welzl. Efficient parallel computation of arrangements of hyperplanes in $d$ dimensions. In *Proceedings of the Second Symposium on Parallel Algorithms and Architectures*, pages 290–297, 1990.

42. D. Haussler and E. Welzl. Epsilon-nets and simplex range queries. *Discrete and Computational Geometry*, 2:127–151, 1987.

43. G. Kalai. A subexponential randomized simplex algorithm. Extended abstract, 1991.

44. N. Kiryati, Y. Eldar, and A. M. Bruckstein. A probabilistic Hough transform. *Pattern Recognition*, 24, 1991.

45. V. Klee and G. J. Minty. How good is the simplex algorithm? In *Inequalities III*, pages 159–175. Academic Press, 1972.

46. R. Klein. Abstract Voronoi diagrams and their applications. In *Computational Geometry and its Applications (CG '89)*, pages 148–157. LNCS 344, 1988.

47. J. Komlós. Manuscript, 1990.

48. N. Littlestone. Learning quickly when irrelevant attributes abound: a new linear-threshold algorithm. In *Proc. 28th IEEE Symp. on Foundations of Computer Science*, pages 68–77, 1987.

49. J. Matoušek. Cutting hyperplane arrangements. *Discrete and Computational Geometry*, pages 385–406, 1991.

50. J. Matoušek. Efficient partition trees. *Proc. Seventh ACM Symp. on Comp. Geometry*, pages 1–9, 1991.

51. J. Matoušek. Reporting points in halfspaces. In *Proc. 32nd IEEE Symp. on Foundations of Computer Science*, pages 207–215, 1991.

52. P. McMullen. The maximum number of faces of a convex polytope. *Mathematika*, 17:179–184, 1970.

53. N. Megiddo. Linear programming in linear time when the dimension is fixed. *Journal of the ACM*, 31:114–127, 1984.

54. K. Mehlhorn, St. Meiser, and C. Ò'Dunlaing. On the construction of abstract Voronoi diagrams. *Discrete and Computational Geometry*, 6:211 - 224, 1991.

55. K. Mulmuley. A fast planar point location algorithm: part I. In *Proc. 29th IEEE Symp. on Foundations of Computer Science*, pages 580–589, 1988. Full version to appear, *J. Symb. Logic.*

56. K. Mulmuley. On obstructions in relation to a fixed viewpoint. In *Proc. 30th IEEE Symp. on Foundations of Computer Science*, pages 592–597, 1989.

57. K. Mulmuley. On levels in arrangements and Voronoi diagrams. *Discrete and Computational Geometry*, 6:307–338, 1990.

58. K. Mulmuley. A fast planar point location algorithm, II. *Journal of the ACM*, 38:74–103, 1991.

59. K. Mulmuley. Randomized multidimensional search trees: Dynamic sampling. In *Proc. Seventh ACM Symp. on Comp. Geometry*, 1991.

60. K. Mulmuley. Randomized multidimensional search trees: Further results in dynamic sampling. In *Proc. 32nd IEEE Symp. on Foundations of Computer Science*, pages 216 - 227, 1991.

61. K. Mulmuley. Randomized multidimensional search trees: Lazy balancing and dynamic shuffling. In *Proc. 32nd IEEE Symp. on Foundations of Computer Science*, pages 180 - 194, 1991.

62. J. Pach and G. Wöginger. Some new bounds for $\epsilon$-nets. In *Proc. Sixth ACM Symp. on Comp. Geometry*, pages 10–15, 1990.

63. M. S. Paterson and F. F. Yao. Efficient binary space partitions for hidden-surface removal and solid modeling. *Discrete and Computational Geometry*, pages 485–504, 1990.

64. M. Pellegrini. Stabbing and ray shooting in 3 dimensional space. In *Proc. Seventh ACM Symp. on Comp. Geometry*, pages 177–186, 1990.

65. J. Reif and S. Sen. Optimal parallel algorithms for computational geometry. In *Proc. 16th International Conference on Parallel Processing*, 1987.

66. J. Reif and S. Sen. Polling: A new randomized sampling technique for computational geometry. In *Proc. 21st Annual SIGACT Symp.*, 1989.

67. R. Reischuk. A fast probabilistic parallel sorting algorithm. In *Proc. 22nd IEEE Symp. on Foundations of Computer Science*, pages 212–219, 1981.

68. L. A. Santalo. *Integral Geometry and Geometric Probability*. Addison-Wesley, Reading, 1976.

69. N. Sauer. On the density of families of sets. *J. Combinatorial Theory, (A)*, 13, 1972.

70. A. Schrijver. *Theory of Linear and Integer Programming.* Wiley, New York, 1986.

71. O. Schwarzkopf. Dynamic maintenance of geometric structures made easy. *Proc. 32nd IEEE Symp. on Foundations of Computer Science*, pages 197 - 206, 1991.

72. R. Seidel. A simple and fast incremental randomized algorithm for computing trapezoidal decompositions and for triangulating polygons. *Computational Geometry: Theory and Applications*, 1:51 - 64, 1991.

73. R. Seidel. Small-dimensional linear programming and convex hulls made easy. *Discrete and Computational Geometry*, pages 423–433, 1991.

74. M. Sharir. The $k$-set problem for arrangements of curves and surfaces. *Discrete and Computational Geometry*, to appear.

75. S. Shelah. A combinatorial problem; stability and order for models and theories of infinitary languages. *Pacific J. Math.*, 41, 1972.

76. J. Spencer. Puncture sets. *J. Combinatorial Theory A*, 17:329–336, 1974.

77. J. Spencer. *Ten Lectures on the Probabilistic Method.* SIAM, Philadelphia, 1987.

78. R. E. Stone and C. A. Tovey. The simplex and projective scaling methods as iteratively reweighted least squares methods. *SIAM Review*, 33:220–237, 1991.

79. V. N. Vapnik. *Estimation of Dependencies Based on Empirical Data.* Springer-Verlag, New York, 1982.

80. V. N. Vapnik and A. Y. A. Chervonenkis. On the uniform convergence of relative frequencies of events to their probabilities. *Th. Prob. and its Appl.*, 16, 1971.

81. E. Welzl. Personal communication.

82. E. Welzl. Partition trees for triangle counting and other range searching problems. In *Proc. Fourth ACM Symp. on Comp. Geometry*, pages 23–33, 1988.

83. L. Xu, E. Oja, and P. Kultanen. A new curve detection method: Randomized Hough transform (RHT). *Pattern Recognition Lett.*, 5, 1990.

*Computing in Euclidean Geometry*, pp. 163–191
Eds. D.-Z. Du and F. K. Hwang


# THE STATE OF ART ON STEINER RATIO PROBLEMS

D.-Z. DU*
*Department of Computer Science, Princeton University*
*Princeton, NJ, USA*

and

F. K. HWANG
*Bell Laboratories, AT&T, Murray Hill, New Jersey 07974, USA*

ABSTRACT

We give a more complete account of two recent important developments on Steiner ratio problems. The first is Du and Hwang's proof of the Gilbert-Pollak conjecture that the Steiner ratio is $\sqrt{3}/2$ for points in the Euclidean plane. The second is Zelikovsky's demonstration of a heuristic for points on a graph with a performance ratio better than the Steiner ratio, and its extension to the rectilinear plane by Berman and Ramaiyer, and its further extension to an arbitrary metric space by Du, Zhang and Feng.

## 1. Introduction

Let $P$ be a set of $n$ points in a metric space or a graph. A Steiner minimal tree on $P$ is a shortest network interconnecting $P$. Since the problem to find a Steiner minimal tree has been shown to be NP-hard for $P$ in Euclidean spaces, rectilinear spaces and graphs, and no polynomial algorithm is known for any space, it is important to study heuristics for Steiner minimal trees. For a heuristic $h$, its *performance ratio* $\rho_M(h)$ in the space $M$ is defined as

$$\rho_M(h) = \inf_{P \in M} \frac{L_s(P)}{L_h(P)},$$

where $L_s(P)$ and $L_h(P)$ are the lengths of a Steiner minimal tree and of the heuristic $h$ on a set $P$ in $M$.

A *minimum spanning tree* on $P$ is a shortest tree interconnecting $P$ with all edges between points of $P$. Since MSTs can be quickly constructed and their lengths are not exceedingly long, they have served as yardsticks for Steiner minimal tree heuristics. The performance ratios of minimal spanning trees, denoted simply by $\rho_M$, are also known as *Steiner ratios*.

There have been two outstanding problems with respect to Steiner ratios. One is to determine them, in particular, the Steiner ratio for the Euclidean plane

*Supported by DIMAS (Center for Discrete Mathematics and Theoretical Computer Science), a National Science Foundation Science and Technology Center - NSF-STC88-09648 and NSF-CCR89-20505. Also from Institute of Applied Mathematics, Chinese Academy of Sciences, and Supported in part by the National Natural Science Foundation of China.

was conjectured to be $\sqrt{3}/2$ by Gilbert and Pollak[18] and only recently proved by Du and Hwang.[13] The second problem is to find other heuristics with proven better performance ratios than the Steiner ratio. Such a heuristic will be called a *better heuristic*. The existence of a better heuristic was only recently shown by Zelikovsky[32] for the graph version, then extended to the rectilinear plane by Berman and Ramaiyer,[1] and finally extended to all metric spaces by Du, Zhang and Feng.[16]

Although the proof of Du and Hwang has appeared,[12,13] the format of journal articles has imposed some constraints on the organization and presentation of the material. On the other hand, the papers of Zelikovsky, Berman and Ramaiyer, and Du, Zhang and Feng are either conference papers or preprints and thus not easily accessible. The purpose of this review is to give a friendlier and more thorough recapitulation of these important results; hopefully, making them more transparent and insightful.

## 2. Background of the $\sqrt{3}/2$ Conjecture

There is a long history of mathematical research on shortest networks. Possibly, Fermat was the first person who studied these kinds of problems. Given three points on the Euclidean plane, Fermat raised the problem of finding a point such that the sum of three distances from this point to three given points achieves minimum. The solution of the Fermat problem is as shown in Fig. 1.

Figure 1. Solution of the Fermat problem.

When the given points $A, B$ and $C$ form a triangle with all inner angles less than $120°$, then the solution $S$ lies inside of triangle $ABC$ and satisfies the condition $\angle ASB = \angle BSC = \angle CSA = 120°$. If triangle $ABC$ has an angle not less than $120°$, say $\angle B \geq 120°$, then the solution is the point $B$.

There are two ways to generalize the Fermat problem. The first generalization is the problem of finding a point such that the sum of $n$ distances from the point to $n$ given points achieves minimum. This generalization is still called the *Fermat problem*. The second is the problem of finding a shortest network interconnecting $n$ given points on the euclidean plane. This is a generalization of the Fermat problem in the sense that the shortest network interconnecting three points is exactly the solution of the Fermat problem. The second generalization is called the *Steiner tree problem*. In fact, such a shortest network must be a tree, which is called a *Steiner minimum tree (SMT)*, for the given set of points.

An SMT for a point set $P$ may contain vertices not in $P$. Such vertices are called *Steiner points* while vertices in $P$ are called *regular points*. It is clear that

(1) all leaves are regular points.

From the properties of SMT for three points, we can also see that

(2) any two edges meet at an angle of at least 120°,

(3) every Steiner point has degree exactly three, and every angle at a Steiner point equals 120°.

In fact, if two edges, say $AB$ and $BC$, meet at an angle $\angle ABC$ of less than 120°, then by the solution of the Fermat problem, we see that $AB$ and $BC$ cannot form an SMT for $\{A, B, C\}$. Namely, we can shorten the total length of the network by replacing $AB$ and $BC$ with an SMT for $\{A, B, C\}$. This proves (2). (3) follows immediately from (2).

The Fermat problem and the Steiner tree problem coincide only for sets of three points. In general, their solutions are very different. We refer the reader to Kuhn[24] for an introduction to the Fermat problem. Who invented the Steiner tree problem? As far as we know Jarnik and Kössler[22] in 1934 raised the Steiner tree problem in a Czeckoslovakia journal. However, Courant and Robbins[7] in their famous 1941 book "What is Mathematics" referred to it as the Steiner problem. The popularity of their book was responsible for bringing the Steiner tree problem to people's attention. Two important papers in the 1960's further laid a solid groundwork for future study. Melzak[26] first gave a finite algorithm for the Steiner tree problem. Gilbert and Pollak[18] produced an excellent survey of the problem, raised many new topics including the Gilbert-Pollak conjecture, and extended the problem to other spaces.

A tree connecting the point set $P$ and satisfying (1), (2) and (3) is called a *Steiner tree (ST)*. Its topology (the graph structure of the network) is called a *Steiner topology*.

A leaf in an ST must be a regular point. However, the inverse is false. If in an ST every regular point is a leaf, then the ST is called a *full* ST. An ST for $n$ points can contain at most $n-2$ Steiner points. It contains exactly $n-2$ Steiner points if and only if it is a full ST. If a regular point is not a leaf, then we can split the ST at this regular point. In this way, the ST can be decomposed into edge-disjoint full sub-STs. Such full sub-STs are called *full components* of the ST. The topology of a full ST is also called a *full topology*.

Let $m$ denote a minimum spanning tree (MST). To get some feeling on the Steiner ratio, let us look at three points $A, B$ and $C$ forming an equilateral triangle with unit edge length. Clearly, $L_s(A, B, C) = \sqrt{3}$ and $L_m(A, B, C) = 2$. So, $L_s(A, B, C)/L_m(A, B, C) = \sqrt{3}/2$. By the definition of the Steiner ratio, $\rho \geq \sqrt{3}/2$.

Gilbert and Pollak[18] conjectured $\rho = \frac{\sqrt{3}}{2}$. That is, an equilateral triangle gives the worse case for using MST to approximate SMT. They verified the inequality for $n = 3$ in the following way (see Fig. 2).

Let $A, B$, and $C$ be three points. If the triangle $ABC$ has an inner angle not less than 120°, then $L_s(A,B,C) = L_m(A,B,C) > (\sqrt{3}/2)L_m(A,B,C)$. Otherwise, the SMT has a Steiner point $S$. Assume that $A$ is as close to $S$ as $B$ and $C$. Let $B'(C')$ be on line $(S,B)$ (line $(S,C)$) such that $\Delta AB'C'$ is equilateral.

$$\begin{aligned} L_s(A,B,C) &= d(B,B') + d(C,C') + d(A,S) + d(B',S) + d(C',S) \\ &= d(B,B') + d(C,C') + \frac{\sqrt{3}}{2}(d(A,B') + d(A,C')) \\ &\geq \frac{\sqrt{3}}{2}(d(A,B) + d(A,C)) \\ &\geq \frac{\sqrt{3}}{2} l_m(A,B,C). \end{aligned}$$

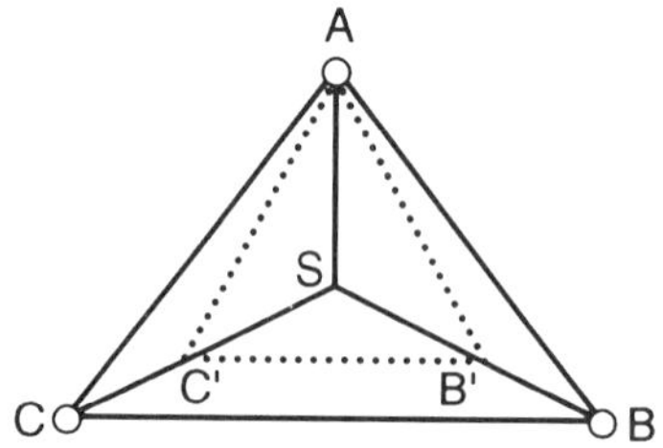


Figure 2. Gilbert and Pollak's proof for $n = 3$.

Gilbert-Pollak's conjecture was then verified by Pollak[27] for $n = 4$, by Du, Hwang and Yao[14] for $n = 5$ and by Rubinstein and Thomas[29] for $n = 6$. Along another line, the lower bound of $\rho$ for general $n$ has been pushed up from 0.5 (by Moore as reported in [18]) to 0.57 by Graham and Hwang,[19] to 0.74 by Chung and Hwang,[6] to 0.8 by Du and Hwang[11] and to 0.824 by Chung and Graham.[5] Finally, Du and Hwang[13] proved the conjecture by employing a new approach.

## 3. A Minimax Theorem

We introduce a minimax theorem in this section. This theorem plays an important role in the proof of Gilbert-Pollak's conjecture.

Let $f(x)$ be a concave function on an interval $[a,b]$. Then the minimum value of $f(x)$ is achieved at $a$ or $b$, one of the two endpoints of the interval.

Now, let $\{g_i(x)\}_{i\in I}$ be a family of finitely many continuous concave functions on an interval $[a,b]$ and let $f(x) = \max_{i\in I} g_i(x)$. Then $f(x)$ is piece-wise concave. The minimum value of $f(x)$ is achieved at $a$ or $b$ or some boundary points between the pieces on which $f(x)$ is concave.

In general, let $\{g_i(x)\}_{i\in I}$ be a family of finitely many continuous concave functions on a polytope $X$ and let $f(x) = \max_{\in I} g_i(x)$ on $X$. Define $X_i = \{x \in X \mid g_i(x) = f(x)\}$. Then in each $X_i$, $f(x)$ is concave. If every $X_i$ is a polytope, then we can conclude immediately that the minimum value of $f(x)$ on $X$ is achieved at some vertex of $X_i$. In general $X_i$ is not a polytope (see Fig. 3). We will employ a new method to prove a similar result.

A subset $Y$ of the polytope $X$ is called an *extreme set* of $X$ if $x, y \in X$ and $\lambda x + (1-\lambda)y \in Y$ for some $\lambda$ in the interval (0,1) imply $x, y \in Y$. For example, every vertex is an extreme set and the set $X$ itself is also an extreme set. For any $x \in X$, define $M(x) = \{i \mid x \in X_i\}$. A point $x$ in $X$ is called a *critical point* if there exists an extreme set $Y$ such that $x \in Y$ and that $y \in Y$ and $M(x) \subseteq M(y)$ imply $M(x) = M(y)$. In other words, a critical point is a point with maximal $M(x)$ in some extreme set $Y$. We can view critical points as vertices of $X_i$'s.

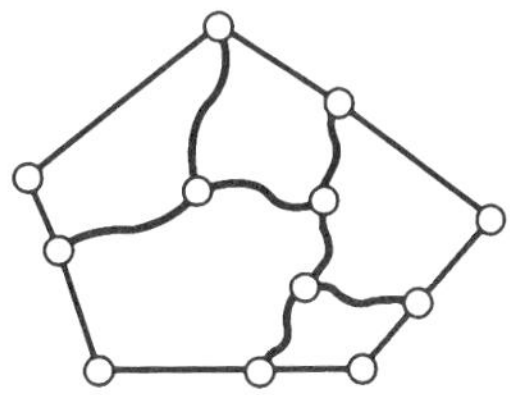

Figure 3. $X_i$ not a polygon.

Since $X_i$ may not be polytopes, the following theorem is nontrivial.

**Theorem 3.1** *Suppose that $f(x) = \max_{i \in I} g_i(x)$ where $I$ is finite and every $g_i(x)$ is a continuous, concave function in a polytope $X$. Then the minimum value of $f(x)$ over the polytope $X$ is achieved at some critical point.*

*Proof.* Suppose that $x^*$ is a point minimizing $f(x)$. Since all $g_i(x)$ are continuous, there is a neighborhood $V$ of $x^*$ such that for any $x \in V$, $M(x) \subseteq M(x^*)$. For $i \in M(x^*)$ define $A(i) = \{x \in X \mid g_i(x) \geq f(x^*)\}$. Then, each $A(i)$ is a convex set and $V$ is contained in the union of $A(i)$ for $i \in M(x^*)$. Let $Y$ be an extreme set such that $x^*$ is a *relative interior point* of $Y$, that is, for any $x \in Y$ and for sufficiently small positive number $\lambda$, $x^* + \lambda(x^* - x) \in Y$. Consider a critical point $\hat{x}$ in $Y$ such that $M(x^*) \subseteq M(\hat{x})$. We will show that $\hat{x}$ is also a point minimizing $f(x)$. Therefore, the theorem is proved.

For contradiction, suppose that $x$ is not a point minimizing $f(x)$. Then, for every $i \in M(x^*)$, $\hat{x}$ is a point of $A(i)$. Denote $x(\lambda) = x^* + \lambda(x^* - \hat{x})$. We claim that for any positive number $\lambda$, $x(\lambda)$ is not in $A(i)$ for every $i \in M(x^*)$. In fact, if the point $x(\lambda)$ for some positive $\lambda$ is in $A(i)$, then the point $x^*$ as an interior point of the segment $[\hat{x}, x(\lambda)]$ can be written as $x^* = c\hat{x} + (1-c)x(\lambda)$ where $0 < c = \lambda/(1+\lambda) < 1$. Thus, we have $g_i(x^*) \geq cg_i(\hat{x}) + (1-c)g_i(x(\lambda) > f(x^*)$, contradicting the fact that $i \in M(x^*)$. Since $x(\lambda)$ is in $Y$ for sufficiently small positive number $\lambda$, the fact that $x(\lambda)$ is not in any $A(i)$ for $i \in M(x^*)$ contradicts the fact that $V$ is covered by the union of $A(i)$'s.□

From the proof of the theorem we can also conclude the following.

**Corollary 3.2** *If $I(x) \subseteq I(y)$, then $f(x) \geq f(y)$.*

We now translate the Gilbert-Pollak's conjecture into a minimax problem as described above.

Given a Steiner topology $t$ and a point set $P$, the ST of topology $t$ on $P$ is denoted by $t(P)$. Similarly, given a spanning tree topology $s$ and a point set $P$, the spanning tree of topology $s$ on $P$ is denoted by $s(P)$. Let $I$ be the set of spanning tree topologies. Since any ST can be decomposed into an edge-disjoint union of smaller full STs, to prove Gilbert-Pollak's conjecture, it suffices to prove that for any full topology $t$, if $t(P)$ exists then $\ell(t(P)) \geq (\sqrt{3}/2)\min_{s\in I}\ell(s(P))$, where $\ell(T)$ denotes the length of a tree $T$. It is equivalent to say that for any full topology $t$,

$$\min_{p\in E_t}\max_{s\in I} g_{t,s}(P) \geq 0, \qquad (\star)$$

where $E_t = \{P \mid t(P) \text{ exists}\}$ and $g_{t,s}(P) = \ell(t(P)) - (\sqrt{3}/2)\ell(s(P))$ . Note that a point set can be determined by a full ST and a full ST can be determined by its topology and all edge lengths. Let $t(x)$ denote the ST with full topology $t$ and $(2n-3)$-dimensional *parameter vector* $x$ whose components are the edge lengths of the ST. Let $P(t;x)$ denote the set of regular points determined by $t(x)$. Then $E_t = \{P(t;x) \mid x > 0\}$. Denote $X = \{x = (x_1 \dots x_{2n-3}) \mid x \geq 0 \text{ and } x_1 + \cdots + x_{2n-3} = 1\}$. For simplicity of notations, we write $g_{t,s}(x)$ and $s(t;x)$ instead of $g_{t,s}(P(t;x))$ and $s(P(t;x))$, respectively. By similarity, it is easy to know that $(\star)$ holds if and only if for every $t$,

$$\min_{x\in X}\max_{x\in I} g_{t,s}(x) \geq 0.$$

Here, we have allowed the edge length to be zero; in this case, $t(x)$ can be seen as a limiting ST with topology $t$ and a positive edge-length vector $y$ as $y$ goes to $x$. Note that throughout this paper, a parameter vector is said to exist for a point set and a topology if either an ST with that topology or a limit of STs with that topology, exists for the point set. The following lemma shows the concavity of the function $g_{t,s}(x)$.

**Lemma 3.3** *$\ell(s(t;x))$ is a convex function with respect to $x$ .*

*Proof.* Let $A$ and $B$ be two regular points. We show that the distance between $A$ and $B$, $d(A,B)$ is a convex function of $x$. Find a path in $t(x)$ which connects points $A$ and $B$. Suppose the path has $k$ edges with lengths $x_{1'} \dots x_{k'}$ and with directions $e_1 \dots e_k$, respectively, where $e_1 \dots e_k$ are unit vectors. Then $d(A,B) = \| x_{1'}e_1 + \cdots + x_{k'}e_k \|$ where $\| \cdot \|$ is the euclidean norm. Note that a norm is a convex function and the part inside the norm is linear with respect to $x$. Thus, $d(A,B)$ is a convex function with respect to $x$. Finally, we notice that a sum of convex functions is also a convex function.□

The basic idea in the proof of Gilbert-Pollak's conjecture is the following: First, we translate the problem into a minimax problem $(\star)$. By the minimax theorem, we know that to prove $(\star)$, it suffices to prove that at every critical point $x$, $f_t(x) = \max_{s\in I} g_{t,s}(x) \geq 0$ . Note that $I(x) = \{s \mid s(t;x) \text{ is an MST on } P(t;x)\}$. Thus, a critical point $x$ in the interior of $X$ corresponds to a point set $P(t;x)$ with maximal number of MSTs. According to this property, we can find a geometric characteristic of $P(t;x)$. So, the problem reduces to checking the conjecture on some special point sets. To develop some intuition for this approach let us use this idea to verify the conjecture for $n = 4$.

## 4. Four Points

There are several ways to verify the conjecture for four points.[15,27,28] The method we use here is not the simplest one. However, it contains the seed for the general case.

Suppose that there is a set $P$ of four points such that $L_s(P) < (\sqrt{3}/2)L_m(P)$. Let $T$ be an SMT for $P$. We will find a contradiction in the following. First, we claim that $T$ must be a full ST. In fact, if $T$ is not a full ST, then $T$ can be decomposed into two trees $T'$ and $T''$ which are STs for point sets $P'$ and $P''$, respectively. $P'$ and $P''$ have 2 or 3 points. So, $L_s(P') \geq (\sqrt{3}/2)L_s(P')$ and $L_s(P'') \geq (\sqrt{3}/2)L_m(P'')$. Adding them together, we obtain that $L_s(P) \geq (\sqrt{3}/2)(L_m(P') + L_m(P'')) \geq (\sqrt{3}/2)L_m(P)$, a contradiction.

Now, suppose that $t$ is the topology of $T$. Then $t$ is full. Clearly, there exits $x \in X$ such that $t(x)$ is similar to $T$, so that $\max_{s\in I} g_{t,s}(x) = \ell(t(x)) - (\sqrt{3}/2)L_m(P(t;x)) = 1 - (\sqrt{3}/2)L_m(P(t;x)) < 0$. Denote $f_t(x) = \max_{s\in I} g_{t,s}(x)$ and $F(t) = \min_{x\in X} f_t(x)$. Then $F(t) < 0$. A point $x^* \in X$ is called a *minimum point* if $f_t(x^*) = F(t)$.

**Lemma 4.1** *If $f(T) < 0$, then every minimum point $x^*$ is an interior point of $X$.*

*Proof.* Note that $t(x^*)$ has five edges. Four of them are incident to regular points; one of them is between two Steiner points. $x^*$ is an interior point of $X$ if and only if $t(x^*)$ has no edge of length 0. Since $f_t(x^*) = F(t) < 0, \ell(t(x^*)) < (\sqrt{3}/2)L_m(P(t;x^*))$. By the argument in the second paragraph of this section, the four edges incident to regular points cannot have length 0. So, we need only to prove that the edge between two Steiner points cannot be of length 0. Suppose to the contrary that this edge has length 0. Then $t(x^*)$ contains a Steiner point of degree four and hence cannot be an SMT on $P(t;x^*)$. Suppose that $T$ is an SMT on $P(t;x^*)$. Then $\ell(T) < \ell(t(x^*)) = 1 < (\sqrt{3}/2)L_m(P(t;x^*))$. So, $T$ must be a full ST. Let $t'$ be the topology of $T$. Let $x' \in X$ be such that $t'(x')$ is similar to $T$. Then

$$\begin{aligned} f_{t'} &= 1 - (\sqrt{3}/2)L_m(P(t';x')) \\ &= 1 - (\sqrt{3}/2)L_m(P(t;x^*))/\ell(T) \\ &< 1 - (\sqrt{3}/2)L_m(P(t;x^*)) \\ &= f_t(x^*). \end{aligned}$$

Thus, $F(t') < F(t)$. However, there are only two full topologies for four points. They are homeomorphic to each other. So, $F(t) = F(t')$, a contradiction.□

By Theorem 3.1, a minimum point $x^*$ must be a critical point. So, the set $I(x^*)$ of MST topologies is maximal. Now, we want to discover the geometric structure of point set $P(t;x^*)$ by using this property. First, we prove some properties of MSTs. Those properties were discovered by Rubinstein and Thomas.[29]

**Lemma 4.2** *Two MSTs can never cross, i.e., edges meet only at vertices.*

*Proof.* Suppose that $AB$ and $CD$ are two edges crossing at the point $E$ (Fig. 4) and they belong to two MSTs $T$ and $T'$, respectively. Without loss of generality, assume that $EA$ has a smallest length among the four segments $EA, EB, EC$ and $ED$. Removing the edge $CD$ from the tree $T'$, the remainder has two connected components containing $C$ and $D$, respectively. Without loss of generality, assume that $A$ and $C$ are in the same component. Note that $\ell(AD) < \ell(EA) + \ell(ED) \leq \ell(CD)$. Connecting the two components by $AD$ results in a shorter spanning tree, contradicting the minimality of $T'$. □

Let $M(t;x) = \{m \mid m$ is the topology of an MST on $P(t;x)\}$. Let $\Gamma(t;x)$ be the union of all MSTs on $P(t;x)$. By Lemma 4.2, $\Gamma(t;x)$ is a planar graph.

**Lemma 4.3** *Every polygon of $\Gamma(t;x)$ has at least 2 equal longest edges.*

*Proof.* Suppose to the contrary that $\Gamma(t;x)$ has a polygon $Q$ with the unique longest edge $e$. Let $m$ be the MST containing $e$. For every edge $e'$ of $Q$ not in $m$, the union of $m$ and $e'$ contains a cycle. If this cycle contains $e$, then $(m\backslash\{e\})\cup\{e'\}$ is a spanning tree shorter than $m$, a contradiction.

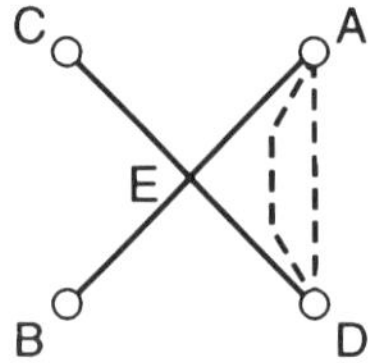


Figure 4. AB and CD intersect at E.

Thus, such a cycle does not contain $e$. Hence, for every $e'$ in $Q$ not in $m$, $m$ has a path connecting two endpoints of $e'$. These paths and the edges of $Q \cap m$ form a cycle in $m$, a contradiction. □

Let $C(t;x^*)$ be the convex hull of the set $P(t;x^*)$ of four points. Now, we are ready to determine the geometric structure of $P(t;x^*)$.

**Lemma 4.4** *If $f(T) < 0$, then $x^*$ is a minimum point, then $\Gamma(t;x^*)$ divides $C(t;x^*)$ into two equilateral triangles.*

*Proof.* If the lemma is false, then one of the following must occur: (a) $\Gamma(t;x^*)$ has a *free* edge, an edge not on any polygon of $\Gamma(t;x^*)$. (b) $\Gamma(t;x^*)$ has a polygon of more than three edges. (c) Neither (a) nor (b) occurs and $\Gamma(t;x^*)$ has a nonequilateral triangle. We will show that in each case, the number of MSTs can be increased.

First, assume that (a) occurs. Embedding $\Gamma(t;x^*)$ into a triangulation of $C(t;x^*)$, we can find a triangle $\Delta$ containing the free edge $e$. Let $T$ be an MST containing $e$. Since $e$ is a free edge, there is an edge $e'$ of $\Delta$, which is not in $\Gamma(t;x^*)$, such that $T' = T\backslash\{e\}\cup\{e'\}$ is a spanning tree. (Such an edge $e'$ must exist, for if $\Delta$ has only one edge not in $\Gamma(t;x^*)$, then this edge must have the desired property; if $\Delta$ has two edges not in $\Gamma(t;x^*)$, then the one which lies between two connected components of $T\backslash\{e\}$ meets the requirement.) Clearly, $\ell(e) < \ell(e')$. By Lemma 4.1, $x^*$ is an interior point of $X$. Hence we can decrease the length of $e'$ and fix all other edge-lengths in the triangulation, so that every edge in $\Gamma(t;x^*)$ does not change its length. Let $\ell$ be the length of the shrinking $e'$. At the beginning, $\ell = \ell(e')$. At the end, $\ell = \ell(e) < \ell(e')$. For each $\ell$, denote by $\bar{P}(\ell)$ the corresponding set of regular points. Then $P(\ell(e')) = P(t; x^*)$. Consider the set $L$ of all $\ell \in [\ell(e), \ell(e')]$ satisfying the condition that there is a minimum point $y$ in $X$ such that $\bar{P}(\ell) = P(t; y)$. Since $\ell(e') \in L$, $L$ is nonempty. Moreover, $L$ is a closed set since all minimum points form a closed set. Now, we consider the minimal element $\ell^*$ of $L$. Suppose $\bar{P}(\ell^*) + P(t; y)$. Then for any $m \in M(t; x^*), \ell(m(t; y)) = \ell(m(t; x^*))$. Since both $x^*$ and $y$ are minimum points, we have $f_t(x^*) = f_t(y)$ , that is, the length of an MST for $P(t; x^*)$ equals that of $P(t; y)$. Hence $M(t; x^*) \subseteq M(t; y)$. However, $x$ is a critical point. Thus, $M(t; x^*) = M(t; y)$. By Lemma 4.1, $y$ is an interior point of $X$. This means that there exists a neighborhood of $\ell^*$ such that for $\ell$ in it, the ST of topology $t$ exists for the point set $\bar{P}(\ell)$. Thus, there exists $\ell < \ell^*$ such that $\bar{P}(\ell) = P(t; z)$ for some vector $z$ (not-necessarily in $X$ ). Since $\ell(m(t; x))$ is continuous with respect to $x$, there is a neighborhood of $y$ such that for every point $y'$ in the neighborhood, $M(t; y') \subseteq M(t; y)$. So, $\ell$ can be chosen to make $z$ satisfy $M(t; z) \subseteq M(t; y)$. Note that $M(t; x^*) = M(t; y)$ and for every $m \in M(t; x^*), \ell(m(t; z)) = \ell(m(t; x^*))$. It follows that for every $m \in M(t; x^*), m(t; z)$ is an MST for $P(t; z)$. Thus, $M(t; z) = M(t; x^*)$ and $f_t(x^*) = f(z)$. Suppose $hz \in X$ where $h$ is a positive number. By Corollary 3.2, $f_t(x^*) = f_t(hz) = hf_t(z)$. So, $h = 1$, i.e., $z \in X$. Hence, $z$ is a minimum point, contradicting the minimality of $\ell^*$.

For the other two cases we can give similar proofs by decreasing the length of an edge not in $\Gamma(t; x^*)$ in case (b) and by increasing the length of all shortest edges in $\Gamma(t; x^*)$ in case (c).□

Now consider four regular points forming a rhombus (see Fig. 5 where an SMT is in solid lines and an MST in dashed lines).

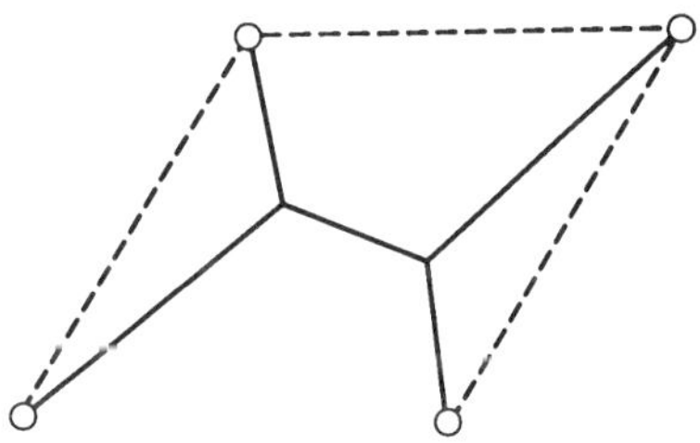

Figure 5. SMT and MST for a rhombus.

It is easily verified that $L_s(P(t; x^*))/L_m(P(t; x^*)) = \sqrt{7}/3 > \sqrt{3}/2$ . So, $f_t(x^*) > 0$, contradicting the assumption $F(t) < 0$.

From the structure of $\Gamma(t; x^*)$ for four points, we may expect that in general, $\Gamma(t; x^*)$ for a minimum point $x^*$ is a union of equilateral triangles. Thus we have to deal with two questions: (A) Suppose that the expectation is right. How do we verify the conjecture on such special point set $P(t; x^*)$? (B) Is our expectation right?

In the next two sections, we will answer these two questions.

## 5. Minimum Hexagonal Trees

In order to answer the first question at the end of the last section, we now study a different kind of tree. Given three directions (Fig. 6) each two of which meet at an angle of 120°, a shortest network interconnecting a given set $P$ of points and having edges all parallel to the three directions is called a *minimum hexagonal tree* on $P$. Let $L_h(P)$ denote the length of the minimum hexagonal tree for $P$. Weng[31] observed the following lemma.

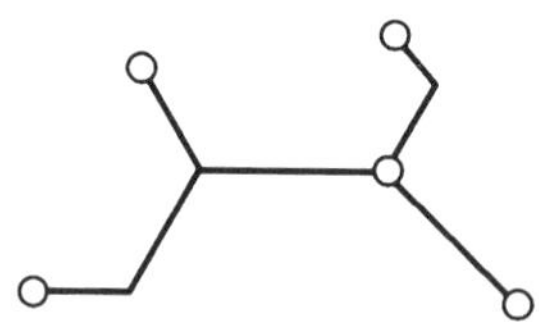

Figure 6. A hexagonal tree.

**Lemma 5.1** $L_s(P) \geq (\sqrt{3}/2)L_h(P)$

*Proof.* First, we note that if a triangle $ABC$ has the angle at $A$ not less than 120°, then $\ell(BC) \geq (\sqrt{3}/2)(\ell(AB) + \ell(AC))$. Now, each edge of an SMT can be replaced by two edges meeting at an angle of 120° and parallel to the given directions. Therefore, the lemma holds.□

Let $T$ be a minimum hexagonal tree for a given set $P$ of points. A point of $T$ but not in $P$ is called a *junction* if the point is incident to at least three lines. Since only three possible slopes exist for lines, there are exactly three lines meeting at a junction. In a hexagonal tree, an edge is referred to as a path between two vertices (regular points or junctions). Thus, an edge can contain several straight segments. An edge is called a *straight edge* if it contains only one straight segment, and is called a *nonstraight edge* if it is not a straight edge. Note that an edge with more than two straight segments can always be replaced by an edge with at most two straight segments. Thus, we assume in the following that every edge has at most two straight segments. Any two nonparallel segments adjacent to each other meet at an angle of 120° since if they meet at an angle of 60°, then we can shorten the tree easily. When we talk about an edge of a junction, its *first segment* is the segment incident to the junction. The other segment, if it exists, is the *second segment* of the edge. A hexagonal tree for $n$ points is said to be *full* if it contains exactly $n-2$ junctions. Any hexagonal tree can be decomposed into an edge-disjoint union of smaller full hexagonal trees. Such a smaller full hexagonal tree will be said to be a full hexagonal subtree of the hexagonal tree.

Consider a minimum hexagonal tree. Suppose a junction has two nonstraight edges. Then these two edges have segments in the same direction. Flip the edges if necessary to line up these two segments, then the second segments of these two edges as well as the first segment of the third edge are three segments each lying completely on one side of the line just constructed. Therefore one side has the majority of the three segments and we can move the line either to shorten the tree or to decrease the number of nonstraight edges. Hence, there exists a minimum hexagonal tree such that each of its junctions has at most one nonstraight edge.

Consider an equilateral triangular lattice. Let the three directions of a hexagonal tree be parallel to edges of an equilateral triangle in the lattice, respectively. Then, we have the following.

**Lemma 5.2** *For any set of n lattice points, there is a minimum hexagonal tree whose junctions are all lattice points.*

*Proof.* Suppose that the lemma is false. Then there exists a set of points such that every minimum hexagonal tree contains a junction which is not a lattice point. Call such a set *bad*. Let $P$ be a smallest bad set. Then every minimum hexagonal tree for $P$ must be a full tree with no junction on a lattice point. (Otherwise, a smaller bad set exists.) Consider a minimum hexagonal tree $T$ with the property that each of its junction has at most one nonstraight edge. There exists a junction $J$ adjacent to two regular points. (Otherwise, $T$ contains a cycle.) Let $C$ be the third vertex adjacent to $J$. Since $C$ being a regular point implies $J$ being a lattice point, we see that $C$ is a junction. We will show that one of the following occurs: (1) $J$ is a lattice point. (2) A junction can be moved either to a regular point or to another junction. (3) $C$ can be moved further away from $J$. Since the last movement cannot last forever and two junctions joining together would result in a shortening of the tree, we obtain a contradiction.

Let us first consider the case that both edges $AJ$ and $JB$ are straight. If $AJ$ and $JB$ are in different directions, then $J$ is a lattice point. Hence, they are in the same direction. Let $e$ be a line through $C$, parallel to $AB$. If $C$ has a nonstraight edge (other than $CJ$) with an endpoint lying in the $AB$-side of $e$, then we can shorten the tree by flipping $e$. If $C$ has a straight edge overlapping $e$, then we can move edge $JC$ such that either $J$ or $C$ meets a regular point or a junction other than $J$ and $C$. Hence, we may assume that the two vertices adjacent to $C$ are on the other side of $e$ away from $AB$. Now, we can move $C$ further away from $J$ (Fig. 7).

Secondly, we consider the case that $AJ$ is a straight edge and $JB$ is a nonstraight edge with a segment in the same direction as $AJ$. Flip $JB$, if necessary, to line up the two first segments of $AJ$ and $JB$. Let $BD$ be the first segment of $JB$. Then $D$ must be a lattice point. Now, we use $D$ to replace $B$ to go back to the first case.

Thirdly, if $AJ$ is a straight edge and $JB$ is a nonstraight edge without a segment in the same direction as $AJ$, then $J$ can be moved either to $A$ or to a lattice point (Fig. 8). Since other cases are symmetric to the above three, the lemma is proved. □

Suppose that every two adjacent lattice points is of distance one. Let $P$ be a subset of $n$ lattice points such that $L_m(P) = n - 1$. By Lemma 5.2, a minimum hexagonal tree for $P$ is of length at least $(n - 1)$, the length of a minimum spanning tree for $P$. By Lemma 5.1, we have

$$L_s(P) \geq (\sqrt{3}/2)L_m(P)).$$

Thus, we give a positive answer to the first question (A).

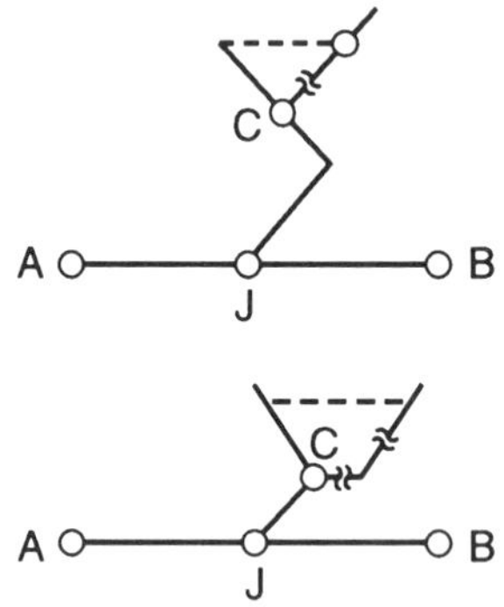


Figure 7. C is moved further away from AB.

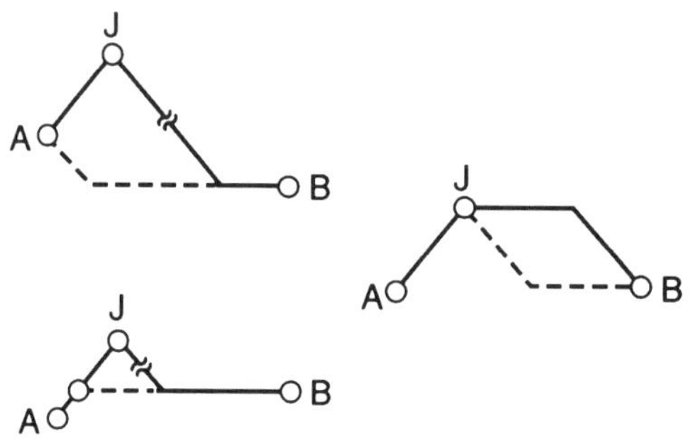


Figure 8. J can be moved either to A or to a lattice point.

## 6. Inner Spanning Trees

In this section, we study question (B) in section 4. At first glance, our answer is negative. That is, in general the union $\Gamma(t; x^*)$ of MSTs at a minimum point $x^*$ may not be a subset of lattice points in an equilateral triangle lattice. To explain this, let us examine the proof of Lemma 4.4. In the proof, we need to vary an edge-length, under the triangle inequality constraint, without disturbing lengths of the other edges in $\Gamma(t; x^*)$. If this is possible for every length, we will say that the edge-lengths are *independent*.

If the point set $P(t; x^*)$ is the vertex set of a convex polygon, then the edge-lengths of $\Gamma(t; x^*)$ are independent. In fact, $\Gamma(t; x^*)$ must be a subgraph of a triangulation of the convex hull of $P(t; x^*)$. Since $P(t; x^*)$ lies on the boundary of the triangulation, there does not exist a cycle in the dual graph of the triangulation. The triangles can be linearly ordered and drawn one by one in the given order. Note that each triangle (except the first one) to be drawn has only one side fixed. Thus the other two sides can have any lengths. This means that the edge-lengths in the triangulation are independent.

For the set $P(t;x^*)$ of four points, if $t(x^*)$ is a full ST, then $P(t;x^*)$ must be the vertex set of a convex quadrilateral.[27] So, the proof of Lemma 4.4 is valid. However, in general, $P(t;x^*)$ may not lie on the boundary of the convex hull of $P(t;x^*)$ (Fig. 9). In this case, the edge-lengths of $\Gamma(t;x^*)$ may not be independent. So, a straightforward generalization of Lemma 4.4 would not work.

From the above explanation, we see that the fact that $P(t;x^*)$ lies on the boundary of the convex hull of $P(t;x^*)$ plays an important role in the proof for four points. An idea motivated from this observation is that although this fact does not hold in general for convex hull, we may use the boundary of a different type of area. Define a *Steiner hull* for a given set of points $P(t;x^*)$ as an area in which an SMT lies. We will consider only spanning trees lying inside the area.

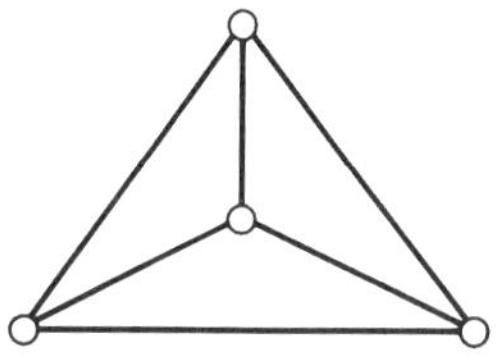

Figure 9. $P(t;x^*)$ not on the boundary.

Consider a full topology $t$ and a full ST $T$ of topology $t$. If we inflate the edges of $T$ to have an $\varepsilon$ width, then $T$ is a polygonal region with a boundary (Fig. 10).

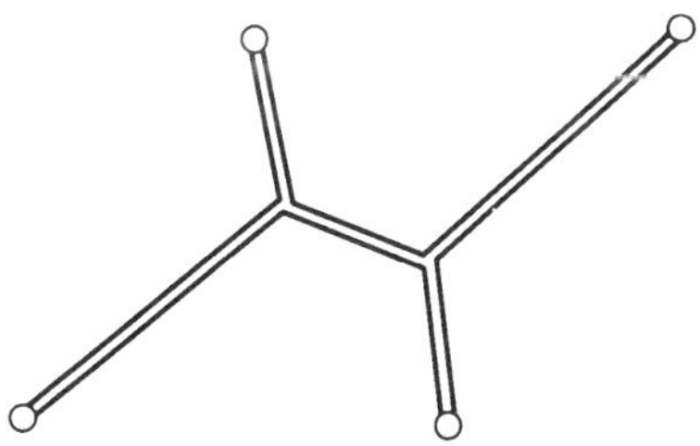

Figure 10. Boundary of an inflated SMT.

Two regular points are called *adjacent* in $t$ if they are consecutive on the boundary of $T$. For a Steiner tree $t(x)$ with a full topology $t$ and a parameter vector $x$, connecting all pairs of adjacent regular points by straight lines, we obtain a polygon. We call such a polygon *the characteristic area* of $t$ at $x$, denoted by $C(t;x)$. Note that $t(x)$ lies inside of $C(t;x)$ and $P(t;x)$ lies on the boundary of $C(t;x)$. A spanning tree on $P(t;x)$ is called an *inner spanning tree* for $t$ at $x$ if it lies inside of $C(t;x)$. Clearly, Gilbert-Pollak's conjecture is a corollary of the following theorem.

**Theorem 6.1** *For every full topology $t$ and parameter vector $x$, there is an inner spanning tree $N$ for $t$ at $x$ such that $\ell(t(x)) \geq (\sqrt{3}/2)\ell(N)$.*

Let $I(t;x)$ denote the set of inner spanning tree topologies for $t$ at $x$ and define $g_{t,s(x)} = \ell(t(x)) - (\sqrt{3}/2)l(s(P(t;x)))$ for $s \in I(t;x)$. To prove Theorem 6.1, it suffices to prove the following inequality.

$$\min_{x \in X} \max_{s \in I(t;x)} g_{t,s}(x) \geq 0.$$

This is a minimax problem different from the one we discussed in section 2 because the index set $I(t;x)$ depends on $x$. So, we need to modify Theorem 2.1 as follows.

**Theorem 6.2** *Suppose that $f(x) = max_{i \in I(x)} g_i(x)$ where $I(x)$ is a subset of a finite set $I$ and every $g_i(x)$ is a continuous and concave function. Denote $M(x) = \{i \in I(x) \mid f(x) = g_i(x)\}$. If for any $x$, there is a neighborhood of $x$ such that for any point $y$ in the neighborhood, $M(y) \subseteq M(x)$, then the minimum value of $f(x)$ over a simplex is achieved at some critical point.*

The proof of Theorem 6.2 is similar to that of Theorem 2.1. Let $MI(t;x)$ denote the set of topologies of minimum inner spanning trees on $P(t;x)$. To see that Theorem 6.2 is applicable to our problem, let us prove the following lemma.

**Lemma 6.3** *For every $x$, there is a neighborhood of $x$ such that for any $y$ in the neighborhood, $MI(t;y) \subseteq MI(t;x)$.*

*Proof.* First, we show that for any $m \in MI(t;x)$ there exists a neighborhood $U$ of $x$ such that for any $y \in U$, $m$ is in $MI(t;y)$. For contradiction, suppose such a neighborhood does not exist. Then there is a sequence of points $y_k$ converging to $x$ such that $m \notin MI(t;y_k)$. Thus every $m(t;y_k)$ has at least one edge not in the characteristic area $C(t(y_k))$. Since the number of edges is finite, there exists a subsequence of $m(t;y_k)$ each of which contains an edge not in $C(t(x))$, but these edges converge to an edge $AB$ in $m(t;x)$. It is easy to see that $AB$ is on the boundary of the area $C(t(x))$ and that $A$ and $B$ are not adjacent. (An edge between two adjacent regular points always lies in the characteristic area.) Since all vertices in an inner spanning tree lie on the boundary of $C(t(x))$, there is a regular point lying in the interior of the segment $AB$, contradicting the minimality of $m(t;x)$.

Now, we prove the lemma by contradiction. Suppose that there is a sequence of points $y_k$ converging to $x$ such that for each $y_k$, a spanning tree topology $m_k$ exists such that $m_k \in MI(t;y_k) \backslash MI(t;x)$. Since the number of spanning tree topologies is finite, there is a subsequence of points $y_{k'}$ with the same $m_{k'}$, denoted by $m$. We can also assume that this subsequence lies inside of the neighborhood $U$ of $x$. Thus, for every $k', \ell(m(t;y_{k'}) \leq \ell(m'(t;y_{k'}))$ for all $m' \in MI(t;x)$ since $MI(t;x) \subseteq I(t;y_{k'})$. Letting $k' \to \infty$, we obtain that $\ell(m(t;x)) \leq \ell(m'(t;x))$ for $m' \in MI(t;x)$. Since $m \notin MI(t;x), m(t;x)$ must not be an inner spanning tree. It follows that there exists a neighborhood of $x$ such that for any point $y$ in the neighborhood, $m(t;y)$ is not an inner spanning tree for $t(y)$, contradicting the existence of the subsequence of points $y_{k'}$.□

Denote $f_t(x) = \max_{s \in I(x)} g_{t,s}(x)$. An immediate consequence of Lemma 6.3 is that $f_t(x)$ is continuous over $X$. Denote $F(t) = \min_{x \in X} f_t(x)$. By Theorem 6.2 and

Lemmas 3.1 and 6.3, $F(t)$ is achieved at some critical point. Choose a full topology $t^*$ such that $F(t^*) = \min_t F(t)$ where $t$ is over all full topologies. We prove Theorem 6.1 by contradiction. Suppose that Theorem 6.1 is false, i.e., $F(t^*) < 0$, and that $n$ is the smallest number of regular points such that $F(t^*) < 0$. From now on, a point $x$ in $X$ is called a *minimum point* if and only if $f_{t^*}(x) = F(t^*)$. Since we use the inner spanning tree instead of the spanning tree, we have to give different proofs for the following lemmas.

**Lemma 6.4** *If $F(t^*) < 0$, then every minimum point is an interior point of $X$.*

*Proof.* Suppose to the contrary that there exists a minimum point $x$ on the boundary of $X$, that is, $t^*(x)$ have some edges vanished. If there is a vanished edge incident to a regular point, then $t^*(x)$ can be decomposed into several edge-disjoint smaller ST's. Since each smaller ST has fewer regular points, we can apply Theorem 6.1 to these ST. Note that a union of inner spanning trees for the smaller ST's is an inner spanning tree for $t^*(x)$. We find a contradiction to $F(t^*) < 0$ by summing all inequalities. So, every vanished edge is between two Steiner points. In this case, it is easy to find a topology $t$ satisfying the following conditions (Fig. 11).

(1) Two regular points are adjacent in $t$ if and only if they are adjacent in $t^*$. ($t$ is called a *companion* of $t^*$ when $t$ satisfies this condition.)

(2) There is a tree $T$ interconnecting the $n$ points in $P(t^*; x)$, with the topology $t$ and with length less than $\ell(t^*(x))$.

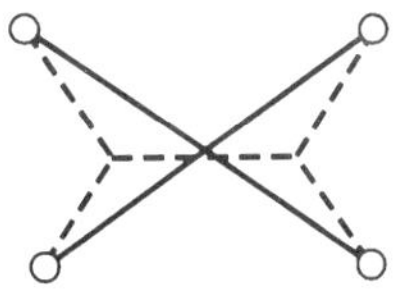

Figure 11. Finding a companion.

If the ST of topology $t$ for $P(t; x)$ exists, then there exists a parameter vector $y$ such that $P(t; y) = P(t^*; x)$. Let $h = 1/\ell(t(y))$. Since $\ell(t(y)) \leq \ell(T) < \ell(t^*) = 1$, $h > 1$. Note that $t(hy)$ is similar to $t(y)$. Hence $f_t(hy) = 1 - (\sqrt{3}/2)L_t(hy) = 1 - (\sqrt{3}/2)hL_t(y) = 1 - (\sqrt{3}/2)hL_{t^*}(x) < f_{t^*}(x) = F(t^*)$ where $L_t(x)$ denotes the length of a minimum inner spanning tree on $P(t; x)$. Since $hy \in X$, we have $F(t) \leq f_t(hy) < F(t^*)$, contradicting the minimality of $F(t^*)$.

If the ST of topology $t$ for $P(t^*; x)$ does not exist, then we cannot use the above argument directly since $f_t(y)$ is undefined. (Remember that $F(t^*)$ is a minimum over all full Steiner topologies. So even though $T$ is a shorter tree, there is no contradiction to the minimality of $F(t^*)$.) Now, we consider any tree of topology $t$. Such a tree can be determined by edge lengths and angles at every Steiner point. Write the lengths into a length vector $y$ and the angles into an angle vector $\theta$. Denote such a tree by $t(y, \theta)$. Two regular points are said to be adjacent in $t(y, \theta)$ if in an ST of topology $t$, the corresponding two regular points are adjacent. Constructing the characteristic area for $t(y, \theta)$ by connecting every pair of adjacent regular points, we can define an inner spanning tree and a minimum inner spanning tree for $t(y,\theta)$ in a similar way. Let $L_t(y,\theta)$ denote the length of a minimum inner spanning tree for $t(y,\theta)$. We can also show the continuity of $L_t(y,\theta)$. Restrict all angles to be between $0°$ and $360°$ and the sum of any three angles at the same Steiner point to equal $360°$. Let $Y_t$ be the set of vectors $(y,\theta)$ with the described restrictions on $\theta$ and the restrictions $\sum y_i = 1$ and $y \geq 0$. Then $Y_t$ is compact. So, the function $h$ defined by $h_t(y,\theta) = 1-(\sqrt{3}/2)L_t(y,\theta)$ reaches its minimum in $Y_t$. We denote this minimum value by $H(t)$. By an argument similar to that in the last paragraph, we can prove that $H(t) < F(t^*)$. Thus, $H(t) < F(t)$.

Suppose that $h_t(y,\theta) = H(t)$. Then $t(y,\theta)$ is neither a full ST nor a limit of full STs; for if it is, then $F(t) \leq h_t(y,\theta)$. Thus, $y$ has a zero component corresponding to an edge incident to a regular point, or $\theta$ has a component less than $120°$. In the former case, we decompose $h(y,\theta)$ and find a full topology $t'$ with fewer regular points such that $H(t') < 0$. In the latter case, we may assume that the former case does not occur. Then at least one edge of two sides of the angle corresponding to a smallest component of $\theta$ must be vanished. Otherwise, we can shorten the tree without changing the topology. Now, a vanished edge between two Steiner points is found again (and every vanished edge is between Steiner points). So, we can find a new companion $t'$ of $t$ such that $H(t') < H(t)$. Repeating the above argument, we will obtain infinitely many full topologies with at most $n$ regular points, contradicting the finiteness of the number of topologies. □

**Lemma 6.5** *Two minimum inner spanning trees can never cross, i.e., edges meet only at vertices.*

*Proof.* Recall the proof of Lemma 4.2 (see Fig. 4) in which we established that $\ell(AD) \leq \ell(CD)$. If $AD$ lies in the characteristic area, then connecting the two components by $AD$ results in a shorter inner spanning tree, contradicting the minimality of $T'$. If $AD$ does not lie in the characteristic area, there must exist some regular points lying inside of the triangle $EAD$. Consider the convex hull of those regular points and two points $A$ and $D$. The boundary of the convex hull other than the edge $AD$ must lie in the characteristic area. This boundary contains a path from $A$ to $D$. In this path there exists two adjacent vertices which belong to different connected components of $T'\setminus\{CD\}$. Connecting two such adjacent vertices also results in an inner spanning tree shorter than $T'$, a contradiction. □

Now, we define $\Gamma(t;x)$ to be the union of minimum inner spanning trees on $P(t;x)$. From Lemma 6.5, we see that $\Gamma(t;x)$ divides the characteristic area $C(t;x)$ into smaller areas each of which is bounded by a polygon with vertices all being regular points. Such a polygon is called a polygon of $\Gamma(t;x)$ if it is a subgraph of $\Gamma(t;x)$.

**Lemma 6.6** *Every polygon of* $\Gamma(t;x)$ *has at least 2 equal longest edges.*

*Proof.* Similar to the proof of Lemma 4.3. □

Note that the characteristic area of $t(x)$ is bounded by a polygon of $n$ edges. Partitioning the area into $n-2$ triangles by adding $n-3$ edges, we will obtain a network with $n$ vertices and $2n-3$ edges. This network will be called a triangulation of $C(t;x)$. Let us first ignore the full ST $t(x)$ and consider the relationship between the vertex set and the length of edges. Note that in the previous discussion, when we say that a set $P$ of points is given, we really mean that the distance between every two points in the set is given; that is, relative positions between those points have been given. With this understanding, we make the following observations.

(1) The vertex set (i.e., the set of regular points, $P(T;x)$) can be determined by $2n-3$ edge lengths of the network.

(2) The $2n-3$ edge-lengths are independent variables; that is, the network could vary by changing any edge-length and fixing all others.

Note that every $\Gamma(t;x)$ can be imbedded in a triangulation of $C(t;x)$. Thus, the edge-length in $\Gamma(t;x)$ are independent. The following lemma shows that $\Gamma(t;x)$ has a critical structure.

**Lemma 6.7** *If $f(T)<0$, $\Gamma(t;x)$ partitions $C(t;x)$ into exactly $n-2$ equilateral triangles.*

*Proof.* Similar to the proof of Lemma 4.4. □

Let $a$ be the length of an edge of the equilateral triangles in the $\Gamma(t;x)$. If we divide the plane into a union of disjoint equilateral triangles with edge length $a$, then all regular points in $\Gamma(t;x)$ can be placed on the lattice points. The following lemma is easy to prove.

**Lemma 6.8** *An MST for n lattice points has length at least $(n-1)a$. For the point set $P(t;x)$, the minimum inner spanning tree has length exactly $(n-1)a$.*

By Lemma 6.7, we see that for the point set $P(t;x)$, every minimum inner spanning tree is an MST in the plane. Thus, in order to derive a contradiction to $f_t(x)=F(t)<0$, it suffices to verify the truth of Gilbert-Pollak's conjecture for the point set $P(t;x)$, which is a set of $n$ lattice points. This has been done in Section 5.

## 7. A Final Touch For A Rigorous Proof

In previous sections, we ignored the following two problems:

(1) As $x$ varies freely in $X$, the full ST $t(x)$ may have two non-adjacent edges crossing each other.

(2) When we construct a characteristic area from a full ST without crossing edges it is still possible to obtain a polygon with lines crossing each other.

In both cases, the characteristic polygon of such an ST is not well defined.

To overcome the second problem, it suffices to view the polygon, obtained by connecting adjacent regular points, as a Rieman surface such that all regular points are on the boundary consisting of the straight segments between adjacent regular points without crossing (Fig. 12).

The same strategy may also work for the first problem. However, in some cases it could be hard to determine the surface on which the characteristic area is defined. So, we suggest an alternative treatment.

First, we notice that for any two edges on $t(x)$, we can give a system of linear inequalities to describe the sufficient and necessary condition for these two edges to intersect each other. For example, suppose that $A_1 \cdots A_7$ is a convex path in $t(x)$ and $x_i = d(A_i, A_{i+1})$. Then $A_1A_2$ and $A_6A_7$ intersect each other iff $x_1+x_2 \geq x_4+x_5 \geq x_2$ and $x_6+x_5 \geq x_2+x_3 \geq x_5$.

Thus, if we delete the $x$, at which $t(x)$ has non-adjacent edges intersecting, from $X$, then the closure $\hat{X}_t$ of remaining points is still a polytope, but not necessarily convex. However, the following fact holds.

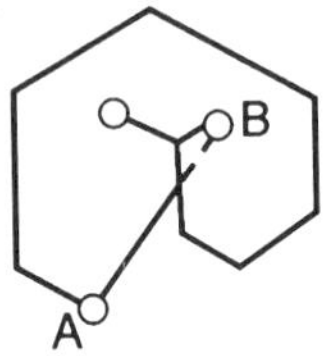


Figure 12. Characteristic area.

**Lemma 7.1** *For any point $x$ in the interior of $\hat{X}_t$ $t(x)$ cannot have non-adjacent edges intersecting. However, for $x$ on the boundary of $\hat{X}_t$ $t(x)$ can have non-adjacent edges intersecting; this happens only if there is a regular point touching an edge or touching another regular point.*

Noting this fact, we can modify our former proof to meet the current case. We now give a rigorous proof of Gilbert-Pollak's conjecture.

First, we note that to prove Gilbert-Pollak's conjecture, it suffices to prove that for every full topology $t$ and every vector $x \in \hat{X}_t$, there exists an inner spanning tree $N$ on $P(t;x)$ such that $\ell(t(x)) \geq (\sqrt{3}/2)\ell(N)$. Suppose that the latter one is false. Let $n$ be the smallest number such that there exists a set $P$ of $n$ points such that there exists a full topology $t$ for $n$ points and a vector $x$ such that $\ell(t(x)) < (\sqrt{3}/2)\ell(N)$ for every inner spanning tree $N$ on $P(t;x)$. Define $\hat{F}(t) = \min_{x \in \hat{X}} f(x)$. Choose a full topology $t^*$ such that $\hat{F}(t^*) = \min_t \hat{F}(t)$ where $t$ is over all full topology. A point $x^*$ is called a *minimum point* if $f_{t^*}(x^*) = \hat{F}(t^*)$.

By Lemma 6.3, $f_{t^*}(x)$ is a continuous function, so a minimum point always exists. Consider a minimum point $x$. The following lemma gives the properties of $x$.

**Lemma 7.2** *If $f(T) < 0$, then a minimum point $x$ must be an interior point of $\hat{X}_{t^*}$.*

*Proof.* By the proof of Lemma 6.4, we need to consider only the case that $x$ is on the new boundary; i.e. the part of the boundary of $\hat{X}_{t^*}$ which is not on the boundary of $X$. Note that if $x$ is a new boundary point, then $t(x)$ has a regular point touching an edge or another regular point. In the former case, we can decompose $t(x)$ at the touching point to obtain two trees each with less than $n$ regular points. In the latter case, we can reduce the number of regular points by one. In either case, a contradiction is achieved by an argument similar to the one used in the proof of Lemma 6.4. □

Since $\hat{X}_t$ may be nonconvex, the minimax theorem (Theorem 3.1) does not apply. Instead we use the following lemma.

**Lemma 7.3** *Suppose that $x$ is a minimum point and that $y$ is a point in $\hat{X}_{t^*}$ such that the segment $[x, y] \subset \hat{X}_t$ and $MI(t^*; x) \subseteq MI(t^*; y)$. Then $y$ is also a minimum point.*

*Proof.* For any $m$ in $MI(t^*, x)$, define $A(m) = \{z \in X_{t^*} \mid \ell(m(t^*; z)) \leq L_{t^*}(x)\}$. Since $\ell(m(t^*; z))$ is a convex function, $A(m)$ is a convex region. By the proof of Theorem 3.1, we know that the union of all $A(m)$ for $m$ in $MI(t^*; x)$ covers a neighborhood of $x$.

Now, we show that $f_{t^*}(y) = F(t^*)$. Suppose to the contrary that $f_{t^*}(y) > F(t^*)$. Note that $MI(t^*; x) \subseteq MI(t^*; y)$. Thus, for every $m \in MI(t^*; x)$, $\ell(m(t^*; y)) < f_{t^*}(x)$. We claim that for all positive number $c$, the point $x + c(x - y)$ is not in $A(m)$ for every $m \in MI(t^*; x)$. In fact, if the point $x + c(x - y)$ for some positive $c$ is in $A(m)$, then the point $x$ as an interior point of the segment $[y, z]$ where $z = x + c(x - y)$ can be written as $x = \lambda y + (1 - \lambda)z$ where $0 < \lambda - c/(1 + c) < 1$. Thus,

$$\begin{aligned} \ell(m(t^*; x)) &\leq \lambda \ell(m(t^*; y)) + (1 - \lambda)\ell(m(t^*; z)) \\ &< f_{t^*}(x), \end{aligned}$$

contradicting that $m \in MI(t^*; x)$. Finally, the fact that $x + c(x - y)$ for all $c > 0$ is not in every $A(m)$ for $m \in MI(t^*; x)$ contradicts that the union of all $A(m)$'s covers a neighborhood of $x$. □

**Lemma 7.4** *Suppose that $f(T) < 0$. Let $x$ be a minimum point. Let $y$ be a parameter vector for $t^*$ but not necessarily in $X_{t^*}$, such that $MI(t^*; x) \subseteq MI(t^*; y)$ and $\ell(m(t^*; x)) = \ell(m(t^*; y))$ for $m \in MI(t^*, x)$. Suppose that for any $w$ in $[x, y], t^*(w)$ exists and does not have non-adjacent edges intersecting. Then $y$ is also a minimum point.*

*Proof.* To see this, note that there always exists a positive number $h$ such that $hy \in X_{t^*}$. By Lemma 7.3, we have that $f_{t^*}(x) = f_{t^*}(hy) = h \cdot f_{t^*}(y)$. Thus $h = 1$, so that $y \in \hat{X}_{t^*}$. Therefore, $y$ is a minimum point by using Lemma 7.3 again. □

Using the above lemmas, we can prove that the union $\Gamma(t; x)$ of minimum inner spanning trees divides the characteristic area $C(t; x)$ into $n - 2$ equilateral triangles. The proof is similar to the proof of Lemma 4.4 except that the point $z$ in the proof needs to be chosen sufficiently close to $y$ such that for any $w$ in $[y, z], t(x)$ exists and does not have non-adjacent edges intersecting.

Finally, the proof proceeds to its completion analogous to its counterpart in Section 6.

## 8. Background of the Better Heuristics Problem

Over the years numerous heuristics for SMT have been proposed for regular points in various metric spaces and graphs. Often their superiority over MSTs were claimed and demonstrated by simulations. But no mathematical proof of superiority was ever given. One exception is that for the rectilinear metric and Poisson distributed regular points, Bern[3] proved that a heuristic tree obtained by a very simple improvement over an MST has a shorter average length. Later, Hwang and Yao[21] extended this result to the usual case when the number of regular points is fixed. But still there has been no proof that any heuristic has a better performance ratio than the Stein ratio.

In 1990, Zelikovsky[31] finally brought a breakthrough to the problem by giving a better heuristic for the graph problem. Unfortunately, his results were published in a conference proceedings not easily accessible, and his writing was very obscure. Berman and Ramaiyer[1] came to the rescue by restating Zelikovsky's results in a much more comprehensible way, giving some improvement for the graph problem and also giving a better heuristic for the rectilinear space. Du, Zhang and Feng[16] made further improvements and gave a better heuristic for an arbitrary metric space, including the Euclidean plane as a special case. In the next two sections we will mostly review the general results of Du, Zhang and Feng, and review Zelikovsky and Berman and Ramaiyers' more specific heuristics only when they are different from the general case. As the methodology for getting better heuristics is now available, the pressing problem for future research is how to push the method to its tilt so that even better heuristics can be obtained.

## 9. A Better Heuristic for an Arbitrary Metric Space

A *k-size quasi ST* is a tree with a Steiner topology and each of its full component contains at most $k$ terminals. A *minimum k-size quasi* ST, denoted by $Q_k$, is a shortest such tree. No polynomial algorithm is known to construct $Q_k$ for $k \geq 3$. However, a heuristic $Z_k$ for $Q_k$ exists such that $\| Z_k \| / \| Q_k \|$ is bounded, where $\| \cdot \|$ denotes the length.

The concept of MST can be extended to the case where some of the regular points are point-sets. For a given set $P$ of $n$ points and subset $P_i \subset P, i = 1, \ldots, s$, define $P' = \cup_{i=1}^{s} P_i$. An MST $T(P; P_1, \ldots, P_s)$ is a set of $s+ \mid P \mid - \mid P' \mid -1$ edges connecting the $s$ point sets and the $\mid P \mid - \mid P' \mid$ singletons where a point-set is considered connected if any of its points is connected. The ordinary MST $T(P)$ can be viewed as the special case $s = 0$. To assure that the connected graph is a tree, it is required that no set (including singleton) is contained in another set, $\mid P_i \cap P_j \mid \leq 1$ and that there does not exist a cyclic sequence of subscripts such that $\mid P_i \cap P_j \mid = 1$ for all $i$ and $j$ consecutive in that cyclic sequence. Such a set $\{P_i\}$ is called a *cycle-free set*.

Let $S_i$ denote a full SMT for $P_i$, if one exists. If $S_i$ exists for $P_i$, $i = 1, \cdots s,$, define

$$W(P; P_1, \cdots, P_s) = \| T(P) \| - \| T(P; P_1, \cdots, P_s) \| - \sum_{i=1}^{s} \| S_i \| .$$

Then $W$ is the saving of length over an MST by using $S_i$ to connect points in $P_i$.

Du, Zhang and Feng, extending an idea of Zelikovsky for constructing $Z_3$, gave the following greedy algorithm for constructing $Z_k$ in an arbitrary metric space $M$:

Step 1. For each $i = 2, \cdots, k$, construct an SMT for each $\binom{n}{i}$ subsets of $i$ regular points. Store the SMT in the list $L$ if it is a full ST.

Step 2. For $j = 1, 2, \cdots, n-1$ select $S_j$ from $L$ such that $S_1, \cdots, S_j$ is cycle-free and

$W(P; P_1, \cdots, P_j)$ maximizes $W(P; P_1, \cdots, P_{j-1}, P_j^*)$ over all $S_j^*$ from $L$. Delete $S_j^*$ from $L$ if $P_1, \cdots, P_{j-1}, P_j^*$ is not cycle-free.

Step 3. Stop when $L$ is empty.

Since each edge is in $L$, the output $Z_k$ is always a tree interconnecting $N$. However, the efficiency of the algorithm depends on the existence of an efficient construction of $k$-point SMT in $M$. Given that there exists such a procedure which requires $f(k)$ time, then step 1 takes $O(kn^k f(k))$ time to generate the $O(kn^k)$ SMTs and step 2 takes $O(n \log n)$ time to generate on MST for $\{P; P_1, \ldots, P_{j-1}, P_j^*\}$ for each $S_j^* \in L$. Since there are $O(kn^k)$ $S_j^*$ and step 2 is repeated $O(n)$ times, the $Z_k$ algorithm requires $O(kn^k(n^2 \log n + f(k)))$ time.

For $\{P_1, \cdots, P_s\}$ a cycle-free set, define

$$\begin{aligned} W(P; P_{t+1}, \cdots, P_s \mid P_1, \cdots, P_t) &= W(P; P_1, \cdots, P_s) - W(P; P_1, \cdots, P_t) \\ &= \| T(P; P_1, \cdots, P_t) \| - \| T(P; P_1, \cdots, P_s) \| . \end{aligned}$$

**Theorem 9.1** $\| Z_k(P) \| \leq \frac{k-2}{k-1} \| Q_2(P) \| + \frac{1}{k-1} \| Q_k(N) \|$ *for* $k \geq 3$.

*Proof.* This inequality is proved by induction on $s$, where $s$ is the number of $P_i$ in $Z_k(N)$ such that $| P_i | \geq 3$. Let $Q_k = \cup_{i=1}^t S_i^*$ where each $S_i^*$ is a full SMT, and let $\{S_1, \cdots, S_t\}$ be the full SMTs chosen by the algorithm. If $s = 0$, then each $S_i^*$ must be an edge for otherwise $W(P; P_1) \geq W(P, P_i^*) > 0$ and $s \geq 1$. Therefore $W(P; P_1, \cdots, P_s) = 0 = W(P; P_1^*)$, and the inequality is trivially true. So, assume $s \geq 1$. Subtracting from $\| Q_2(N) \| = \| Z_2(N) \|$ both sides of the inequality in Theorem 9.1, then

$$W(P; P_1, \cdots, P_s) \geq \frac{1}{k-1} W(P; P_1^*, \cdots, P_t^*).$$

Suppose that $S_1$ contains $x$ regular points. Then $S_1 \cup Q_k$ contains $x - 1$ cycles since $S_1$ contributes $x - 2$ new vertices and $2x - 3$ new edges. Without loss of generality, assume that $S_1^*, \cdots, S_y^*$, $1 \leq y \leq x - 1$, are the full STs whose deletions from $Q_k$ break all the $x - 1$ cycles. Then from the choice of $S_1$ in the algorithm,

$$(k-1)W(P; P_1) \geq (x-1)W(P; P_1) \geq \sum_{i=1}^{y} W(P; P_i^*) \geq W(P; P_1^*, \cdots, P_y^*).$$

Since

$$W(P; P_1, \cdots, P_s) = W(P; P_1) + W(P; P_2, \cdots, P_s \mid P_1)$$

and

$$W(P; P_1^*, \cdots, P_t^*) = W(P; P_1^*, \cdots, P_y^*) + W(P; P_{y+1}^*, \cdots, P_t^* \mid P_1^*, \cdots, P_y^*),$$

it suffices to prove

$$(k-1)W(P; P_2, \cdots, P_s \mid P_1) \geq W(P; P_{y+1}^*, \cdots, P_t^* \mid P_1^*, \cdots, P_y^*).$$

Let $T(P; P'_2, \cdots, P'_i \mid P_1)$ denote a minimum $k$-size quasi ST given that $P_1$ is chosen. Then

$$(k-1)W(P; P_2, \cdots, P_s \mid P_1) \geq W(P; P'_2, \cdots, P'_s \mid P_1) \geq W(P; P^*_{y+1}, \cdots, P^*_t \mid P_1),$$

where the first inequality is due to the inductive assumption and the second due to the minimality of $T(P; P'_2, \cdots, P'_s \mid P_1)$,

Let $P_{1j}, j = 1, \cdots, J$, denote the pairs of adjacent vertices in $S_1$. Then

$$\begin{aligned} W(P; P^*_{y+1}, \cdots, P^*_t \mid P_1) &= W(P; P^*_{y+1}, \cdots, P^*_t \mid P_{11}, \cdots, P_{1J}) \\ &\geq W(P; P^*_{y+1}, \cdots, P^*_t \mid P^*_1, \cdots, P^*_y), \end{aligned}$$

since each $P_{ij}$ is contained in some $P^*_i, i = 1, \cdots, y$. Therefore

$$(k-1)W(P; P_2, \cdots, P_s \mid P_1) \geq W(P; P^*_{y+1}, \cdots, P^*_t \mid P^*_1, \cdots, P^*_y). \square$$

For a tree $T$ interconnecting $P$ in $M$, define

$$\rho(T) = \inf_{N \in M} \frac{\| SMT(N) \|}{\| T(N) \|}.$$

**Corollary 9.2** $\rho^{-1}(Z_k) = \frac{k-2}{k-1}\rho^{-1}(Q_2) + \frac{1}{k-1}\rho^{-1}(Q_k)$ *for* $k \geq 3$.

Therefore, if $\rho(Q_2) < \rho(Q_k)$ for some $k \geq 3$, then $Z_k$ is a better heuristic. Proving this inequality will be the goal of the rest of this section. Berman and Ramaiyer recently claimed the following stronger result.

**Theorem 9.3** $\rho^{-1}(Z_k) \leq \rho^{-1}(Q_2) - \sum_{i=3}^{k} \frac{\rho^{-1}(Q_{i-1}) - \rho^{-1}(Q_i)}{i-1}$.

## 10. Lower Bounds of $Q_k$

Consider a full SMT $T$. Turn $T$ into a rooted binary tree $T_r$ by choosing an arbitrary edge and setting its middle point $r$ as the root for $T_r$. The root is considered at the first level of $T_r$, and a vertex is at the $i^{\text{th}}$ level if the path from the root to the vertex contains $i$ vertices (including both the root and the vertex).

**Lemma 10.1** *For any rooted binary tree* $T$, *there exists a one-to-one mapping* $f$ *from internal nodes to leaves such that for* $u$ *an internal node*

(i) $f(u)$ *is a descendant of* $u$

(ii) *all paths from* $f(u)$ *to* $u$ *for all* $u$ *are edge disjoint.*

*Proof.* The proof is by induction on the number $i$ of internal nodes. If $i = 1$, Lemma 10.1 is trivially true. Suppose that $i > 1$. Let $T'$ be the tree obtained from $T$ by deleting two leaves $x$ and $y$ having the same parent node $z$. By the inductive hypothesis, there exists a mapping $f'$ on $T'$ satisfying (i) and (ii). Suppose that $f'(w) = z$. Define $f(u) = f'(u)$ except $f(w) = x$ and $f(z) = y$.

It is easily verified that $f$ satisfies (i) and (ii). $\square$

**Theorem 10.2** $\rho(Q_k) \geq \frac{K}{K+1}$ *where* $K = \lfloor \log_2 k \rfloor$.

*Proof.* The theorem is trivially true for $n \leq k$. The general $n > k$ case is proved by induction on $n$. It suffices to prove the theorem for a full SMT $T$. Let $T_r$ be the rooted binary tree obtain from $T$ with the one-to-one mapping $f$. The length of an edge is its length in $M$.

Let $I_\ell$ be the set of interval nodes at the $\ell^{\text{th}}$ level. Define $U_d = \cup_{i \equiv d (\text{mod } K)} I_i$. Then $U_d$ for $d = 1, \cdots, K$ are disjoint. Let $p(u)$ denote the path from an interval node $u$ to $f(u)$. Define

$$L_d = \sum_{u \in U_d} \| p(u) \| .$$

By Lemma 10.1 (ii),

$$\| T_r \| \geq \sum_{d=1}^{K} L_d.$$

Therefore there exists a $d$ such that $L_d \leq \| T_r \| /K$. Consider $U_d$ for such a $d$. Decomposing $T_r$ at nodes of $U_d$ yields a collection of rooted binary trees each rooted at a node of $U_d$ and having leaves either in $U_d$ or in the leaf-set of $T_r$. Denote such a tree by $T_x$ if it is rooted at $x$. Clearly, each $T_x$ has at most $k$ leaves.

For each leave $u$ of $T_x$ which is an internal node of $T_r$, connect $u$ to $f(u)$ with an edge. Thus $T_x$ is transformed to a quasi ST $Q(x)$ on at most $k$ terminals. Clearly, the union of $Q(r)$ and $Q(x), x \in U_d$ is connected, hence it is a $k$-size quasi ST. Thus

$$Q_k \leq \| Q(r) \| + \sum_{x \in U_d} \| Q(x) \| \leq \| T_r \| + L_d \leq \frac{K+1}{K} \| T \| . \square$$

The case $k = 4$ was also obtained by Berman and Ramaiyer. Since $\rho(Q_k) \to 1$ as $k \to \infty$,

**Corollary 10.3** *For every $M$ such that $\| SMT(N) \| \neq \| MST(N) \|$ for all $N \in M$, there exists a $k$ large enough such that $Z_k$ is a better heuristic.*

**Corollary 10.4** $\rho(Z_{128}) > \sqrt{3}/2$ *for the Euclidean metric.*

For $k = 3$ Zelikovsky gave a better result.

**Theorem 10.5** $\rho(Q_3) > \frac{3}{5}$.

*Proof.* Let $T_r$ be defined as in Theorem 10.2 and let $T_r$ be a subtree of $T_r$ rooted at $u$. Let $I_u$ denote the set of terminal nodes of $T_u$. For each $u \in I_r$ let $g(u)$ denote a leaf closest to $u$, and let $p(u)$ denote the path from $u$ to $g(u)$. Then $g(u)$ can be chosen to equal $g(v)$ where $v$ is a children node of $u$. Suppose that $v$ and $w$ are the two children nodes of $u$. Let $a(u)$ be the unique simple path connecting $g(v)$ and $g(u)$ in $T_r$. Define $A_u = \cup_{i \in I_u} a(i)$. Then $A_u$ is a tree interconnecting $I_u$. It can be easily verified that $\| A_i \| \leq 2 \| T_r \| - 2 \| p(r) \|$. Define $\Delta(u) = \| A_u \| - \| T_u \|$. Then

$$\Delta(u) = \begin{cases} 0 & \text{if } u \text{ is a leaf,} \\ \Delta(v) + p(v) + \Delta(w) + p(w) & \text{otherwise.} \end{cases}$$

It is easily solved that $\Delta(u) = \sum_{i \in I_u - \{u\}} p(i)$.

Let $e = [u, v]$ where $v$ is a child node of $u$, and $v$ has two children nodes $x$ and $y$. Let $P_e$ denote the three terminals $g(w)$, $g(x)$ and $g(y)$, and let $S(e)$ be an SMT on

$P_e$. Since $p(u) \cap p(v) = p(v)$,

$$\| S_e \| \leq \| a(u) \| + \| a(v) \| - \| p(v) \| .$$

Let $A(T_1 \mid P_{e_1}, \cdots, P_{e_t})$ denote a tree obtained from $A_r$ by substituting $S(e_i)$ for $a(u_i)$ and $a(v_i)$, $i = 1, \cdots, t$. Then

$$\| A_r(P_e) \| \leq \| A_r \| - \| p(v) \| .$$

Furthermore, if $\{P_{e_1}, \cdots, P_{e_t}\}$ is a cycle-free set, then

$$\| A_r(P_{e_1}, \cdots, P_{e_t}) \| \leq \| A(r) \| - \sum_{i=1}^{t} \| p(v_i) \| .$$

It can be verified that $\{P_{e_1}, \cdots, P_{e_t}\}$ is a cycle-free set if and only if $e_i \cap e_j = \phi$ for all $1 \leq i < j \leq t$. Since the edge set $E$ of a rooted binary tree can be easily partitioned into three disjointed subsets where edges in each subset do not intersect, one subset, say, $E_1$ must satisfy

$$\sum_{e_i \in E_1} \| p(v_i) \| \geq \frac{1}{3} \sum_{e_i \in E} p(v_i).$$

Thus

$$\begin{aligned} \| A_r(\{P_{e_i} : e_i \in E_1\}) \| &\leq \| A_r \| - \frac{1}{3} \sum_{e_i \in E} p(v_i) \\ &\leq \| A_r \| - \frac{1}{3} \Delta(r) \\ &\leq \| A_r \| - \frac{1}{3} \{\| A_r \| - \| T_r \|\} \\ &= \frac{2}{3} \| A_r \| + \frac{1}{3} \| T_r \| . \\ &\leq \frac{5}{3} \| T_r \| - \frac{4}{3} \| p(r) \| . \end{aligned}$$

It follows immediately,

$$\rho(Q_3) \geq \frac{\| T_r \|}{\| A_r(\{P_{e_i} : e_i \in E_1\}) \|} > \frac{3}{5} . \square$$

**Corollary 10.6** $\rho(Z_3) \geq \frac{6}{11}$.

Note that the Steiner ratio is bounded by $n/2(n-1)$ for the graph problem.[23] Therefore, for $n > 12$, the heuristic $Z_3$ has a better lower bound for the performance ratio.

## 11. Lower Bounds of $\rho_k$ for the Rectilinear Metric

For the rectilinear metric Berman and Ramaiyer obtained better lower bounds of $\rho(Q_3)$.

**Theorem 11.1** $\rho(Q_3) \geq 4/5$ *for the rectilinear metric.*

*Proof.* (sketch): It suffices to prove Theorem 11.6 for full SMTs, i.e. each leaf is of degree one. Let $T$ denote a full SMT. As proved by Hwang[20] we may assume that $T$ is of the following two types (see Fig. 13).

**Type 1.** All Steiner points lie on one line, say, $x = 0$. Each regular point is connected to the $x = 0$ line by a horizontal edge which rotates to the left and to the right. The first and the last such edges can have length zero. Let the $i^{\text{th}}$ regular point $p_i$ have the Cartesian coordinates $(x_i, y_i)$, $i = 0, \cdots, n-1$, where $y_0 \le y_1 \le \cdots \le y_{n-1}$. Then

$$\| T \| = y_{n-1} - y_0 + \sum_{i=0}^{n-1} x_i.$$

Define $P_i = \{p_{2i}, p_{2i+1}, p_{2i+2}\}$ for $i = 0, 1, \cdots, n'$ where $n' \equiv \lceil n/2 \rceil - 2$. Let $S_i$ denote an SMT for $P_i$. Then

$$\| S_i \| = y_{2i+2} - y_{2i} + x_{2i+1} + \max\{x_{2i}, x_{2i+2}\}.$$

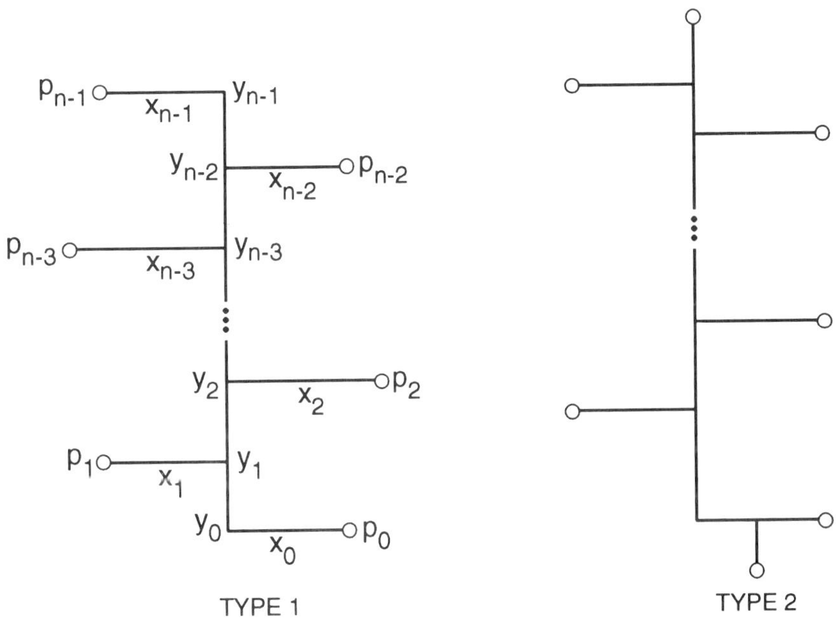


Figure 13. Two types of full SMTs

It is also easily verified that

$$\| T(P; P_0, \cdots, P_{n'}) \| = y_{n-1} - y_{n-2} + x_{n-1} + x_{n-2}.$$

Since $\{P_0, \cdots, P_{n'}\}$ is a cycle-free set, the tree

$$T_1 \equiv T(P; P_0, \cdots, P_{n'}) \cup \left( \sum_{i=0}^{n'} S_i \right)$$

is a 3-size quasi ST. Define $n^* = \lfloor n/2 \rfloor - 1$ and $x_i = 0$ for $i \geq n$. Then

$$\| T_1 \| = y_{n-1} - y_0 + \sum_{i=0}^{n^*} x_{2i+1} + \sum_{i=0}^{n^*} \max\{x_{2i}, x_{2i+2}\}.$$

We now consider a second cycle-free set which is obtained from $\{P_0, \cdots, P_{n'}\}$ by substituting $P'_{i-1} \neq \{p_{2i-2}, p_{2i-1}, p_{2i+1}\}$ for $P_{i-1}$ if $x_{2i-1} > x_{2i+1}$. Note that

$$\| S'_{i-1} \| \leq y_{2i+1} - y_{2i-2} + x_{2i-1} + x2i - 2.$$

By further considering two more cycle-free sets obtained similarly as the second set, Berman and Ramaiyer were able to show that the four trees together use up at most five copies of each edges of $T$. Therefore, one of the four trees, say, $T_1$, has a length at most $5/4 \| T \|$. Since $\| Q_3 \| \leq \| T_1 \|$, Theorem 11.6 is proved.
**Type 2.** This type differs from Type 1 in that $n$ must be even and the first Steiner point is connected to the second through a corner. The proof is similar to that for Type 1. □

**Corollary 1.2** *$\rho(Z_3) \geq 8/11$ for the rectilinear metric.*

Since Hwang[20] has proved that $\rho = 2/3 < 8/11$ for the rectilinear metric, $Z_3$ is a better heuristic.

Berman and Ramaiyer also claimed the following result:

**Theorem 11.3** *$\rho(Q_4) \geq 6/7$ and $\rho(Z_4) \geq 7/12$.*

In a later extended abstract Berman and Ramaiyer[2] generalized Theorem 11.1 to the following theorem.

**Theorem 11.4** *$\rho(Q_k) \geq (2k-2)/(2k-1)$ for $k \geq 2$ and the rectilinear metric.*

They also remarked that the $k = 3$ case was concurrently obtained by Zelikovsky.

## 12. Open Problems

We proved the existence of better polynomial-time heuristics. However, the running time of the heuristic is bounded by a polynomial of high degree. If we could prove that $\rho_3 > \sqrt{3}/2$, then we can get down the degree. Could $\rho_3$ be determined by the same approach as determining $\rho_2$? It is an open question right now. Some results along this line were given by Du, Zhang and Feng.[16] However, they have not characterized the set of $P_s$ on which the number of MSTs is maximized. They conjectured that $\rho_3 = (1+\sqrt{3})\sqrt{2}/(1+\sqrt{2}+\sqrt{3})$. In a space of dimension more than two, such a set is even harder to determine. For example, in the three-dimensional Euclidean space, all polytopes with only equilaterally triangular facets are candidates. Recently, W. D. Smith[30] showed that in the euclidean space of dimension $d, 3 \leq d \leq 9$, the Steiner ratio is not achieved by the vertex set of a regular simplex. This suggests that determining the Steiner ratio in a higher dimensional space is much more difficult but an interesting topic for further research.

Let $\rho(d)$ denote the Steiner ratio in a $d$-dimensional space. Chung and Gilbert[4] proved that $\lim_{d\to\infty} \rho(d) \leq \sqrt{3}/(4-\sqrt{2}) = 0.6698$ and conjectured that the equality sign holds for the inequality. Graham and Hwang[19] showed that $\lim_{d\to\infty} \rho(d) \geq \frac{1}{\sqrt{3}} = 0.57$. Du[8] pushed up this lower bound to 0.615.

In a normed plane, Du, Graham and Liu[10] conjectured that the Steiner ratio is between $2/3$ and $\sqrt{3}/2$. However, they could only prove an upper bound $(1+\sqrt{19})/6 = 0.8931$. Liu and Du[25] showed that in the $L_p$-normed plane, the Steiner ratio is upper bounded by $\sqrt{3}/2$.

If Du, Graham and Liu's conjecture is right, then the smallest Steiner ratio happens in the $L_1$-normed plane, i.e., the plane with rectilinear distance. Hwang[18] proved that in the $L_1$-normed plane, the Steiner ratio is exactly $2/3$. Graham and Hwang[19] conjectured that in a $d$-dimensional space with rectilinear distance, the Steiner ratio is $d/(2d-1)$.

Du and Graham[9] conjectured that the Steiner ratio in a normed space equals to that in its dual space.

**Acknowledgement** The first author wishes to thank Professors R. L. Graham, R. Tarjan, A. C. Yao and F. F. Yao for their encouragement in his research.

## References

1. P. Berman and V. Ramaiyer, An approximation algorithm for the Steiner tree problem, preprint.
2. P. Berman and V. Ramaiyer, Improved approximations for the Steiner tree problem, preprint.
3. M. Bern, Two probabilistic results on rectilinear Steiner trees, *Proceedings, 18th STOC* 1986) 433-441.
4. F. R. K. Chung and E. N. Gilbert, Steiner trees for the regular simples, *Bull. Inst. Math. Acad. Sinica*, **4** (1976) 313-325.
5. F. R. K. Chung and R. L. Graham, A new bound for euclidean Steiner minimum trees, *Ann. N.Y. Acad. Sci.*, **440** (1985) 328-346.
6. F. R. K. Chung and F. K. Hwang, A lower bound for the Steiner tree problem, *SIAM J. Appl. Math.*, **34** (1978) 27-36.
7. R. Crourant and H. Robbins, *What is Mathematics?*, (Oxford Univ. Press, New York, 1941).
8. D. Z. Du, On Steiner ratio conjectures, *Annals of Operation Research*, to appear.
9. D. Z. Du and R. L. Graham, The Steiner ratio in dual normed plane, in preparation.
10. D. Z. Du, R. L. Graham, and Z. C. Liu, Minimum Steiner trees in normed planes, preprint.
11. D. Z. Du and F. K. Hwang, A new bound for the Steiner ratio, *Trans. Amer. Math. Soc.*, **278** (1983) 137-148.
12. D. Z. Du and F. K. Hwang, The Steiner ratio conjecture of Gilbert-Pollak is

true, *Proc. Nat. Acad. Sci.* **87** (1990) 9464-9466.

13. D. Z. Du and F. K. Hwang, A proof of Gilbert-Pollak's conjecture on the Steiner ratio, *Algorithmica*, to appear.
14. D. Z. Du, F. K. Hwang and E. N. Yao, The Steiner ratio conjecture is true for five points, *J. Combinatorial Theory, Ser. A*, **32** (1982) 396-400.
15. D. Z. Du, E. N. Yao and F. K. Hwang, A short proof of a result of Pollak on Steiner minimal trees, *J. Combinatorial Theory, Ser. A*, **32** (1982) 396-400.
16. D. Z. Du, Y. J. Zhang, and Q. Feng, On better heuristic for euclidean Steiner minimum trees, *Proceedings 32nd FOCS* (1991).
17. M. R. Garey, R. L. Graham and D. S. Johnson, The complexity of computing Steiner minimal trees, *SIAM J. Appl. Math.*, **32** (1977) 835-859.
18. E. N. Gilbert and H. O. Pollak, Steiner minimal trees, *SIAM J. Appl. Math.*, **16** (1968) 1-29.
19. R. L. Graham and F. K. Hwang, Remarks on Steiner minimal trees, *Bull. Inst. Math. Acad. Sinica*, **4** (1976) 177-182.
20. F. K. Hwang, On Steiner minimal trees with rectilinear distance, *SIAM J. Appl. Math*, **30** (1976) 104-114.
21. F. K. Hwang and Y. C. Yao, Comments on Bern's probabilistic results on rectilinear Steiner trees, *Algorithmica* **5** (1990) 591-598.
22. V. Jarnik and O. Kossler, O minimalnich gratech obsahuijicich n danych bodu (in Czech), *Casopis Pesk. Mat. Fry.*, **63** (1934) 223-235.
23. R. M. Karp, Reducibility among combinatorial problems, in R. E. Miller and J. W. Tatcher (ed.), *Complexity of Computer Computation* (Plenum Press, New York, 1972) 85-103.
24. H. W. Kuhn, Steiner's problem revisited, in *Studies in Optimization*, G. B. Dantzig and B. C. Eaves (editor), (Math Assoc. of America, 1975) 98-107.
25. Z. C. Liu and D. Z. Du, On Steiner minimal trees with $L_p$ distance, to appear in *Algorithmica*.
26. Z. A. Melzak, On the problem of Steiner, *Canada. Math. Bull.*, **4** (1961) 143-148.
27. H. O. Pollak, Some remarks on the Steiner problem, *J. Combinatorial Theory, Ser. A*, **24** (1978) 278-295.
28. J. H. Rubinstein and D. A. Thomas, A variational approach to the Steiner network problem, *Proceedings of NATO Workshop on Topological Networks*, Copenhagen, Denmark, 1989.
29. J. H. Rubinstein and D. A. Thomas, The Steiner ratio conjecture for six points, to appear in *J. of Combinatorial Theory*.
30. W. D. Smith, How to find Steiner minimal trees in euclidea $d$-space, to appear in *Algorithmica*.
31. J. F. Weng, Steiner problem in hexagonal metric, unpublished manuscript.

32. A. Z. Zelikovsky, The 11/6-approximation algorithm for the Steiner problem on networks, manuscripts.

*Computing in Euclidean Geometry*, pp. 193–233
Eds. D.-Z. Du and F. K. Hwang


# VORONOI DIAGRAMS and DELAUNAY TRIANGULATIONS

STEVEN FORTUNE
*AT&T Bell Laboratories, 600 Mountain Avenue*
*Murray Hill, NJ 07974, USA*

ABSTRACT

The Voronoi diagram is a fundamental structure in computational geometry and arises naturally in many different fields. This chapter surveys properties of the Voronoi diagram and its geometric dual, the Delaunay triangulation. The emphasis is on practical algorithms for the construction of Voronoi diagrams.

## 1 Introduction

Let $S$ be a set of $n$ points in $d$-dimensional euclidean space $E^d$. The points of $S$ are called *sites*. The *Voronoi diagram* of $S$ splits $E^d$ into regions with one region for each site, so that the points in the region for site $s \in S$ are closer to $s$ than to any other site in $S$.

The *Delaunay triangulation* of $S$ is the unique triangulation of $S$ so that there are no elements of $S$ inside the circumsphere of any triangle. Here 'triangulation' is extended from the planar usage to arbitrary dimension: a triangulation decomposes the convex hull of $S$ into simplices using elements of $S$ as vertices. The existence and uniqueness of the Delaunay triangulation are perhaps not obvious, but these properties follow easily since it is the dual of the Voronoi diagram.

The Voronoi diagram and Delaunay triangulation of the same 250 points appear side by side in figure 1. The points are chosen uniformly in the unit square.

The Voronoi diagram is ubiquitous, arising independently in many different fields. The eponymous Voronoi in 1908 gave a careful definition of the structure, motivated by the study of quadratic forms as initiated by Gauss and Dirichlet. Crystalographers including Delaunay studied the problem of filling space with congruent copies of a set of crystals, starting in the 1920s. Metallurgists call Voronoi regions Wigner-Seitz zones after Wigner and Seitz, who in 1933 used them to study equilibrium properties of alloys. A name in geography for Voronoi regions is Thiessen polygons, following Thiessen who in 1911 used them to improve the estimation of precipitation. Blum suggested Voronoi diagrams as a descriptor for the shape of a set of objects in 1967, hence the use of the Blum transform in pattern recognition.

More recently, Voronoi diagrams have become a central subject in computational geometry. The motivating problem is nearest-site search. Given a fixed set of point sites, this problem is to be able to answer repeated queries of the form "which site is closest to point $q$?" This problem can be solved by computing the Voronoi

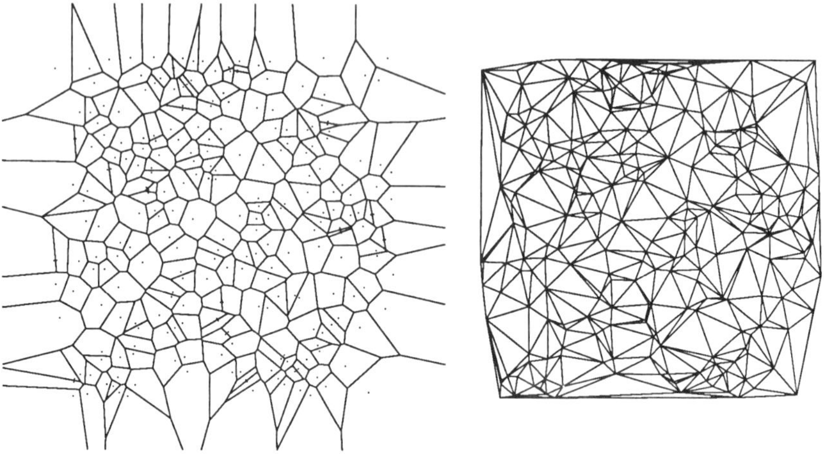

Figure 1: The Voronoi diagram (on the left) and the Delaunay triangulation (on the right).

diagram of the set of sites, and for each query locating the region of the Voronoi diagram containing the query point.

In an influential early paper, Shamos and Hoey[56] sketched a divide-and-conquer algorithm to compute the Voronoi diagram of a set of $n$ planar points in time $O(n \log n)$. Independently, Green and Sibson[32] gave an incremental algorithm for the same problem with $O(n^{3/2})$ expected running time. The incremental algorithm was generalized to arbitrary dimension by Bowyer[7] and Watson.[58] Brown[10] observed that a Voronoi diagram in dimension $d$ can be obtained from an appropriate convex hull in dimension $d+1$, so any convex hull algorithm can be used for Delaunay triangulations. Subsequently, algorithms were obtained for an amazing collection of generalizations of the Voronoi diagram, too numerous to mention except briefly. An easy generalization is to change the metric, either uniformly, for example by using a general $L_p$ metric or a convex distance function, or locally, for example by additive or multiplicative weights on sites. Another generalization is to change the shape of sites; circles and line segments have been considered. 'Higher order' Voronoi diagrams result by partitioning space on the basis of the $k$ nearest neighbors, for some fixed $k$; the 'furthest point' Voronoi diagram partitions space on the basis of the $n$th closest site, i.e., the furthest site. 'Constrained' Delaunay triangulations allow some edges to be specified before the triangulation is constructed. 'Geodesic' Voronoi diagrams are defined inside a polygonal environment, where the distance between two points is given by the length of the shortest path connecting the two points lying entirely inside the polygonal environment. 'Abstract' Voronoi diagrams can be used to study Voronoi diagram algorithms given only a set of axioms describing the behavior of the diagram.

Aurenhammer[3] gives a very inclusive survey of Voronoi diagrams, with more than 200 references. He summarizes the early history of Voronoi diagrams, gives many of the applications in mathematics and the natural sciences, and outlines many of the generalizations of Voronoi diagrams and algorithms for constructing them. References for all the topics just mentioned can be found in his paper.

This chapter also surveys Voronoi diagrams and Delaunay triangulations. The focus is narrow; the only topic is sets of point sites with the euclidean metric in $E^d$. This is the most important case for applications. This chapter gives a careful definition of the Voronoi diagram and the Delaunay triangulation and discusses elementary properties.

The emphasis of this chapter is on algorithms. One goal is to demonstrate that there are usable and efficient algorithms for computing Delaunay triangulations, both in two dimensions and in higher dimension. For, say, a set of points chosen from a uniform distribution inside a sphere, the combinatorial complexity of the Delaunay triangulation grows linearly with the number of sites. The random-incremental algorithm discussed below can compute the Delaunay triangulation of such a set of sites in time $O(n \log n)$, and the running time constant is small. Hence it is perfectly feasible to compute the Delaunay triangulation of "large" sets of points, say $10^5$ or $10^6$ sites in dimension two. A caveat is that the combinatorial complexity of the Delaunay triangulation grows exponentially with dimension. Hence the running time of any algorithm also grows exponentially with dimension. For small dimension, say dimension less than five, the Delaunay triangulation is still a useful tool.

Four fundamental algorithms are discussed in some detail. These are the flipping algorithm, the incremental algorithm, the random incremental algorithm, and the plane-sweep algorithm. These algorithms are fundamental because they are both theoretically important and reasonable implementation choices. The flipping algorithm is the simplest imaginable algorithm for computing the Delaunay triangulation in two dimensions. It starts with an arbitrary triangulation of the set of points and produces the Delaunay triangulation by a sequence of local modifications. The incremental algorithm is slightly more complicated. Sites are added one by one, with the Delaunay triangulation updated to include the new site after each addition. The advantage of the incremental algorithm is that it generalizes readily to higher dimension. The random incremental algorithm a recent innovation. It is the incremental algorithm with the insertion order chosen at random with all permutations equally likely. The random insertion order guarantees good worst-case performance. The plane-sweep algorithm is based on the plane-sweep paradigm. It provides a contrast because its analysis is based on different principles from the others.

All of the algorithms have been implemented; at least one has been used fairly extensively. The implementation of these algorithms is neither trivial nor excessively complicated. Some attention is given in the chapter to the geometric data structures and geometric primitives required to implement the algorithms, as well as implementation perils. A brief empirical comparison of the algorithms is included.

The background required for this chapter is a basic familiarity with algorithms and with geometry. Exposure to computational geometry is certainly helpful but is not essential. Standard references in computational geometry are books by Edelsbunner[23] and by Preparata and Shamos.[50] Many concepts from the theory of convex polyhedra are used throughout the chapter. Appendix 1 defines the terms that are used; it is adequate for reference and review but is certainly inadequate as a primer. Standard references on convex polyhedra are books by Grunbaum[33] and Brøndsted.[9]

The table of contents is given below. Section 2 defines the Voronoi diagram and the Delaunay triangulation and gives basic properties. One such property is a connection with convex polyhedra in dimension one higher. This connection is fundamental to the analysis of many of the algorithms and is used throughout the chapter. Section 3 of the chapter is an eclectic enumeration of results known about the Voronoi diagram and Delaunay triangulation. For example, the Delaunay triangulation has 'optimality' properties that makes it attractive for use in finite element analysis. The bulk of the chapter is the section 4 on algorithms; it is independent of section 3.

**Contents**

## 2 Definition of the Voronoi diagram and Delaunay triangulation

The Voronoi diagram of a set of sites partitions euclidean space $E^d$ on the basis of the site closest to a particular point. To handle the cases where a point is closest to several equidistant sites, the Voronoi diagram is best defined as a cell complex. A *cell* is the intersection of a finite number of hyperplanes and open halfspaces, and a *cell complex* is a finite collection of pairwise disjoint cells so that every face of every cell is in the collection. Similarly, the Delaunay triangulation is a cell complex decomposing the convex hull of the set of sites and having the sites as vertices.

The next section gives a careful definition of the Voronoi diagram and Delaunay triangulation as cell complexes and discusses elementary properties of these two structures. Section 2.2 outlines a fundamental connection between Delaunay triangulations in $E^d$ and convex hulls in $E^{d+1}$, and between Voronoi diagrams in $E^d$ and intersection of halfspaces in $E^{d+1}$. The topic of section 2.3 is the combinatorial complexity of a Voronoi diagram, that is, the number of cells in the Voronoi diagram, as a function of the number of sites. Known upper and lower bounds are given, both in the worst case and in the 'typical' case.

### *2.1 Voronoi diagrams and Delaunay triangulations*

Let $S$ be a set of $n$ point sites in $E^d$. For a nonempty subset $R$ of $S$, the *Voronoi cell* of $R$, $V(R)$, is the set of all points of $E^d$ equidistant from all sites in $R$ and closer to every site of $R$ than to any site not in $R$. Clearly, for a singleton set $R = \{r\}$, $V(R)$ is the set of all points strictly closer to $r$ than to any other site. Any point of $E^d$ lies in $V(R')$ for some $R' \subset S$, though $R'$ may have more than one element. For $R \subseteq S$, $V(R)$ may be empty, either because there is no point equidistant from all $r \in R$ or because any point equidistant from all $r \in R$ is also equidistant from some $r' \in S - R$. The *Voronoi diagram* $V$ is the collection of all nonempty Voronoi cells $V(R)$, for $R$ a subset of $S$.

Figure 2(a) depicts a Voronoi diagram of a set $S$ of 11 sites in two dimensions. For each site there is an open two-dimensional cell consisting of the points in the plane for which the site is the strict closest among all the sites. Each such cell has a boundary consisting alternately of Voronoi edges (one-dimensional cells) and Voronoi vertices (zero-dimensional cells). Voronoi edges are equidistant from two sites and Voronoi vertices are equidistant from at least three sites. The unbounded Voronoi cells are exactly the cells corresponding to sites on the convex hull of $S$. Figure 3(a) depicts the Voronoi diagram of a set $S$ of 7 sites in three dimensions. The boundary of each 3-cell consists of facets (2-cells), edges, and vertices.

If $R \subset S$ and $V(R)$ is a nonempty Voronoi cell, then the *Delaunay cell* $D(R)$ is cell($R$) (the relative interior of the convex hull of $R$). The *Delaunay triangulation* $D$ is the collection of Delaunay cells $D(R)$, where $R$ varies over subsets of $S$ with $V(R)$ nonempty. An immediate property of the Delaunay triangulation is that it has *empty circumspheres*: for $R \subseteq S$, $D(R)$ is a Delaunay cell exactly if there is a circumsphere of $R$ that contains no site of $S-R$ in its interior. Such a circumsphere

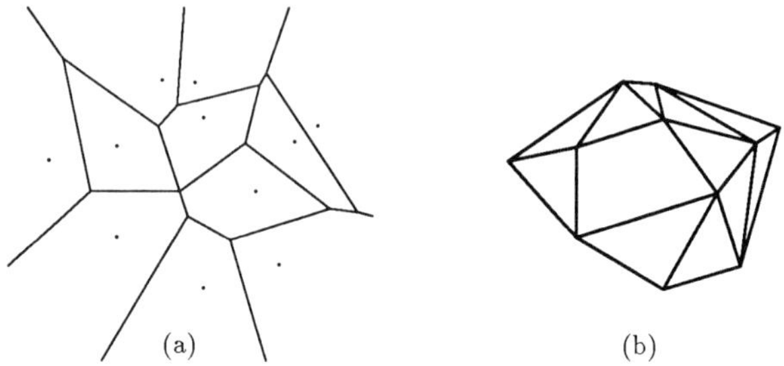

(a) (b)

Figure 2: (a) A Voronoi diagram in dimension 2; (b) the Delaunay triangulation of the same sites.

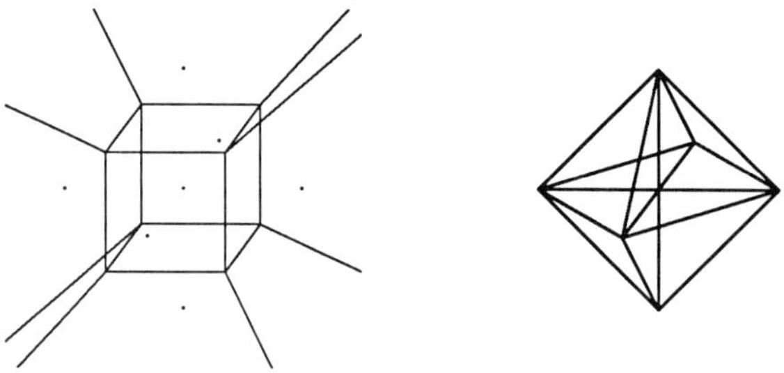

(b)

(a)

Figure 3: (a) A Voronoi diagram in dimension 3; (b) the Delaunay triangulation of the same sites.

can be obtained with center any point in the Voronoi cell $V(R)$, so the circumsphere is unique exactly if $V(R)$ is a vertex.

Figure 2(b) gives the two-dimensional Delaunay triangulation of the same sites as in figure 2(a). Notice the correspondence between a Delaunay 2-cell and a Voronoi vertex (the cell need not contain the vertex), between a Delaunay edge and a Voronoi edge, and between a Delaunay vertex, i.e. a site, and a Voronoi 2-cell. In figure 2(b) one of the Delaunay 2-cells is a quadrilateral with four sites as vertices; this is because the four sites are cocircular. Thus the Delaunay triangulation need not be a proper triangulation, i.e., its cells need not all be simplices. The Delaunay quadrilateral corresponds to the vertex of degree four in the Voronoi diagram in figure 2(a). Figure 3(b) gives the three-dimensional Delaunay triangulation of the same sites as in figure 3(a). The correspondence is now between a Delaunay 3-cell and a Voronoi vertex, between a Delaunay facet and a Voronoi edge, between a Delaunay edge and a Voronoi facet, and between a site and a Voronoi 3-cell.

The following theorem gives fundamental properties of the Delaunay triangulation and the Voronoi diagram. One proof can be obtained using standard properties of convex polyhedra together with the connection to convex polyhedra discussed in the next section. A direct proof, possibly more illuminating, is sketched in Appendix 2.

**Theorem 2.1** *Let $S$ be a set of $n$ points in $E^d$ with Voronoi diagram $V$ and Delaunay triangulation $D$.*

1. *$V$ is a cell complex that partitions $E^d$.*
2. *$D$ is a triangulation of $S$.*
3. *$V$ and $D$ are dual, that is, for $R, R' \subseteq S$, $V(R)$ is a face of $V(R')$ iff $D(R')$ is a face of $D(R)$.*
4. *If $R \subseteq S$, $V(R)$ is unbounded iff every site of $R$ is on the boundary of the convex hull of $S$.*

A nonempty Voronoi cell $V(R)$ has the same dimension as the flat of points equidistant from all sites in $R$, since it is a relatively open subset of the flat. It is not hard to check that the flat has dimension $d$ minus the dimension of the convex hull of $R$. Hence the dimensions of $V(R)$ and $D(R)$ always sum to $d$. Since the dimension of $D(R)$ is at most $|R|-1$, the dimension of $V(R)$ is at least $d+1-|R|$.

If $V(R)$ is not empty and the dimension of $V(R)$ exceeds $d+1-|R|$; then $V(R)$ and $D(R)$ are *degenerate*; this happens only if the sites of $R$ lie on a common sphere and also on a common flat of dimension less than $|R|-1$. If $|R| = 1, 2, 3$, then $V(R)$ and $D(R)$ are never degenerate: this is obvious for $|R| = 1$ and $|R| = 2$, and for $|R| = 3$ follows from the observation that three sites cannot be both collinear and cospherical.

In low dimensions, degenerate Voronoi cells can be enumerated explicitly. In dimension 2, the only degenerate Voronoi cell is a vertex that is equidistant from four or more sites, as in Figure 2. The corresponding Delaunay cell is the interior of the convex hull of the sites. In dimension 3, a Voronoi vertex can be equidistant from five or more sites, and the corresponding Delaunay cell is the interior of a convex polyhedron. A second degeneracy in dimension 3 is a Voronoi edge equidistant from four or more sites. In this case the four sites are cocircular and also lie on a common plane, so the Delaunay cell is the interior of a circularly-inscribed polygon sitting in 3-space.

It can happen that $k+2$ sites lie on a common $k$-flat on the boundary of the convex hull of the set of sites, for some $k < d$. Then some boundary face of the convex hull is the union of several Delaunay cells.

$S$ is in *general position* if it is affinely independent and if no $d+2$ points lie on a common sphere. Clearly, if $S$ is in general position, then all Delaunay and Voronoi cells are nondegenerate, so every Delaunay cell $D(R)$ is a simplex, every Voronoi cell $V(R)$ has dimension $d+1-|R|$, and every face of the convex hull of $S$ is a Delaunay cell.

It is convenient to have a local characterization of the Delaunay triangulation. Two opposite $d$-cells cell$(R)$ and cell$(R')$ are *locally Delaunay* if $R$ and $R'$ both have circumspheres $C$ and $C'$, and $R'-R$ is outside $C$ (equivalently, $R-R'$ is outside $C'$). A triangulation $T$ is *locally Delaunay* if every pair of opposite $d$-cells is locally Delaunay.

**Lemma 2.2** *A triangulation is Delaunay iff it is locally Delaunay.*

**Proof.** (Sketch) If $T$ is Delaunay, then easily $T$ is locally Delaunay. Suppose $T$ is locally Delaunay. Choose a $d$-cell cell$(R)$ and site $s \in S-R$; it suffices to show that $s$ is outside the circumsphere of $R$. We can choose a point $r \in$ cell$(R)$ so that open segment $rs$ avoids all $k$-cells of $T$, for $k \leq d-2$, and intersects $(d-1)$-cells in a single point. Intersection with segment $rs$ gives a linear order on an alternating sequence of $d$- and $(d-1)$-cells cell$(R) = P_0, Q_1, P_1, \ldots, P_{m-1}, Q_m, P_m$ ending at a $d$-cell $P_m$ incident to $s$. An easy downwards induction, $i = m, \ldots, 1$, using the local Delaunay property shows that the hyperplane through $Q_i$ separates $s$ from $P_{i-1}$ and that $s$ lies outside the circumsphere of $P_{i-1}$. The lemma follows using the property that the Delaunay triangulation is the unique triangulation having empty circumspheres. □

## *2.2 Connection with convex polyhedra*

The Delaunay triangulation of a set $S$ of point sites in dimension $d$ can be obtained by appropriately lifting the sites to $d+1$ dimensions and then projecting the lower convex hull of the lifted sites back to $d$ dimensions. The Voronoi diagram in dimension $d$ can be obtained by appropriately mapping sites to hyperplanes in dimension $d+1$, taking the intersection of the upper halfspaces, and projecting back to $d$ dimensions. Much of the material in this section is from Edelsbrunner and Seidel[22]; they extend earlier work of Brown.[10] Edelsbrunner's book[23] is another source.

We now describe the appropriate mappings. To make the geometry easier to visualize, we consider the case when the set of sites $S$ is in dimension 2. We identify

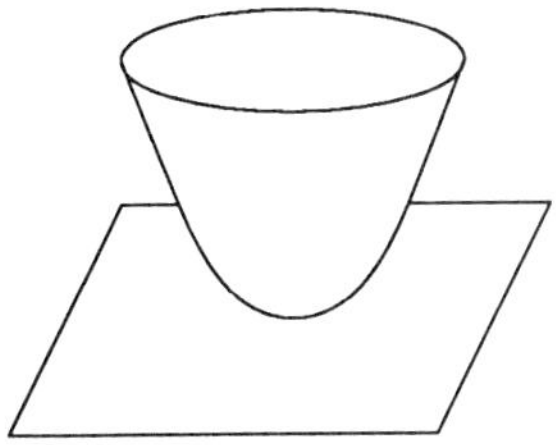

Figure 4: The paraboloid $\Lambda$.

$E^2$ with the plane spanned by the first 2 coordinate axes of $E^3$; it is the *base plane*. We think of the third coordinate axis as the vertical direction. We write for example $x$ for the point $(x_1, x_2)$ and we use dot products, thus $x \cdot p$ is $x_1p_1+x_2p_2$. If $x = (x_1, x_2)$ and $r$ is a real, then $(x, r)$ is $(x_1, x_2, r)$. This notation makes the generalization to higher dimension obvious.

For Delaunay triangulations, the appropriate *lifting map* $\lambda : E^2 \to E^3$ is $\lambda(x) = (x_1, x_2, x_1^2 + x_2^2) = (x, x \cdot x)$. The image of $\lambda$ is

$$\Lambda = \{(x, z) : x \in E^2, z = x \cdot x\}.$$

$\Lambda$ is a paraboloid of revolution (about the vertical axis). See Figure 4.

A plane is *nonvertical* if it is not parallel to the vertical axis. Clearly, if $P$ is a nonvertical plane, then any point in $E^3$ is either above, on, or below $P$. The crucial property of $\Lambda$ is the following. Any nonvertical plane $P$ either misses $\Lambda$, touches it at a tangent point, or the projection of $P \cap \Lambda$ onto the base plane is a circle. In the last case, the portion of $\Lambda$ below the plane projects inside the circle, and the portion above projects outside. Conversely, given any circle in $E^2$ there is a plane that intersects $\Lambda$ in a set whose projection is the circle. To see this, note that for any nonvertical plane $P$ there is a real $c$ and a point $p \in E^2$ so that

$$P = \{(x, z) : x \in E^2, z = 2x \cdot p - p \cdot p + c\}.$$

The vertical distance from the point $(x, 2x \cdot p - p \cdot p + c)$ of $P$ to the point $(x, x \cdot x)$ of $\Lambda$ is

$$x \cdot x - 2x \cdot p + p \cdot p - c = r^2 - c,$$

where $r$ is the distance from $x$ to $p$. If $c < 0$, this is positive for all $x$, and $P$ is below $\Lambda$. If $c = 0$, then this is zero exactly at $x = p$, and it is easy to see that $P$ is tangent to $\Lambda$ at $(p, p \cdot p)$. If $c > 0$, then the projection of $P \cap \Lambda$ on the base plane is the circle with center $p$ and radius $\sqrt{c}$. A point $(x, x \cdot x)$ of $\Lambda$ is above $P$ iff $r^2 > c$, i.e., iff $x$ is outside the circle. The converse direction, obtaining a plane from a circle, is similar.

Suppose $S$ is a set of sites in the base plane. Consider the convex hull $H$ of $\lambda(S)$. A face of $H$ is a *lower* face if it has a nonvertical supporting plane $P$ so that $H$ lies on or above $P$. Equivalently, a lower face is visible to an observer positioned at

(a) (b)

Figure 5: (a) A Delaunay triangulation; (b) the convex hull of the lifted sites.

a very negative vertical coordinate without looking through the interior of $H$. The Delaunay triangulation of a set of sites $S$ is shown in figure 5(a); 5(b) shows the convex hull $H$ of $\lambda(S)$.

**Lemma 2.3** *Let $H$ be the convex hull of $\lambda(S)$ and $D$ the Delaunay triangulation of $S$. Then $D$ is the set of projections of lower faces of $H$ onto the base hyperplane.*

**Proof.** Suppose $F$ is a lower face of $H$, then $F = \text{cell}(\lambda(R))$, for some nonempty $R \subseteq S$. Since $F$ is lower, there is some nonvertical supporting plane $P$ with $\lambda(R) \subseteq P$ and $\lambda(S-R)$ above $P$. The projection of $P \cap \Lambda$ onto the base plane is a circle $C$. Since $\lambda(R) \subseteq P \cap \Lambda$, $R$ lies on $C$; since $\lambda(S-R)$ lies above $P$, $S-R$ is outside $C$. Hence cell($R$) is Delaunay, but cell($R$) is also the projection of $F = \text{cell}(\lambda(R))$ onto the base plane. The converse argument is essentially the reverse. □

A result that is similar to lemma 2.3 can also be obtained using stereographic projection.[10] Place a unit-radius 2-sphere $\Sigma$ so that it is tangent to the base plane, and let $t$ be the point of $\Sigma$ diametrically opposite the point of tangency. The *stereographic projection* $\sigma(x)$ of a point $x$ in the base plane is the point of intersection of $\Sigma$ with the line segment $tx$. The crucial property of stereographic projection[50] is that a circle in the base plane maps to a circle on $\Sigma$. Let $H_\sigma$ be the convex hull of $\{t\} \cup \sigma(S)$. An argument similar to lemma 2.3 shows that $\sigma$ is a bijection between cells of $D$ and the faces of $H_\sigma$ that are not incident to $t$.

Now we consider Voronoi diagrams. The appropriate mapping sends the site $s$ to the plane

$$P_s = \{(x, z) : x \in E^2, z = 2x \cdot s - s \cdot s\}.$$

Clearly $P_s$ is tangent to $\Lambda$ at $(s, s \cdot s)$. Consider also the surface

$$U_s = \{(x, z) : x \in E^2, z = -x \cdot x + 2x \cdot s - s \cdot s\}.$$

At a point $x$ in the base plane, the vertical height of $U_s$ is the negative square of the distance from $x$ to $s$. Thus $U_s$ is an upside-down paraboloid touching the base plane at $s$ and below the base plane elsewhere. See Figure 6 for the situation in dimension 1. If we think of $P_s$, $U_s$ and $\Lambda$ as functions from the base plane to the vertical direction,

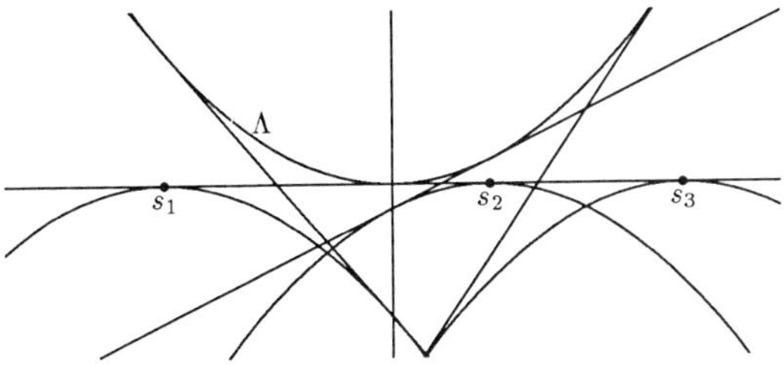


Figure 6: The relation among $\Lambda$, $P_s$, and $U_s$.

then obviously $U_s = P_s - \Lambda$. The crucial point is that $\Lambda$ is independent of $s$: for sites $s, s'$ and any $x$ in the base plane, we have $P_s(x) > P_{s'}(x)$ iff $U_s(x) > U_{s'}(x)$ iff $x$ is further from $s$ than from $s'$. Similarly, $P_s(x) = P_{s'}(x)$ iff $x$ is equidistant from $s$ and $s'$.

**Lemma 2.4** *Let $I$ be the intersection of the closed halfspaces above the planes $\{P_s\}$. The Voronoi diagram $V$ is the set of projections of faces of $I$ onto the base plane.*

**Proof.** $F$ is a face of $I$ exactly if there is a subset $R$ of $S$ so that $F \subseteq P_r$ for every $r \in R$ and $F$ is above $P_{r'}$ for every $r' \in S - R$. Hence if $(x, z) \in F$ for some $x$ in the base plane, then $x$ is equidistant from every $r \in R$ and closer to every $r \in R$ than to any $r' \in S - R$. Hence $x \in V(R)$. The converse is similar. □

It is possible to give a visual interpretation of lemma 2.4. Use opaque paint to give each surface $U_s$ a different color. An observer sitting at a very positive vertical coordinate will see the *upper envelope* of the collection of surfaces, that is, the portion of each surface not obscured by some surface above it. The pattern visible to the observer is the Voronoi diagram of $S$, since a surface $U_s$ obscures all other surfaces exactly at points $x$ of the base plane for which $x$ is closer to $s$ than to any other site. Now instead paint each plane $P_s$ using corresponding colors. The observer sitting inside $I$ at a very positive vertical coordinate sees exactly the same pattern, since for any $x$, the vertical order of the planes $\{P_s\}$ is the same as the vertical order of the surfaces $\{U_s\}$.

### *2.3 Combinatorial complexity*

Let $S$ be a set of $n$ sites in $E^d$. The *combinatorial complexity* of the Voronoi diagram $V$ is the number of cells in $V$. Clearly, the number of cells of $V$ is the same as the number of cells of the Delaunay triangulation $D$.

If $S$ is in general position, then the total number of cells of $D$ is at most the number of $d$-cells of $D$ times a constant depending exponentially on $d$. This follows because any cell of $D$ is a face of some $d$-cell of $D$, and if $S$ is in general position, then each $d$-cell is a simplex and has at most $\binom{d+1}{k+1}$ $k$-faces. If $S$ is not in general position, then for example the sites could all be cospherical, and there is only a single $d$-cell but many $k$-cells for $k < d$.

In dimension 2 it is easy to show using planarity that the number of vertices of $V$ (or triangles of $D$) is at most $3n - 6$ and the number of edges of $V$ (or edges of $D$) is at most $2n - 4$. In dimension higher than 2, it is possible to bound the number of cells of $V$ using the upper bound theorem for convex polyhedra.[9,23] The upper bound theorem states that the number of $j$-faces of a convex polyhedron in dimension $d$ with $n$ facets is asymptotically $O(n^{\min(d-j,\lfloor d/2 \rfloor)})$. ($\lfloor x \rfloor$ and $\lceil x \rceil$ are the integers resulting from $x$ by rounding down and up, respectively.) Using the correspondence of section 2.2 between the Voronoi diagram $V$ in dimension $d$ and the convex polyhedron $I$ in dimension $d+1$, the number of $j$-faces of $V$ is asymptotically at most $O(n^{\min(d+1-j,\lceil d/2 \rceil)})$. Hence the combinatorial complexity of $V$ is at most $O(n^{\lceil d/2 \rceil})$.

A simple example attains this worst-case bound, at least within a constant factor independent of $n$. First consider dimension 3, with coördinate axes $x_1$, $x_2$, and $x_3$. Let $P_1$ be the segment along the $x_1$ axis with endpoints $(0,0,0)$ and $(1,0,0)$ and $P_2$ the segment in the $x_1 = 0$ plane parallel to the $x_2$ axis with endpoints $(0,0,1)$ and $(0,1,1)$ Put $n/2$ sites equally spaced along $P_1$ and $n/2$ sites equally spaced along $P_2$. It is easy to see that there is an empty sphere through any two adjacent sites on $P_1$ and any two adjacent sites on $P_2$. This gives a total of $\theta(n^2)$ Delaunay triangles. Notice that each site can be perturbed within a small ball while maintaining the $\theta(n^2)$ complexity. This example can be extended to give $\theta(n^{\lceil d/2 \rceil})$ cells in any odd dimension $d$.

Using known exact bounds on the number of faces of a convex polyhedron and the connection between convex polyhedra and Voronoi diagrams, Seidel[54] gives *exact* bounds on the worst-case number of $j$-faces of a Voronoi diagram in $d$ dimensions, for all $j$ and $d$.

For a set of point sites chosen 'at random', the combinatorial complexity of the Voronoi diagram is probably much smaller. However, the available results are not as general as might be desired. One model of 'random sites' is an infinite set of point sites given by a unit-intensity Poisson process[53] in $E^d$; then the expected number of sites in a subset of $E^d$ is equal to the measure of the subset. Meijering[45] shows that the expected number of Voronoi neighbors of a site is a constant depending only on the dimension $d$. The constant is 6 for $d = 2$ and about 15.54 for $d = 3$.

A second model of 'random' sites is point sites chosen with a uniform distribution inside a bounded set $H \subset E^d$. The situation is more complicated because of edge effects near the boundary of $H$. Dwyer[20] shows that the expected number of Delaunay $d$-cells for a set of $n$ sites chosen uniformly within a sphere in $E^d$ is $\sim c_d n$. The constant $c_d$ agrees with the constants given by Meijering for $d = 2$ and $d = 3$. For $d > 5$, an estimate[21] of the constant accurate to within 1% is

$$c_d \approx \frac{\sqrt[4]{e}\sqrt{2/\pi}}{d(d+1)}(2\pi d)^{d/2}.$$

Dwyer conjectures that the bound holds for point sets chosen uniformly from within an arbitrary convex set $H$. If only a small fraction of the points lie near the boundary of $H$, then edge effects are negligible.

## 3 Properties of the Voronoi diagram and Delaunay triangulation

### *3.1 Optimality of the Delaunay triangulation*

Good finite element meshes avoid long, skinny triangles as much as possible. A classic result of Lawson[42] is that two-dimensional Delaunay triangulations optimize one formalization of this criterion.

**Lemma 3.1** *Over all proper triangulations of a set of sites $S \subset R^2$ in general position, the Delaunay triangulation maximizes the minimum angle of any triangle.*

**Proof.** Let angles($T$) be the sorted sequence of angles of triangles in $T$. The lemma follows by showing that the Delaunay triangulation $D$ is the triangulation that lexicographically maximizes angles($T$), where smaller angles are more significant in the ordering. We show that if $T$ is not the Delaunay triangulation, then there is a lexicographically larger triangulation $T'$. If $T$ is not Delaunay, then by Lemma 2.2 it is not locally Delaunay, so there must be two triangles $abc$ and $acd$ so that $d$ is inside the circumcircle of triangle $abc$ (and $b$ is inside the circumcircle of $acd$). Necessarily $abcd$ is a convex quadrilateral. Some easy elementary geometry shows that the minimum angle of triangles $abd$ and $bcd$ is larger than the minimum angle of triangles $abc$ and $acd$. Hence if $T'$ is the triangulation obtained from $T$ by replacing edge $ac$ with edge $bd$, then angles($T'$) is larger in the lexicographic ordering than angles($T$). □

If the set of sites $S$ is not in general position, then some sites may be cocircular and a Delaunay cell may be the interior of a circularly-inscribed convex polygon with more than three sides. In this case the Delaunay triangulation $D$ can be completed to a proper triangulation by adding diagonals to the convex polygon. The minimum angle in the resulting triangulation is independent of how diagonals are added, since it is determined by the shortest side of the convex polygon. Essentially the same argument as lemma 3.1 shows that any such completion of the Delaunay triangulation maximizes the minimum angle over all triangulations. However, one might require that the triangulation lexicographically maximize the sorted sequence of angles, in which case it matters very much how diagonals are added. Mount and Saalfeld[47] give an efficient algorithm that determines how to add diagonals so that the resulting triangulation is the lexicographic maximum.

No generalization of lemma 3.1 is known for higher dimensions, at least in terms of some angular measure of simplices. Rajan[51] gives a different optimality criterion that does hold in any dimension.

The *min-containment sphere* of a simplex is the minimum-radius sphere that contains the simplex. If the center of the circumsphere of the simplex lies inside the simplex, then the circumsphere is the min-containment sphere. Otherwise, the center of the min-containment sphere lies on some face of the simplex. For example, in two dimensions, if a triangle is acute, then its min-containment circle is its circumcircle, and if the triangle is obtuse, then its min-containment circle has center at the midpoint of the longest edge. For $T$ a proper triangulation, let maxrad$(T)$ be the maximum radius of any min-containment sphere of any simplex in $T$.

**Lemma 3.2** *Over all proper triangulations $T$ of a set of sites $S \subset E^d$ in general position, the Delaunay triangulation minimizes* maxrad$(T)$.

**Proof.** (Sketch) Recall the lifting map $\lambda$ from section 2.2. A proper triangulation $T$ in $E^d$ can be lifted to a polyhedral surface $\hat{T}$ in $E^{d+1}$ by mapping cell$(R)$ to cell$(\lambda(R))$. An easily verified fact is that the square of the radius of the min-containment sphere of a cell $C$ of $T$ is the maximum over $x$ in $C$ of the vertical distance between the paraboloid $\Lambda$ and the polyhedral surface $\hat{T}$ above $x$. The maximum vertical distance between $\Lambda$ and the polyhedral surface $\hat{T}$ is minimized if $\hat{T}$ is the convex hull of $\lambda(S)$, i.e. if $T$ is the Delaunay triangulation $D$. □

### *3.2 Geometric graph properties*

A Delaunay triangulation or Voronoi diagram can be thought of as a geometrically embedded graph by restricting attention to just its vertices and edges. A number of results exploit this geometric graph structure.

Several classical graphs with vertex set $S \subset E^d$ capture some form of the neighborhood structure of $S$. The *euclidean minimum spanning tree EMST* of $S$ is the spanning tree of $S$ that minimizes the sum of the euclidean lengths of the edges in the tree. The *relative neighborhood graph*[57] *RNG* of $S$ has an edge connecting $s_i$ and $s_j$ if there are no sites in the interior of the intersection of the two discs centered at $s_i$ and $s_j$ with radius the distance between $s_i$ and $s_j$. The *Gabriel graph*[31] *GG* of $S$ has an edge connecting $s_i$ and $s_j$ if there are no sites in the interior of the disc having as diameter the segment $s_i s_j$. It is easy to see that the edge sets of these graphs satisfy

$$EMST \subseteq RNG \subseteq GG \subseteq D.$$

In two dimensions, these containment relations can form the basis of algorithms that compute these graphs.[50]

A graph $G$ with vertex set $S \subset E^d$ has *dilation c* if for any $s_i, s_j \in S$, the length of the shortest path between $s_i$ and $s_j$ along edges of $G$ is at most $c$ times the euclidean distance between $s_i$ and $s_j$. Here the length of an edge is just the euclidean distance between its endpoints. Clearly the complete graph has dilation 1, but it has $O(n^2)$ edges. In two dimensions, Delaunay triangulations give planar graphs with small dilation but only $O(n)$ edges. For any set $S$, Chew[12] gives a planar graph $\Delta = \Delta(S)$ with dilation $c_\Delta = 2$; the edges of $\Delta$ are edges of the Delaunay triangulation of $S$ using a convex distance function whose unit ball is an equilateral triangle. Naively it would seem that the dilation $c_D$ of the graph with edges from the regular Delaunay triangulation $D$ would be smaller, since a circle is much more symmetric than a triangle. However this has not been shown; Keil and Gutwin[40] show that $c_D \leq 2\pi/(3\cos(\pi/6)) \approx 2.42$. It is not hard to see that $c_D \geq \pi/2$. Chew[12] conjectures that the true bound for $c_D$ is close to $\pi/2$. Using other methods, it is possible to get graphs with small dilation. For any set of $n$ points in $E^d$ and any $\epsilon > 0$, Vaidya[59] is able to construct a graph with dilation $1+\epsilon$ and $O(n/\epsilon^d)$ edges.

Given a cell complex whose union is the plane, one can ask if it is the Voronoi diagram of some set of sites $S$. Ash and Bolker[1] give a criterion that can decide this question. Let $G$ be the edge graph of the cell complex and assume that each vertex has three edges incident. For each vertex $v$ of $G$ there are three rays with endpoint $v$ so that the edges incident to $v$ bisect the angles formed by the rays. Each ray enters one of the planar regions incident to $v$; the ray entering planar region $R$ is the *central ray* from $v$ into $R$. Ash and Bolker show that the cell complex is a Voronoi diagram iff for each region the central rays entering the region have a point in common. If the central rays entering the region intersect in a unique point, then it is the site determining the region (the point is unique in all but extremely degenerate situations). Ash and Bolker[2] give generalizations for other forms of planar Voronoi diagrams. Apparently no comparable condition is known for Voronoi diagrams in higher dimensions. The analogous question of whether a cell complex is a Delaunay triangulation is easy to answer: lemma 2.2 gives an easy local characterization of Delaunay triangulations.

### *3.3 Inscribability*

Steinitz's theorem[33] states that the edge graphs of convex polytopes in three dimensions are exactly the three-connected planar graphs. A long-standing open question of Steiner was to characterize which of these graphs are *inscribable*, that is, which graphs are actually the edge graphs of convex polytopes with all vertices lying on a common sphere. Rivin and Smith[52] give the following answer. A three-connected planar graph $G$ is inscribable iff the edges of $G$ can be assigned real number weights so that the weight of a circuit around a face is exactly one and the weight of a circuit not bounding a face is strictly greater than one. Note that three-connected planar graphs have combinatorially unique embeddings, so the faces of $G$ are uniquely determined. A variant of the ellipsoid algorithm for linear programming can determine whether the weights exist, in time polynomial in the number of vertices of $G$.

Rivin and Smith's theorem has some interesting consequences for Delaunay triangulations in two dimensions. One is a test to decide if an undirected graph is the edge graph of a Delaunay triangulation. Suppose $G$ is a three-connected planar graph with a specified 'external' face and specified 'extreme' vertices incident to the external face. $G$ has a *Delaunay realization* if it can be embedded in the plane so that its edges become Delaunay edges, the 'external' face is unbounded, and 'extreme' vertices are extreme on the convex hull of the embedded vertices. Notice that four or more vertices around an internal face must be embedded on a common circle and a chain of vertices bounding the external face between extreme vertices must be embedded on a common line. Let $G'$ be the graph obtained from $G$ by inserting a new vertex in the external face with edges to the extreme vertices. Using stereographic projection

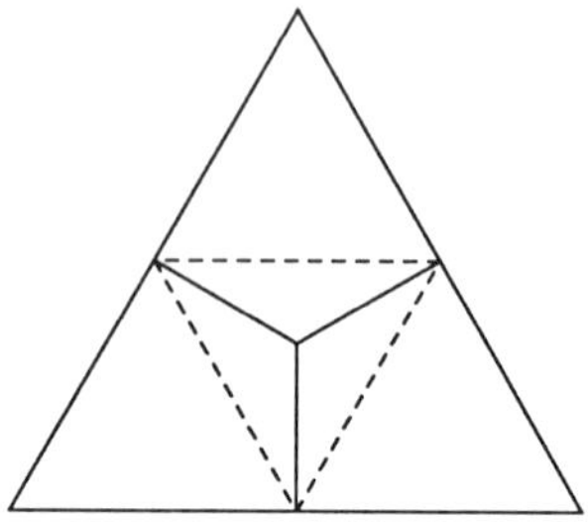

Figure 7: If the dashed edges are added, the resulting graph is not Delaunay

as described in section 2.2, it is easy to see that $G$ has a Delaunay realization exactly if $G'$ is inscribable (the new vertex of $G'$ gets mapped to the point $t$ used to define the stereographic projection).

Another consequence is a condition characterizing when the sites of an improper Delaunay triangulation can be perturbed to obtain a particular completion of the triangulation. Suppose $D$ is the Delaunay triangulation of a set $S$ with edge graph $G$. Let $G'$ be the graph as before, using the true extreme sites of $S$. Complete $G'$ to a proper triangulation, then delete the new vertex and all edges incident to it, and call the new graph $G''$. $G''$ is like a completion of $G$ with the additional property that nonextreme sites on the convex hull of $S$ have been decided to be either extreme or inside the convex hull. Dillencourt, Rivin and Smith[18] show that $G''$ always has a Delaunay realization if $G'$ is not bipartite. In the case that $G'$ is bipartite, say with color classes red and blue, then $G''$ is still Delaunay realizable as long as both red-red and blue-blue edges were added to $G$ to get $G''$. Notice that $G'$ is bipartite only in an extremely degenerate situation: there must be an even number of edges between extreme vertices on the convex hull, and an even number of edges around each internal face; even in this degenerate situation, most completions are still Delaunay realizable. If $G'$ is bipartite and only red-red or blue-blue edges are added, then $G''$ is not Delaunay realizable. For example, in figure 7, if the dashed edges are added, then the resulting graph is not Delaunay realizable. Dillencourt, Rivin, and Smith are also able to show that any triangulation without chords or nonfacial triangles is Delaunay realizable; a chord is an edge connecting two nonconsecutive vertices on the outer face, which is the only face that does not have to be a triangle.

Dillencourt[17] gives a necessary condition for a graph to be a Delaunay triangulation. For specific graphs this condition may be easier to apply than the circuit condition given by Rivin and Smith. A graph is *1-tough* if deleting any $k$ vertices (and incident edges) splits the graph into at most $k$ components. Dillencourt shows that the edge graph $G$ of a nondegenerate Delaunay triangulation is 1-tough. If the Delaunay triangulation is degenerate, then $G$ is 'almost' 1-tough: deleting any $k$ vertices results in at most $k+1$ components. In either case a consequence is that $G$ has a perfect matching, using a classic theorem of Tutte. Dillencourt offers the 1-toughness of Delaunay graphs as a partial explanation for the observed phenomenon that 'most' Delaunay graphs have Hamiltonian cycles.

## 4 Algorithms

This section outlines some of the fundamental algorithms that are known for Delaunay triangulations and Voronoi diagrams. For various reasons, it is much more convenient to express algorithms in terms of the Delaunay triangulation than the Voronoi diagram, so we do so here. The Voronoi diagram is easily obtained from the data structure suggested below for Delaunay triangulations; the only additional information needed is the location of vertices, edges, etc.

The algorithms are described using the 'real random access machine' (real RAM) model.[50] Thus memory can contain real numbers; operations $+$, $-$, $\times$, $/$, $=$ and $>$ on reals are exact and take unit time. The algorithm descriptions also assume that the set $S$ of sites is in general position. These two assumptions are discussed more fully in section 4.7.

The 'figure of merit' of an algorithm is usually its worst-case running time. In $E^2$, a worst-case lower bound of $\Omega(n \log n)$ is known for computing the Delaunay triangulation of $n$ sites in the real-RAM model,[50] and in $E^d$, $d \geq 3$, a lower bound of $\Omega(n^{\lceil d/2 \rceil})$ follows because the Delaunay triangulation can have that complexity. Algorithms that match these lower bounds are known, so in a sense there is no reason to consider any other algorithms. However, there are other more subjective criteria. One is difficulty, both conceptual and in terms of implementation. Another is running time 'in practice.' The probabilistic estimates of section 2.3 suggest that the combinatorial complexity of the Delaunay triangulation is linear for 'random' sites, so worst-case estimates are far too pessimistic. However, relatively little is known about expected running time on 'random' sites, and there are few empirical comparisons of Delaunay triangulation algorithms in the literature.

The next section discusses some basic tools used to implement the algorithms. It outlines a graph data structure for representing triangulations and convex hulls, and gives geometric primitives used by many of the algorithms. Perhaps the simplest Delaunay triangulation algorithm in two dimensions is the "flipping algorithm," discussed in section 4.2. It attempts to produce the Delaunay triangulation by local modification of an arbitrarily-chosen triangulation. Naively, it has no connection with convex hulls, but a good way to understand it turns out to be with the lifting map of section 2.2. Section 4.3 gives an incremental algorithm for Delaunay triangulations, where the sites are added one by one. The incremental algorithm generalizes nicely to arbitrary dimension; the best way of understanding it is as an incremental convex hull algorithm on the set of lifted sites, that is, as the beneath-beyond algorithm.[50] If the sites are inserted in random order, then it turns out that the expected worst-case running time of the incremental algorithm is improved; this is discussed in section 4.4. The plane-sweep algorithm in section 4.5 is one of the few algorithms that has no apparent connection with convex hulls. However, there is a way of understanding it using a different mapping to three dimensions. Section 4.6 briefly discusses other algorithms. The general position assumption and the real RAM model are reconsidered in section 4.7. Finally, some implementation issues and results are given in section 4.8.

## *4.1 Primitives for Delaunay triangulation algorithms*

This section describes two geometric primitives, the "orientation test" and the "incircle test," that are sufficient for implementing many Delaunay triangulation algorithms. Also, there is a description of a simple graph data structure that can be used to represent a triangulation or convex hull.

The *orientation* of a sequence of $d+1$ points $(p_1, \ldots, p_{d+1})$ in $E^d$ is either $-1$, $0$, or $+1$; it is 0 if the set of $d+1$ points lie on a common $k$-flat for $k < d$. Otherwise it is defined recursively as follows: for $d = 1$, $(p_1, p_2)$ has positive orientation if $p_1 < p_2$ and negative otherwise; for $d > 1$, identify $E^d$ with the hyperplane spanned by points $p_2, \ldots, p_{d+1}$ so that $p_1$ lies in the positive halfspace of $E^d$, then the orientation of $(p_1, \ldots, p_{d+1})$ is the orientation of $(p_2, \ldots, p_{d+1})$ in $E^d$. Thus for $d = 2$, $(p_1, p_2, p_3)$ has positive orientation if $p_1$ lies to the left of the line directed from $p_2$ to $p_3$, or equivalently if $p_1, p_2, p_3$ are in counterclockwise order. For $d = 3$, $(p_1, p_2, p_3, p_4)$ has positive orientation if looking from $p_1$ to the plane spanned by $p_2, p_3, p_4$, $(p_2, p_3, p_4)$ has positive orientation. Orientation is preserved under an even permutation of the points and negated under an odd permutation. If $p_i = (p_{i1}, p_{i2}, \ldots, p_{id})$, then the orientation of $p_1, \ldots, p_{d+1}$ is given by the sign of the determinant

$$\begin{vmatrix} p_{11} & p_{12} & \cdots & p_{1d} & 1 \\ p_{21} & p_{22} & \cdots & p_{2d} & 1 \\ \vdots & \vdots & & \vdots & \vdots \\ p_{d+1,1} & p_{d+1,2} & \cdots & p_{d+1,d} & 1 \end{vmatrix}.$$

(The value of this determinant is the signed volume of the simplex spanned by $p_1, \ldots, p_{d+1}$, up to a constant factor depending only on $d$). The *orientation test* on $d+1$ points is just to evaluate the sign of this determinant. The orientation test is sufficient to determine 'convexity properties' of a set of points; Edelsbrunner and Mücke[24] give a number of examples of its use.

Suppose the sequence of $d+1$ points $(p_2, \ldots, p_{d+2})$ in $E^d$ has positive orientation. By the results of section 2.2, $p_1$ is outside the circumsphere of $p_2, \ldots, p_{d+2}$ exactly if $\lambda(p_1)$ is above the hyperplane spanned by $\lambda(p_2), \ldots, \lambda(p_{d+2})$, which happens exactly if the orientation of $(\lambda(p_1), \ldots, \lambda(p_{d+2}))$ is positive, which in turn happens exactly if the determinant

$$\begin{vmatrix} p_{11} & \cdots & p_{1d} & p_{11}^2 + \ldots + p_{1d}^2 & 1 \\ p_{21} & \cdots & p_{2d} & p_{21}^2 + \ldots + p_{2d}^2 & 1 \\ \vdots & & \vdots & \vdots & \vdots \\ p_{d+2,1} & \cdots & p_{d+2,d} & p_{d+2,1}^2 + \ldots + p_{d+2,d}^2 & 1 \end{vmatrix}$$

is positive. If the determinant is zero, then it is easy to see that $p_1, \ldots, p_{d+2}$ are cospherical, and if the determinant is negative, then $p_1$ is inside the circumsphere of $p_2, \ldots, p_{d+2}$. The evaluation of the sign of this determinant is the *incircle test*[35] for points $p_1, \ldots, p_{d+2}$.

A triangulation algorithm requires some data structure to represent the triangulation. There are known data structures that will represent arbitrary $d$-dimensional cell complexes.[8,43] However, because of the general position assumption made earlier the Delaunay triangulation $D$ of the set of sites $S$ is proper, that is, all cells are simplices. Hence a fairly simple graph data structure can represent the triangulation. Each $d$-cell $C$ of the triangulation should be represented by a node in memory; the node contains a vector of the indices of the $d+1$ sites that determine the cell and a vector of $d+1$ pointers to the nodes for the $d$-cells that are opposite $C$ in the triangulation. A facet common to two $d$-cells is thus represented by two pointers, one in each direction between the nodes representing the two $d$-cells. A convenient correspondence is that the $k$th pointer stored in a node corresponds to the facet obtained by deleting the $k$th site from the node. Facets on the boundary of the triangulation could be distinguished with null pointers. A perhaps preferable alternative is to give the triangulation the topology of a sphere. One way to do this is to create a dummy vertex $\nu$ (thought of as somewhere in the unbounded complement of the triangulation) and to triangulate the complement topologically using $\nu$, that is, to add a $d$-cell formed from $\nu$ and each boundary facet of the triangulation. Then the boundary facets are obtained by examining the $d$-cells containing $\nu$.

Cells of dimension $d-2$ or less are not represented explicitly; each such cell is determined by a subset of the sites of some $d$-cell. Some care is needed to exploit the structure implied by lower-dimensional cells. For example, the facets and $d$-cells incident to a ridge form a cyclic order around the ridge. Traversal of this cyclic order is possible at the expense of some manipulation of facet pointers. Suppose cell($R$) is a $d$-cell with ridge cell($Q$). A convenient representation of cell($Q$) is the pair of sites in $R - Q$. Each such site $s$ determines a facet cell($R - \{s\}$); these are the two facets on either side of cell($R$) in the cyclic ordering. If the pointer for facet cell($R - \{s\}$) is traversed to reach a $d$-cell cell($R'$), then $R' - Q$ is obtained from $R - Q$ by deleting $s$ and adding a site $s'$; $s'$ can be deduced by examining all facet pointers of cell($R'$), since the pointer for facet cell($R' - \{s'\}$) points to cell($R$). Similarly, the ridges, facets, and $d$-cells incident to a $(d-3)$-cell have the topology of a sphere, which can be traversed in an analogous though more cumbersome fashion.

Essentially the same data structure can be used to represent the boundary of the convex hull of a lifted set of sites, since all the faces are still simplices. Notice that the boundary of the convex hull already has the topology of a sphere. Clearly, the Delaunay triangulation can be obtained by just selecting the lower faces of the convex hull and ignoring the vertical coordinate of a lifted site.

### 4.2 Flipping

Let $S$ be a set of $n \geq 3$ sites in $E^2$. The simplest algorithm for computing the Delaunay triangulation of $S$ is edge flipping:

1. Determine some triangulation of $S$.

2. While there are two opposite triangles $abc$ and $acd$ that are not locally Delaunay, flip the diagonal, that is, replace the two triangles with triangles $abd$ and $bcd$. This flip is a *Delaunay diagonal flip.*

A simple way to determine a triangulation of $S$ is to add the sites one by one in $x$-sorted order (using $y$-order to break ties). The triangles for a new site $s_i$ are constructed using boundary edges of the current triangulation visible from $s_i$; these can be determined by traversing the boundary of the triangulation first clockwise, then counterclockwise, from the previous last site $s_{i-1}$. For the second step, the incircle test from section 4.1 can be used to determine if two opposite triangles are not locally Delaunay. Notice that if two opposite triangles are not locally Delaunay, they together form a convex quadrilateral, so indeed it is possible to flip the diagonal. To save scanning the triangulation at every iteration, a queue of edges can be maintained between opposite triangles that are not locally Delaunay. The queue is initialized once at the beginning, and updated as necessary using the bounding edges of a quadrilateral involved in a flip. Working out the details, it is easy to see that the algorithm can be made to run in time $O(n \log n + f)$, where $f$ is the number of flips.

Why is the number of flips finite? A triangulation $T$ in $E^2$ lifts to a polyhedral surface $\hat{T}$ in $E^3$ by mapping a face $F = \text{cell}(R)$ into a face $\hat{F} = \text{cell}(\lambda(R))$. Suppose that two opposite triangles $\Delta_1 = abc$ and $\Delta_2 = acd$ are not locally Delaunay. An observer sitting at a negative vertical position looking up at the surface will see that the two lifted triangles $\hat{\Delta}_1$ and $\hat{\Delta}_2$ form a concave dihedral angle. Flipping the diagonal results in a triangulation $T'$ with triangles $\Delta'_1 = abd$ and $\Delta'_2 = bcd$ substituted for $\Delta_1$ and $\Delta_2$. The lifted surface $\hat{T}'$ now has two triangles $\hat{\Delta}'_1$ and $\hat{\Delta}'_2$ forming a convex dihedral angle. The symmetric difference between $\hat{T}$ and $\hat{T}'$ is essentially the boundary of the tetrahedron $\lambda(a)\lambda(b)\lambda(c)\lambda(d)$; $\hat{T}'$ has resulted from $T$ by substituting the two faces of the tetrahedron visible from below for the two faces obscured by the interior of the tetrahedron. Thus $\hat{T}'$ is everywhere equal to or below $\hat{T}$. In particular, the edge $\lambda(a)\lambda(c)$ lies above $\hat{T}'$, and above every subsequent lifted triangulation. Hence edge $ac$ will never reappear in any subsequent triangulation. Since there are only $\binom{n}{2}$ possible edges among the $n$ sites of $S$, and some edge disappears forever after every flip, the number of flips is at most $\binom{n}{2}$.

This quadratic bound can be achieved in the worst case. Consider the triangulation $T$ in Figure 8a; the Delaunay triangulation $D$ of the same set of sites is depicted in Figure 8b. At least $\Omega(n^2)$ Delaunay flips are required to transform $T$ into $D$, no matter in what order the flips are performed.[27] Notice that only $O(n)$ diagonal flips are necessary to transform $T$ into $D$–this bound holds for the flip distance be-

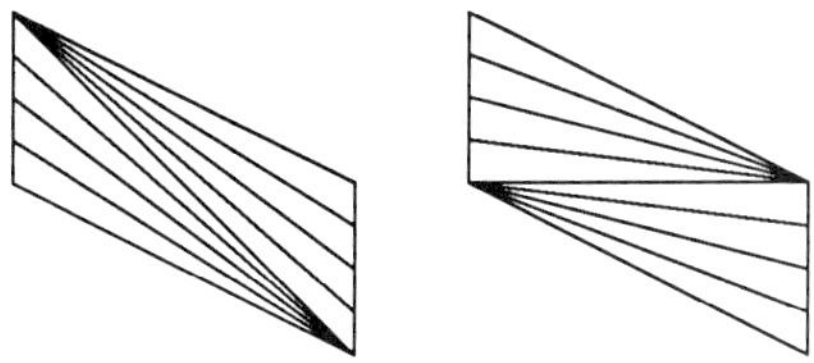

Figure 8: $\Omega(n^2)$ Delaunay flips are necessary to transform triangulation $T$ in (a) into $D$ in (b).

tween any two triangulations of a convex polygon–but not all the flips are Delaunay diagonal flips.

Not much is known about the number of flips required to transform a 'random' triangulation into the Delaunay triangulation, in part because of the lack of a satisfactory definition of a 'random' triangulation. Empirical evidence suggests that edge flipping is a reasonable algorithm in practice; see section 4.8. If the sites are added in random order, and flips are performed to update the triangulation to the Delaunay triangulation after each addition, then Guibas, Knuth and Sharir[34] show that the expected number of flips is linear. The expectation is over the permutation used to order insertions and holds for any set of sites.

Naively, it seems that the flipping algorithm should extend to higher dimension. The notion of a 'flip' is now more complicated; it does not suffice to consider just two adjacent simplices. Let $S$ be a set of sites in $E^d$, and consider a set $R \subset S$ of $d+2$ sites so that all sites are on the boundary of the convex hull of $R$. The set $\lambda(R)$ forms a simplex in $E^{d+1}$. Viewed from below, some facets are visible and some facets are obscured by the interior of the simplex. A *Delaunay flip* in a triangulation $T$ of $S$ substitutes a set of cells $V$ for a set of cells $O$ if $V$ and $O$ are projections of the visible and obscured facets of such a simplex $\lambda(R)$. For example, for $S \subset E^3$, a simplex of $d+2$ points in $E^4$ has five tetrahedral faces, either two or three of which can be visible; hence a Delaunay flip either replaces two tetrahedra with three, or three tetrahedra with two. Clearly, if $T'$ results from $T$ by a Delaunay flip, then $\hat{T}'$ is everywhere at or below $T$. Since some simplex disappears forever at each flip, and there are only a finite number of simplices with vertices chosen from $S$, it would seem that the Delaunay triangulation results after a finite number of flips. Unfortunately, flipping can get stuck. Edelsbrunner and Shah[25] show that there are non-Delaunay triangulations in $E^3$ for which no Delaunay flip applies. Roughly speaking, the problem is that two opposite simplices can be locally nonDelaunay, but the union of the two is not convex, so no flip is possible. This is possible for the first time in $E^3$; in $E^2$ if two opposite triangles do not form a convex quadrilateral, then they are locally Delaunay. Edelsbrunner and Shah show that an incremental form of flipping does work in higher dimension; the incremental flipping is similar to the incremental algorithm, discussed below.

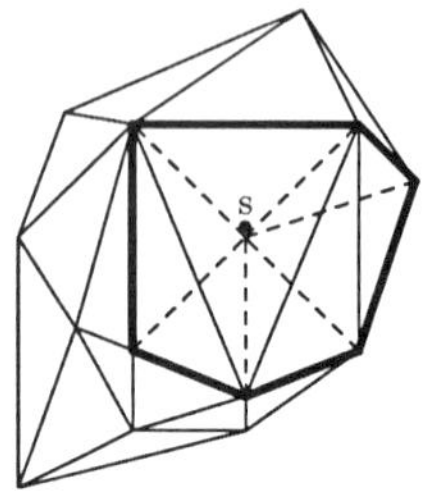


Figure 9: $P$ is given by dark edges and its triangulation using dashed edges.

### *4.3 The incremental algorithm*

The incremental algorithm adds sites one by one; at each addition, the Delaunay triangulation of the previous set of sites is updated to include the new site.

In two dimensions, the incremental algorithm is easy. For simplicity, choose an initial triangulation so that all subsequent sites lie inside the triangulation. Given a new site $s$, find the set $T_v$ of triangles of the current triangulation whose circumcircles contains $s$. The boundary of the union of triangles $T_v$ is a polygon $P$ star-shaped from $s$ (that is, for any point $p$ in $P$, segment $sp$ is contained in $P$). See Figure 9. Let $T_s$ be the triangulation of $P$ whose triangles are each formed from an edge of $P$ and $s$, and replace $T_v$ by $T_s$. The resulting triangulation is Delaunay since it is locally Delaunay: each edge lies between two old triangles, or between an old triangle whose circumcircle does not contain $s$ and a triangle formed using the edge and $s$, or between two new triangles formed from $s$ and the vertices of an old triangle whose circumcircle contains $s$. In the worst case, every triangle of $T$ could appear in $T_v$, so adding a single site takes linear time, and adding all sites takes quadratic time. The non-worst-case analysis is given below.

This algorithm generalizes to higher dimension. However, the generalization is best understood as an incremental convex hull algorithm, applied to the set of lifted sites $\lambda(S)$. Thus the incremental Delaunay triangulation algorithm is just a version of the beneath-beyond algorithm.[50]

A useful observation concerns the changes to the convex hull of a point set when a new point is added. Suppose $R \cup \{r\}$ is a finite affinely independent point set in $E^d$, $H$ is the convex hull of $R$, $r$ is outside $H$, and $H'$ is the convex hull of $R \cup \{r\}$. Then the faces of $H'$ are exactly the faces of $H$ invisible from $r$ and on the horizon from $r$, faces of the form cell$(F \cup \{r\})$ for some horizon face $F$ of $H$, and $r$ itself.[44] See Figure 10. (This characterization requires affine independence, in particular, that $r$ is not contained in the affine span of any horizon face. The general position assumption implies that the set of lifted sites $\lambda(S)$ is affinely independent. Figure 10 is drawn with nontriangular faces to make it appear more like a three-dimensional polyhedron.) Notice that any face of $H$ visible from $r$ lies in the interior of $H'$.

Let $S$ be a set of at least $d+2$ sites in $E^d$. The incremental algorithm will

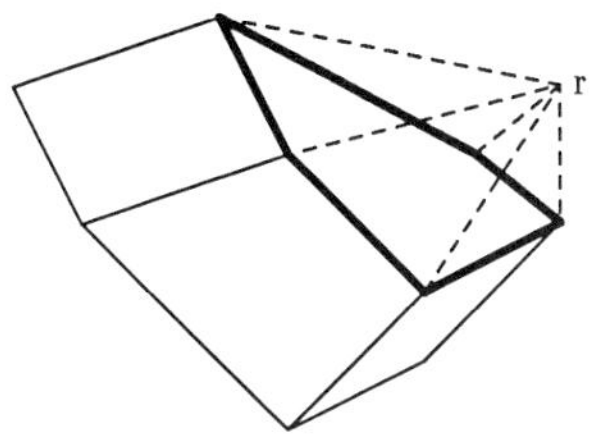


Figure 10: Obtaining $H'$ from $H$. The heavy lines are the horizon faces.

produce the convex hull of the lifted sites $\lambda(S)$. The convex hull is represented using a graph data structure $G$, as described in section 4.1. The first step of the algorithm is to choose a subset of $d+2$ sites. The $d+2$ lifted sites form a simplex in $E^{d+1}$ with $d+1$ $d$-faces, so $G$ starts with $d+1$ nodes.

Suppose $R \subset S$ and the convex hull $H$ of $\lambda(R)$ has been constructed. The iterative step of the algorithm is to choose a new site $s \in S-R$ and construct the convex hull $H'$ of $\lambda(R \cup \{s\})$. Using the observation given before, adding $s$ can be split into two steps:

1. Find the set $H_v$ of $d$-cells of $H$ visible from $\lambda(s)$.
2. Replace $H_v$ with the set of $d$-faces $H_s = \{F_s : F_s = \text{cell}(F \cup \{\lambda(s)\}),\ F \in H_h\}$, where $H_h$ is the set of horizon $(d-1)$-faces of $H$ from $\lambda(s)$.

(It should be clear that the algorithm in two dimensions given earlier is essentially the same, but described in terms of the triangulation itself.) For the first step, it suffices to find a single face in $H_v$, since as described in lemma 4.1 below, the remaining faces can then be found by a graph search in $G$.

Sets $H_v$ and $H_s$ form topological $d$-balls with the same boundary $H_h$. However, their worst-case sizes are different. $H_v$ can be essentially all of $H$. Since $H$ is a convex polytope in $d+1$ dimensions, by the upper bound theorem it has at most size $O(m^{\lceil d/2 \rceil})$, where $m$ is the number of sites determining $H$. $H_s$ is the set of $d$-faces incident to a single vertex of a convex polytope in $d+1$ dimensions. This set has at most the size of a convex polytope in dimension $d$, i.e. $O(m^{\lfloor d/2 \rfloor})$. One way to see this is to push a hyperplane supporting $H'$ at $\lambda(s)$ into $H'$ slightly. The resulting intersection is a convex polytope in $d$ dimensions with faces in one-one correspondence with the faces of $H_s$ (and also $H_h$).

**Lemma 4.1** *Given a single face in $H_v$, it is possible to determine $H_s$ in time $O(|H_v|)$.*

**Proof.** Here is a sketch of one algorithm. The data structure nodes for the $d$-faces in $H_s$ can be created if the horizon $(d-1)$-faces $H_h$ are known. These are exactly the $(d-1)$-faces between faces in $H_v$ and faces not in $H_v$, and can be enumerated by a depth-first search through the graph data structure $G$ starting at any face in $H_v$, backtracking at a transition from a face in $H_v$ to a face not in $H_v$. The adjacencies among the faces in $H_s$ can be discovered as follows. There is a $(d-1)$-face between $d$-faces $F_s$ and $F_s'$ of $H_s$ exactly if $(d-1)$-faces $F$ and $F'$ in $H_h$ have a common $(d-2)$-face. Furthermore, any $(d-2)$-face $f$ of a face in $H_h$ is incident to exactly two faces of $H_h$, since $H_h$ is topologically a $(d-1)$-sphere. The $d$-cells of $H$ incident to such a $(d-2)$-face $f$ are cyclically ordered around $f$, and in particular the $d$-cells of $H_v$ form a connected portion of the cyclic order. Hence given a $(d-2)$-face $f$ of a face in $H_h$, the other face of $H_h$ incident to $f$ can be found by walking in this cyclic order through $H_v$ until a $d$-cell not in $H_v$ is discovered. The total time spent to construct all the cells $H_s$ and to discover adjacencies is clearly $O(|H_v| + |H_h|) = O(|H_h|)$ plus the time spent spent traversing cyclic orders around $(d-2)$-faces. However, a $d$-cell of $H_v$ is visited at most once for each of its $(d-2)$-faces, so this time is also $O(|H_v|)$. □

**Theorem 4.2** *Let $S$ be a set of $n$ sites in $E^d$, $d \geq 2$. The incremental algorithm to compute the Delaunay triangulation of $S$ can be implemented so that in the the worst case it uses time $O(n^{\lceil (d+1)/2 \rceil})$ and space $O(n^{\lceil d/2 \rceil})$.*

**Proof.** The sites should be added in coordinate-sorted order (using lexicographic order to break ties). Then the next lifted site always sees one of the $d$-cells incident to the last lifted site added, so simply examining all such $d$-cells determines a $d$-cell visible to the new lifted site. The total time spent in the algorithm is determined by the number of $d$-cells ever created, since each $d$-cell gets charged once for being created, at most once for being deleted, and at most once for being examined. If the current hull is determined by $m$ lifted sites, then adding a new lifted site creates at most $O(m^{\lfloor d/2 \rfloor})$ new $d$-cells. Summing from 1 to $n$ yields $O(n^{\lceil (d+1)/2 \rceil})$. The space usage is determined by the maximum number of $d$-cells of the convex hull, which is $O(n^{\lceil d/2 \rceil})$. □

Edelsbrunner[23] gives a careful description of this algorithm without assuming that the set of sites is in general position. Unfortunately, the details of the algorithm become much more complicated.

If the set of sites is 'random' the running time bound of theorem 4.2 is unduly pessimistic. The probabilistic estimates in section 2.3 suggest that $|H|$ is linear; hence one expects quadratic behavior from even a naive implementation that examines every face of $H$ to find the faces of $H_v$. The probabilistic estimates also suggest that for a new 'random' site, $|H_v|$ is constant, so examining every face of $H$ is unduly expensive. However, sorting the points as in theorem 4.2 destroys randomness (at least as far as is known), and the points may not all be available at the beginning of the algorithm. Hence an efficient way of finding a face in $H_v$ is needed. Various heuristics have been suggested. One is to start at a site of the current Delaunay triangulation and walk through the triangulation in the direction of a new site $s$. The walk will find either a cell containing $s$ or leave the triangulation; either case yields a face of $H_v$. Bowyer[7] suggests that a walk starting from a site near the centroid of the Delaunay triangulation should visit about $O(n^{1/d})$ cells, although no rigorous analysis is known.

An improvement is to subdivide space with a regular grid. Before the algorithm starts, the set of sites in each grid cell is determined. The walk for a new site can then start at an already inserted site lying in the same or a nearby grid cell. In two dimensions, Ohya, Iri and Murota[48,49] give a 'quaternary incremental' algorithm that uses a hierarchy of grids in order to maintain an approximately uniform distribution of sites throughout the insertion process. They report linear performance for a variety of distributions. An alternative method of finding a face of $H_v$ is presented in the next section.

### *4.4 The random incremental algorithm*

The *random incremental algorithm* inserts sites in random order, with each permutation equally likely. One reason for considering the random incremental algorithm is that its expected worst-case running time is better than the worst-case running time of the incremental algorithm. Here the expectation is over the permutation used to order insertion; 'expected worst case' means the expected running time for the worst set of sites.

A requirement of the random incremental algorithm is a data structure to find a face of the current hull seen by a new lifted site. One such search data structure is based on a triangulation of the current hull obtained in a straightforward fashion from the sequence of intermediate convex hulls.

The initial triangulation is just the $(d+1)$-simplex $H_0$ formed by the first $d+2$ lifted sites. Suppose $s$ is a new site, $R$ is the current set of lifted sites, and $H$, $H'$, $H_v$ and $H_s$ are as in the incremental algorithm. The triangulation of $H'$ is obtained from the triangulation of $H$ by adding, for each $d$-face $F$ in $H_v$, the $(d+1)$-simplex $G_{Fs}$ formed by $F$ and $\lambda(s)$. The simplices $\{G_{Fs}\}$ fill out the volume between $H$ and $H'$ in the obvious way. Each simplex $G_{Fs}$ has a single *parent* $d$-face $F$ in $H_v$, possibly *child* $d$-faces in $H_s$, and possibly *sibling* $d$-faces shared with other simplices $G_{F's}$ just added. The triangulation can be represented using the data structure described in section 4.1, in one dimension higher. In fact, the triangulation simultaneously serves as the search data structure and as the representation of the boundary of the current convex hull.

Given a new site $s$, depth-first search in the triangulation of $H$ can be used to find a face of $H$ seen by $\lambda(s)$. The triangulation should be thought of as a directed graph. The nodes are the $(d+1)$-simplices and arcs correspond to $d$-faces. There is an arc directed from parent to child, and an arc in each direction between siblings. All of the faces of the initial simplex $H_0$ are child faces, directed away from $H_0$. Depth-first search starts at $H_0$; the search proceeds from a simplex $G$ through a child or sibling pointer if $\lambda(s)$ is beyond the corresponding $d$-face of $G$, that is, if the hyperplane through the face separates $G$ from $\lambda(s)$. The search stops if $\lambda(s)$ is beyond some $d$-face of a simplex $G$ and there is no opposite simplex; this happens only if the $d$-face is part of the boundary of the current hull $H$. There is one subtlety to the search: if the search enters a simplex, the search should continue only if the simplex is *active*, that is, if $\lambda(s)$ and the simplex are on the same side of the hyperplane through the parent

$d$-face of the simplex. It may happen that an inactive simplex is entered through a sibling pointer; the search should backtrack immediately to save unnecessary work (and to make the analysis below go through).

Why is the search guaranteed to find a $d$-face of $H$ seen by $\lambda(s)$? Choose a point $h$ in the initial simplex $H_0$ so that the segment from $h$ to $\lambda(s)$ intersects only $d$- and $(d+1)$-simplices, and those in vertices and open segments, respectively. The sequence of $(d+1)$-simplices intersected by the segment in order from $h$ to $\lambda(s)$ can be traversed using sibling $d$-faces and $d$-faces from parent to child, since $\lambda(s)$ is beyond every intermediate convex hull. Any $(d+1)$-simplex in the sequence is active because the hyperplane through its father face separates $h$ from some point on the segment and hence separates $h$ from $\lambda(s)$. It follows that $\lambda(s)$ and the $(d+1)$-simplex are on the same side of the hyperplane. Clearly the last $d$-face intersected by the segment is a $d$-face of $H$ seen by $\lambda(s)$, so there is at least one path through which the depth-first search can succeed.

**Theorem 4.3** *Let $S$ be a set of $n$ sites in $E^d$. The expected running time and space of the random-incremental algorithm are in $O(n^{\lceil d/2 \rceil})$ if $d \geq 3$. If $d = 2$, then the expected running time is in $O(n \log n)$ and the expected space is in $O(n)$. This bound holds for the worst-case set of sites; the expectation is over the permutation used to order insertions.*

**Proof.** Let $\mathcal{B}$ be the subsets of $S$ of size $d+1$. An element $B \in \mathcal{B}$ has *scope* $k$ if there are exactly $k$ sites in the interior of the circumsphere of $B$. Let $b_k$ and $b_{\leq k}$ be the number of elements of $\mathcal{B}$ of scope $k$ and of scope between 0 and $k$, respectively. Clearly, if $B$ has scope 0, then cell($B$) is a Delaunay $d$-cell; hence $b_0$ is the number of Delaunay $d$-cells and $b_0$ is in $O(n^{\lceil d/2 \rceil})$. A fundamental and not especially obvious fact[13,14] is that for $k \geq 1$,

$$b_{\leq k} \in O(n^{\lceil d/2 \rceil} k^{\lceil (d+1)/2 \rceil}).$$

Surprisingly, good bounds on $b_k$ are not known.

The sites are inserted in some random order. We first bound the expectation of the total number of $d$-faces that appear on any intermediate convex hull. For $B \in \mathcal{B}$ of scope $k$, cell($\lambda(B)$) appears on some intermediate convex hull exactly if the $d+1$ elements of $B$ appear in the random order before any of the $k$ elements lying inside the circumsphere of $B$. The probability of this happening is

$$p_k = \frac{(d+1)!k!}{(k+d+1)!} = \frac{(d+1)!}{(k+1)(k+2)\cdots(k+d+1)}.$$

Thus the expected number of $d$-faces is

$$\sum_{k=0}^{n-(d+1)} p_k b_k \quad = \quad b_0 + \sum_{k=1}^{n-(d+1)} p_k (b_{\leq k} - b_{\leq k-1})$$

$$\begin{aligned} &= b_0 + \sum_{k=1}^{n-(d+1)} (p_k - p_{k+1}) b_{\leq k} \\ &\in O(n^{\lceil d/2 \rceil}) + \sum_{k=1}^{n-(d+1)} O(k^{-(d+2)} n^{\lceil d/2 \rceil} k^{\lceil (d+1)/2 \rceil}) = O(n^{\lceil d/2 \rceil}). \end{aligned}$$

The space usage of the algorithm is determined by the total number of $d$-faces and has expectation $O(n^{\lceil d/2 \rceil})$.

Now we bound the total time spent traversing the search data structure for all sites. Consider the time spent at a single non-root simplex for a single site $s$. We charge the time spent leaving the simplex through child and sibling faces to the parent face $F = \mathrm{cell}(\lambda(B))$ of the simplex. The total charge is constant, since there are only $d$ child and sibling faces. Such a charge happens only if the simplex is active, that is, if $\lambda(s)$ and the simplex are on the same side of the hyperplane through the parent face, or equivalently if $s$ is inside the circumsphere of $B$. Hence the total charge to a parent face $F = \mathrm{cell}(\lambda(B))$ is a constant times the scope of $B$. The expected time spent traversing the data structure is at most the sum over all potential faces of the probability that the face gets created times the total possible charge to the face:

$$\sum_{k=1}^{n-(d+1)} k p_k b_k.$$

If $d \geq 3$, essentially the same analysis as above shows that the sum is $O(n^{\lceil d/2 \rceil})$. The case $d=2$ is special: the sum becomes

$$\sum_{k=1}^{n-(d+1)} O(k^{-(d+1)} n^{\lceil d/2 \rceil} k^{\lceil (d+1)/2 \rceil}) = O(n \sum_{k=1}^{n-3} k^{-1}) = O(n \log n).$$

In either case it is easy to see that the sum bounds the total expected running time of the algorithm. □

Other analyses of this algorithm are possible. Clarkson, Mehlhorn and Seidel[13,15] give an elegant self-contained analysis that does not require the bound on the magic value $b_{\leq k}$; Boissonnat *et al.*[6] give another. The analysis given here follows the outline given by Guibas, Knuth, and Sharir[34] for a random-incremental diagonal-flipping algorithm in two dimensions. The original random-incremental algorithm for convex hulls is due to Clarkson and Shor.[14] That algorithm maintains a "conflict graph" that has an edge connecting each uninserted site with each ridge of the current hull that would be destroyed by the addition of the new site. The conflict graph makes finding the faces visible from an uninserted site trivial; the work is in maintaining the conflict graph. The data structure described here is very similar to one suggested by Boissonnat and Teillaud[5]; the idea of using intermediate Voronoi diagrams to aid searching can be traced to Green and Sibson.[32]

Chazelle[11] shows that it is possible to 'derandomize' this algorithm, that is, to choose deterministically an insertion order that guarantees the same asymptotic running time. Unfortunately, the algorithm that computes the insertion order is quite complicated.

As in the case of the incremental algorithm, the worst-case bound of theorem 4.3 is unduly pessimistic. Suppose the set of sites $S$ has the property that the expected size of the Delaunay triangulation of a random subset $R \subseteq S$ is linear in $|R|$. A set of points drawn uniformly in the unit sphere satisfies this property, as discussed in section 2.3. Then the expected running time of the random incremental algorithm is $O(n \log n)$.[13,15]

### 4.5 The plane-sweep algorithm

The plane-sweep method constructs a cell complex in the plane by observing the intersection of the cell complex with a line, the 'sweepline', as the line sweeps across the plane. For a general discussion of the plane-sweep method, see for example the book by Preparata and Shamos.[50] We use the plane-sweep method to construct the Voronoi diagram or Delaunay triangulation of a set of points in two dimensions.

It is perhaps not immediately obvious how to use the plane-sweep method to compute a Voronoi diagram $V$ efficiently in two dimensions. Suppose the sweepline is parallel to the $x$-axis and is moving upwards, that is in the plus $y$-direction. The vertices of $V$ can be split into two categories: those with two edges incident from below and one from above, and those with one edge incident from below and two from above. It is easy to predict the first kind of vertex, because the two edges incident from below are adjacent on the sweepline data structure. It is not so easy to predict the second type of event. In effect, it is caused by an edge intersected by the sweepline and a site lying above the sweepline. However, it is undesirable to consider all quadratically many such pairs.

In fact, there is a plane-sweep algorithm that computes the Voronoi diagram of a set $S$ of $n$ sites in the plane in time $O(n \log n)$.[28] It is convenient to describe the algorithm as producing the Delaunay triangulation $D$. In addition to the general position assumption, assume that no two sites of $S$ lie on a common horizontal line and that no site is the topmost point of the circumcircle of any Delaunay triangle.

Let $l_y$ be the horizontal line with $y$-coordinate $y$ ($l_y$ will be the sweepline). Let $y_{\min}$ be the minimum $y$-coordinate of a site in $S$ and assume $y > y_{\min}$. For a point $p \in l_y$, grow a circle $C_p$ with topmost point $p$ until it touches a site in $S$. In general, $C_p$ touches a single site $s$, and in fact for every point $p$ in some open interval $I_y(s) \subseteq l_y$, $C_p$ touches exactly $s$. See figure 11. At an endpoint $p$ of $I_y(s)$, $C_p$ also touches a site $t$ with interval $I_y(t)$ also incident to $p$; possibly $C_p$ touches another site as well. The sequence of intervals $I_y(s_1), I_y(s_2) \ldots, I_y(s_k)$ along $l_y$ determines a path of Delaunay edges $F_y = s_1s_2, s_2s_3, \ldots, s_{k-1}s_k$. $F_y$ need not be simple: a vertex can appear arbitrarily many times (though an edge can appear at most twice). $F_y$ can be thought of as the 'frontier' of the set of Delaunay cells that have some circumcircle with interior below $l_y$.

What happens as $y$ increases and $l_y$ moves upwards? Path $F_y$ can change in two ways. First, $l_y$ can become tangent to the top of the circumcircle of three sites

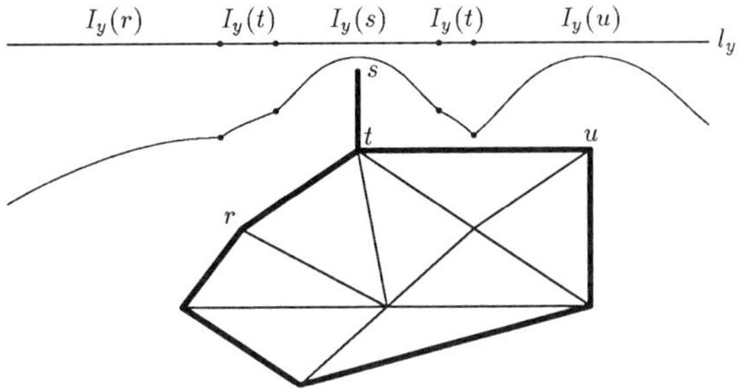


Figure 11: The heavy path is $F_y$; the curved path is the center of circles $C_p$ for $p \in l_y$.

$s, t, u$ that form two consecutive edges $st$ and $tu$ along $F_y$. Edges $st$ and $tu$ form a concave angle in the sense that $s, t, u$ and the topmost point of the circle appear in that order around the circle. In this case, the interval $I_y(t)$ disappears and edge $su$ replaces edges $st$ and $tu$ along $F_y$. Notice that $stu$ must be a Delaunay triangle and $su$ a Delaunay edge. Second, $l_y$ can meet some site $s$. If $s$ were not a site, then point $s$ would lie in some interval $I_y(t)$ with intervals $I_y(u_1)$ before and $I_y(u_2)$ after. Since $s$ is a site, for slightly larger $y$ there will be an open interval $I_y(s)$, and the subpath of $F_y$ consisting of $u_1t$, $tu_2$ is replaced with $u_1t$, $ts$, $st$, $tu_2$. Notice that $st$ is a Delaunay edge, and that this is the first time site $s$ has appeared. Eventually all sites are passed and path $F_y$ forms the convex hull of $S$; then there are no further changes to $F_y$.

Using this analysis, a plane-sweep algorithm can be designed that computes the Delaunay triangulation. The algorithm maintains the current path $F_y$ and an event queue ordered by $y$-coordinate. The two events are passing a site and passing the top of a circle. Initially, the sweep line is at $y = y_{\min}$, the event queue contains all sites, and the path $F_y$ consists only of the lowest site. An invariant assertion of the algorithm is that the event queue contains an event for each circle through sites $s, t, u$ so that edges $st$ and $tu$ are consecutive on $F_y$ and form a concave angle. If the next event is to pass the top of circle formed by edges $st$ and $tu$, then the action is to replace these edges with $su$, and to record $stu$ as a Delaunay cell and $tu$ as a Delaunay edge. If the next event is a site $s$, then the site $t$ to be connected to $s$ must be discovered. Site $t$ can be found by binary search in path $F_y$; the necessary primitive decides if $s_x$ lies to the left or right of the topmost point of a circle tangent to $l_y$ and touching two sites $u$ and $v$ so that $uv$ is an edge of $F_y$. Once $t$ is found, the path is augmented with edges $ts$ and $st$, and edge $st$ recorded as a Delaunay edge. For either event, path $F_y$ changes, so insertions and deletions to the event queue are necessary to maintain the invariant. However, since the changes to $F_y$ are local, only a constant number of event queue operations are required.

**Theorem 4.4** *The sweepline algorithm can compute the Delaunay triangulation of $n$ sites $S \subset E^2$ in time $O(n \log n)$ and $O(n)$ space.*

**Proof.** The path data structure can be implemented using a balanced binary tree, so that insertions, deletions, and searches each take time $O(\log n)$. Similarly, the event queue can be implemented as a heap, so insertions, deletions, and determining the next event also each take time $O(\log n)$. The number of events is equal to the number of sites plus the number of Delaunay triangles, which is $O(n)$. Each event uses only a constant number of operations on the path data structure and the event queue, so the total running time is $O(n \log n)$. The total space is determined by the maximum length of the path $F_y$, which is linear, since the number of edges in the triangulation is linear. □

There is an interpretation in three dimensions that may clarify the algorithm. We sketch the interpretation here; Guibas and Stolfi[36] discuss it at greater length. Call the plane spanned by the $x$- and $y$-axes the base plane, as usual, and now let the vertical direction be the $z$-direction. Put a three-dimensional cone $C(s)$ over each site $s$; the height of the cone over a point $x$ in the plane is equal to the distance from $x$ to $s$. Paint each cone a different opaque color; then an observer sitting below the plane looking up sees the lower envelope of the cones. Clearly the vertical projection of the lower envelope of the cones on the base plane is the Voronoi diagram.

Consider a plane $P$ that makes the same angle with the base plane as the cones and that intersects the base plane in a line $l$ parallel to the $x$-axis. The intersection of $P$ with a cone projects onto the base plane as a parabola; the intersection of $P$ with the lower envelope of the cones projects to a sequence of a parabolic arcs. In fact, the sequence is just the locus of centers of the circles $C_p$, as $p$ varies along $l$. See figure 11.

Now imagine sweeping the plane $P$ in the positive $y$-direction, so that $l$ moves in the positive $y$-direction as well. For small motions of $P$, $P$ intersects the same sequence of cones on the lower envelope; however, the parabolic arcs obtained by projecting on the base plane become less sharply curved. The intersection of $P$ and the lower envelope can change combinatorially in two ways. First, $l$ can hit a new site and $P$ the corresponding cone; notice that in fact $P$ becomes tangent to the cone. After a slight additional motion of $l$ and $P$, the new cone yields a new parabolic arc in the projection of the intersection of $P$ and the lower envelope. Second, $P$ can pass through the point of intersection of three cones on the lower envelope; this corresponds to the sweepline discovering a vertex of the Voronoi diagram (or a Delaunay circle). One of the cones now disappears from the intersection of the lower envelope and $P$, and the projection onto the base plane loses a parabolic arc.

What happens if the set of sites $S$ lies in three dimensions? One might hope that a 'space-sweep' algorithm would be output-sensitive, that is, its running time would be approximately proportional to the number of Delaunay tetrahedra. The algorithm does generalize to three dimensions. It is perfectly possible to define the 'frontier' $F_z$ as the set of Delaunay cells that have a circumsphere touching a sweep plane parallel to the $xy$-plane at height $z$. $F_z$ can be given the structure of a plane, just as in two dimensions the frontier has a linear structure. There are now three kinds of events: discovering a new site (and connecting the site by a Delaunay edge to a previously discovered site), discovering a triangle, and discovering a tetrahedron.

The first event is of course easy to predict; finding the connecting edge requires point location in a dynamic planar subdivision, but there are reasonably efficient algorithms for that. The third event can also be predicted, because at least two triangles sharing a common edge are involved. The difficult event to predict is the creation of a triangle. Conceivably the current frontier has a vertex with $O(n)$ Delaunay edges incident to it. The next Delaunay triangle may be formed from two of those edges. However, there are $O(n^2)$ possible pairs, and it seems necessary to consider all of them. Hence, as far as is known, space-sweep algorithms require at least quadratic complexity, even if the Delaunay triangulation has linear complexity.

### *4.6 Other algorithms*

The divide-and-conquer algorithm was the first worst-case optimal algorithm for Voronoi diagrams in two dimensions.[56] It splits the set of sites into two halves of approximately the same size with a vertical line, recursively computes the Voronoi diagram of the two halves, and then merges the two triangulations. The merge takes linear time in the worst case, so the worst-case running time of the algorithm on $n$ sites is $O(n \log n)$. Guibas and Stolfi give a careful description of the algorithm in terms of the Delaunay triangulation.[35]

If the set of sites is uniformly distributed in the unit square, then the expected running time can be improved. Bentley, Weide, and Yao[4] show that a combination of bucketing techniques and any $O(n \log n)$ algorithm leads to an algorithm with expected running time $O(n)$. Dwyer shows that a simpler algorithm, essentially first splitting with a vertical line, then a horizontal line, has expected running time $O(n \log \log n)$. This is further improved to $O(n)$ by Katajainen and Koppinen.[38]

A Delaunay triangulation can always be computed by lifting the set of sites and using a convex hull algorithm. One alternative to the beneath-beyond algorithm is the 'gift wrapping' algorithm.[50] The gift-wrapping algorithm takes time $O(ns)$, where $n$ is the number of points and $s$ is the number of facets on the convex hull of the set of points. Hence, in the worst case it is slower than the beneath-beyond algorithm. However, Dwyer[20] is able to show that if the set of sites is uniformly distributed in a sphere, then a version of the gift-wrapping algorithm applied to the lifted sites takes linear time. The linear time results from the use of bucketing techniques in addition to the linear expected combinatorial complexity of the Delaunay triangulation. A second convex hull algorithm is due to Seidel[55]; it takes time $O(n^2 + s \log n)$, where $s$ and $n$ are as before. Conceivably there are situations where its use could lead to an efficient algorithm for Delaunay triangulations.

### *4.7 The real RAM and the general position assumption*

The 'real RAM' model and the general position assumption for sites considerably simplify the description of geometric algorithms. However, the real RAM is unimplementable and the set of sites of interest may not be in general position. It is not clear how best to proceed without these assumptions; this section discusses some of the issues. This discussion is appropriate for all geometric algorithms, not just Delaunay triangulation algorithms.

A plausible approach is to use integers instead of reals. Thus an input point is specified by a vector of integers; scaling of the input may be necessary to approximate the input adequately on the integer grid. Most of the Delaunay triangulation algorithms can be implemented using the geometric primitives in section 4.1 (the sweepline algorithm has a different set of primitives). Each of these primitives is a sign-evaluation of a determinant formed from the coordinates of the input points and can be evaluated using integer arithmetic. The issue is the bit complexity of the arithmetic. As far as is known, it is necessary to evaluate the determinant in order to determine its sign. A coarse estimate of the bit complexity of the determinant evaluation can be obtained by viewing the determinant as a polynomial in the input coordinates. If a polynomial has degree $d$ and the variables have bit length $b$, then approximately $bd$ bits are required to evaluate the polynomial. In dimension $d$, the orientation test has degree $d+1$ and the incircle test degree $d+2$. Thus with only standard 32-bit integer arithmetic, in dimension $d = 2$ the input coordinates can have bit length at most $b = 7$, i.e. the input sites have to lie on a $128 \times 128$ grid. It is reasonable to take the dimension $d$ to be constant, so extended-precision integer arithmetic adds only a constant multiplicative factor of overhead.

Unfortunately, at least with naive implementations of extended-precision integer arithmetic, the constant is large. Consider Gaussian elimination to evaluate the determinant of a $d \times d$ matrix with integer entries. The usual estimate is that Gaussian elimination takes $O(d^3)$ arithmetic steps. However, to implement Gaussian elimination using integers without division, the bit length of matrix entries increases. A better estimate of the number of $b$-bit primitive arithmetic operations needed to evaluate a $d \times d$ determinant with $b$-bit entries is exponential in $d$.[39] An alternative is to use modular arithmetic over a collection of finite fields, so division is possible, and then use the Chinese remainder theorem to reconstruct the determinant.[41] This results in about $O(d^4)$ $b$-bit arithmetic operations to evaluate a $d \times d$ determinant with $b$-bit entries. No extensive comparison of determinant evaluation methods appears to be available, at least in the context of geometric algorithms.

It is a considerable simplification to allow only integer coordinates. Rational coordinates arise naturally; for example, the coordinates of the point of intersection of two line segments are rational even if the endpoint coordinates are integer. However, the bit-complexity estimates go up dramatically if the input coordinates are rational. Even a single addition can double bit complexity: the appropriate measure of the bit-complexity of $p/q$ is the sum of the bit lengths of $p$ and $q$, thus the bit-complexity of $p/q + r/s = (ps + qr)/(qs)$ is the sum of the bit-complexities of $p/q$ and $r/s$.

Exact arithmetic has an advantage that certain symbolic perturbation schemes guarantee that the input can be treated as if it were in general position.[24,26,60,61] Edelsbrunner and Mücke[24] suggest a scheme called 'simulation of simplicity;' they work out the details for algorithms that use the orientation test and the incircle test. Suppose that the input sites are $S = \{s_1, s_2, \ldots, s_n\} \subset E^d$, where $s_i = (s_{i1}, s_{i2}, \ldots, s_{id})$. Imagine perturbing $s_{ij}$ by an amount $\epsilon^{2^{id-j}}$, for $\epsilon$ very small. For sufficiently small $\epsilon > 0$ the resulting set of sites is guaranteed to be in general position. Edelsbrunner and Mücke show that this perturbation can be simulated symbolically, without any computation involving $\epsilon$ or even knowing $\epsilon$. For the orientation test on $d+1$ points, they give a sequence of determinants constructed using the coordinates of the unperturbed points. They show that the sign of the first nonzero determinant in the sequence is the same as the sign of the determinant obtained from the perturbed points. The first determinant in the sequence is just the determinant obtained from the unperturbed points, so overhead is incurred only in the case that it is zero, i.e., in the degenerate case. A possible disadvantage of the method is the length of the sequence of determinants; for the incircle test in dimension 2, the length is already 15. At least in principle the determinants can be generated as needed. A second disadvantage is that there is no control over the perturbation. Thus a nonextreme site on the convex hull of the triangulation may be perturbed to become extreme, in effect adding a flat triangle.

The alternative to exact arithmetic is floating point arithmetic, and the bit-complexity issue no longer arises. However, the general position assumption becomes much more stringent. Standard error analysis techniques from numerical analysis can be applied to the geometric primitives.[16] Consider the orientation-test determinant obtained from a set of sites $s_1, \ldots, s_{d+1}$. In floating point arithmetic using, say, Gaussian elimination, it is possible to compute an approximate determinant $\hat{\Delta}$ and an error bound $\delta > 0$. If $|\hat{\Delta}| \geq \delta$, then the sign of $\hat{\Delta}$ is correct; if $|\hat{\Delta}| < \delta$, then the true sign is unknown, but there are slightly perturbed sites $s'_1, \ldots, s'_{d+1}$ so that the true determinant is zero. The 'slight perturbation' moves sites by an amount depending upon the relative error $\epsilon$ of floating point arithmetic, the dimension $d$, and the distance of the sites from the origin. Hence if the set $S$ of sites is in 'sufficiently general' position, that is, no determinant changes sign when the sites are slightly perturbed, then the floating-point implementation is correct.

What happens if the set of sites is not in 'sufficiently general' position? A naive implementation might interpret the 'can't tell' answer as implying that the determinant can be taken as zero. Unfortunately, this can be disastrous: choose three sites $s_1, s_2, s_4$ in the plane that form a right angle, and then add a site $s_3$ very close to $s_2$. The implementation might take $s_1, s_2, s_3$ as collinear and $s_2, s_3, s_4$ as collinear, implying the collinearity of $s_1, s_2, s_4$, which is manifestly false. Such problems can lead to catastrophic failure of algorithms implemented naively with floating point arithmetic.

There has been some work applying the techniques of numerical analysis to geometric algorithms as a whole.[29,37,46] For Delaunay triangulations and Voronoi diagrams in two dimensions, the following results can be obtained for any set of sites, nearly degenerate or not.[30] The flipping, incremental, and sweepline algorithms can each be implemented to produce an 'approximate Delaunay triangulation'. This means that the triangulation has almost-empty circumcircles; any site lying inside a circumcircle is only 'slightly' inside. From the 'approximate Delaunay triangulation' an 'approximate Voronoi diagram' can be produced: if a point-location algorithm reports that a query point $p$ lies in the approximate Voronoi cell of $s$, then $s$ may not be the true closest site to $p$, but any other site $s'$ is only a small amount closer to $p$. The meaning of 'small' and 'slightly' depends upon the relative error $\epsilon$ of floating point arithmetic, the number $n$ of sites, and the distance of the sites from the origin, but not on any geometric properties of the configuration of the sites.

### *4.8 Implementation issues*

It is not trivial to obtain a reliable and complete implementation of any of the Delaunay triangulation algorithms. This section discusses a few implementation issues, offers some advice, and reports experience with four of the algorithms.

A fundamental implementation decision is the choice of exact or floating point arithmetic. This choice likely influences every other implementation decision. Exact arithmetic is the only known way to obtain a completely reliable implementation, at least in general. The disadvantage is the inconvenience and potential cost of exact arithmetic. Edelsbrunner and Mücke[24] use multiple-word integer arithmetic to implement orientation-test primitives on points on an integer grid. They report approximately a factor of 10 degradation in performance as compared to using single-word arithmetic on points on an integer grid drastically reduced in size. They suggest the use of special-purpose hardware to accelerate integer determinant calculations. Karasick, Lieber, and Nackman[39] implement the divide-and-conquer algorithm in two dimensions using adaptive-precision rational arithmetic. Their effort was principally addressed at accelerating the performance of rational arithmetic. Their ultimate implementation was at least three orders of magnitude faster than a naive implementation of rational arithmetic, and only a factor of about 4 slower than floating point arithmetic.

An implementation using floating point arithmetic is not guaranteed to be completely reliable (except as cited before). However, floating point arithmetic is the typical implementation choice and may be reliable enough in some circumstances. A minimal requirement is to raise a flag if some primitive cannot reliably deduce the correct answer. (This is often not done, perhaps because the required interval analysis techniques are not understood.) If some primitive does give an ambiguous answer, a simple trick is to repeat the algorithm after a small random perturbation of the input data. If perturbation is not acceptable, an alternative is to examine each use of a primitive test. Some ambiguous answers truly are 'don't cares' and are easily handled; others may require localized use of higher precision. Such analysis is still a black art.

Whether exact or floating point arithmetic is required, it is better to construct the Delaunay triangulation than the Voronoi diagram. One reason is to avoid manipulating computed values: in exact arithmetic, Voronoi vertices are rationals with high bit complexity; in floating point arithmetic, Voronoi vertices may be poorly determined if the defining sites are close to being affinely dependent. Another reason is that it is simpler to manipulate a cell complex with regular bounded cells than one with irregular unbounded cells. The Voronoi diagram can be obtained in linear time from the completed Delaunay triangulation, if required.

| | lines | $d$ | $n$=500 | $n$=1000 | $n$=5000 | $n$=10000 | $n$=50000 |
|---|---|---|---|---|---|---|---|
| Flipping | 300 | 2 | 2.1 | 4.5 | 26.2 | 59.3 | 316.9 |
| Divide-and-conquer | 900 | 2 | 2.0 | 3.8 | 21.5 | 42.3 | 222.0 |
| Plane sweep | 900 | 2 | 1.2 | 2.5 | 12.1 | 24.1 | 128.0 |
| Random-incremental | 1500 | 2 | 2.2 | 4.5 | 24.2 | 51.0 | 285.0 |
| | | 3 | 8.4 | 18.5 | 102.4 | 210.0 | - |
| | | 4 | 45.3 | 103.0 | 638.0 | - | - |
| | | 5 | 287.0 | - | - | - | - |

Figure 12: Performance of Delaunay triangulation algorithms, in seconds.

The worst-case $O(n^{\lceil d/2 \rceil})$ complexity of the Delaunay triangulation should not be taken too seriously. The linear complexity $c_d n$ suggested by the probabilistic estimates of section 2.3 is more realistic in practice. However, the constant $c_d$ indeed does grow exponentially with $d$, so Delaunay triangulations in dimension 5 or bigger should be approached with caution.

Figure 12 gives timings for various algorithms. All algorithms are implemented in the C programming language and run on a VAX 8550. The primitives are implemented using floating point arithmetic, in some cases with heuristics to improve reliability. The input points are chosen in the unit cube using the library pseudo-random number generator. Both the timings and the number of lines of source code should be taken as suggestive only, since the implementations were done by different people and with different levels of optimization. For example, an unoptimized version of the plane-sweep algorithm requires only half as many lines of code but is considerably slower. The flipping algorithm sorts points by angle rather than by coordinate. The divide-and-conquer algorithm takes advantage of the uniform distribution of input points by using both horizontal and vertical splitting directions.[19] The plane-sweep algorithm uses bucketing for both the priority queue and sweepline data structure, giving good performance on uniform data. An important optimization for the random-incremental algorithm is to use a dot-product rather than an orientation test in the search data structure; the child face normal is calculated when a triangle is added to the triangulation. The running times include input and output; at least for the plane-sweep algorithm, the time for conversion between character and floating point representations is comparable to the time actually required to compute the Delaunay triangulation.

In two dimensions, all of the algorithms are reasonable. The times are sufficiently close that any one of the algorithms could be made the fastest by sufficient optimization. The times are also sufficiently small that optimization may be irrelevant. In higher dimension, the random-incremental algorithm is attractive because it is relatively simple to implement and has good running time, both worst-case and expected-case. One alternative is to use the incremental algorithm with a different search data structure to find a face seen by a new lifted site; another alternative is the variant of gift-wrapping suggested by Dwyer.[20]

For large problems storage may be more relevant than running time. The 5000 point example in four dimensions required roughly 70 megabytes of main storage. Because of paging overhead, the actual clock time for the example was about a factor of 10 larger than the running time. The implementation of the random-incremental algorithm was not optimized to minimize storage. The search data structure is the predominant use of storage. A variant of the random-incremental algorithm that uses a conflict graph rather than than search data structure might require much less storage.[14,13]

*Appendix 1: Definitions from the theory of polyhedra*

We review some basic definitions from the theory of convex polyhedra. The definitions are more or less standard, except perhaps for the definition of a cell complex as consisting of relatively open sets. For more extensive discussion, see books by Brøndsted,[9] Edelsbrunner,[23] and Grunbaum.[33]

$E^d$ is $d$-dimensional euclidean space. A *k-flat* is an affinely closed subset of $E^d$ of dimension $k$; *lines, planes*, and *hyperplanes* are flats of dimension 1,2, and $d-1$, respectively. A set of points in $E^d$ is *affinely independent* if no $k+2$ points lie on a common $k$-flat, for $k<d$. An *open halfspace* is the set of points to one side of a hyperplane; a *closed halfspace* includes the hyperplane. A hyperplane *separates* two sets if one set is contained in one of the closed halfspaces and the other is contained in the complementary open halfspace. A *convex polyhedron* is the intersection of a finite number of closed halfspaces; if it is bounded it is a *convex polytope*. A *cell* is the intersection of a finite set of flats and open halfspaces; equivalently a cell is the relative interior of a convex polyhedron. If $R \subseteq E^d$, then cell($R$) is the relative interior of the convex hull of $R$. A *simplex* is cell($R$) for an affinely independent set of points $R$; thus a two-dimensional simplex is the interior of a triangle and a three-dimensional simplex is the interior of a tetrahedron. A *sphere* in $E^d$ is the set of points in $E^d$ at a fixed distance from its center. A point of $E^d$ is either inside, on, or outside a sphere; a *circumsphere* of a set $R \subset E^d$ is a sphere with all points of $R$ on the sphere.

A closed halfspace *supports* a cell or convex polyhedron $P$ if it contains $P$ and its bounding hyperplane intersects the closure of $P$. A hyperplane *supports* $P$ if one of its closed halfspaces supports $P$. A *face* of $P$ is the relative interior of the intersection of a hyperplane supporting $P$ with the closure of $P$. Basic facts are that a face is a cell and that a face of a face of $P$ is a face of $P$. If $F$ is a face of $G$, then $F$ and $G$ are *incident*. A face is a *k-face* if its affine closure has dimension $k$; *vertices, edges, ridges*, and *facets* are faces of dimension 0,1,$d-2$, and $d-1$, respectively. An *extreme point* of a polyhedron is a vertex of the polyhedron. A face $F$ of a convex polyhedron $P$ is *visible* from a point $x$ not in $P$ if every hyperplane supporting $P$ through $F$ separates $x$ from $P$; $F$ is a *horizon face* from $x$ if there is a hyperplane supporting $P$ containing $x$ and $F$; $F$ is *invisible* from $x$ if no hyperplane supporting $P$ through $F$ separates $x$ from $P$.

A *cell complex* is a finite collection of pairwise disjoint cells so that the face of every cell is in the collection. Two $k$-cells of a cell complex are *opposite* if they have a common $(k-1)$-face. A *triangulation* $T$ of a finite point set $S$ is a cell complex whose union is the convex hull of $S$ and whose vertex set is $S$. Any cell of a triangulation is cell$(R)$ for some $R \subseteq S$. This is a convenient but nonstandard definition because cells are not required to be simplices. A *proper triangulation* is a triangulation all of whose cells are simplices. Any triangulation can be *completed* to a proper triangulation by appropriately subdividing nonsimplicial cells. A convex polyhedron can be partitioned into a cell complex, in fact, a triangulation: one cell is the relative interior of the polyhedron and the other cells are its faces.

*Appendix 2: Proof of theorem 2.1*

We first show that if $V(R)$ and $V(R')$ are both Voronoi cells, then $V(R')$ is a face of $V(R)$ iff $R \subset R'$. Suppose $R \subset R'$. For $r \in R'$, $s \in S-R'$, let $H_{rs}$ be the hyperplane equidistant from $r$ and $s$ with positive halfspace containing $R'$. Let $I$ be $\cap H_{rs}$ and $Q$ the closure of $V(R)$. We can choose a hyperplane $H$ so that $Q \cap I = Q \cap H$ by choosing $H$ through $B$ avoiding the intersection of the positive halfspaces and the intersection of the negative halfspaces. Thus the relative interior of $Q \cap I$ is a face. However the relative interior of $Q \cap I$ is $V(R')$, since $V(R')$ is a relatively open subset of $I$, and any point of $Q$ is equidistant from all sites in $R'$. The converse direction, that $V(R')$ a face of $V(R)$ implies $R \subset R'$, is easy: any point in a face of $V(R)$ is in the closure of $V(R)$, and hence is in $V(R')$ for some $R' \supset R$.

The first claim is that $V(R)$ is a cell complex with union $E^d$. Clearly, $V(R)$ is a cell and any point of $E^d$ lies in some Voronoi cell. If $F$ is a face of $V(R)$, then any point $x \in F$ is on the boundary of $V(R)$, so $x \in V(R')$ for some $R' \supset R$. Therefore $F = V(R')$ since $V(R')$ is a face and $x$ lies in exactly one face of $V(R)$.

The second claim is that the Delaunay triangulation is a triangulation. We must show that Delaunay cells are pairwise disjoint, that the face of a Delaunay cell is a Delaunay cell, and that the union of the Delaunay cells is the convex hull of $S$. The disjointness of distinct Delaunay cells $D(R)$ and $D(R')$ follows by considering circumspheres $C$ and $C'$ of $R$ and $R'$, respectively, and observing that either $C$ and $C'$ have disjoint interiors, or if not, then $R \cap R' \subset C \cap C'$ and by the empty-sphere property, $R-R'$ and $R'-R$ lie on opposite sides of the hyperplane through $C \cap C'$. A face of a Delaunay cell $D(R')$ is cell$(R)$ for $R = R' \cap H$ and $H$ a hyperplane supporting the convex hull of $R'$; since $R$ lies on a hyperplane, it is easy to see that the empty circumsphere of $R'$ can be perturbed to be an empty circumsphere of $R$, and so cell$(R)$ is a Delaunay cell. Let $q$ be a point in the convex hull of $S$; we show that $q$ is in some Delaunay cell. Assume that there is some Delaunay $d$-cell $D(R)$ (if not, then $S$ lies in some lower-dimensional flat, and the proof works in that flat). Choose a point $p \in D(R)$ so that $pq$ avoids all $k$-cells for $k < d-1$ and so that $pq$ intersects any Delaunay $(d-1)$-cell in at most a point. The sequence of Delaunay $d$- and $(d-1)$-cells intersected by $pq$ in order from $p$ to $q$ can end only at $q$: a Delaunay $d$-cell intersecting $pq$ has a $(d-1)$-face that is also Delaunay and further along $pq$;

a Delaunay $(d-1)$-cell intersecting $pq$ must have an incident Delaunay $d$-cell further along $pq$, since $q$ is in the convex hull of $S$. Hence $q$ is in the closure of some Delaunay $d$-cell, and in some Delaunay cell.

The third claim is duality. Suppose $V(R)$ and $V(R')$ are both Voronoi cells, so both $D(R)$ and $D(R')$ are Delaunay cells. Then all the following steps are easy or have been established: $V(R')$ is a face of $V(R)$ iff $R \subset R'$ iff $R = R' \cap H$ for some hyperplane $H$ supporting the convex hull of $R'$ iff $D(R)$ is a face of $D(R')$.

The fourth claim is an easy exercise.

*Acknowledgements.*

I would like to thank Jon Bentley, Paul Chew, Ken Clarkson, Michael Dillencourt, Rex Dwyer, Herbert Edelsbrunner, John Hobby, Allan Wilks, and Margaret Wright for helpful comments.

## References

1. P.F. Ash, E.D. Bolker, Recognizing Dirichlet tessellations, *Geometriae Dedicata* **19**:175-206, 1985.
2. P.F. Ash, E.D. Bolker, Generalized Dirichlet tessellations, *Geometriae Dedicata* **20**:209-243, 1986.
3. F. Aurenhammer, Voronoi diagrams – a survey of a fundamental geometric data structure, *ACM Computing Surveys* **23**:345-405, 1991.
4. J. Bentley, B. Weide, A. Yao, Optimal expected-time algorithms for closest-point problems, *ACM Trans. Math. Software* **6**, 563-580, 1980.
5. J.-D. Boissonnat, M. Teillaud, An hierarchical representation of objects: the Delaunay tree, *Proc. 2nd Ann. Symp. Comp. Geom.* 260-268, 1989.
6. J.-D. Boissonnat, O. Devillers, R. Schott, M. Teillaud, M. Yvinec, Applications of random sampling to online algorithms in computational geometry, research report 1285, INRIA-Sophia Antipolis, 1990.
7. A. Bowyer, Computing Dirichlet tessellations, *The Comp. Journal* **24**(2):162-166, 1981.
8. E. Brisson, Representing geometric structures in $d$-dimensions: topology and order, *Proc. of the Fifth Annual Symp. on Comp. Geom.*, 218:227, 1989.
9. A. Brøndsted, *An Introduction to Convex Polytopes*, Springer-Verlag, Berlin, 1983.
10. K.Q. Brown, Voronoi diagrams from convex hulls, *Information Processing Letters* **9**:223-228, 1979.
11. B. Chazelle, An optimal convex hull algorithm and new results on cuttings, *Proc. 32nd Ann. Symp. Found. of Comp. Sci.* 29-38, 1991.

12. P. Chew, There are planar graphs almost as good as the complete graph, *J. Comp. System Science* **39**(2):205-219, 1989.
13. K.L. Clarkson, Randomized geometric algorithms, this volume.
14. K.L. Clarkson, P.W. Shor, Applications of random sampling in computational geometry, II, *Disc. Comp. Geom.* **4**:387-421, 1989.
15. K.L. Clarkson, K. Mehlhorn, R. Seidel, Four results on randomized incremental constructions, to appear, *Symp. Theor. Aspects of Comp. Sci.*, 1992.
16. G. Dahlquist, Å. Björck, N. Anderson, *Numerical Methods*, Prentice-Hall, 1974.
17. M.B. Dillencourt, Toughness and Delaunay triangulations, *Disc. and Comp. Geom.* **5**(6):575-601, 1990.
18. M.B. Dillencourt, I. Rivin, W.D. Smith, Combinatorial structure of Delaunay triangulations in the plane, manuscript, 1991.
19. R. A. Dwyer, A faster divide-and-conquer algorithm for constructing Delaunay triangulations, *Algorithmica* **2**(2):137-151, 1987.
20. R. A. Dwyer, Higher-dimensional Voronoi diagrams in linear expected time, *Disc. and Comp. Geometry*, **6**:343-367, 1991.
21. R. Dwyer, private communication.
22. H. Edelsbrunner, R. Seidel, Voronoi diagrams and arrangements, *Disc. and Comp. Geom.* **8**(1):25-44, 1986.
23. H. Edelsbrunner, *Algorithms in Combinatorial Geometry*, Springer-Verlag, Berlin, 1987.
24. H. Edelsbrunner, E.P. Mücke, Simulation of Simplicity, a technique to cope with degenerate cases in geometric computations, *ACM Trans. Graphics* **9**(1):66-104, 1990.
25. H. Edelsbrunner, Flipping works for regular triangulations, manuscript, in preparation, U. Illinois, 1991.
26. I. Emiris, J. Canny, A general approach to removing degeneracies, *32nd Annual Symp. on Found. of Comp. Sci.* 405-413, 1991.
27. S. Fortune, A note on Delaunay diagonal flips, unpublished manuscript, AT&T Bell Labs, 1987.
28. S. Fortune, Sweepline algorithms for Voronoi diagrams, *Algorithmica* **2**:153-174, 1987.
29. S. Fortune, Stable maintenance of point-set triangulation in two dimensions, *30th Ann. Symp. on the Found. of Comp. Sci.*, 494-499, 1989.
30. S. Fortune, Numerical stability of algorithms for Delaunay triangulations and Voronoi diagrams in two dimensions, manuscript, in preparation, AT&T Bell Labs, 1991.
31. K.R. Gabriel, R.R. Sokal, A new statistical approach to geographic variation analysis, *Systematic Zoology* **18**:259-278, 1969.

32. P.J. Green, R. Sibson, Computing Dirichlet tessellations in the plane, *Computer Journal*, **21**(22):168-173, 1977.

33. B. Grunbaum, *Convex Polytopes*, John Wiley and Sons, New York, 1967.

34. L.J. Guibas, D.E. Knuth, M. Sharir, Randomized incremental construction of Delaunay and Voronoi diagrams, *ICALP 90*, 414-431, Springer-Verlag, Berlin, 1990.

35. L.J. Guibas, J. Stolfi, Primitives for the manipulation of general subdivisions and the computation of Voronoi diagrams, *ACM Trans. Graphics* 4(2):74-123, 1985.

36. L. J. Guibas, J. Stolfi, Ruler, compass, and computer: the design and analysis of geometric algorithms, *Theoretical Foundations of Computer Graphics and CAD* 111-165, R.A. Earnshaw, ed., Springer-Verlag, New York, 1988.

37. C. Hoffmann, The problems of accuracy and robustness in geometric computation. *Computer* **22**:31-42, 1989.

38. J. Katajainen, M. Koppinen, Constructing Delaunay triangulations by merging buckets in quadtree order, manuscript, 1987.

39. M. Karasick, D. Lieber, L. Nackman, Efficient Delaunay triangulation using rational arithmetic, *ACM Trans. Graphics* **10**(1):71-91, 1990.

40. J.M. Keil, C.A. Gutwin, The Delaunay triangulation closely approximates the complete euclidean graph, *Algorithms and Data Structures, Workshop WADS 89* 47-56, F. Dehne, J.-R. Sack, N. Santoro, eds, Springer-Verlag LNCS 382, Berlin, 1989.

41. D. E. Knuth, *The Art of Computer Programming, Vol.2, Seminumerical Algorithms*, Addison-Wesley, 1969.

42. C.L. Lawson, Software for $C^1$ surface interpolation, *Mathematical Software III* 161-194, J.R. Rice, ed., Academic Press, 1977.

43. P. Lienhardt, Subdivisions of $n$-dimensional spaces and $n$-dimensional generalized maps, *Proc. of the Fifth Annual Symp. on Comp. Geom.*, 228-236, 1989.

44. P. McMullen, G.C. Shephard, *Convex Polytopes and the Upper Bound Conjecture*, Cambridge University Press, 1971.

45. J.L. Meijering, Interface area, edge length and number of vertices in crystal aggregates with random nucleation, *Phillips Res. Rep* 8:270-290, 1953.

46. V. J. Milenkovic. *Verifiable implementations of geometric algorithms using finite precision arithmetic*, Ph.D. Thesis, Carnegie-Mellon, 1988.

47. D.M. Mount, A. Saalfeld, Globally-equiangular triangulations of cocircular points in $O(n \log n)$ time, *Proc. of the Fourth Annual Symp. on Comp. Geom.*, 143-152, 1988.

48. T. Ohya, M. Iri, K. Murota, A fast Voronoi-diagram with quaternary tree bucketing, *Inf. Proc. Letters* **18**(4):227-231, 1984.

49. T. Ohya, M. Iri, K. Murota, Improvements of the incremental method for the Voronoi diagram with computational comparison of various algorithms, *J. of the O.R. Res. Society of Japan* **27**:306-337, 1984.

50. F.P. Preparata, M.I. Shamos, *Computational Geometry*, Springer-Verlag, 1985.

51. V.T. Rajan, Optimality of the Delaunay triangulation in $R^d$, *Proc. of the Seventh Annual Symp. on Comp. Geom*, 357-363, 1991.

52. I. Rivin, W.D. Smith, Inscribable graphs, manuscript, NEC Corporation, 1991.

53. L.A. Santaló, *Integral Geometry and Geometric Probability*, Encyclopedia of Mathematics and its Applications, Addison-Wesley, 1976.

54. R. Seidel, The complexity of Voronoi diagrams in higher dimension, *Proc. of the 20th Allerton Conference on Communication, Control, and Computing*, 1982.

55. R. Seidel, Constructing higher-dimensional convex hull algorithms at logarithmic cost per face, *Proc. 18th Ann. Symp. Theory. Comp.* 404-413, 1986.

56. M. Shamos, D. Hoey, Closest-point problems, *Proc. 16th Ann. Symp. Found. Comp. Sci.* 151-162, 1975.

57. G.T. Toussaint, The relative neighborhood graph of a finite planar set, *Pattern Recognition* **12**(4):262-268, 1980. *The Comp. Journal* **24**(2):167-172, 1981.

58. D.F. Watson, Computing the $n$-dimensional Delaunay tessellation with application to Voronoi polytopes, *The Computer Journal* **24**(2):167-172, 1981.

59. P. Vaidya, A sparse graph almost as good as the complete graph on points in $k$ dimensions, *Disc. Comp. Geom* **64**:369-381, 1991.

60. C. Yap, Symbolic treatment of geometric degeneracies, *J. Symb. Comp.* **10**:349-370, 1990.

61. C. Yap, A geometric consistency theorem for a symbolic perturbation scheme *J. Comp. Sys. Sci.* **40**(1):2-18, 1990.

*Computing in Euclidean Geometry*, pp. 235–286
Eds. D.-Z. Du and F. K. Hwang


# Polar Forms and Triangular B-Spline Surfaces

Hans-Peter Seidel *

Department of Computer Science, University of Waterloo
Waterloo, Ontario, N2L 3G1, Canada

## 1 Introduction

Computer Aided Geometric Design (CAGD) is concerned with the construction of complex 3-dimensional objects on a computer. Typical examples are cars, ships, and air planes. Most of these objects are best described in terms of their boundary surfaces. Research in CAGD has therefore largely focused on the development of powerful surface schemes.

In order to be useful for CAGD applications, a surface scheme has to satisfy several criteria: Given an outline of the desired shape by means of a so-called control mesh, the surface scheme should produce a smooth surface that roughly mimics the shape of this given control mesh. Furthermore, one would like to be able to control the shape of a surface not just globally, but also locally. Moreover, the construction should be independent of a specific coordinate system: In particular, the relationship between a surface and its underlying control mesh has to be invariant under transformations such as scalings, rotations, and translations. Finally, one has to be able to interrogate the final surface for quality control and manufacturing.

Although there has been significant research on the use of algebraic, piecewise algebraic, and analytic surfaces, over the last decade, the most popular class of surfaces is still the class of parametric piecewise polynomials.

Most surface schemes using parametric piecewise polynomial surfaces are based on tensor product surfaces: Given a curve scheme $F(u) = \sum_i B_i(u)\mathbf{b}_i$ with vector valued control points $\mathbf{b}_i \in \mathbb{R}^3$, the corresponding tensor product scheme is defined as $F(u,v) = \sum_{i,j} B_i(u)B_j(v)\mathbf{b}_{i,j}$ with control points $\mathbf{b}_{i,j} \in \mathbb{R}^3$. If piecewise polynomial curves are represented in the Bernstein-Bézier [21, 5], or in the B-spline representation [18, 60], one then obtains Bézier or B-spline tensor product surfaces, in their rational version called NURBS surfaces (non uniform rational B-splines).

*This work has been partly supported by the Natural Sciences and Engineering Research Council of Canada through Personal Operating Grant OGP0105573 and Strategic Operating Grant STR0040527

Unfortunately, tensor product surfaces are inherently rectangular in nature. While tensor products have proven themselves an excellent tool for the modeling of largely regular surfaces, e.g. outer car bodies, they have well-known draw-backs if the modeling of largely irregular objects, e.g. inner car panels, is required. Therefore, not surprisingly, the need for B-spline like surfaces over non-rectangular regions has been expressed quite early [61].

This paper presents a new triangular B-spline scheme that allows to construct piecewise polynomial surfaces over arbitrary triangulations of the parameter plane. The development of this scheme is based on the study of polar forms. Polar forms have originally been a tool from classical mathematics [80]. They have first been introduced to CAGD by P. de Faget de Casteljau [21, 22, 23] and by L. Ramshaw [57, 58, 59]. The author has subsequently extended this theory to more general surface representations and has used polar forms for the development of B-patches [69, 71, 75]. Further extensions to simplex splines have finally led to the new triangular B-spline scheme described in this paper [17, 33, 34, 70, 72].

While previous approaches to the construction of B-spline like surfaces over irregular domains have been based on subdivision [10, 26], interpolation [52], and on the use of multisided patches [47, 48, 62], the new scheme is based on blending functions and control points. The resulting surfaces are defined as linear combinations of the blending functions and are parametric piecewise polynomials over an arbitrary triangulation of the parameter plane, whose shape is determined by their control points.

The paper is organized as follows: After a brief introduction to bivariate polynomials and polar forms (Section 2) we discuss triangular Bézier patches (Section 3) and introduce a new surface representation for bivariate polynomials, the B-patch (Section 4). In connection with simplex splines (Section 5) this finally leads to the construction of the new triangular B-spline scheme (Section 6). We hope that our presentation will provide a thorough unterstanding of the polar form of a polynomial surface, of triangular Bézier patches, and of some of the main issues that are involved in the construction of triangular B-spline surfaces over arbitrary triangulations.

# 2 Polynomials and Polar Forms

In this section we introduce the polar form of a surface and show that polynomials and polar forms are equivalent to each other. The material in this section is essentially taken from [57], although our proofs are slightly different.

**Definition 2.1 (Affine Map)** *A map $f : \mathbb{R}^2 \to \mathbb{R}^t$ is* affine *iff it preserves affine combinations, i.e., iff $f$ satisfies $f(\sum_{i=0}^m \alpha_i \mathbf{u}_i) = \sum_{i=0}^m \alpha_i f(\mathbf{u}_i)$ whenever $\sum_{i=0}^m \alpha_i = 1$.* ♣

Every affine map $f$ has a *linearization* $f_*$. In order to define $f_*$ we consider the embeddings

$$\epsilon_0 : \mathbb{R}^2 \to \mathbb{R}^3 : \mathbf{u} = (u, v) \mapsto \hat{\mathbf{u}} = (0, u, v) \tag{2.1}$$

and

$$\epsilon_1 : \mathbb{R}^2 \to \mathbb{R}^3 : \mathbf{u} = (u, v) \mapsto \bar{\mathbf{u}} = (1, u, v). \tag{2.2}$$

It is easily checked that $\epsilon_1$ is an affine map.

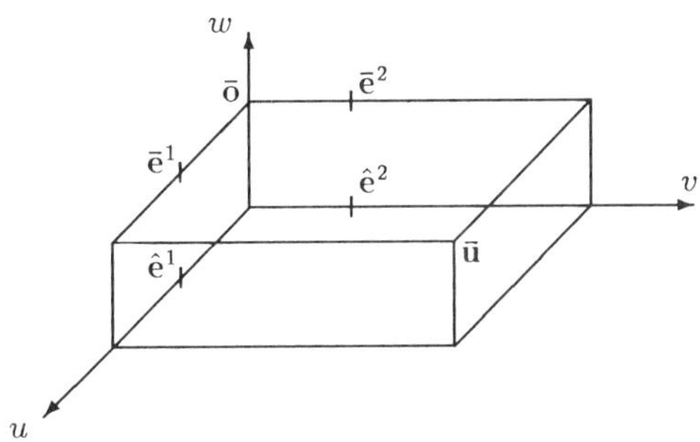


Figure 1: The embeddings $\epsilon_0 : \mathbb{R}^2 \to \mathbb{R}^3 : \mathbf{u} \mapsto \hat{\mathbf{u}}$ and $\epsilon_1 : \mathbb{R}^2 \to \mathbb{R}^3 : \mathbf{u} \mapsto \bar{\mathbf{u}}$.

**Proposition 2.2 (Linearization)** *Affine maps $f : \mathbb{R}^2 \to \mathbb{R}^t$ and linear maps $f_* : \mathbb{R}^3 \to \mathbb{R}^t$ are equivalent. In particular, given a map of either type, a unique map of the other type exists that satisfies the identity $f(\mathbf{u}) = f_*(\bar{\mathbf{u}})$. In this situation $f_*$ is called the linearization of $f$.*

**Proof:** Let $\mathbf{o} \in \mathbb{R}^2$ be the origin and let $\mathcal{B} = \{\mathbf{e}_1, \mathbf{e}_2\}$ be a basis for $\mathbb{R}^2$. Since $\bar{\mathcal{B}}_o = \{\bar{\mathbf{o}}, \bar{\mathbf{e}}_1, \bar{\mathbf{e}}_2\}$ is a basis of $\mathbb{R}^3$ there exists a unique linear map $f_*$ satisfying $f_*(\bar{\mathbf{o}}) = f(\mathbf{o})$, $f_*(\bar{\mathbf{e}}_1) = f(\mathbf{e}_1)$, and $f_*(\bar{\mathbf{e}}_2) = f(\mathbf{e}_2)$. Writing $\mathbf{u} = (1 - u_1 - u_2)\mathbf{o} + u_1\mathbf{e}_1 + u_2\mathbf{e}_2$ as an affine combination of $\mathbf{o}$, $\mathbf{e}_1$, and $\mathbf{e}_2$, it is easily checked that $f(\mathbf{u}) = f_*(\bar{\mathbf{u}})$. The converse is straightforward. ♣

**Example 2.3 (Linearization)** *Consider the affine map $f : \mathbb{R}^2 \to \mathbb{R}$ defined by*

$$f(\mathbf{u}) = 7 + 3u + v$$

*where $\mathbf{u} = u\mathbf{e}_1 + v\mathbf{e}_2 \in \mathbb{R}^2$. Instead of $\bar{\mathcal{B}}_o$ we consider the basis $\mathcal{B}_0 = \{\bar{\mathbf{o}}, \hat{\mathbf{e}}_1, \hat{\mathbf{e}}_2\}$. The linearization $f_*$ of $f$ w.r.t. $\mathcal{B}_0$ is given as*

$$f_*(\mathbf{v}) = 7w + 3u + v$$

*where $\mathbf{v} = w\bar{\mathbf{o}} + u\hat{\mathbf{e}}_1 + v\hat{\mathbf{e}}_2 \in \mathbb{R}^3$. This follows from $f_*(\bar{\mathbf{o}}) = 7$, $f_*(\bar{\mathbf{e}}_1) = f_*(\bar{\mathbf{o}} + \hat{\mathbf{e}}_1) = f(\mathbf{e}_1)$ and $f_*(\bar{\mathbf{e}}_2) = f_*(\bar{\mathbf{o}} + \hat{\mathbf{e}}_2) = f(\mathbf{e}_2)$. ♣*

**Definition 2.4 (symmetric, multilinear, multiaffine)** *Let $\phi$ be an $n$-variate map. Then $\phi$ is* symmetric *iff $\phi(\mathbf{u}_1, \ldots, \mathbf{u}_n) = \phi(\mathbf{u}_{\pi(1)}, \ldots, \mathbf{u}_{\pi(n)})$ for all permutations $\pi \in \Sigma_n$. The map $\phi$ is* multilinear *(*multiaffine*) iff $\phi$ is linear (affine) in each argument if the others are held fixed. ♣*

The following proposition generalizes Proposition 2.2 to the symmetric, $n$-variate case.

**Proposition 2.5 (Multilinearization)** *Symmetric multiaffine maps $f : (\mathbb{R}^2)^n \to \mathbb{R}^t$ and symmetric multilinear maps $f_* : (\mathbb{R}^3)^n \to \mathbb{R}^t$ are equivalent. In particular, given a map of either type, a unique map of the other type exists that satisfies the identity $f(\mathbf{u}_1, \ldots, \mathbf{u}_n) = f_*(\bar{\mathbf{u}}_1, \ldots, \bar{\mathbf{u}}_n)$.*

**Proof:** The proof is almost identical to the proof of Proposition 2.2. Let $\mathbf{o} \in \mathbb{R}^2$ be the origin and let $\mathcal{B} = \{\mathbf{e}_1, \mathbf{e}_2\}$ be a basis of $\mathbb{R}^2$. Since $\bar{\mathcal{B}}_o = \{\bar{\mathbf{o}}, \bar{\mathbf{e}}_1, \bar{\mathbf{e}}_2\}$ is a basis of $\mathbb{R}^3$, a symmetric multilinear map $f_* : (\mathbb{R}^3)^n \to \mathbb{R}^t$ is defined by the $\binom{n+2}{2}$ values $f_*(\bar{\mathbf{o}}, \ldots, \bar{\mathbf{o}}, \bar{\mathbf{e}}_1, \ldots, \bar{\mathbf{e}}_1, \bar{\mathbf{e}}_2, \ldots, \bar{\mathbf{e}}_2)$. Setting $f_*(\bar{\mathbf{o}}, \ldots, \bar{\mathbf{o}}, \bar{\mathbf{e}}_1, \ldots, \bar{\mathbf{e}}_1, \bar{\mathbf{e}}_2, \ldots, \bar{\mathbf{e}}_2) = f(\mathbf{o}, \ldots, \mathbf{o}, \mathbf{e}_1, \ldots, \mathbf{e}_1, \mathbf{e}_2, \ldots, \mathbf{e}_2)$ we thus obtain a symmetric multilinear map $f_*$ that satisfies the identity $f(\mathbf{u}_1, \ldots, \mathbf{u}_n) = f_*(\bar{\mathbf{u}}_1, \ldots, \bar{\mathbf{u}}_n)$. Again, the converse is trivial. ♣

We are now ready to apply the concepts that we have developed so far to the study of polynomials.

**Theorem 2.6 (Blossoming Principle)** *Polynomials $F : \mathbb{R}^2 \to \mathbb{R}^t$ of degree $n$, symmetric multiaffine maps $f : (\mathbb{R}^2)^n \to \mathbb{R}^t$, and symmetric multilinear maps $f_* : (\mathbb{R}^3)^n \to \mathbb{R}^t$ are equivalent. In particular, given a map of either type, unique maps of the other two types exist that satisfy the identity $F(\mathbf{u}) = f(\mathbf{u}, \ldots, \mathbf{u}) = f_*(\bar{\mathbf{u}}, \ldots, \bar{\mathbf{u}})$. In this situation $f$ and $f_*$ are called the* multiaffine *and* multilinear polar form *or* blossom *of $F$. Furthermore, the $q$-th directional derivative of $F$ with respect to vectors $\xi_1, \ldots, \xi_q \in \mathbb{R}^2$ is given as*

$$D_{\xi_1,\ldots,\xi_q} F(\mathbf{u}) = \frac{n!}{(n-q)!} f_*(\bar{\mathbf{u}}, \ldots, \bar{\mathbf{u}}, \hat{\xi}_1, \ldots, \hat{\xi}_q) \tag{2.3}$$

*where $\hat{\xi} = \epsilon_0(\xi) \in \mathbb{R}^3$. ♣*

**Proof:** We start with existence of $f_*$. Since $\mathcal{B}_0 = \{\bar{o}, \hat{e}_1, \hat{e}_2\}$ is a basis of $\mathbb{R}^3$, a symmetric multilinear map $f_* : (\mathbb{R}^3)^n \to \mathbb{R}^t$ is completely defined by the $\binom{n+2}{2}$ values $f_*(\bar{o}, \ldots, \bar{o}, \hat{e}_1, \ldots, \hat{e}_1, \hat{e}_2, \ldots, \hat{e}_2)$. Setting

$$f_*(\underbrace{\bar{o}, \ldots, \bar{o}}_{i}, \underbrace{\hat{e}_1, \ldots, \hat{e}_1}_{j}, \underbrace{\hat{e}_2, \ldots, \hat{e}_2}_{k}) = \frac{i!}{n!} \frac{\partial^{j+k} F(\mathbf{o})}{\partial_u^j \partial_v^k} \tag{2.4}$$

we thus obtain a multilinear map $f_*$, and writing $\bar{\mathbf{u}} = \bar{o} + u\hat{e}_1 + v\hat{e}_2$ we see that $f_*$ satisfies

$$\begin{aligned} f_*(\bar{\mathbf{u}}, \ldots, \bar{\mathbf{u}}) &= \sum_{i+j+k=n} \binom{n}{i\,j\,k} f_*(\underbrace{\bar{o}, \ldots, \bar{o}}_{i}, \underbrace{\hat{e}_1, \ldots, \hat{e}_1}_{j}, \underbrace{\hat{e}_2, \ldots, \hat{e}_2}_{k}) u^j v^k \\ &= \sum_{i+j+k=n} \frac{\partial^{j+k} F(\mathbf{o})}{\partial_u^j \partial_v^k} \frac{u^j v^k}{j!k!} \\ &= F(\mathbf{u}) \end{aligned}$$

where the last step follows from Taylor's Theorem.

The above argument also shows that any symmetric multiaffine map $f_*$ satisfying $f_*(\bar{\mathbf{u}}, \ldots, \bar{\mathbf{u}}) = F(\mathbf{u})$ must satisfy (2.4), hence $f_*$ is unique. Finally, (2.3) follows from the fact that the differential operator $D_\xi$ is linear in $\xi$. Together with Proposition 2.5 this completes the proof of Theorem 2.6. ♣

**Corollary 2.7 (Monomials)** *Consider a monomial*

$$F(\mathbf{u}) = \sum_{j+k \le n} \mathbf{a}_{jk} u^j v^k, \quad \mathbf{a}_{jk} \in \mathbb{R}^t. \tag{2.5}$$

*Then the coefficients* $\mathbf{a}_{jk}$ *satisfy*

$$\mathbf{a}_{jk} = \binom{n}{i\,j\,k} f_*(\underbrace{\bar{o}, \ldots, \bar{o}}_{i}, \underbrace{\hat{e}_1, \ldots, \hat{e}_1}_{j}, \underbrace{\hat{e}_2, \ldots, \hat{e}_2}_{k}). \tag{2.6}$$

*The multilinear polar form* $f_*$ *of* $F$ *with respect to* $\mathcal{B}_0 = \{\bar{o}, \hat{e}_1, \hat{e}_2\}$ *is given as*

$$f_*(\mathbf{v}_1, \ldots, \mathbf{v}_n) = \sum_{\mathcal{P}_n = \{I,J,K\}} \binom{n}{|I|\,|J|\,|K|}^{-1} \mathbf{a}_{|J||K|} \prod_{i \in I} w_i \prod_{j \in J} u_j \prod_{k \in K} v_k \tag{2.7}$$

*where* $\mathbf{v}_i = w_i\bar{o} + u_i\hat{e}_1 + v_i\hat{e}_2 \in \mathbb{R}^3$. *The multiaffine polar form* $f$ *of* $F$ *with respect to* $\mathcal{B} = \{e_1, e_2\}$ *is given as*

$$f(\mathbf{u}_1, \ldots, \mathbf{u}_n) = \sum_{\mathcal{P}_n = \{I,J,K\}} \binom{n}{|I|\,|J|\,|K|}^{-1} \mathbf{a}_{|J||K|} \prod_{j \in J} u_j \prod_{k \in K} v_k. \tag{2.8}$$

*where* $\mathbf{u}_i = u_i\mathbf{e}_1 + v_i\mathbf{e}_2 \in \mathbb{R}^2$. *Here* $\mathcal{P}_n = \{I, J, K\}$ *denotes a partition of* $\{1, \ldots, n\}$ *into the sets* $I, J, K$.

**Proof:** Using Taylor's Theorem, Formula (5) follows directly from (2.3). Writing $\mathbf{v}_i = w_i\bar{\mathbf{o}} + u_i\hat{\mathbf{e}}_1 + v_i\hat{\mathbf{e}}_2$, the multinomial theorem then shows

$$f_*(\mathbf{v}_1,\ldots,\mathbf{v}_n) = \sum_{\mathcal{P}_n=\{I,J,K\}} f_*(\underbrace{\bar{\mathbf{o}},\ldots,\bar{\mathbf{o}}}_{i},\underbrace{\hat{\mathbf{e}}_1,\ldots,\hat{\mathbf{e}}_1}_{j},\underbrace{\hat{\mathbf{e}}_2,\ldots,\hat{\mathbf{e}}_2}_{k})\prod_{i\in I} w_i \prod_{j\in J} u_j \prod_{k\in K} v_k,$$

and (2.7) follows. Finally, specialization of (2.7) to $\mathbf{v}_i = \bar{\mathbf{u}}_i = (1, u_i, v_i)$ yields (2.8). ♣

**Example 2.8 (Quadratics)** *Consider the real-valued polynomial* $F : \mathbb{R}^2 \to \mathbb{R}$ *given by*

$$F(\mathbf{u}) = 1 + 4u + 6v - u^2 - 2uv - v^2$$

*where* $\mathbf{u} = u\mathbf{e}_1 + v\mathbf{e}_2 \in \mathbb{R}^2$. *The multiaffine polar form* $f : \mathbb{R}^2 \times \mathbb{R}^2 \to \mathbb{R}$ *of* $F$ *w.r.t.* $\mathcal{B} = \{\mathbf{e}_1, \mathbf{e}_2\}$ *is given as*

$$f(\mathbf{u}_1,\mathbf{u}_2) = 1 + 2(u_1+u_2) + 3(v_1+v_2) - u_1u_2 - (u_1v_2 + v_1u_2) - v_1v_2,$$

*while the multilinear polar form* $f_* : \mathbb{R}^3 \times \mathbb{R}^3 \to \mathbb{R}$ *of* $F$ *w.r.t.* $\mathcal{B}_0 = \{\bar{\mathbf{o}}, \hat{\mathbf{e}}_1, \hat{\mathbf{e}}_2\}$ *is given as*

$$f_*(\mathbf{v}_1,\mathbf{v}_2) = w_1w_2 + 2(w_1u_2 + u_1w_2) + 3(w_1v_2 + v_1w_2) - u_1u_2 - (u_1v_2 + v_1u_2) - v_1v_2.$$

*Finally, the monomial coefficients* $\mathbf{a}_{jk}$ *can be reconstructed as*

$$\mathbf{a}_{00} = f_*(\bar{\mathbf{o}},\bar{\mathbf{o}}) = 1; \quad \mathbf{a}_{10} = 2f_*(\bar{\mathbf{o}},\hat{\mathbf{e}}_1) = 4; \quad \mathbf{a}_{01} = 2f_*(\bar{\mathbf{o}},\hat{\mathbf{e}}_2) = 6;$$

$$\mathbf{a}_{20} = f_*(\hat{\mathbf{e}}_1,\hat{\mathbf{e}}_1) = -1; \quad \mathbf{a}_{11} = 2f_*(\hat{\mathbf{e}}_1,\hat{\mathbf{e}}_2) = -2; \quad \mathbf{a}_{02} = f_*(\hat{\mathbf{e}}_2,\hat{\mathbf{e}}_2) = -1;$$

*where* $\bar{\mathbf{o}} = (1,0,0) \in \mathbb{R}^3$, $\hat{\mathbf{e}}^1 = (0,1,0) \in \mathbb{R}^3$, *and* $\hat{\mathbf{e}}^2 = (0,0,1) \in \mathbb{R}^3$. ♣

**Example 2.9 (Cubics)** *Consider the real-valued polynomial* $F : \mathbb{R}^2 \to \mathbb{R}$ *given by*

$$F(\mathbf{u}) = 1 + 3u + 6v + 3u^2 + 6uv - 3v^2 - 5u^3 - 15u^2v - 12uv^2 - 3v^3$$

*where* $\mathbf{u} = u\mathbf{e}_1 + v\mathbf{e}_2 \in \mathbb{R}^2$. *The multiaffine polar form* $f : \mathbb{R}^2 \times \mathbb{R}^2 \times \mathbb{R}^2 \to \mathbb{R}$ *of* $F$ *w.r.t.* $\mathcal{B} = \{\mathbf{e}_1, \mathbf{e}_2\}$ *is given as*

$$\begin{aligned} f(\mathbf{u}_1,\mathbf{u}_2,\mathbf{u}_3) &= 1 + (u_1+u_2+u_3) + 2(v_1+v_2+v_3) + (u_1u_2 + u_2u_3 + u_3u_1)\\ &+(u_1v_2 + v_1u_2 + u_2v_3 + v_2u_3 + u_3v_1 + v_3u_1) - (v_1v_2 + v_2v_3 + v_3v_1)\\ &-5u_1u_2u_3 - 5(u_1u_2v_3 + u_1v_2u_3 + v_1u_2u_3) - 4(u_1v_2v_3 + v_1u_2v_3 + v_1v_2u_3)\\ &-3v_1v_2v_3, \end{aligned}$$

*while the he multilinear polar form* $f_* : \mathbb{R}^3 \times \mathbb{R}^3 \times \mathbb{R}^3 \to \mathbb{R}$ *of* $F$ *w.r.t.* $\mathcal{B}_0 = \{\bar{\mathbf{o}}, \hat{\mathbf{e}}_1, \hat{\mathbf{e}}_2\}$ *is given as*

$$\begin{aligned} f(\mathbf{v}_1,\mathbf{v}_2,\mathbf{v}_3) &= w_1w_2w_3 + (u_1w_2w_3 + w_1u_2w_3 + w_1w_2u_3)\\ &+2(v_1w_2w_3 + w_1v_2w_3 + w_1w_2v_3) + (u_1u_2w_3 + w_1u_2u_3 + u_1w_2u_3)\\ &+(u_1v_2w_3 + v_1u_2w_3 + w_1u_2v_3 + w_1v_2u_3 + v_1w_2u_3 + u_1w_2v_3)\\ &-(v_1v_2w_3 + w_1v_2v_3 + v_1w_2v_3) - 5u_1u_2u_3\\ &-5(u_1u_2v_3 + u_1v_2u_3 + v_1u_2u_3) - 4(u_1v_2v_3 + v_1u_2v_3 + v_1v_2u_3)\\ &-3v_1v_2v_3. \end{aligned}$$

*Finally, the monomial coefficients* $\mathbf{a}_{ij}$ *can be reconstructed as*

$$\mathbf{a}_{00} = f_*(\bar{\mathbf{o}}, \bar{\mathbf{o}}, \bar{\mathbf{o}}) = 1; \quad \mathbf{a}_{10} = 3f_*(\bar{\mathbf{o}}, \bar{\mathbf{o}}, \hat{\mathbf{e}}_1) = 3; \quad \mathbf{a}_{01} = 3f_*(\bar{\mathbf{o}}, \bar{\mathbf{o}}, \hat{\mathbf{e}}_2) = 6;$$

$$\mathbf{a}_{20} = 3f_*(\bar{\mathbf{o}}, \hat{\mathbf{e}}_1, \hat{\mathbf{e}}_1) = 3; \quad \mathbf{a}_{11} = 6f_*(\bar{\mathbf{o}}, \hat{\mathbf{e}}_1, \hat{\mathbf{e}}_2) = 6; \quad \mathbf{a}_{02} = 3f_*(\bar{\mathbf{o}}, \hat{\mathbf{e}}_2, \hat{\mathbf{e}}_2) = -3;$$

$$\mathbf{a}_{30} = f_*(\hat{\mathbf{e}}_1, \hat{\mathbf{e}}_1, \hat{\mathbf{e}}_1) = -5; \quad \mathbf{a}_{21} = f_*(\hat{\mathbf{e}}_1, \hat{\mathbf{e}}_1, \hat{\mathbf{e}}_2) = -15;$$

$$\mathbf{a}_{12} = f_*(\hat{\mathbf{e}}_1, \hat{\mathbf{e}}_2, \hat{\mathbf{e}}_2) = -12; \quad \mathbf{a}_{03} = f_*(\hat{\mathbf{e}}_2, \hat{\mathbf{e}}_2, \hat{\mathbf{e}}_2) = -3;$$

*where* $\bar{\mathbf{o}} = (1,0,0) \in \mathbb{R}^3$, $\hat{\mathbf{e}}_1 = (0,1,0) \in \mathbb{R}^3$, *and* $\hat{\mathbf{e}}_2 = (0,0,1) \in \mathbb{R}^3$. ♣

The following Theorem expresses the continuity conditions between two polynomials in terms of polar forms:

**Theorem 2.10 ($C^q$-Conditions)** *Let* $F : \mathbb{R}^2 \to \mathbb{R}^t$ *and* $G : \mathbb{R}^2 \to \mathbb{R}^t$ *be two polynomials of degree* $n$, *and let* $\mathbf{u} \in \mathbb{R}^2$. *Then the following three statements are equivalent:*

- $F$ *and* $G$ *are* $C^q$*-continuous at* $\mathbf{u}$.
- $f_*(\bar{\mathbf{u}}, \ldots, \bar{\mathbf{u}}, \hat{\xi}_1, \ldots, \hat{\xi}_l) = g_*(\bar{\mathbf{u}}, \ldots, \bar{\mathbf{u}}, \hat{\xi}_1, \ldots, \hat{\xi}_l)$ *for* $l = 0, \ldots, q$, *and* $\xi_1, \ldots, \xi_q \in \mathbb{R}^2$.
- $f(\mathbf{u}, \ldots, \mathbf{u}, \mathbf{u}_1, \ldots, \mathbf{u}_q) = g(\mathbf{u}, \ldots, \mathbf{u}, \mathbf{u}_1, \ldots, \mathbf{u}_q)$ *for* $\mathbf{u}_1, \ldots, \mathbf{u}_q \in \mathbb{R}^2$.

**Proof:** The equivalence between the first two statements follows from (2.3). The equivalence between the last two statemments follows from Proposition 2.5 and from the identity $\bar{\mathbf{u}} + \hat{\xi} = \overline{\mathbf{u} + \xi}$. ♣

**Notation 2.11 (Multiplicative Notation)** *It is cumbersome to indicate the number of arguments in a polar form by underbracing. This notational inconvenience can easily be overcome by switching from additive to* multiplicative *notation. This means that from now on we will write*

$$f(\mathbf{u}_1 \cdots \mathbf{u}_n) \quad \text{instead of} \quad f(\mathbf{u}_1, \ldots, \mathbf{u}_n). \tag{2.9}$$

*If all arguments coincide, our notation then simplifies to*

$$f(\mathbf{u}^n) \quad \text{instead of} \quad f(\underbrace{\mathbf{u}, \ldots, \mathbf{u}}_{n}). \tag{2.10}$$

*Terms with several multiple arguments can then be written, e.g., as*

$$f(\mathbf{u}_0{}^i \mathbf{u}_1{}^j \mathbf{u}_2{}^k) \quad \text{instead of} \quad f(\underbrace{\mathbf{u}_0, \ldots, \mathbf{u}_0}_{i}, \underbrace{\mathbf{u}_1, \ldots, \mathbf{u}_1}_{j}, \underbrace{\mathbf{u}_2, \ldots, \mathbf{u}_2}_{k}). \tag{2.11}$$

*If nothing else, multiplicative notation avoids the writing of brackets and indices, and also saves some space. A deeper mathematical justification for this notation, using tensors, is given in [57].* ♣

# 3 Triangular Bézier Patches

Triangular Bézier patches have originally been introduced by de Casteljau [21, 22] using polar forms. They have since been studied widely and are completely understood by now. Several excellent accounts on the subject are available [27, 28, 29, 14, 20]. Our notes essentially follow the outline in [29] with proofs in the spirit of [57].

## 3.1 The Bernstein-Bézier Representation

Consider a triangle $\Delta = [\mathbf{t}_0, \mathbf{t}_1, \mathbf{t}_2]$ in $\mathbb{R}^2$. Given a point $\mathbf{u} \in \mathbb{R}^2$ there exist unique scalars $\lambda_0(\mathbf{u})$, $\lambda_1(\mathbf{u})$, and $\lambda_2(\mathbf{u})$ such that

$$\mathbf{u} = \sum_{i=0}^{2} \lambda_i(\mathbf{u})\mathbf{t}_i \quad \text{and} \quad \sum_{i=0}^{2} \lambda_i(\mathbf{u}) = 1. \tag{3.1}$$

The coefficients $\lambda = (\lambda_0, \lambda_1, \lambda_2)$ are called *barycentric coordinates.*

It is easy to check that the barycentric coordinates satisfy

$$\lambda_0(\mathbf{u}) = \frac{d(\mathbf{u}, \mathbf{t}_1, \mathbf{t}_2)}{d(\mathbf{t}_0, \mathbf{t}_1, \mathbf{t}_2)}, \; \lambda_1(\mathbf{u}) = \frac{d(\mathbf{t}_0, \mathbf{u}, \mathbf{t}_2)}{d(\mathbf{t}_0, \mathbf{t}_1, \mathbf{t}_2)}, \; \lambda_2(\mathbf{u}) = \frac{d(\mathbf{t}_0, \mathbf{t}_1, \mathbf{u})}{d(\mathbf{t}_0, \mathbf{t}_1, \mathbf{t}_2)} \tag{3.2}$$

where

$$d(\mathbf{u}_0, \mathbf{u}_1, \mathbf{u}_2) = \det \begin{pmatrix} 1 & 1 & 1 \\ u_0 & u_1 & u_2 \\ v_0 & v_1 & v_2 \end{pmatrix}. \tag{3.3}$$

Note that $d(\mathbf{u}_0, \mathbf{u}_1, \mathbf{u}_2)$ is twice the area of the triangle $\Delta(\mathbf{u}_0, \mathbf{u}_1, \mathbf{u}_2)$.

**Definition 3.1 (Bernstein Polynomials)** *Let $\Delta = [\mathbf{t}_0, \mathbf{t}_1, \mathbf{t}_2]$ be a triangle in $\mathbb{R}^2$, and let $\lambda_d(\mathbf{u})$, $d = 0, 1, 2$ be the barycentric coordinates of $\mathbf{u}$ with respect to $\Delta$. The polynomials*

$$B^{\Delta}_{ijk}(\mathbf{u}) = \binom{n}{ijk} \lambda_0(\mathbf{u})^i \lambda_1(\mathbf{u})^j \lambda_2(\mathbf{u})^k, \quad i + j + k = n \tag{3.4}$$

*are called* Bernstein polynomials *with respect to $\Delta$.* ♣

It follows immediately from this definition that the Bernstein Polynomials satisfy the following recurrence relation:

**Corollary 3.2 (Recurrence)** *Let $\Delta = [\mathbf{t}_0, \mathbf{t}_1, \mathbf{t}_2]$ be a triangle in $\mathbb{R}^2$, and let $\lambda_d(\mathbf{u})$, $d = 0, 1, 2$, be the barytcentric coordinates of $\mathbf{u}$ w.r.t. $\Delta$. The Bernstein polynomials satisfy the recurrence*

$$B^{\Delta}_{000}(\mathbf{u}) = 1 \tag{3.5}$$

*and*

$$B^{\Delta}_{ijk}(\mathbf{u}) = \lambda_0(\mathbf{u})\; B^{\Delta}_{i-1,j,k}(\mathbf{u}) + \lambda_1(\mathbf{u})\; B^{\Delta}_{i,j-1,k}(\mathbf{u}) + \lambda_2(\mathbf{u})\; B^{\Delta}_{i,j,k-1}(\mathbf{u}), \tag{3.6}$$

*where terms with negative indices are set to zero.* ♣

**Definition 3.3 (Bézier Polynomials)** *Let $\Delta = [\mathbf{t}_0, \mathbf{t}_1, \mathbf{t}_2]$ be a triangle in $\mathbb{R}^2$, and let $F : \mathbb{R}^2 \rightarrow \mathbb{R}^t$ be a polynomial. The representation*

$$F(\mathbf{u}) = \sum_{i+j+k=n} B^{\Delta}_{ijk}(\mathbf{u})\mathbf{b}_{ijk} \tag{3.7}$$

*is called* Bernstein-Bézier representation *of $F$ with respect to $\Delta$. The restriction of $F$ to $\Delta$ is called a* triangular Bézier patch. *The coefficients $\mathbf{b}_{ijk} \in \mathbb{R}^t$ are called* Bézier points. *They form the* Bézier net.

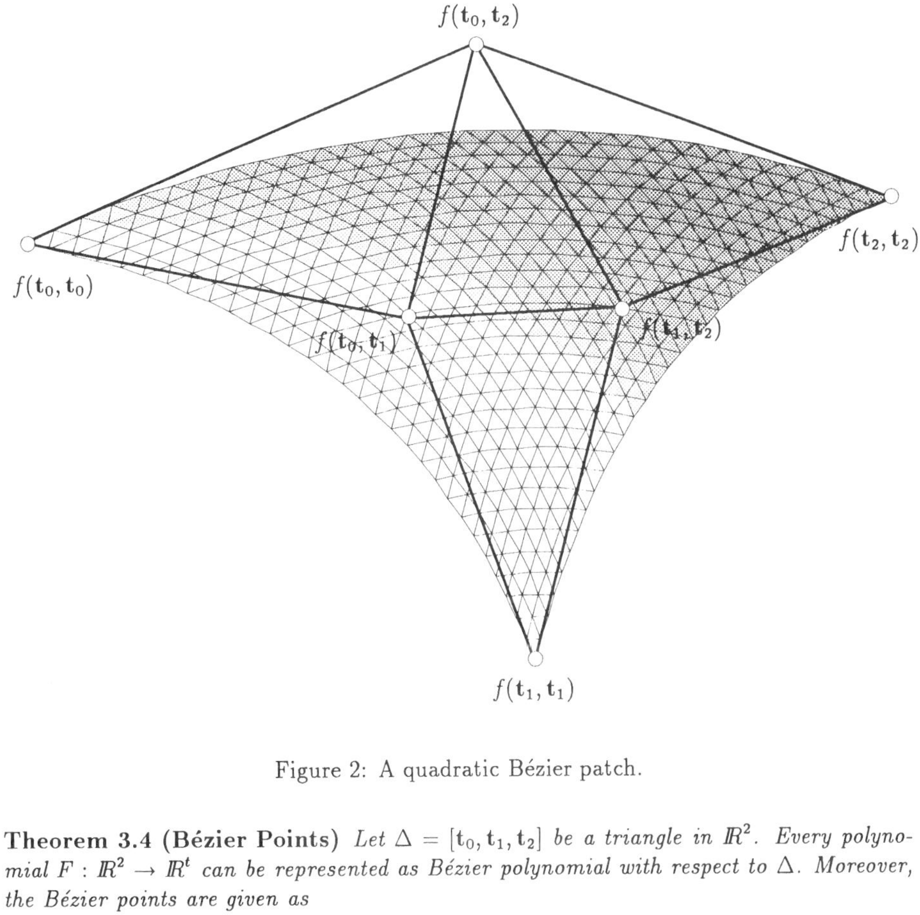


Figure 2: A quadratic Bézier patch.

**Theorem 3.4 (Bézier Points)** *Let $\Delta = [\mathbf{t}_0, \mathbf{t}_1, \mathbf{t}_2]$ be a triangle in $\mathbb{R}^2$. Every polynomial $F : \mathbb{R}^2 \rightarrow \mathbb{R}^t$ can be represented as Bézier polynomial with respect to $\Delta$. Moreover, the Bézier points are given as*

$$\mathbf{b}_{ijk} = f(\underbrace{\mathbf{t}_0, \ldots, \mathbf{t}_0}_{i}, \underbrace{\mathbf{t}_1, \ldots, \mathbf{t}_1}_{j}, \underbrace{\mathbf{t}_2, \ldots, \mathbf{t}_2}_{k}) = f({\mathbf{t}_0}^i {\mathbf{t}_1}^j {\mathbf{t}_2}^k) \tag{3.8}$$

*where $f$ is the multiaffine polar form of $F$.*

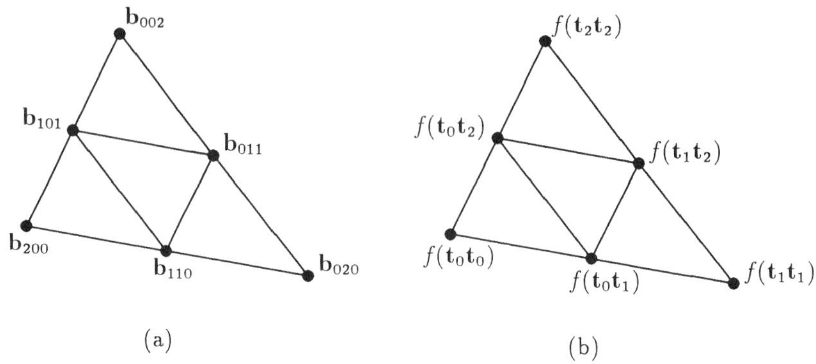


Figure 3: A quadratic Bézier patch: Bézier points (a) and polar values (b) (schematic).

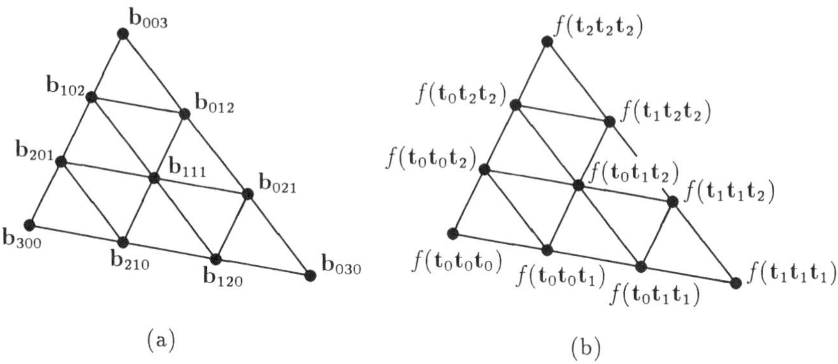


Figure 4: A cubic Bézier patch: Bézier points (a) and polar values (b) (schematic).

**Proof:** Let $F : \mathbb{R}^2 \to \mathbb{R}^t$ be given, and let $f$ be its multiaffine blossom. Representing $\mathbf{u} = \sum_{i=0}^{2} \lambda_i(\mathbf{u})\mathbf{t}_i$ in barycentric coordinates with respect to $\Delta = [\mathbf{t}_0, \mathbf{t}_1, \mathbf{t}_2]$, the standard multinomial theorem yields

$$\begin{aligned} F(\mathbf{u}) &= f(\mathbf{u}^n) \\ &= \sum_{i+j+k=n} \binom{n}{ijk} \lambda_0(\mathbf{u})^i \lambda_1(\mathbf{u})^j \lambda_2(\mathbf{u})^k f({\mathbf{t}_0}^i {\mathbf{t}_1}^j {\mathbf{t}_2}^k) \\ &= \sum_{i+j+k=n} B_{ijk}^{\Delta}(\mathbf{u}) f({\mathbf{t}_0}^i {\mathbf{t}_1}^j {\mathbf{t}_2}^k) \end{aligned}$$

and the assertion follows. ♣

**Remark 3.5 (Linear Independence)** *In particular, Theorem 3.4 shows that every real valued polynomial can be written as linear combination of the Bernstein polynomials. Hence, the Bernstein polynomials* $B_{ijk}^{\Delta}(\mathbf{u})$, $i + j + k = n$, *are a basis of the space* $\Pi^n(\mathbb{R}^2)$ *of real-valued polynomials on* $\mathbb{R}^2$. ♣

**Example 3.6 (Quadratics)** *Consider the real-valued polynomial*

$$F(\mathbf{u}) = 1 + 4u + 6v - u^2 - 2uv - v^2$$

*from Example 2.8. The multiaffine blossom $f$ of $F$ with respect to $\mathcal{B} = \{\mathbf{e}_1, \mathbf{e}_2\}$ is given as*

$$f(\mathbf{u}_1, \mathbf{u}_2) = 1 + 2(u_1 + u_2) + 3(v_1 + v_2) - u_1u_2 - (u_1v_2 + v_1u_2) - v_1v_2.$$

*Then Theorem 3.4 implies that the Bézier points of $F$ with respect to the* standard triangle *$\Delta = [\mathbf{o}, \mathbf{e}_1, \mathbf{e}_2]$ are given as*

$$\mathbf{b}_{200} = f(\mathbf{o}, \mathbf{o}) = 1; \quad \mathbf{b}_{020} = f(\mathbf{e}_1, \mathbf{e}_1) = 4; \quad \mathbf{b}_{002} = f(\mathbf{e}_2, \mathbf{e}_2) = 6;$$

$$\mathbf{b}_{110} = f(\mathbf{o}, \mathbf{e}_1) = 3; \quad \mathbf{b}_{011} = f(\mathbf{e}_1, \mathbf{e}_2) = 5; \quad \mathbf{b}_{101} = f(\mathbf{o}, \mathbf{e}_2) = 4; \quad \clubsuit$$

**Example 3.7 (Cubics)** *Consider the real-valued polynomial*

$$F(\mathbf{u}) = 1 + 3u + 6v + 3u^2 + 6uv - 3v^2 - 5u^3 - 15u^2v - 12uv^2 - 3v^3$$

*from Example 2.9. The multiaffine blossom $f$ of $F$ with respect to $\mathcal{B} = \{\mathbf{e}_1, \mathbf{e}_2\}$ is given as*

$$\begin{aligned} f(\mathbf{u}_1, \mathbf{u}_2, \mathbf{u}_3) = {} & 1 + (u_1 + u_2 + u_3) + 2(v_1 + v_2 + v_3) + (u_1u_2 + u_2u_3 + u_3u_1) \\ & + (u_1v_2 + v_1u_2 + u_2v_3 + v_2u_3 + u_3v_1 + v_3u_1) - (v_1v_2 + v_2v_3 + v_3v_1) \\ & - 5u_1u_2u_3 - 5(u_1u_2v_3 + u_1v_2u_3 + v_1u_2u_3) - 4(u_1v_2v_3 + v_1u_2v_3 + v_1v_2u_3) \\ & - 3v_1v_2v_3. \end{aligned}$$

*Then Theorem 3.4 implies that the Bézier points of $F$ with respect to the* standard triangle *$\Delta = [\mathbf{o}, \mathbf{e}^1, \mathbf{e}^2]$ are given as*

$$\mathbf{b}_{300} = f(\mathbf{o}, \mathbf{o}, \mathbf{o}) = 1; \quad \mathbf{b}_{210} = f(\mathbf{o}, \mathbf{o}, \mathbf{e}_1) = 2; \quad \mathbf{b}_{201} = f(\mathbf{o}, \mathbf{o}, \mathbf{e}_2) = 3;$$

$$\mathbf{b}_{120} = f(\mathbf{o}, \mathbf{e}_1, \mathbf{e}_1) = 4; \quad \mathbf{b}_{111} = f(\mathbf{o}, \mathbf{e}_1, \mathbf{e}_2) = 5; \quad \mathbf{b}_{102} = f(\mathbf{o}, \mathbf{e}_2, \mathbf{e}_2) = 4;$$

$$\mathbf{b}_{030} = f(\mathbf{e}_1, \mathbf{e}_1, \mathbf{e}_1) = 2; \quad \mathbf{b}_{021} = f(\mathbf{e}_1, \mathbf{e}_1, \mathbf{e}_2) = 3;$$

$$\mathbf{b}_{012} = f(\mathbf{e}_1, \mathbf{e}_2, \mathbf{e}_2) = 3; \quad \mathbf{b}_{003} = f(\mathbf{e}_2, \mathbf{e}_2, \mathbf{e}_2) = 1; \quad \clubsuit$$

The following result is sometimes useful for representing the graph of a real-valued polynomial in Bernstein-Bezier form:

**Corollary 3.8 (Greville Abscissae)** *The Bézier points of the polynomial $F(\mathbf{u}) = \mathbf{u}$ with respect to $\Delta = [\mathbf{t}_0, \mathbf{t}_1, \mathbf{t}_2]$ are given as*

$$\mathbf{b}_{ijk} = \frac{1}{n}(i\mathbf{t}_0 + j\mathbf{t}_1 + k\mathbf{t}_2). \tag{3.9}$$

**Proof:** The multiaffine blossom $f$ of $F$ is given by $f(\mathbf{u}_1 \cdots \mathbf{u}_n) = \frac{1}{n}(\mathbf{u}_1 + \cdots + \mathbf{u}_n)$, and Theorem 3.4 yields $\mathbf{b}_{ijk} = f(\mathbf{t}_0{}^i \mathbf{t}_1{}^j \mathbf{t}_2{}^k) = \frac{1}{n}(i\mathbf{t}_0 + j\mathbf{t}_1 + k\mathbf{t}_2)$. ♣

**Corollary 3.9 (Degree Raising)** *Let $F(\mathbf{u}) = \sum B^{\Delta}_{ijk}(\mathbf{u})\mathbf{b}_{ijk}$ be a Bézier polynomial of degree $n$. Then $F$ can also be represented as Bézier polynomial $F(\mathbf{u}) = \sum B^{\Delta}_{ijk}(\mathbf{u})\mathbf{b}^*_{ijk}$ of degree $n+1$. The Bézier points $\mathbf{b}^*_{ijk}$ in this representation are given as*

$$\mathbf{b}^*_{ijk} = \frac{1}{n+1}(i\mathbf{b}_{i-1,j,k} + j\mathbf{b}_{i,j-1,k} + k\mathbf{b}_{i,j,k-1}). \tag{3.10}$$

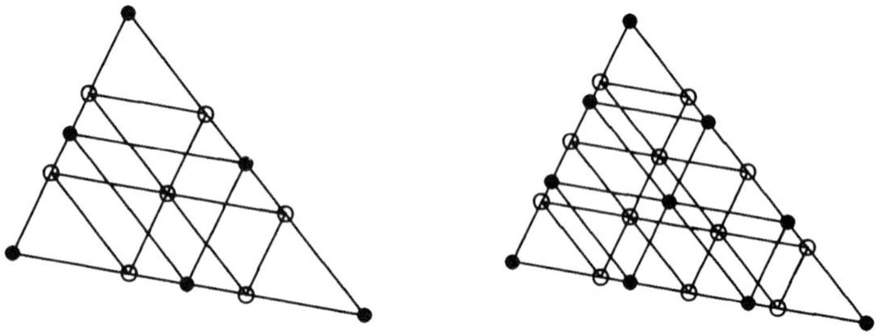

(a) (b)

Figure 5: Degree raising from quadratic to cubic (a), and from cubic to quartic (b).

**Proof:** We follow the proof in [57]. Let $f_n$ be the multiaffine $n$-blossom of $F$. The map $f_{n+1}$ defined by

$$f_{n+1}(\mathbf{u}_1 \cdots \mathbf{u}_{n+1}) = \frac{1}{n+1} \sum_{l=1}^{n+1} f(\mathbf{u}_1 \cdots \mathbf{u}_{l-1}\mathbf{u}_{l+1} \cdots \mathbf{u}_{n+1}) \tag{3.11}$$

is symmetric, multiaffine, and satisfies $f_{n+1}(\mathbf{u}^{n+1}) = F(\mathbf{u})$. Therefore $f_{n+1}$ is the unique multiaffine $(n+1)$-blossom of $F$. Then Theorem 3.4 implies

$$\begin{aligned}
\mathbf{b}^*_{ijk} &= f_{n+1}(\mathbf{t}_0{}^i \mathbf{t}_1{}^j \mathbf{t}_2{}^k) \\
&= \frac{1}{n+1}\Big(\sum_{l=1}^{i} f_n(\mathbf{t}_0{}^{i-1}\mathbf{t}_1{}^j\mathbf{t}_2{}^k) + \sum_{l=i+1}^{i+j} f_n(\mathbf{t}_0{}^i\mathbf{t}_1{}^{j-1}\mathbf{t}_2{}^k) + \sum_{l=i+j+1}^{n} f_n(\mathbf{t}_0{}^i\mathbf{t}_1{}^j\mathbf{t}_2{}^{k-1})\Big) \\
&= \frac{1}{n+1}(i\mathbf{b}_{i-1,j,k} + j\mathbf{b}_{i,j-1,k} + k\mathbf{b}_{i,j,k-1}). \quad ♣
\end{aligned}$$

**Remark 3.10 (Convergence)** *It can be shown that the Bézier nets that are obtained by successive degree raising converge uniformly to the surface. Unfortunately, this convergence is too slow to be useful in practice.* ♣

## 3.2 The de Casteljau Algorithm and its Applications

The most important algorithm for triangular Bézier patches is the de Casteljau Algorithm [21, 22]. Due to the equivalence between polynomials, symmetric multilinear maps, and symmetric multiaffine maps, the de Casteljau Algorithm comes in three different versions: multilinear, multiaffine, and polynomial. Besides from evaluating the polar form, or a point on the surface, the de Casteljau Algorithm can also be used for subdivision, differentiation, and the computation of smooth joints between adjacent patches.

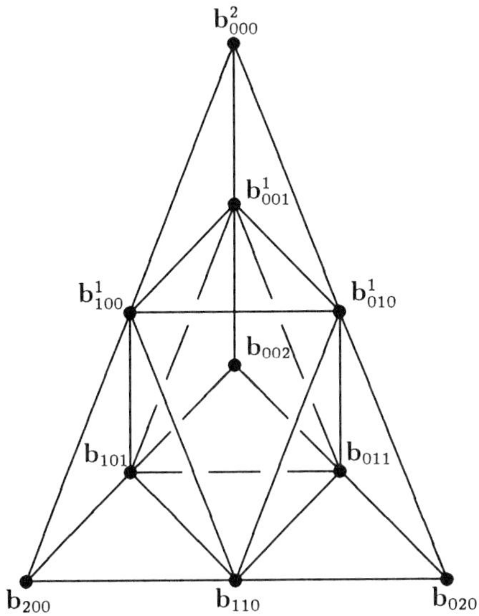


Figure 6: The de Casteljau Algorithm for a quadratic Bézier patch.

**Theorem 3.11 (Multilinear de Casteljau Algorithm)** *The de Casteljau Algorithm can be used to compute the multilinear polar form: Let* $F(\mathbf{u}) = \sum B^{\Delta}_{ijk}(\mathbf{u})\mathbf{b}_{ijk}$ *be a triangular Bézier patch w.r.t.* $\Delta = [\mathbf{t}_0, \mathbf{t}_1, \mathbf{t}_2]$*, and let* $\mathbf{v}_1, \ldots, \mathbf{v}_n \in \mathbb{R}^3$ *be given. Affine independence of* $[\mathbf{t}_0, \mathbf{t}_1, \mathbf{t}_2]$ *implies that* $\bar{\mathcal{B}}_{\Delta} = \{\bar{\mathbf{t}}_0, \bar{\mathbf{t}}_1, \bar{\mathbf{t}}_2\}$ *is a basis of* $\mathbb{R}^3$. *Let* $\mathbf{v}_l = \sum_{d=0}^{2} \lambda_d(\mathbf{v}_l)\bar{\mathbf{t}}_d$ *be the representation of* $\mathbf{v}_l$ *w.r.t.* $\bar{\mathcal{B}}_{\Delta}$. *Set* $\lambda_{d,l} = \lambda_d(\mathbf{v}_l)$ *and compute* $\mathbf{b}^n_{000}$ *recursively according to*

$$\mathbf{b}^0_{ijk} = \mathbf{b}_{ijk}, \tag{3.12}$$

*and*

$$\mathbf{b}^l_{ijk} = \lambda_{0,l}\ \mathbf{b}^{l-1}_{i+1,j,k} + \lambda_{1,l}\ \mathbf{b}^{l-1}_{i,j+1,k} + \lambda_{2,l}\ \mathbf{b}^{l-1}_{i,j,k+1}, \quad l = 1, \ldots, n. \tag{3.13}$$

*Then*

$$f_*(\mathbf{v}_1 \cdots \mathbf{v}_n) = \mathbf{b}^n_{000}. \tag{3.14}$$

*This algorithm is called* multilinear de Casteljau Algorithm. *The simplicial array of points* $\mathcal{S} = \{\mathbf{b}^l_{ijk} \mid i+j+k+l = n\}$ *is called* de Casteljau simplex.

**Proof:** It suffices to show that $\mathbf{b}^l_{ijk} := f_*(\bar{\mathbf{t}}_0{}^i \bar{\mathbf{t}}_1{}^j \bar{\mathbf{t}}_2{}^k \mathbf{v}_1 \cdots \mathbf{v}_l)$ provides a closed form solution to the recurrence (3.12),(3.13). For $l = 0$ this follows directly from Theorem 3.4, and for $l \geq 1$ this follows from the fact that $f_*$ is linear in $\mathbf{v}_l$. ♣

**Remark 3.12 (Coordinates)** *Note that the coordinates $\lambda_{d,l}$ of an arbitrary vector $\mathbf{v} \in \mathbb{R}^3$ w.r.t. $\bar{\mathcal{B}}_\Delta$ do* not *add up to 1 in general. However, if $\mathbf{v} = \bar{\mathbf{u}} = \epsilon_1(\mathbf{u})$ with $\mathbf{u} \in \mathbb{R}^2$, then $\lambda_i(\bar{\mathbf{u}}) = \lambda_i(\mathbf{u})$ are the barycentric coordinates of $\mathbf{u}$ w.r.t. $\Delta = [\mathbf{t}_0, \mathbf{t}_1, \mathbf{t}_2]$.*

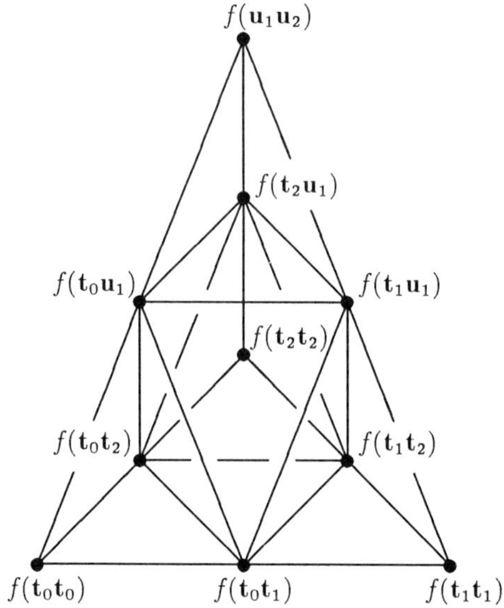


Figure 7: The multiaffine de Casteljau Algorithm for a quadratic Bézier patch.

**Corollary 3.13 (Multiaffine de Casteljau Algorithm)** *The de Casteljau Algorithm can be used to compute the multiaffine polar form: Let $F(\mathbf{u}) = \sum B^\Delta_{ijk}(\mathbf{u})\mathbf{b}_{ijk}$ be a triangular Bézier patch w.r.t. $\Delta = [\mathbf{t}_0, \mathbf{t}_1, \mathbf{t}_2]$, and let $\mathbf{u}_1, \ldots, \mathbf{u}_n \in \mathbb{R}^2$ be given. Let $\mathbf{u}_l = \sum_{d=0}^2 \lambda_d(\mathbf{u}_l)\mathbf{t}_d$ be the representation of $\mathbf{u}_l$ in barycentric coordinates w.r.t. $\Delta = [\mathbf{t}_0, \mathbf{t}_1, \mathbf{t}_2]$. Set $\lambda_{d,l} = \lambda_d(\mathbf{u}_l)$, and compute $\mathbf{b}^n_{000}$ recursively according to the recurrence (3.12),(3.13). Then*

$$f(\mathbf{u}_1 \cdots \mathbf{u}_n) = \mathbf{b}^n_{000}. \tag{3.15}$$

*This algorithm is called* multiaffine de Casteljau Algorithm.

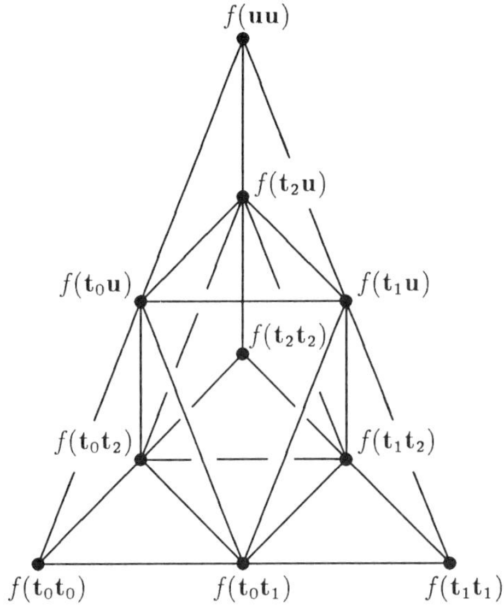


Figure 8: The de Casteljau Algorithm for a quadratic Bézier patch: Evaluation.

**Proof:** Using Theorem 3.11, the proof follows from $f(\mathbf{u}_1 \cdots \mathbf{u}_n) = f_*(\bar{\mathbf{u}}_1 \cdots \bar{\mathbf{u}}_n)$. An independent proof follows from the observation that $\mathbf{b}^l_{ijk} := f(\mathbf{t}_0{}^i \mathbf{t}_1{}^j \mathbf{t}_2{}^k \mathbf{u}_1 \cdots \mathbf{u}_l)$ is again a closed form solution to the recurrence (3.12),(3.13). ♣

**Corollary 3.14 (Evaluation)** *The de Casteljau Algorithm can be used for evaluation: Let* $F(\mathbf{u}) = \sum B^{\Delta}_{ijk}(\mathbf{u})\mathbf{b}_{ijk}$ *be a triangular Bézier patch w.r.t.* $\Delta = [\mathbf{t}_0, \mathbf{t}_1, \mathbf{t}_2]$, *and let* $\mathbf{u} \in I\!R^2$. *Let* $\mathbf{u} = \sum_{d=0}^{2} \lambda_d(\mathbf{u})\mathbf{t}_d$ *be the representation of* $\mathbf{u}$ *in barycentric coordinates w.r.t.* $\Delta = [\mathbf{t}_0, \mathbf{t}_1, \mathbf{t}_2]$. *Set* $\lambda_{d,l} = \lambda_d(\mathbf{u})$, *and compute* $\mathbf{b}^n_{000}$ *recursively according to the recurrence (3.12),(3.13). Then*

$$F(\mathbf{u}) = \mathbf{b}^n_{000}. \tag{3.16}$$

*This algorithm is the original* de Casteljau Algorithm.

**Proof:** Again $\mathbf{b}^l_{ijk} := f(\mathbf{t}_0{}^i \mathbf{t}_1{}^j \mathbf{t}_2{}^k \mathbf{u}^l)$ is a closed form solution to (3.12),(3.13). ♣

**Corollary 3.15 (Subdivision)** *The de Casteljau Algorithm can be used for subdivision: In the situation of Corollary 3.14, the Bézier points of* $F$ *w.r.t. the three subtriangles* $\Delta_0 = [\mathbf{u}, \mathbf{t}_1, \mathbf{t}_2]$, $\Delta_1 = [\mathbf{t}_0, \mathbf{u}, \mathbf{t}_2]$, *and* $\Delta_2 = [\mathbf{t}_0, \mathbf{t}_1, \mathbf{u}]$ *are given by the points on the faces of the de Casteljau simplex* $\mathcal{S}$. *More precisely,* $\mathcal{S}_0 = \{\mathbf{b}^l_{0jk} \mid j + k + l = n\}$ *contains the Bézier points w.r.t.* $\Delta_0$, $\mathcal{S}_1 = \{\mathbf{b}^l_{i0k} \mid j + k + l = n\}$ *contains the Bézier points w.r.t.* $\Delta_1$, *and* $\mathcal{S}_2 = \{\mathbf{b}^l_{ij0} \mid i + j + l = n\}$ *contains the Bézier points w.r.t.* $\Delta_2$.

**Proof:** According to Theorem 3.4 the Bézier points of $F$ w.r.t. $\Delta_0 = [\mathbf{u}, \mathbf{t}_1, \mathbf{t}_2]$ are given by $\mathcal{S}_0 = \{f(\mathbf{t}_1{}^j \mathbf{t}_2{}^k \mathbf{u}^l) \mid j + k + l = n\}$, and the proof follows from the fact that $\mathbf{b}_{0jk}^l = f(\mathbf{t}_1{}^j \mathbf{t}_2{}^k \mathbf{u}^l)$ is a closed form solution of the recurrence (3.12),(3.13). ♣

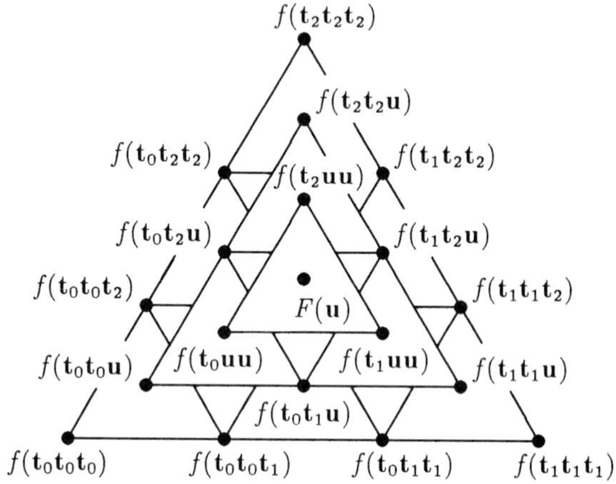


Figure 9: The de Casteljau Algorithm for a cubic Bézier patch: Evaluation.

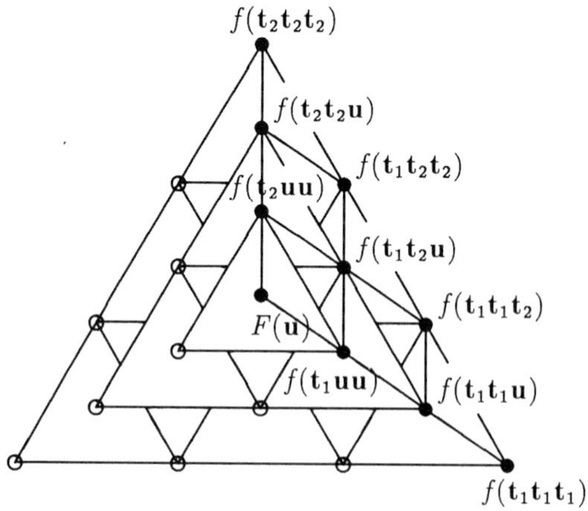


Figure 10: The de Casteljau Algorithm for a cubic Bézier patch: Subdivision.

**Corollary 3.16 (Tangent Plane)** *The de Casteljau Algorithm can be used for computing the tangent plane: In the situation of Corollary 3.14 the tangent plane at $F(\mathbf{u})$ is spanned by the three points $\mathbf{b}^{n-1}_{100}$, $\mathbf{b}^{n-1}_{010}$, and $\mathbf{b}^{n-1}_{001}$.*

**Proof:** Consider the vector $\xi_1 = \mathbf{t}_1 - \mathbf{t}_0$. Then (2.3) shows

$$\begin{aligned} D_\xi F(\mathbf{u}) &= n\ f_*(\hat{\xi}_1 \bar{\mathbf{u}}^{n-1}) = n\ (f_*(\bar{\mathbf{t}}_1 \bar{\mathbf{u}}^{n-1}) - f_*(\bar{\mathbf{t}}_0 \bar{\mathbf{u}}^{n-1})) \\ &= n\ (f(\mathbf{t}_1 \mathbf{u}^{n-1}) - f(\mathbf{t}_0 \mathbf{u}^{n-1})) \\ &= n\ (\mathbf{b}^{n-1}_{010} - \mathbf{b}^{n-1}_{100}). \end{aligned}$$

The same argument holds for $\xi_2 = \mathbf{t}_2 - \mathbf{t}_0$, and the assertion follows. ♣

**Corollary 3.17 (Differentiation)** *The de Casteljau Algorithm can be used for differentiation: Let $F(\mathbf{u}) = \sum B^{\Delta}_{ijk}(\mathbf{u})\mathbf{b}_{ijk}$ be a triangular Bézier patch w.r.t. $\Delta = [\mathbf{t}_0, \mathbf{t}_1, \mathbf{t}_2]$, and let $\mathbf{u} \in \mathbb{R}^2$ and vectors $\xi_1, \ldots, \xi_q \in \mathbb{R}^2$ be given. Let $\mathbf{u} = \sum_{d=0}^{2} \lambda_d(\mathbf{u})\mathbf{t}_d$ be the representation of $\mathbf{u}$ in barycentric coordinates w.r.t. $\Delta = [\mathbf{t}_0, \mathbf{t}_1, \mathbf{t}_2]$, and let $\xi_l = \lambda_1(\xi_l)(\mathbf{t}_1 - \mathbf{t}_0) + \lambda_2(\xi_l)(\mathbf{t}_2 - \mathbf{t}_0)$ be the representation of $\xi_l$ w.r.t. $\mathcal{B} = \{\mathbf{t}_1 - \mathbf{t}_0, \mathbf{t}_2 - \mathbf{t}_0\}$. Let $\lambda_0(\xi_l) = -\lambda_1(\xi_l) - \lambda_2(\xi_l)$. Set $\lambda_{d,l} = \lambda_d(\xi_l)$ for $l = 1, \ldots, q$ and $\lambda_{d,l} = \lambda_d(\mathbf{u})$ for $l = q+1, \ldots, n$, and compute $\mathbf{b}^n_{000}$ recursively according to the recurrence (3.12),(3.13). Then*

$$D_{\xi_1,\ldots,\xi_q} F(\mathbf{u}) = \frac{n!}{(n-q)!}\mathbf{b}^n_{000}. \tag{3.17}$$

*This algorithm is called the* differential de Casteljau Algorithm.

**Proof:** Theorem 2.6 shows $D_{\xi_1,\ldots,\xi_q} F(\mathbf{u}) = \frac{n!}{(n-q)!} f_*(\hat{\xi}_1 \cdots \hat{\xi}_q \mathbf{u}^{n-q})$. Since $\hat{\xi}_l = \lambda_0(\xi)\bar{\mathbf{t}}_0 + \lambda_1(\xi)\bar{\mathbf{t}}_1 + \lambda_2(\xi)\bar{\mathbf{t}}_2$, the assertion then follows from Theorem 3.11. Again we remark that the coordinates $\lambda_{d,l}$ of a vector $\hat{\xi} = \epsilon_0(\xi)$ w.r.t. $\bar{\mathcal{B}}_\Delta$ add up to 0 instead of 1. ♣

**Remark 3.18 (Symmetry)** *Due to the symmetry of the polar form $f$ the various differentiation and evaluation steps in Corollary 3.17 commute. This observation has received a great deal of attention in [27, 28].* ♣

**Corollary 3.19 ($C^q$-Conditions)** *Let $F(\mathbf{u}) = \sum B^{\Delta}_{ijk}(\mathbf{u})\mathbf{b}_{ijk}$ be a triangular Bézier patch w.r.t. $\Delta = [\mathbf{t}_0, \mathbf{t}_1, \mathbf{t}_2]$, and let $G(\mathbf{u}) = \sum B^{\tilde{\Delta}}_{ijk}(\mathbf{u})\tilde{\mathbf{b}}_{ijk}$ be a triangular Bézier patch w.r.t. $\tilde{\Delta} = [\tilde{\mathbf{t}}_0, \mathbf{t}_1, \mathbf{t}_2]$. Let $\tilde{\mathbf{t}}_0 = \sum_{d=0}^{2} \lambda_d(\tilde{\mathbf{t}}_0)\mathbf{t}_d$ be the representation of $\tilde{\mathbf{t}}_0$ in barycentric coordinates w.r.t. $\Delta = [\mathbf{t}_0, \mathbf{t}_1, \mathbf{t}_2]$. Set $\lambda_{d,l} = \lambda_d(\tilde{\mathbf{t}}_0)$, and let $\mathbf{b}^l_{ijk}$ be computed according to the de Casteljau Algorithm. Then $F$ and $G$ are $C^q$-continuous along their common boundary $[\mathbf{t}_1, \mathbf{t}_2]$ if and only if*

$$\tilde{\mathbf{b}}_{ijk} = \mathbf{b}^i_{0jk}, \quad i = 0 \ldots, q. \tag{3.18}$$

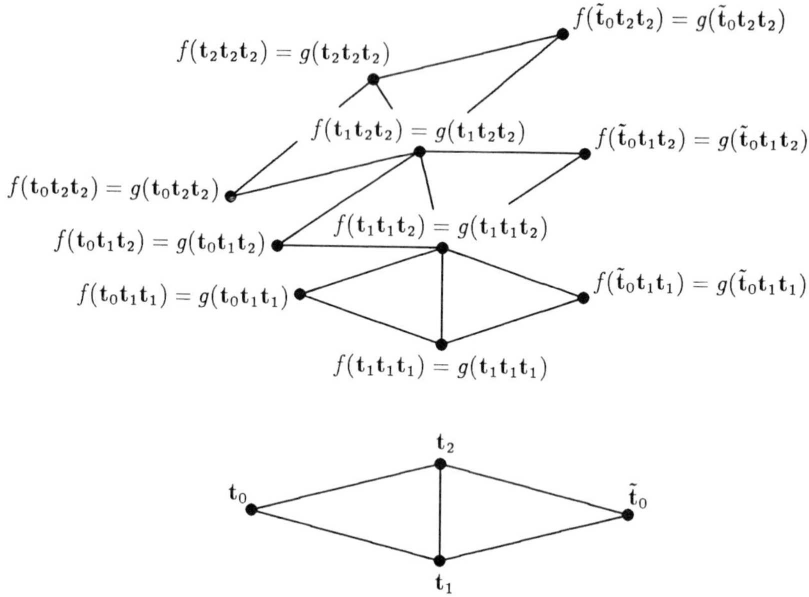


Figure 11: $C^1$-joint between two adjacent cubic Bézier patches: Each subconfiguration is planar and is an affine image of the domain configuration $[\mathbf{t}_0, \mathbf{t}_1, \tilde{\mathbf{t}}_0, \mathbf{t}_2]$.

**Proof:** Using $\mathbf{b}^i_{0jk} = f({\mathbf{t}_1}^j{\mathbf{t}_2}^k\tilde{\mathbf{t}}_0^i)$ and $\tilde{\mathbf{b}}_{ijk} = g({\mathbf{t}_1}^j{\mathbf{t}_2}^k\tilde{\mathbf{t}}_0^i)$, the assertion follows from Theorem 2.10. ♣

**Remark 3.20 ($C^1$-Conditions)** *For $q = 1$, Corollary 3.19 states that $F$ and $G$ are $C^1$-continuous if and only if every tuple of points $(\mathbf{b}_{1,j,k}, \mathbf{b}_{0,j+1,k}, \mathbf{b}_{0,j,k+1}, \mathbf{c}_{1,j,k})$, $j + k = n - 1$, is co-planar, and is an affine image of the quadrilateral $(\mathbf{t}_0, \mathbf{t}_1, \mathbf{t}_2, \mathbf{t}_3)$. This situation is shown in Figure 11.* ♣

We conclude with the following theorem that relates the shape of a triangular Bézier patch to the shape of its control polygon (see Figure 12):

**Theorem 3.21 (Shape Control)** *Let $F(\mathbf{u}) = \sum B^{\Delta}_{ijk}(\mathbf{u})\mathbf{b}_{ijk}$ be a triangular Bézier patch w.r.t. $\Delta = [\mathbf{t}_0, \mathbf{t}_1, \mathbf{t}_2]$. Then the shape of the surface is related to the shape of its control net as follows:*

- **Convex Hull Property:** *If $\mathbf{u} \in \Delta$, then $F(\mathbf{u})$ lies inside the closed convex hull of the control net $\{\mathbf{b}_{ijk} | i + j + k = n\}$.*
- **Corner Vertex Interpolation:** *$F(\mathbf{t}_0) = \mathbf{b}_{n00}$, $F(\mathbf{t}_1) = \mathbf{b}_{0n0}$, and $F(\mathbf{t}_2) = \mathbf{b}_{00n}$, i.e., $F$ interpolates the corner vertices of its control net.*

- **Tangent Plane at Corner Vertices:** *At the corner vertices* $F(\mathbf{t}_0)$, $F(\mathbf{t}_1)$, *and* $F(\mathbf{t}_2)$, *the surface* $F$ *is tangent to its control net.*
- **Boundary Curves:** *The restriction of* $F$ *to the edge* $[\mathbf{t}_1, \mathbf{t}_2]$ *is a Bézier curve with control points* $\mathbf{b}_{0jk}$. *Analogous statements hold for the other boundary curves.*
- **Affine Invariance:** *The relationship between the surface* $F$ *and its control net is affinely invariant: If* $\phi : \mathbb{R}^t \to \mathbb{R}^t$ *is an affine map, then the Bézier points of the image surface* $\tilde{F} = F \circ \phi$ *are the images* $\phi(\mathbf{b}_{ijk})$ *of the Bézier points* $\mathbf{b}_{ijk}$, *i.e., the diagram*

$$\begin{array}{ccc} (\mathbf{b}_{i,j,k})_{i+j+k=n} & \xrightarrow{\phi} & (\phi(\mathbf{b}_{i,j,k}))_{i+j+k=n} \\ \downarrow & & \downarrow \\ F & \xrightarrow{\phi} & \phi \circ F \end{array}$$

*commutes.*

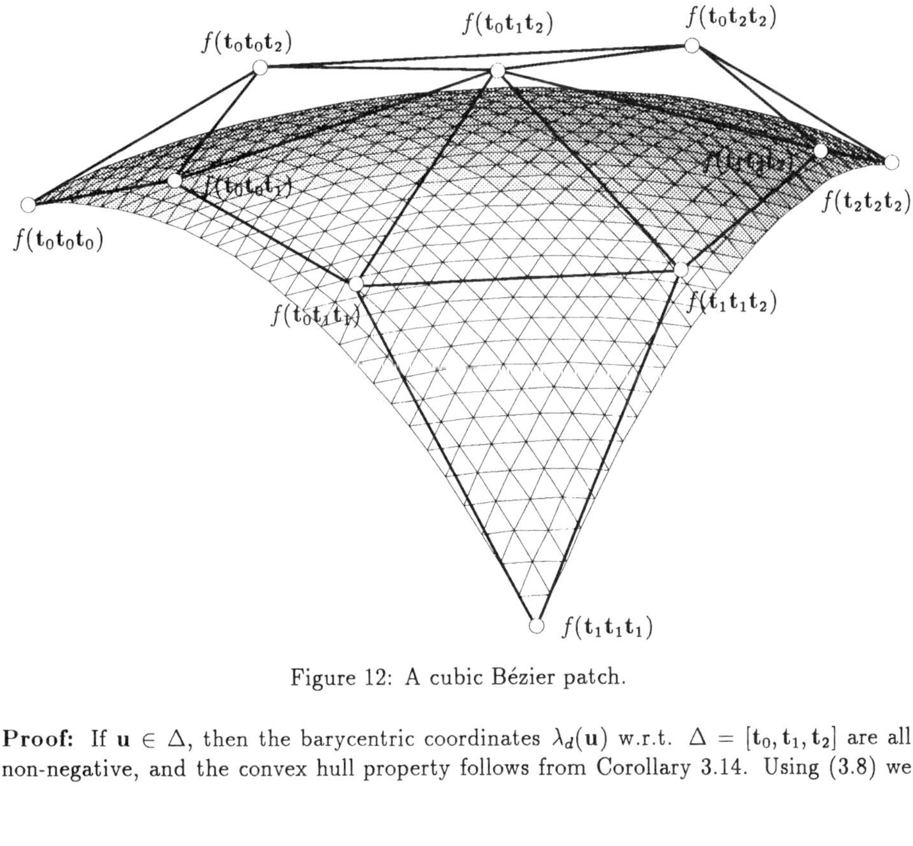


Figure 12: A cubic Bézier patch.

**Proof:** If $\mathbf{u} \in \Delta$, then the barycentric coordinates $\lambda_d(\mathbf{u})$ w.r.t. $\Delta = [\mathbf{t}_0, \mathbf{t}_1, \mathbf{t}_2]$ are all non-negative, and the convex hull property follows from Corollary 3.14. Using (3.8) we get $F(\mathbf{t}_0) = f(\mathbf{t}^n) = \mathbf{b}_{n00}$. Using Corollary 3.16 it also follows that $F$ is tangent to its control net at the corner vertices. The statement on boundary curves directly follows from the fact that the barycentric coordinates $\lambda_0(\mathbf{u})$ of $\mathbf{u} \in [\mathbf{t}_1, \mathbf{t}_2]$ are zero. Finally, in order to prove affine invariance, we remark that the multiaffine polar form $\tilde{f}$ of $\tilde{F}$ is given as $\tilde{f} = \phi \circ f$. This follows from the fact that $\phi \circ f$ is symmetric, multiaffine, and satisfies $\phi \circ f(\mathbf{u}^n) = \tilde{F}(\mathbf{u})$. Theorem 3.4 then shows that the Bézier points $\tilde{\mathbf{b}}_{ijk}$ of $\tilde{F}$ are given as $\tilde{\mathbf{b}}_{ijk} = \tilde{f}(\mathbf{t}_0^i \mathbf{t}_1^j \mathbf{t}_2^k) = \phi(f(\mathbf{t}_0^i \mathbf{t}_1^j \mathbf{t}_2^k) = \phi(\mathbf{b}_{ijk})$, and the proof of Theorem 3.21 is complete. ♣

# 4 B-patches

In the previous section we have seen that a polar form $f$ is uniquely defined by its values $f(\mathbf{t}_0^i\mathbf{t}_1^j\mathbf{t}_2^k)$ on the vertices of a triangle $\Delta = [\mathbf{t}_0, \mathbf{t}_1, \mathbf{t}_2]$. This has led us to the Bézier representation. In this section we generalize this situation by assigning a family of - usually different - knots to each vertex of the triangle $\Delta$. The resulting surface representation is called a *B-patch* [68, 71].

## 4.1 The B-patch Representation

**Definition 4.1 (Knot Arrangement)** *Let* $\mathbf{t}_{0,0}, \ldots, \mathbf{t}_{0,n-1}, \mathbf{t}_{1,0}, \ldots, \mathbf{t}_{1,n-1}, \mathbf{t}_{2,0}, \ldots, \mathbf{t}_{2,n-1} \in \mathbb{R}^2$ *be given and assume that every triple of knots* $(\mathbf{t}_{0,i}, \mathbf{t}_{1,j}, \mathbf{t}_{2,k})$, $0 \le i+j+k \le n-1$ *is affinely independent, i.e. that* $\Delta_{i,j,k} = [\mathbf{t}_{0,i}, \mathbf{t}_{1,j}, \mathbf{t}_{2,k}]$ *forms a proper triangle. Then* $\mathcal{A} = \{\mathbf{t}_{0,0}, \ldots, \mathbf{t}_{0,n-1},\ \mathbf{t}_{1,0}, \ldots, \mathbf{t}_{1,n-1},\ \mathbf{t}_{2,0}, \ldots,\ \mathbf{t}_{2,n-1}\}$ *is called a* knot arrangement *associated with the triangle* $\Delta = [\mathbf{t}_0, \mathbf{t}_1, \mathbf{t}_2] := [\mathbf{t}_{0,0}, \mathbf{t}_{1,0}, \mathbf{t}_{2,0}]$. ♣

A typical knot arrangement is shown in Figure 13.

**Definition 4.2 (Normalized B-weights)** *Let* $\mathcal{A}$ *be a knot arrangement associated with a triangle* $\Delta = [\mathbf{t}_0, \mathbf{t}_1, \mathbf{t}_2]$, *and let* $\lambda_{ijk,d}(\mathbf{u})$, $d = 0,1,2$, *be the barycentric coordinates of* $\mathbf{u}$ *w.r.t.* $\Delta_{ijk} = [\mathbf{t}_{0,i}, \mathbf{t}_{1,j}, \mathbf{t}_{2,k}]$. *The polynomials* $B_{ijk}^{\mathcal{A}}(\mathbf{u})$, $i+j+k=n$, *that are defined recursively as*

$$B_{000}^{\mathcal{A}}(\mathbf{u}) = 1 \tag{4.1}$$

*and*

$$B_{ijk}^{\mathcal{A}}(\mathbf{u}) = \lambda_{i-1,j,k,0} B_{i-1,j,k}^{\mathcal{A}}(\mathbf{u}) + \lambda_{i,j-1,k,1} B_{i,j-1,k}^{\mathcal{A}}(\mathbf{u}) + \lambda_{i,j,k-1,2} B_{i,j,k-1}^{\mathcal{A}}(\mathbf{u}), \tag{4.2}$$

*where terms with negative indices are set to zero, are called the* normalized B-weights *over* $\mathcal{A}$. ♣

**Definition 4.3 (B-patches)** *Let* $\mathcal{A}$ *be a knot arrangement associated with a triangle* $\Delta = [\mathbf{t}_0, \mathbf{t}_1, \mathbf{t}_2]$, *and let* $F : \mathbb{R}^2 \to \mathbb{R}^t$ *be a polynomial. The representation*

$$F(\mathbf{u}) = \sum_{i+j+k=n} B_{ijk}^{\mathcal{A}}(\mathbf{u}) \mathbf{c}_{ijk} \tag{4.3}$$

*is called* B-patch representation *of* $F$ *over* $\mathcal{A}$. *The restriction of* $F$ *to* $\Delta$ *is called a* B-patch. *The coefficients* $\mathbf{c}_{ijk} \in \mathbb{R}^t$ *are called* B-patch control points. *They form the* B-patch control net. ♣

**Theorem 4.4 (B-patch Control Points)** *Let* $\mathcal{A}$ *be a knot arrangement associated with a triangle* $\Delta = [\mathbf{t}_0, \mathbf{t}_1, \mathbf{t}_2]$. *Every polynomial* $F : \mathbb{R}^2 \to \mathbb{R}^t$ *can be represented as B-patch over* $\mathcal{A}$. *Moreover, the B-patch control points are given as*

$$\mathbf{c}_{ijk} = f(\mathbf{t}_{0,0} \cdots \mathbf{t}_{0,i-1} \mathbf{t}_{1,0} \cdots \mathbf{t}_{1,j-1} \mathbf{t}_{2,0} \cdots \mathbf{t}_{2,k-1}) \tag{4.4}$$

*where* $f$ *is the multiaffine polar form of* $F$.

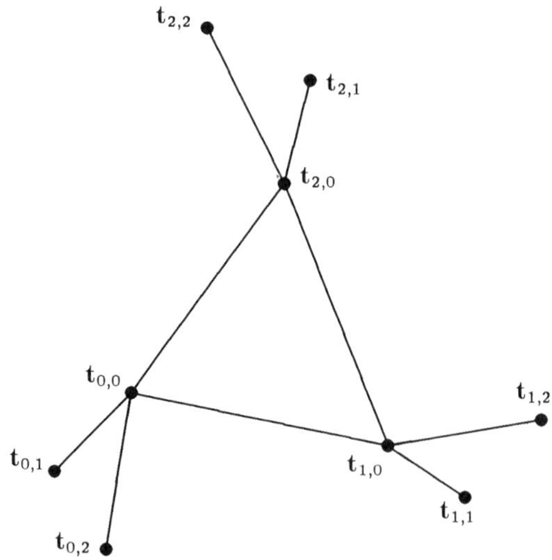


Figure 13: A knot arrangement $\mathcal{A}$ for a cubic B-patch.

**Proof:** Let $F : \mathbb{R}^2 \to \mathbb{R}^t$ be given, and let $f$ be its multiaffine polar form. Define

$$c^l_{ijk}(\mathbf{u}) = f(\mathbf{t}_{0,0} \cdots \mathbf{t}_{0,i-1}\mathbf{t}_{1,0} \cdots \mathbf{t}_{1,j-1}\mathbf{t}_{2,0} \cdots \mathbf{t}_{2,k-1}\mathbf{u}^l), \quad i+j+k+l = n.$$

We show that

$$F(\mathbf{u}) = \sum_{i+j+k+l=n} B^{\mathcal{A}}_{ijk}(\mathbf{u})\mathbf{c}^l_{ijk}(\mathbf{u})$$

for $l = 0, \ldots, n$. We use induction on $\nu = n - l$. The case $\nu = 0$ is trivial. Suppose that our assertion holds for $\nu = n-l$. Representing $\mathbf{u} = \lambda_{ijk,0}(\mathbf{u})\mathbf{t}_{0,i} + \lambda_{ijk,1}(\mathbf{u})\mathbf{t}_{1,j} + \lambda_{ijk,2}(\mathbf{u})\mathbf{t}_{2,k}$ in barycentric coordinates w.r.t. $\Delta_{ijk} = [\mathbf{t}_{0,i}, \mathbf{t}_{1,j}, \mathbf{t}_{2,k}]$ we obtain

$$\mathbf{c}^l_{ijk}(\mathbf{u}) = \lambda_{ijk,0}(\mathbf{u})\mathbf{c}^{l-1}_{i+1,j,k} + \lambda_{ijk,1}(\mathbf{u})\mathbf{c}^{l-1}_{i,j+1,k} + \lambda_{ijk,2}(\mathbf{u})\mathbf{c}^{l-1}_{i,j,k+1},$$

and using the induction hypothesis we get

$$\begin{aligned} F(\mathbf{u}) &= \sum_{i+j+k+l=n} B^{\mathcal{A}}_{ijk}(\mathbf{u})\mathbf{c}^l_{ijk}(\mathbf{u}) \\ &= \sum_{i+j+k+l-1=n} (\lambda_{i-1,j,k,0}(\mathbf{u})B^{\mathcal{A}}_{i-1,j,k}(\mathbf{u}) + \lambda_{i,j-1,k,1}(\mathbf{u})B^{\mathcal{A}}_{i,j-1,k}(\mathbf{u}) \\ &\qquad\qquad + \lambda_{i,j,k-1,2}(\mathbf{u})B^{\mathcal{A}}_{i,j,k-1}(\mathbf{u}))\ \mathbf{c}^{l-1}_{ijk}(\mathbf{u}) \\ &= \sum_{i+j+k+l-1=n} B^{\mathcal{A}}_{ijk}(\mathbf{u})\mathbf{c}^{l-1}_{ijk}(\mathbf{u}) \end{aligned}$$

where the last step follows from (4.2). This proves our claim. In the special case $l = 0$ we thus obtain $F(\mathbf{u}) = \sum B^{\mathcal{A}}_{ijk}(\mathbf{u})\mathbf{c}_{ijk}$, and the proof of Theorem 4.4 is complete. ♣

**Remark 4.5 (Linear Independence)** *In particular, Theorem 4.4 shows that every real valued polynomial can be written as linear combination of the normalized B-weights* $B_{ijk}^{A}(\mathbf{u})$. *Hence, the normalized B-weights* $B_{ijk}^{A}(\mathbf{u})$, $i+j+k=n$, *are a basis of the space* $\Pi^n(\mathbb{R}^2)$ *of real-valued polynomials on* $\mathbb{R}^2$. ♣

The following result is sometimes useful for representing the graph of a real-valued polynomial in B-patch form:

**Corollary 4.6 (Greville Abscissae)** *The B-patch control points of the polynomial* $F(\mathbf{u}) = \mathbf{u}$ *over a knot arrangement* $\mathcal{A}$ *are given as*

$$\mathbf{c}_{ijk} = \frac{1}{n}(\mathbf{t}_{0,0} + \ldots + \mathbf{t}_{0,i-1} + \mathbf{t}_{1,0} + \ldots + \mathbf{t}_{1,j-1} + \mathbf{t}_{2,0} + \ldots + \mathbf{t}_{2,k-1}). \tag{4.5}$$

**Proof:** The multiaffine blossom $f$ of $F$ is given by $f(\mathbf{u}_1 \cdots \mathbf{u}_n) = \frac{1}{n}(\mathbf{u}_1 + \cdots + \mathbf{u}_n)$, and Theorem 4.4 yields $\mathbf{c}_{ijk} = f(\mathbf{t}_{0,0} \cdots \mathbf{t}_{2,k-1}) = \frac{1}{n}(\mathbf{t}_{0,0} + \ldots + \mathbf{t}_{2,k-1})$. ♣

We conclude this section with a brief discussion of the multiple knot case.

**Remark 4.7 (Multiple Knots)** *Suppose that all knots* $\mathbf{t}_{0,0} = \ldots = \mathbf{t}_{0,n-1} = \mathbf{t}_0$, $\mathbf{t}_{1,0} = \ldots = \mathbf{t}_{1,n-1} = \mathbf{t}_1$, *and* $\mathbf{t}_{2,0} = \ldots = \mathbf{t}_{2,n-1} = \mathbf{t}_2$ *appear with multiplicity* $n$. *Then the recurrence (4.1), (4.2) for the normalized B-weights* $B_{ijk}^{A}(\mathbf{u})$ *specializes to the recurrence (3.5), (3.6) for the Bernstein polynomials, and the B-patch representation (4.3) specializes to the Bézier representation (3.7). Accordingly, (4.4) specializes to (3.8), and (4.5) specializes to (3.9).* ♣

The previous remark shows that B-patches subsume triangular Bézier patches in much the same way as the segments of a B-spline curve subsume Bézier curves. Therefore B-patches may be considered the analogue to the segments of a B-spline curve for surfaces.

## 4.2 The de Boor Algorithm and its Applications

In the previous subsection we have introduced B-patches and demonstrated that B-patches are a generalization of triangular Bézier patches in much the same way as B-spline segments are a generalization of Bézier curves.

We now wish to consider the generalization of the de Casteljau Algorithm to the B-patch setting. Due to its analogy to the curve case (where the extension of the de Casteljau Algorithm to B-spline segments is the de Boor Algorithm), we call the resulting symmetric simplicial algorithm *de Boor Algorithm* for polynomials over triangles [68, 71]. Again, the de Boor Algorithm can be used for computing the polar form, for evaluation, for subdivision, and for differentiation.

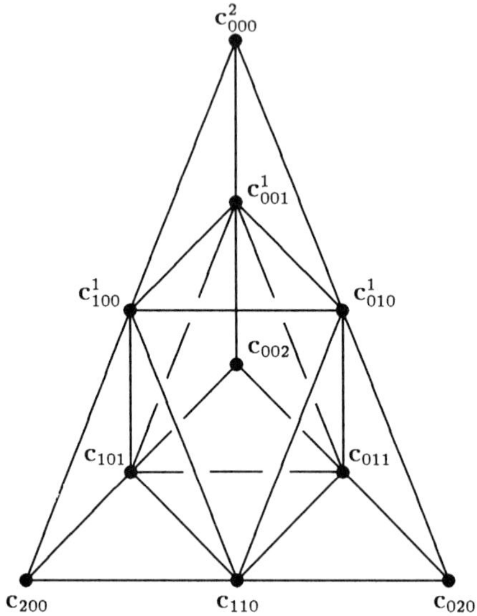


Figure 14: The de Boor Algorithm for a quadratic B-patch.

**Theorem 4.8 (Multilinear de Boor Algorithm)** *The de Boor Algorithm can be used to compute the multilinear polar form: Let* $F(\mathbf{u}) = \sum B^{\mathcal{A}}_{ijk}(\mathbf{u})\mathbf{c}_{ijk}$ *be a B-patch over* $\mathcal{A}$, *and let* $\mathbf{v}_1, \ldots, \mathbf{v}_n \in \mathbb{R}^3$ *be given. Affine independence of* $[\mathbf{t}_{0,i}, \mathbf{t}_{1,j}, \mathbf{t}_{2,k}]$ *implies that* $\bar{\mathcal{B}}_{\Delta_{ijk}} = \{\bar{\mathbf{t}}_{0,i}, \bar{\mathbf{t}}_{1,j}, \bar{\mathbf{t}}_{2,k}\}$ *is a basis of* $\mathbb{R}^3$. *Let*

$$\mathbf{v}_l = \lambda_{ijk,0}(\mathbf{v}_l)\bar{\mathbf{t}}_{0,i} + \lambda_{ijk,1}(\mathbf{v}_l)\bar{\mathbf{t}}_{1,j} + \lambda_{ijk,2}(\mathbf{v}_l)\bar{\mathbf{t}}_{2,k}$$

*be the representation of* $\mathbf{v}_l$ *w.r.t.* $\bar{\mathcal{B}}_{\Delta_{ijk}}$. *Set* $\lambda_{ijk,d,l} = \lambda_{ijk,d}(\mathbf{v}_l)$, *and compute* $\mathbf{c}^n_{000}$ *recursively according to*

$$\mathbf{c}^0_{ijk} = \mathbf{c}_{ijk}, \tag{4.6}$$

*and*

$$\mathbf{c}^l_{ijk} = \lambda_{ijk,0,l}\ \mathbf{c}^{l-1}_{i+1,j,k} + \lambda_{ijk,1,l}\ \mathbf{c}^{l-1}_{i,j+1,k} + \lambda_{ijk,2,l}\ \mathbf{c}^{l-1}_{i,j,k+1}, \quad l = 1, \ldots, n, \tag{4.7}$$

*Then*

$$f_*(\mathbf{v}_1 \cdots \mathbf{v}_n) = \mathbf{c}^n_{000}. \tag{4.8}$$

*This algorithm is called* multilinear de Boor Algorithm. *The simplicial array of points* $\mathcal{S} = \{\mathbf{c}^l_{ijk} \mid i + j + k + l = n\}$ *is called* de Boor simplex.

**Proof:** Again, it suffices to show that $\mathbf{c}_{ijk}^l := f_*(\bar{\mathbf{t}}_{0,0} \cdots \bar{\mathbf{t}}_{0,i-1}\bar{\mathbf{t}}_{1,0} \cdots \bar{\mathbf{t}}_{1,j-1}\bar{\mathbf{t}}_{2,0} \cdots \bar{\mathbf{t}}_{2,k-1} \mathbf{v}_1 \cdots \mathbf{v}_l)$ provides a closed form solution to the recurrence (4.6),(4.7). For $l = 0$ this follows directly from Theorem 4.4, and for $l \geq 1$ this follows from the fact that $f_*$ is linear in $\mathbf{v}_l$. ♣

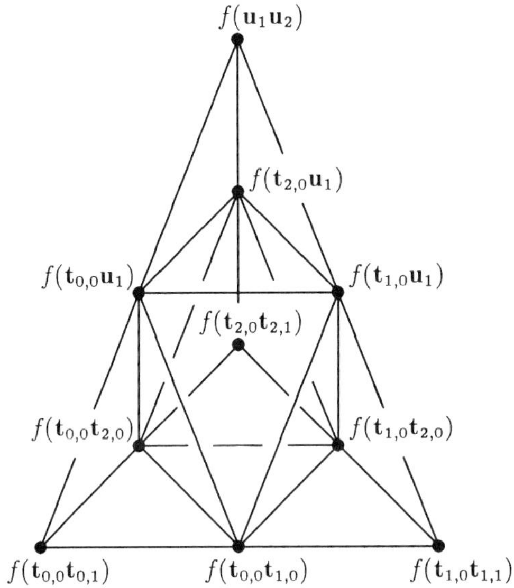


Figure 15: The multiaffine de Boor Algorithm for a quadratic B-patch.

**Corollary 4.9 (Multiaffine de Boor Algorithm)** *The de Boor Algorithm can be used to compute the multiaffine polar form: Let $F(\mathbf{u}) = \sum B_{ijk}^{\mathcal{A}}(\mathbf{u})\mathbf{c}_{ijk}$ be a B-patch over $\mathcal{A}$, and let $\mathbf{u}_1, \ldots, \mathbf{u}_n \in \mathbb{R}^2$ be given. Let*

$$\mathbf{u}_l = \lambda_{ijk,0}(\mathbf{u}_l)\mathbf{t}_{0,i} + \lambda_{ijk,1}(\mathbf{u}_l)\mathbf{t}_{1,j} + \lambda_{ijk,2}(\mathbf{u}_l)\mathbf{t}_{2,k}$$

*be the representation of $\mathbf{u}_l$ in barycentric coordinates w.r.t. $\Delta_{ijk} = [\mathbf{t}_{0,i}, \mathbf{t}_{1,j}, \mathbf{t}_{2,k}]$. Set $\lambda_{ijk,d,l} = \lambda_{ijk,d}(\mathbf{u}_l)$, and compute $\mathbf{c}_{000}^n$ recursively according to the recurrence (4.6),(4.7). Then*

$$f(\mathbf{u}_1 \cdots \mathbf{u}_n) = \mathbf{c}_{000}^n. \tag{4.9}$$

*This algorithm is called* multiaffine de Boor Algorithm.

**Proof:** Using Theorem 4.8, the proof follows from $f(\mathbf{u}_1 \cdots \mathbf{u}_n) = f_*(\bar{\mathbf{u}}_1 \cdots \bar{\mathbf{u}}_n)$. An independent proof follows from the observation that $\mathbf{c}_{ijk}^l := f(\mathbf{t}_{0,0} \cdots \mathbf{t}_{2,k-1}\mathbf{u}_1 \cdots \mathbf{u}_l)$ is again a closed form solution to the recurrence (4.6),(4.7). ♣

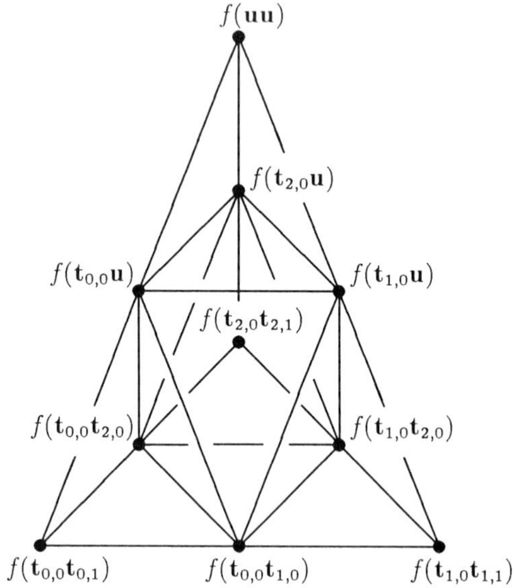


Figure 16: The de Boor Algorithm for a quadratic B-patch: Evaluation.

**Corollary 4.10 (Evaluation)** *The de Boor Algorithm can be used for evaluation: Let* $F(\mathbf{u}) = \sum B_{ijk}^{\mathcal{A}}(\mathbf{u})\mathbf{c}_{ijk}$ *be a B-patch over* $\mathcal{A}$, *and let* $\mathbf{u} \in I\!R^2$. *Let*

$$\mathbf{u} = \lambda_{ijk,0}(\mathbf{u})\mathbf{t}_{0,i} + \lambda_{ijk,1}(\mathbf{u})\mathbf{t}_{1,j} + \lambda_{ijk,2}(\mathbf{u})\mathbf{t}_{2,k}$$

*be the representation of* $\mathbf{u}$ *in barycentric coordinates w.r.t.* $\Delta_{ijk} = [\mathbf{t}_{0,i}, \mathbf{t}_{1,j}, \mathbf{t}_{2,k}]$. *Set* $\lambda_{ijk,d,l} = \lambda_{ijk,d}(\mathbf{u})$, *and compute* $\mathbf{c}_{000}^n$ *recursively according to the recurrence (4.6),(4.7). Then*

$$F(\mathbf{u}) = \mathbf{c}_{000}^n. \tag{4.10}$$

*This algorithm is called* de Boor Algorithm.

**Proof:** Again, $\mathbf{c}_{ijk}^l := f(\mathbf{t}_{0,0} \cdots \mathbf{t}_{2,k-1}\mathbf{u}^l)$ is a closed form solution to (4.6),(4.7). ♣

**Corollary 4.11 (Subdivision)** *The de Boor Algorithm can be used for subdivision: Let* $\mathbf{u}_0, \ldots, \mathbf{u}_{n-1} \in I\!R^2$ *such that each of the sets* $\mathcal{A}_0 = \{\mathbf{u}_0, \ldots, \mathbf{u}_{n-1}, \mathbf{t}_{1,0}, \ldots, \mathbf{t}_{1,n-1}, \mathbf{t}_{2,0}, \ldots, \mathbf{t}_{2,n-1}\}$, $\mathcal{A}_1 = \{\mathbf{t}_{0,0}, \ldots, \mathbf{t}_{0,n-1}, \mathbf{u}_0, \ldots, \mathbf{u}_{n-1}\ \mathbf{t}_{1,0}, \ldots, \mathbf{t}_{1,n-1}\}$, *and* $\mathcal{A}_2 = \{\mathbf{t}_{0,0}, \ldots, \mathbf{t}_{0,n-1}, \mathbf{t}_{1,0}, \ldots, \mathbf{t}_{1,n-1}, \mathbf{u}_0, \ldots, \mathbf{u}_{n-1}\}$ *forms a proper knot arrangement. Then the B-patch representation of* $F$ *over the knot arrangements* $\mathcal{A}_1$, $\mathcal{A}_2$, *and* $\mathcal{A}_3$ *is given by the points on the faces of the de Boor simplex* $\mathcal{S}$. *More precisely,* $\mathcal{S}_0 = \{\mathbf{c}_{0jk}^l \mid j + k + l = n\}$ *contains the*

*B-patch control points w.r.t.* $\mathcal{A}_0$, $\mathcal{S}_1 = \{\mathbf{c}_{i0k}^l \mid j+k+l=n\}$ *contains the B-patch control points w.r.t.* $\mathcal{A}_1$, *and* $\mathcal{S}_2 = \{\mathbf{c}_{ij0}^l \mid i+j+l=n\}$ *contains the B-patch control points w.r.t.* $\mathcal{A}_2$.

**Proof:** According to Theorem 4.4 the B-patch control points of $F$ w.r.t. $\mathcal{A}_0$ are given by $\mathcal{S}_0 = \{f(\mathbf{u}_0 \cdots \mathbf{u}_{i-1}\mathbf{t}_{1,0} \cdots \mathbf{t}_{1,j-1}\mathbf{t}_{2,0} \cdots \mathbf{t}_{2,k-1}) \mid j+k+l=n\}$, and the proof follows from the fact that $\mathbf{c}_{0jk}^l = f(\mathbf{t}_1{}^j\mathbf{t}_2{}^k\mathbf{u}^l)$ is a closed form solution of the recurrence (4.6),(4.7). ♣

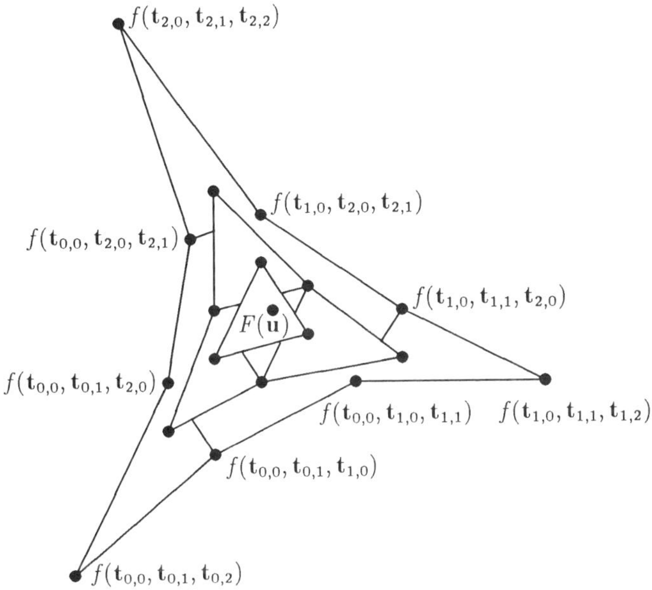


Figure 17: The de Boor Algorithm for a cubic B-patch: Evaluation.

**Corollary 4.12 (Tangent Plane)** *The de Boor Algorithm can be used for computing the tangent plane: In the situation of Corollary 4.10 the tangent plane at* $F(\mathbf{u})$ *is spanned by the three points* $\mathbf{c}_{100}^{n-1}, \mathbf{c}_{010}^{n-1}$, *and* $\mathbf{c}_{001}^{n-1}$.

**Proof:** The proof is completely similar to the proof of Corollary 4.12. ♣

**Corollary 4.13 (Differentiation)** *The de Boor Algorithm can be used for differentiation: Let* $F(\mathbf{u}) = \sum B_{ijk}^{\mathcal{A}}(\mathbf{u})\mathbf{c}_{ijk}$ *be a B-patch over* $\mathcal{A}$, *and let* $\mathbf{u} \in \mathbb{R}^2$ *and vectors* $\xi_1, \ldots, \xi_q \in \mathbb{R}^2$ *be given. Let*

$$\mathbf{u} = \lambda_{ijk,0}(\mathbf{u})\mathbf{t}_{0,i} + \lambda_{ijk,1}(\mathbf{u})\mathbf{t}_{1,j} + \lambda_{ijk,2}(\mathbf{u})\mathbf{t}_{2,k}$$

*be the representation of* $\mathbf{u}$ *in barycentric coordinates w.r.t.* $\Delta_{ijk} = [\mathbf{t}_{0,i}, \mathbf{t}_{1,j}, \mathbf{t}_{2,k}]$, *and let* $\xi_l = \lambda_{ijk,1}(\xi_l)(\mathbf{t}_{1,j} - \mathbf{t}_{0,i}) + \lambda_{ijk,2}(\xi_l)(\mathbf{t}_{2,k} - \mathbf{t}_{0,i})$ *be the representation of* $\xi_l$ *w.r.t.* $\mathcal{B}_{ijk} = \{\mathbf{t}_{1,j} - \mathbf{t}_{0,i}, \mathbf{t}_{2,k} - \mathbf{t}_{0,i}\}$. *Let* $\lambda_{ijk,0}(\xi_l) = -\lambda_{ijk,1}(\xi_l) - \lambda_{ijk,2}(\xi_l)$. *Set* $\lambda_{ijk,d,l} = \lambda_{ijk,d}(\xi_l)$ *for* $l = 1, \ldots, q$ *and* $\lambda_{d,l} = \lambda_{ijk,d}(\mathbf{u})$ *for* $l = q+1, \ldots, n$, *and compute* $\mathbf{c}_{000}^n$ *recursively according to the recurrence (4.6),(4.7). Then*

$$D_{\xi_1,\ldots,\xi_q} F(\mathbf{u}) = \frac{n!}{(n-q)!}\mathbf{c}_{000}^n. \tag{4.11}$$

*This algorithm is called the* differential de Boor Algorithm.

**Proof:** Again the proof is almost identical to the proof of Corollary 3.17. ♣

Similar to the results for triangular Bézier patches, the shape of a B-patch is strongly influenced by the shape of its control polygon:

**Theorem 4.14 (Shape Control)** *Let* $F(\mathbf{u}) = \sum B_{ijk}^{\mathcal{A}}(\mathbf{u})\mathbf{c}_{ijk}$ *be a B-patch over* $\mathcal{A}$. *Then the shape of the surface is related to the shape of its control net as follows:*

- **Convex Hull Property:** *Suppose that* $\Delta_{000} \subseteq \Delta_{ijk}$ *for every* $i+j+k \leq n-1$. *If* $\mathbf{u} \in \Delta_{000}$, *then* $F(\mathbf{u})$ *lies inside the closed convex hull of the control net* $\{\mathbf{c}_{ijk} \mid i+j+k = n\}$.

- **Corner Vertex Interpolation and Tangency at Multiple Knots:** *If* $n$ *knots* $\mathbf{t}_{0,0} = \ldots = \mathbf{t}_{0,n-1} = \mathbf{t}_0$ *coincide, then* $F(\mathbf{t}_0) = \mathbf{b}_{n00}$, *and* $F$ *is tangent to its control net at this point.*

- **Affine Invariance:** *The relationship between the surface* $F$ *and its control net is affinely invariant: If* $\phi : \mathbb{R}^t \to \mathbb{R}^t$ *is an affine map, then the B-patch control points of the image surface* $\tilde{F} = F \circ \phi$ *are the images* $\phi(\mathbf{c}_{ijk})$ *of the B-patch control points* $\mathbf{c}_{ijk}$, *i.e., the diagram*

$$\begin{array}{ccc} (\mathbf{c}_{i,j,k})_{i+j+k=n} & \overset{\phi}{\to} & (\phi(\mathbf{c}_{i,j,k}))_{i+j+k=n} \\ \downarrow & & \downarrow \\ F & \overset{\phi}{\to} & \phi \circ F \end{array}$$

  *commutes.*

**Proof:** Due to the definition of B-patches, the proof of Theorem 4.14 is similar to the proof of Theorem 3.21, and is therefore omitted. ♣

**Remark 4.15 (Multiple Knots)** *We have observed in (4.7) that a B-patch with n-fold knots degenerates to a triangular Bézier patch. In this situation, the de Boor Algorithm degenerates to the de Casteljau Algorithm. Again, this behaviour is completely similar to the situation for curves, where the de Boor Algorithm for the evaluation of a B-spline segment with n-fold knots degenerates to the de Casteljau Algorithm for the evaluation of a Bézier curve.* ♣

# 5 Bivariate B-Splines

A single polynomial surface is nice, but is of course not sufficient for any serious application. What we need are piecewise polynomial surfaces instead. However, if at all possible, we would like to avoid stitching together the individual pieces by hand. What we are looking for are blending functions that produce smooth piecewise polynomial surfaces automatically. Such blending functions for piecewise polynomial surfaces over arbitrary triangulations can be obtained using *bivariate B-splines.*

Bivariate B-splines can be defined in a variety of different ways. Their original geometric definition is obtained by projecting a high-dimensional simplex onto the plane and by taking the volume of the fibre over each point. For this reason they are also called *simplex splines.* Another possible definition is by means of recurrence [51]. Althoug this definition runs into trouble for points on the knot lines, these difficulties can easily be overcome by a slight modification of the initialization of the recurrence. All that is required is a slight modification of the area of support for the piecewise constant B-splines. This phenomenon is completely analogous to the univariate case, where the piecewise constant B-spline blending functions are non-zero on the half-open interval $[\mathbf{t}_i, \mathbf{t}_{i+1})$ instead of the closed interval $[\mathbf{t}_i, \mathbf{t}_{i+1}]$. This observation motivates the following definition:

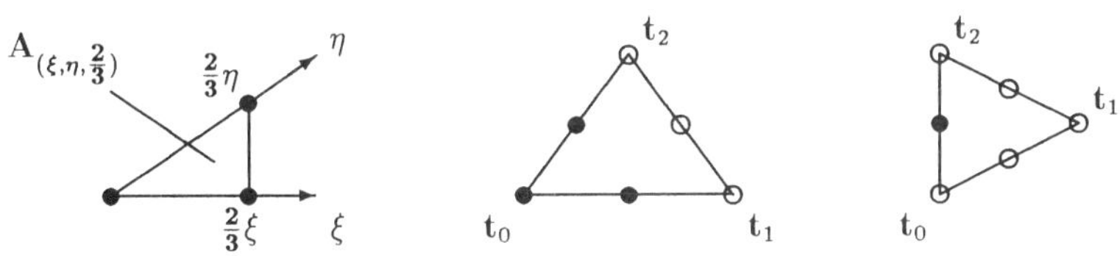


Figure 18: The set $A_{(\xi,\eta,\frac{2}{3})}$ (left), and two half-open convex hulls $[\mathbf{t}_0, \mathbf{t}_1, \mathbf{t}_2)$ (middle, right). The black points belong to the half-open convex hull while the white points do not.

**Definition 5.1 (Half-Open Convex Hull)** *Given* $\mathbf{t}_0, \mathbf{t}_1, \mathbf{t}_2 \in \mathbb{R}^2$, *we denote by* $\Delta = [\mathbf{t}_0, \mathbf{t}_1, \mathbf{t}_2]$ *their* closed convex hull, *and by* $\Delta^\circ = int\ [\mathbf{t}_0, \mathbf{t}_1, \mathbf{t}_2]$ *its* interior. *The* half-open convex hull $[\mathbf{t}_0, \mathbf{t}_1, \mathbf{t}_2)$ *is then defined as follows: Let* $\xi$ *be the horizontal unit vector in* $\mathbb{R}^2$. *A point* $\mathbf{u} \in \mathbb{R}^2$ *belongs to the half-open convex hull* $[\mathbf{t}_0, \mathbf{t}_1, \mathbf{t}_2)$ *if and only if there exists a vector* $\eta$ *with positive slope, and a positive scalar* $\epsilon > 0$, *such that the set*

$$A_{(\xi,\eta,\epsilon)}(\mathbf{u}) = \{s\xi + t\eta \mid 0 < s, t, 0 < s + t < \epsilon\}$$

*is completely contained in* $\Delta^\circ$. *Note that this definition implies* $\Delta^\circ \subset [\mathbf{t}_0, \mathbf{t}_1, \mathbf{t}_2) \subset \Delta$. *The definition of the set* $A_{\xi,\eta,\epsilon}(\mathbf{u})$ *and of the half-open convex hull is illustrated by Figure 18.* ♣

With the definition of the half-open convex hull in place we are ready to define bivariate B-splines by means of the following recurrence relation [51]:

**Definition 5.2 (Bivariate B-splines)** *Let* $V = \{\mathbf{t}_0, \ldots, \mathbf{t}_m\}$ *be a finite set of points in* $\mathbb{R}^2$. *The bivariate B-spline* $M(\mathbf{u}|V) = M(\mathbf{u}|\mathbf{t}_0, \ldots, \mathbf{t}_m)$ *is defined recursively as follows: For* $V = \{\mathbf{t}_0, \mathbf{t}_1, \mathbf{t}_2\}$ *we set*

$$M(\mathbf{u}|\mathbf{t}_0, \mathbf{t}_1, \mathbf{t}_2) = \frac{\chi_{[\mathbf{t}_0, \mathbf{t}_1, \mathbf{t}_2)}}{|d(\mathbf{t}_0, \mathbf{t}_1, \mathbf{t}_2)|}, \tag{5.1}$$

*where* $\chi_{[\mathbf{t}_0,\mathbf{t}_1,\mathbf{t}_2)}$ *is the characteristic function on the half-open convex hull* $[\mathbf{t}_0, \mathbf{t}_1, \mathbf{t}_2)$. *For* $V = \{\mathbf{t}_0, \ldots, \mathbf{t}_m\}$, $m > 2$, *we set*

$$M(\mathbf{u}|\mathbf{t}_0 \ldots, \mathbf{t}_m) = \sum_{d=0}^{2} \lambda_{i_d}(\mathbf{u}) M(\mathbf{u}|\mathbf{t}_0 \ldots, \mathbf{t}_{i_{d-1}}, \mathbf{t}_{i_{d+1}}, \ldots, \mathbf{t}_m) \tag{5.2}$$

*where* $W = \{\mathbf{t}_{i_0}, \mathbf{t}_{i_1}, \mathbf{t}_{i_2}\}$ *is any subset of affinely independent points in* $V$, *and* $\mathbf{u} = \sum_{d=0}^{2} \lambda_{i_d}(\mathbf{u})\mathbf{t}_{i_d}$ *is the representation of* $\mathbf{u}$ *in barycentric coordinates w.r.t.* $W$. *In this situation, the points* $\mathbf{t}_0, \ldots, \mathbf{t}_m$ *are called* knots. ♣

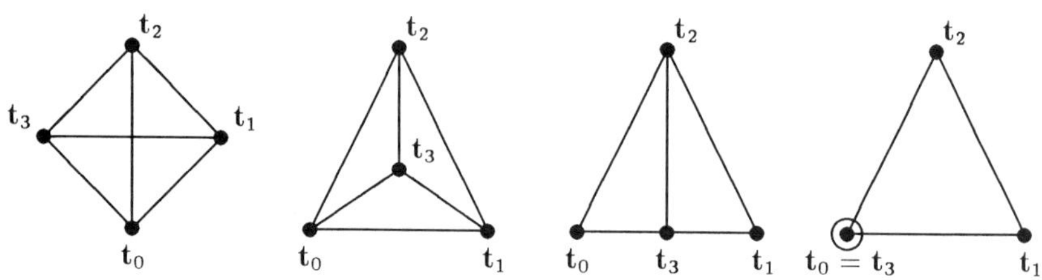


Figure 19: The four essentially different knot configurations for a linear B-spline $M(\mathbf{u}|\mathbf{t}_0, \mathbf{t}_1, \mathbf{t}_2, \mathbf{t}_3)$.

The following example [51] should help to clarify these concepts:

**Example 5.3 (Linear B-splines)** *Let* $\mathbf{t}_0, \mathbf{t}_1, \mathbf{t}_2, \mathbf{t}_3 \in \mathbb{R}^2$ *be given according to Figure 20. We consider evaluation of the linear B-spline* $M(\mathbf{u}|\mathbf{t}_0, \mathbf{t}_1, \mathbf{t}_2, \mathbf{t}_3)$. *Setting* $W = \{\mathbf{t}_0, \mathbf{t}_1, \mathbf{t}_2\}$, *and using (5.2), we get*

$$M(\mathbf{u}|\mathbf{t}_0, \mathbf{t}_1, \mathbf{t}_2, \mathbf{t}_3) = \lambda_0(\mathbf{u})M(\mathbf{u}|\mathbf{t}_1\mathbf{t}_2\mathbf{t}_3) + \lambda_1(\mathbf{u})M(\mathbf{u}|\mathbf{t}_0\mathbf{t}_2\mathbf{t}_3) + \lambda_2(\mathbf{u})M(\mathbf{u}|\mathbf{t}_0\mathbf{t}_1\mathbf{t}_3).$$

*We first consider* $\mathbf{u}_0 \in$ *int* $S_0$. *Then* $M(\mathbf{u}_0|\mathbf{t}_1\mathbf{t}_2\mathbf{t}_3) = M(\mathbf{u}_0|\mathbf{t}_0\mathbf{t}_2\mathbf{t}_3) = 0$, *and we get*

$$M(\mathbf{u}_0|\mathbf{t}_0, \mathbf{t}_1, \mathbf{t}_2, \mathbf{t}_3) = \lambda_2(\mathbf{u}_0)M(\mathbf{u}_0|\mathbf{t}_0\mathbf{t}_1\mathbf{t}_3) = \frac{d(\mathbf{t}_0, \mathbf{t}_1, \mathbf{u}_0)}{d(\mathbf{t}_0, \mathbf{t}_1, \mathbf{t}_2) \cdot d(\mathbf{t}_0, \mathbf{t}_1, \mathbf{t}_3)}.$$

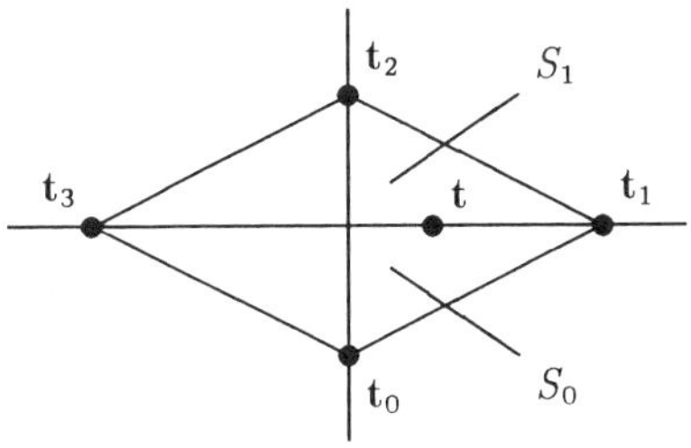


Figure 20: Knot configuration for the linear B-spline in the example.

*Similarly, for* $\mathbf{u}_1 \in int\ S_1$ *we get*

$$M(\mathbf{u}_1|\mathbf{t}_0, \mathbf{t}_1, \mathbf{t}_2, \mathbf{t}_3) = \lambda_0(\mathbf{u}_0)M(\mathbf{u}_0|\mathbf{t}_1\mathbf{t}_2\mathbf{t}_3) = \frac{d(\mathbf{t}_1, \mathbf{t}_2, \mathbf{u}_1)}{d(\mathbf{t}_0, \mathbf{t}_1, \mathbf{t}_2) \cdot d(\mathbf{t}_1, \mathbf{t}_2, \mathbf{t}_3)}.$$

*Let us now consider a point* $\mathbf{t} \in \partial S_0 \cap \partial S_1$ *on the boundary between* $S_0$ *and* $S_1$*, and let*

$$\mathbf{t} = \sigma_1(\mathbf{t}) + \sigma_3(\mathbf{t})$$

*be the representation of* $\mathbf{t}$ *in barycentric coordinates w.r.t.* $\mathbf{t}_1$ *and* $\mathbf{t}_3$. *Note that*

$$\sigma_3(\mathbf{t}) = \frac{d(\mathbf{t}_0, \mathbf{t}_1, \mathbf{t})}{d(\mathbf{t}_0, \mathbf{t}_1, \mathbf{t}_3)} = \frac{d(\mathbf{t}_1, \mathbf{t}_2, \mathbf{t})}{d(\mathbf{t}_1, \mathbf{t}_2, \mathbf{t}_3)}.$$

*Our definition of the half-open convex hull implies that* $\mathbf{t} \in [\mathbf{t}_1, \mathbf{t}_2, \mathbf{t}_3)$*, but* $\mathbf{t} \notin [\mathbf{t}_0, \mathbf{t}_2, \mathbf{t}_3)$. *Therefore*

$$M(\mathbf{t}|\mathbf{t}_0, \mathbf{t}_1, \mathbf{t}_2, \mathbf{t}_3) = \lambda_0(\mathbf{t})M(\mathbf{t}|\mathbf{t}_1\mathbf{t}_2\mathbf{t}_3) = \frac{\sigma_3(\mathbf{t})}{d(\mathbf{t}_0, \mathbf{t}_1, \mathbf{t}_2)}.$$

*On the other hand, we get*

$$\lim_{\mathbf{u}_0 \to \mathbf{t}} M(\mathbf{u}_0|\mathbf{t}_0, \mathbf{t}_1, \mathbf{t}_2, \mathbf{t}_3) = \lim_{\mathbf{u}_1 \to \mathbf{t}} M(\mathbf{u}_1|\mathbf{t}_0, \mathbf{t}_1, \mathbf{t}_2, \mathbf{t}_3) = \frac{\sigma_3(\mathbf{t})}{d(\mathbf{t}_0, \mathbf{t}_1, \mathbf{t}_2)} = M(\mathbf{t}|\mathbf{t}_0, \mathbf{t}_1, \mathbf{t}_2, \mathbf{t}_3).$$

*The example illustrates that a linear B-spline* $M(\mathbf{u}|\mathbf{t}_0, \mathbf{t}_1, \mathbf{t}_2, \mathbf{t}_3)$ *over the knot configuration in Figure 20 as defined by (5.1), (5.2) is well-defined and continuous everywhere.* ♣

Bivariate B-splines offer a variety of features that make them attractive for interactive surface design. Their main properties may be summarized as follows:

**Theorem 5.4 (Bivariate B-splines)** *Let* $V = \{\mathbf{t}_0, \ldots, \mathbf{t}_m\} \subset \mathbb{R}^2$, *and let* $\mathcal{T}_V$ *be the triangulation of* $\mathbb{R}^2$ *that results after connecting every two knots in* $V$ *with a line. The B-spline* $M(\mathbf{u}|\mathbf{t}_0, \ldots, \mathbf{t}_m)$ *satisfies the following properties:*

- **Piecewise Polynomial:** $M(\mathbf{u}|\mathbf{t}_0, \ldots, \mathbf{t}_m)$ *is a piecewise polynomial of degree* $n = m - 2$ *over* $\mathcal{T}_V$.
- **Local Support:** $M(\mathbf{u}|\mathbf{t}_0, \ldots, \mathbf{t}_m)$ *is supported on the closed convex hull* $[V] = [\mathbf{t}_0, \ldots, \mathbf{t}_m]$ *of its knots.*
- **Non-Negativity:** $M(\mathbf{u}|\mathbf{t}_0, \ldots, \mathbf{t}_m) \geq 0$ *for all* $\mathbf{u} \in \mathbb{R}^2$.
- **Smoothness:** *If at most* $\mu$ *knots are collinear,* $2 \leq \mu \leq m$, *then* $M(\mathbf{u}|\mathbf{t}_0, \ldots, \mathbf{t}_m)$ *is* $C^{m-\mu-1}$*-continuous everywhere. Hence, if the knots are in general position (i.e., no three of them are collinear), then* $M(\mathbf{u}|\mathbf{t}_0, \ldots, \mathbf{t}_m)$ *is* $C^{n-1}$*-continuous everywhere.*

**Proof:** The fact that $M(\mathbf{u}|\mathbf{t}_0, \ldots, \mathbf{t}_m)$ is a locally supported piecewise polynomial of degree $n$ over $\mathcal{T}_V$ follows directly from the definition. Since every $\mathbf{t} \in [\mathbf{t}_0, \ldots, \mathbf{t}_m]$ is a convex combination of at most 3 knots, the recursion (5.2) also shows that $M(\mathbf{u}|\mathbf{t}_0, \ldots, \mathbf{t}_m)$ is non-negative. The only property that is harder to prove is the $C^{m-1-\mu}$-continuity. However, since the necessary argument is very technical, we omit this proof here, and refer the reader to some of the references. ♣

Quadratic B-splines over 6 different knot configurations are shown in Figure 26.

# 6 A New B-Spline Scheme

How can we use bivariate B-splines to define piecewise polynomial surfaces over arbitrary triangulations? The previous section demonstrates that lots and lots of bivariate B-splines exist. Which ones should we choose, and how should we normalize?

We shall proceed according to the following recipe: Suppose that a triangulation of $\mathbb{R}^2$ or of a bounded region $D \subset \mathbb{R}^2$ is given (see Fig. 21). With each vertex we associate a *cloud* of $n$ additional knots. We then need a rule for selecting $\binom{n+2}{2}$ subsets of $n+3$ knots from the 3 clouds of a given triangle. Each such subset then gives a bivariate B-spline of degree $n$. The linear span of all these B-splines is the spline space of interest. If the B-splines can be renormalized to form a partition of unity, we then obtain an affinely invariant surface scheme.

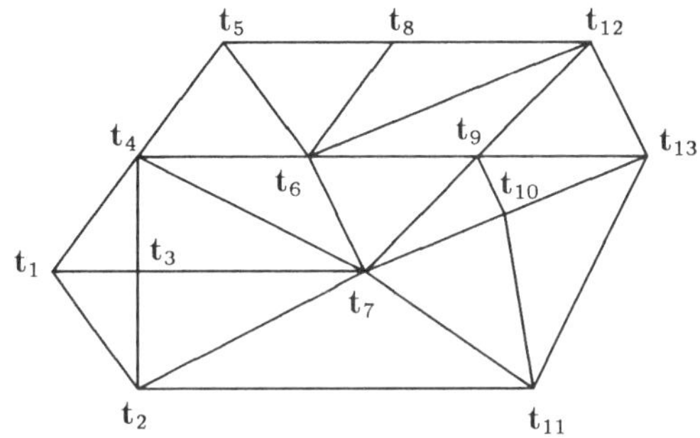


Figure 21: Triangulation of a bounded domain $D \subset \mathbb{R}^2$.

Hence the problem to develop an affinely invariant surface scheme boiles down to

- finding a rule for selecting $\binom{n+2}{2}$ subsets of $n+3$ knots from the 3 clouds of a given triangle, and
- coming up with the right normalization.

## 6.1 A New Spline Space

Surprisingly enough, both the above requirements can be satisfied simultaneously by matching the B-spline recurrence (5.1),(5.2) with the recurrence for the normalized B-weights (4.1), (4.2). In order to be precise, we need the following definition:

**Definition 6.1 (Knot Net)** *Let* $\mathbf{t}_{0,0}, \ldots, \mathbf{t}_{0,n}, \mathbf{t}_{1,0}, \ldots, \mathbf{t}_{1,n}, \mathbf{t}_{2,0}, \ldots, \mathbf{t}_{2,n} \in \mathbb{R}^2$ *be given, and assume that every triple of knots* $(\mathbf{t}_{0,i}, \mathbf{t}_{1,j}, \mathbf{t}_{2,k})$, $0 \le i+j+k \le n$ *is affinely independent, i.e., that* $\Delta_{ijk} = [\mathbf{t}_{0,i}, \mathbf{t}_{1,j}, \mathbf{t}_{2,k}]$ *forms a proper triangle. Define*

$$\Omega_{ijk} = \bigcap_{(i'j'k') \le (ijk)} \Delta_{i'j'k'} \tag{6.1}$$

*and*

$$\Omega_n = int\,(\bigcap_{i+j+k=n} \Omega_{ijk}). \tag{6.2}$$

*If the interior int* $\Omega_n$ *is non-empty, then* $\mathcal{K} = \{\mathbf{t}_{0,0},\ldots,\mathbf{t}_{0,n},\mathbf{t}_{1,0},\ldots,\mathbf{t}_{1,n},\mathbf{t}_{2,0},\ldots,\mathbf{t}_{2,n}\}$ *is called a* knot net associated with the triangle $\Delta = [\mathbf{t}_0,\mathbf{t}_1,\mathbf{t}_2] = [\mathbf{t}_{0,0},\mathbf{t}_{1,0},\mathbf{t}_{2,0}]$. *The set of points* $C_i = \{\mathbf{t}_{i,0},\ldots,\mathbf{t}_{i,n}\}$ *is called the* cloud of knots associated with the vertex $\mathbf{t}_i = \mathbf{t}_{i,0}$. *Loosely speaking, the condition that int* $\Omega_n$ *is non-empty states that the clouds of knots associated with different vertices of a triangle are all separated from each other. In particular, all triangles* $\Delta_{ijk} = [\mathbf{t}_{0,i},\mathbf{t}_{1,j},\mathbf{t}_{2,k}]$ *then have the* same *orientation.* ♣

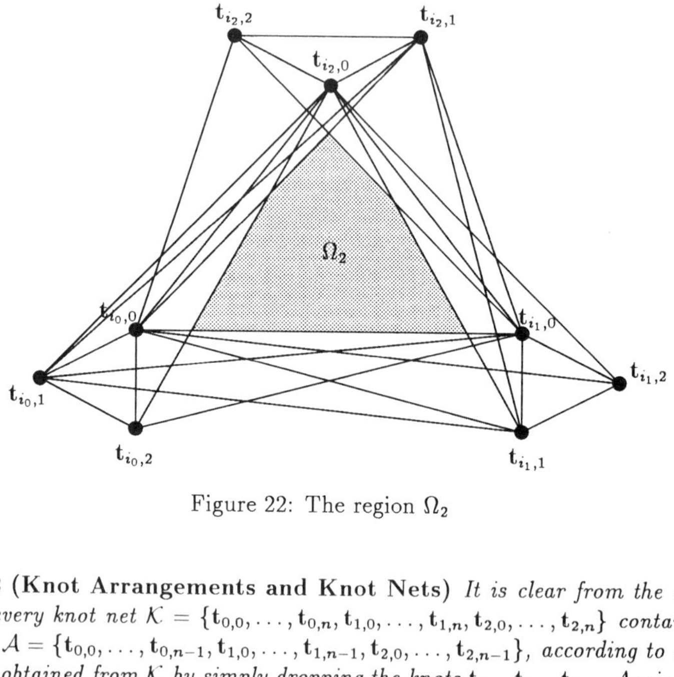


Figure 22: The region $\Omega_2$

**Remark 6.2 (Knot Arrangements and Knot Nets)** *It is clear from the above definition that every knot net* $\mathcal{K} = \{\mathbf{t}_{0,0},\ldots,\mathbf{t}_{0,n},\mathbf{t}_{1,0},\ldots,\mathbf{t}_{1,n},\mathbf{t}_{2,0},\ldots,\mathbf{t}_{2,n}\}$ *contains a knot arrangement* $\mathcal{A} = \{\mathbf{t}_{0,0},\ldots,\mathbf{t}_{0,n-1},\mathbf{t}_{1,0},\ldots,\mathbf{t}_{1,n-1},\mathbf{t}_{2,0},\ldots,\mathbf{t}_{2,n-1}\}$, *according to Definition 4.1, which is obtained from* $\mathcal{K}$ *by simply dropping the knots* $\mathbf{t}_{0,n},\mathbf{t}_{1,n},\mathbf{t}_{2,n}$. *Again, this phenomenon is completely analogous to the curve case, where a B-spline segment of degree* $n$ *is defined by* $2n$ *many knots, while the* $n+1$ *B-spline blending functions that contribute to this segment require* $2n+2$ *knots for their definition.* ♣

**Definition 6.3 (Normalized B-Splines)** *Let* $\mathcal{K} = \{\mathbf{t}_{0,0},\ldots,\mathbf{t}_{0,n},\mathbf{t}_{1,0},\ldots,\mathbf{t}_{1,n},\mathbf{t}_{2,0},\ldots,\mathbf{t}_{2,n}\}$ *be a knot net associated with the triangle* $\Delta = [\mathbf{t}_{0,0},\mathbf{t}_{1,0},\mathbf{t}_{2,0}]$. *We define*

$$V_{ijk}^{\mathcal{K}} = \{\mathbf{t}_{0,0},\ldots,\mathbf{t}_{0,i},\mathbf{t}_{1,0},\ldots,\mathbf{t}_{1,j},\mathbf{t}_{2,0},\ldots,\mathbf{t}_{2,k}\}, \quad i+j+k=n, \tag{6.3}$$

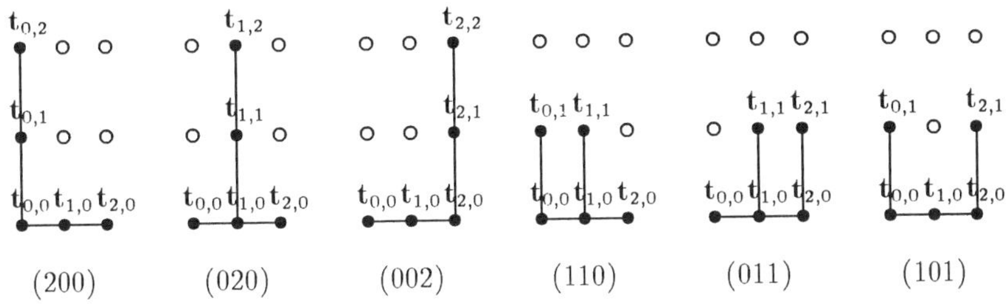


Figure 23: Knot selection for the quadratic normalized B-splines $N_{ijk}^{\mathcal{K}}$ by matching the B-spline recurrence with the recurrence for the normalized B-weights.

*and consider the functions*

$$N_{ijk}^{\mathcal{K}}(\mathbf{u}) = |d(\mathbf{t}_{0,i}, \mathbf{t}_{1,j}, \mathbf{t}_{2,k}| \cdot M(\mathbf{u}|V_{ijk}^{\mathcal{K}}). \tag{6.4}$$

*The functions* $N_{ijk}^{\mathcal{K}}(\mathbf{u})$ *are called* normalized B-splines over $\mathcal{K}$. *Note that (6.3) selects exactly* $\binom{n+2}{2}$ *subsets* $V_{ijk}^{\mathcal{K}}$, $i+j+k=n$, *of* $n+3$ *knots from the 3 clouds of a given triangle* $\Delta$. *For the quadratic case, this selection rule is illustrated by Figure 23.* ♣

The motivation for our selection rule (6.3) and for the normalization (6.4) comes from the following theorem [17]:

**Theorem 6.4 (B-Weights and Normalized B-Splines)** *Let* $\mathcal{K} = \{\mathbf{t}_{0,0}, \ldots, \mathbf{t}_{0,n}, \mathbf{t}_{1,0}, \ldots, \mathbf{t}_{1,n}, \mathbf{t}_{2,0}, \ldots, \mathbf{t}_{2,n}\}$ *be a knot net, and let* $\mathcal{A} = \{\mathbf{t}_{0,0}, \ldots, \mathbf{t}_{0,n-1}, \mathbf{t}_{1,0}, \ldots, \mathbf{t}_{1,n-1}, \mathbf{t}_{2,0}, \ldots, \mathbf{t}_{2,n-1}\}$ *be the corresponding knot arrangement which is obtained from* $\mathcal{K}$ *by dropping the knots* $\{\mathbf{t}_{0,n}, \mathbf{t}_{1,n}, \mathbf{t}_{2,n}\}$. *Let* $N_{ijk}^{\mathcal{K}}(\mathbf{u})$ *be the normalized B-splines over* $\mathcal{K}$, *and let* $B_{ijk}^{\mathcal{A}}$ *be the normalized B-weights over* $\mathcal{A}$. *Then*

$$N_{ijk}^{\mathcal{K}}(\mathbf{u}) = B_{ijk}^{\mathcal{A}}(\mathbf{u}) \quad \text{for all } \mathbf{u} \in \Omega_n, \quad i+j+k=n, \tag{6.5}$$

*where* $\Omega_n$ *is defined according to (6.2).*

**Proof:** We will proceed by induction on $n$. Noting that $\Omega_0 = \text{int}\,\Delta_{000}$ the assertion for $n=0$ readily follows from (4.1) and (5.1).

Suppose it holds for $n-1$. Due to Condition (6.2) on the knot net $\mathcal{K}$, the B-splines $M(\mathbf{u}|V_{i',j',k'}^{\mathcal{K}})$, $i'+j'+k'=n-1$, of degree $n-1$ vanish for $\mathbf{u} \in \Omega_n$ whenever any of the indices $i', j', k'$ is negative. The recursion formula (4.2) for the normalized B-weights thus gives

$$
\begin{aligned}
B^{\mathcal{A}}_{ijk} &= \frac{d(\mathbf{u},\mathbf{t}_{1j},\mathbf{t}_{2k})}{d(\mathbf{t}_{0,i-1},\mathbf{t}_{1j},\mathbf{t}_{2k})}\,N^{\mathcal{A}}_{i-1,j,k}(\mathbf{u}) + \frac{d(\mathbf{t}_{0i},\mathbf{u},\mathbf{t}_{2k})}{d(\mathbf{t}_{0i},\mathbf{t}_{1,j-1},\mathbf{t}_{2k})}\,N^{\mathcal{A}}_{i,j-1,k}(\mathbf{u}) \\
&\quad + \frac{d(\mathbf{t}_{0i},\mathbf{t}_{1j},\mathbf{u})}{d(\mathbf{t}_{0i},\mathbf{t}_{1j},\mathbf{t}_{2,k-1})} N^{\mathcal{A}}_{i,j,k-1}(\mathbf{u}) \\
&= \sigma_{i-1,j,k}\; d(\mathbf{u},\mathbf{t}_{1j},\mathbf{t}_{2k})\; M(\mathbf{u}|V^{\mathcal{K}}_{i-1,j,k}) + \sigma_{i,j-1,k}\; d(\mathbf{t}_{0i},\mathbf{u},\mathbf{t}_{2k})\; M(\mathbf{u}|V^{\mathcal{K}}_{i,j-1,k}) \\
&\quad + \sigma_{i,j,k-1}\; d(\mathbf{t}_{0i},\mathbf{t}_{1j},\mathbf{u})\; M(\mathbf{u}|V^{\mathcal{K}}_{i,j,k-1})
\end{aligned}
$$

where $\sigma_{ijk} = \operatorname{sgn} d(\mathbf{t}_{0,i},\mathbf{t}_{1,j},\mathbf{t}_{2,k})$. Since the triangles $\Delta_{ijk} = [\mathbf{t}_{0i},\mathbf{t}_{1j},\mathbf{t}_{2k}]$ all have the same orientation, we get $\sigma_{ijk} = \sigma_{i-1,j,k} = \sigma_{i,j-1,k} = \sigma_{i,j,k-1}$, whenever $i, j, k > 0$, and therefore

$$
\begin{aligned}
B^{\mathcal{A}}_{ijk}(\mathbf{u}) &= \sigma_{ijk}\; d(\mathbf{t}_{0,i},\mathbf{t}_{1,j},\mathbf{t}_{2,k})\; \{\frac{d(\mathbf{u},\mathbf{t}_{1,j},\mathbf{t}_{2,k})}{d(\mathbf{t}_{0i},\mathbf{t}_{1j},\mathbf{t}_{2k})}\; M(\mathbf{u}|V^{\mathcal{K}}_{i-1,j,k}) \\
&\quad + \frac{d(\mathbf{t}_{0i},\mathbf{u},\mathbf{t}_{2k})}{d(\mathbf{t}_{0i},\mathbf{t}_{1j},\mathbf{t}_{2k})}\; M(\mathbf{u}|V^{\mathcal{K}}_{i,j-1,k}) + \frac{d(\mathbf{t}_{0i},\mathbf{t}_{1j},\mathbf{u})}{d(\mathbf{t}_{0i},\mathbf{t}_{1j},\mathbf{t}_{2k})}\; M(\mathbf{u}|V^{\mathcal{K}}_{i,j,k-1})\} \\
&= |d(\mathbf{t}_{0i},\mathbf{t}_{1j},\mathbf{t}_{2k})|\; \{\lambda_{ijk,0}(\mathbf{u})\; M(\mathbf{u}|V^{\mathcal{K}}_{i-1,j,k}) + \lambda_{ijk,1}(\mathbf{u})\; M(\mathbf{u}|V^{\mathcal{K}}_{i,j-1,k}) \\
&\quad + \lambda_{ijk,2}(\mathbf{u})\; M(\mathbf{u}|V^{\mathcal{K}}_{i,j,k-1})\} \\
&= |d(\mathbf{t}_{0i},\mathbf{t}_{1j}\,,\mathbf{t}_{2k})|\; M(\mathbf{u}|V^{\mathcal{K}}_{i,j,k}) \\
&= N^{\mathcal{K}}_{ijk}(\mathbf{u}),
\end{aligned}
$$

where we have used (5.2). This completes the proof of Theorem 6.4. ♣

Note that Theorem 6.4 implies in particular that the normalized B-splines $N^{\mathcal{K}}_{ijk}(\mathbf{u})$ are linearly independent. The Theorem also implies that the normalized B-splines $N^{\mathcal{K}}_{ijk}(\mathbf{u})$ form a partition of unity on $\Omega_n$, i.e.,

$$
\sum_{i+j+k=n} N^{\mathcal{K}}_{ijk}(\mathbf{u}) = \sum_{i+j+k=n} B^{\mathcal{A}}_{ijk}(\mathbf{u}) = 1 \quad \text{for all } \mathbf{u} \in \Omega_n, \tag{6.6}
$$

We will see below in (6.19) that this property does not just hold *locally*, but also *globally*.

**Remark 6.5 (Multiple Knots)** *Suppose that all knots* $\mathbf{t}_{0,0} = \ldots = \mathbf{t}_{0,n} = \mathbf{t}_0$, $\mathbf{t}_{1,0} = \ldots = \mathbf{t}_{1,n} = \mathbf{t}_1$, *and* $\mathbf{t}_{2,0} = \ldots = \mathbf{t}_{2,n} = \mathbf{t}_2$ *in a knot net* $\mathcal{K}$ *have multiplicity* $n+1$. *Theorem 6.4 shows that the normalized B-splines* $N^{\mathcal{K}}_{ijk}$ *then specialize to the Bernstein polynomials* $B^{\Delta}_{ijk}$ *w.r.t.* $\Delta = [\mathbf{t}_0,\mathbf{t}_1,\mathbf{t}_2]$. ♣

Examples of the quadratic normalized B-splines $N^{I}_{110}(\mathbf{u})$ and $N^{I}_{200}(\mathbf{u})$ over six different knot configurations with different knot multiplicities are shown in Figure 26.

**Remark 6.6 (Knot Selection)** *It should be pointed out that the rule (6.3) for selecting* $\binom{n+2}{2}$ *subsets of* $n+3$ *knots from the* 3 *clouds of a given triangle is completely different from another rule as developed earlier in [15, 45], and subsequently used, e.g., in [1, 37, 79].*

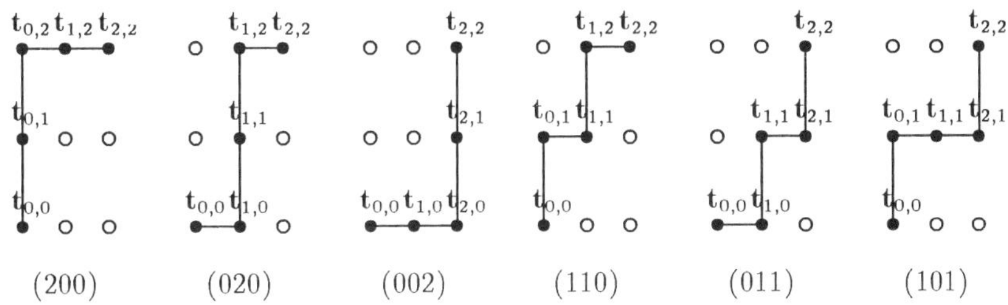


Figure 24: Previous knot selection rule by associating a different knot set with every non descending path from $(0,0)$ to $(2,n)$. This rule is different from our rule as defined above.

*While the rule (6.3) is based on matching the B-spline recurrence with the recurrence for the normalized B-weights, the previous rule is based on the purely combinatorial recipe of associating a different knot set with every non descending path from* $(0,0)$ *to* $(2,n)$ *on the lattice points in* $[0,2] \times [0,n]$. *Note that there are exactly* $\binom{n+2}{n}$ *such different paths.*

*This different knot selection rule is illustrated in Figure 24 for the quadratic case. One of the main differences between the two rules is, e.g., the fact that every knot set according to the rule (6.3) contains all three vertices* $\mathbf{t}_{0,0}, \mathbf{t}_{1,0}, \mathbf{t}_{2,0}$ *of the triangle* $\Delta$, *while this is not the case for the combinatorial recipe.* ♣

Using Definition 6.3 above, the construction of the new spline space may then be completed as follows: Suppose that

$$\mathcal{T} = \{\Delta(I) = [\mathbf{t}_{i_0,0}, \mathbf{t}_{i_1,0}, \mathbf{t}_{i_2,0}] | I = (i_0, i_1, i_2) \in \mathcal{I} \subset \mathbb{Z}_+^3\} \tag{6.7}$$

defines a triangulation of $\mathbb{R}^2$ or of a bounded domain $D \subset \mathbb{R}^2$. We associate with every vertex $\mathbf{t}_i$ a cloud $C_i = \{\mathbf{t}_{i,0}, \ldots, \mathbf{t}_{i,n}\}$ of knots such that $\mathcal{K}_I = \{\mathbf{t}_{i_0,0}, \ldots, \mathbf{t}_{i_0,n}, \mathbf{t}_{i_1,0}, \ldots, \mathbf{t}_{i_1,n}, \mathbf{t}_{i_2,0}, \ldots, \mathbf{t}_{i_2,n}\}$ is a knot net for every $I \in \mathcal{I}$. We then define the sets

$$V_{ijk}^I = \{\mathbf{t}_{i_0,0}, \ldots, \mathbf{t}_{i_0,i}, \mathbf{t}_{i_1,0}, \ldots, \mathbf{t}_{i_1,j}, \mathbf{t}_{i_2,0}, \ldots, \mathbf{t}_{i_2,k}\} \tag{6.8}$$

according to (6.3) and define the normalized B-splines

$$N_{ijk}^I(\mathbf{u}) = |d(\mathbf{t}_{i_0,i}, \mathbf{t}_{i_1,j}, \mathbf{t}_{i_2,k})| \; M(\mathbf{u}|V_{ijk}^I) \tag{6.9}$$

according to (6.4). The linear space

$$\mathcal{S}_V^n = \operatorname{span}\{N_{ijk}^I(\mathbf{u}) | I \in \mathcal{I}\} \tag{6.10}$$

of all these normalized B-splines is the new spline space. Since the normalized B-splines are linearly independent, every surface in this space can then be uniquely represented as a linear combination

$$F(\mathbf{u}) = \sum_{I \in \mathcal{I}} \sum_{i+j+k=n} N_{ijk}^I(\mathbf{u}) \mathbf{c}_{ijk}^I, \quad \mathbf{c}_{ijk} \in \mathbb{R}^t. \tag{6.11}$$

The coefficients $\mathbf{c}_{ijk}^I \in \mathbb{R}^t$, are called *control points*. They form the *control net*.

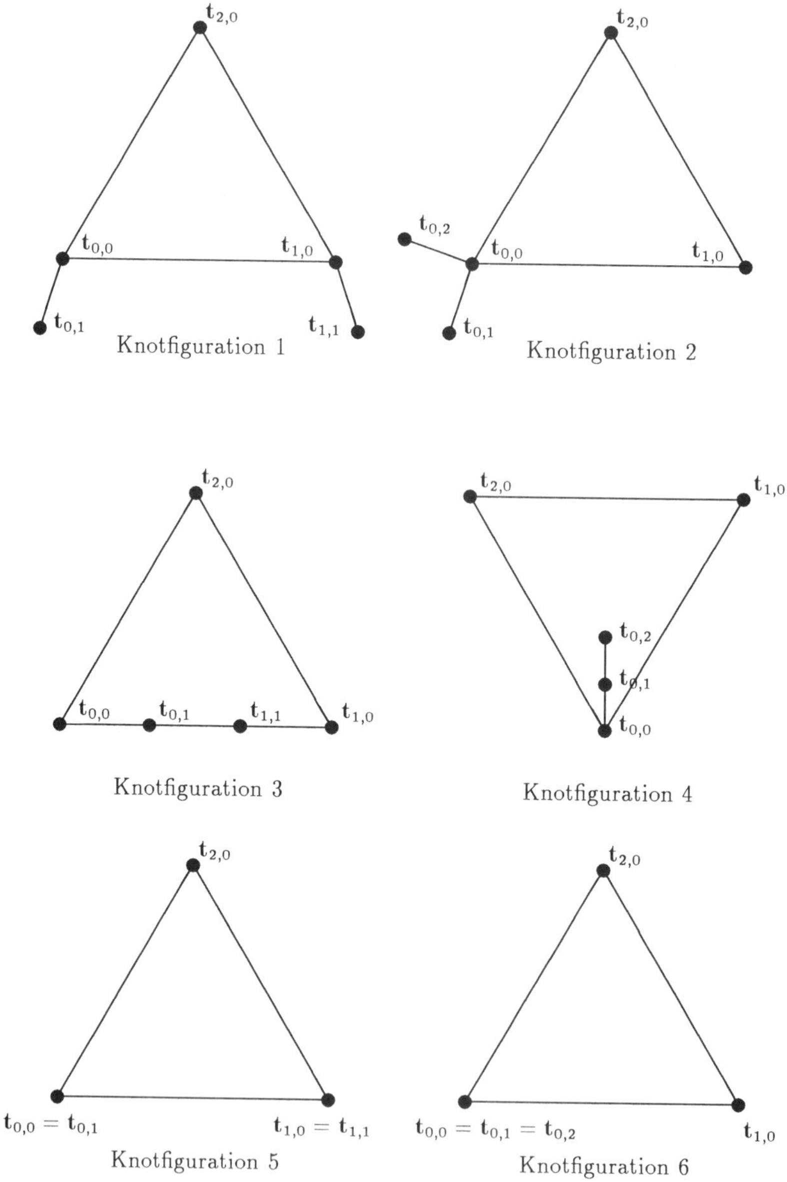


Figure 25: Six different knot configurations with different multiplicities for the quadratic normalized normalized B-splines $N^I_{110}(\mathbf{u})$ and $N^I_{200}(\mathbf{u})$.

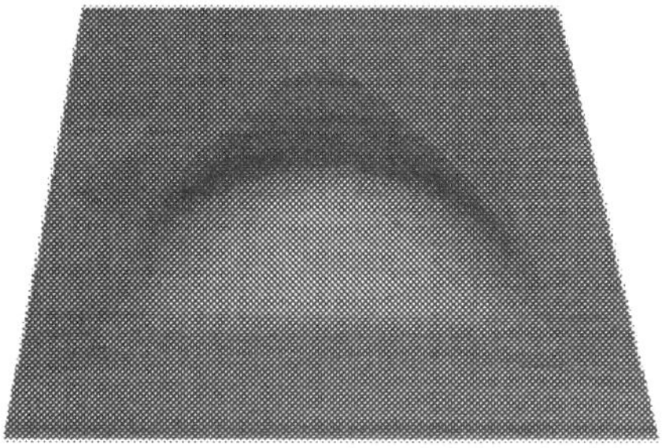

$N^I_{100}$ over Configuration 1

$N^I_{200}$ over Configuration 2

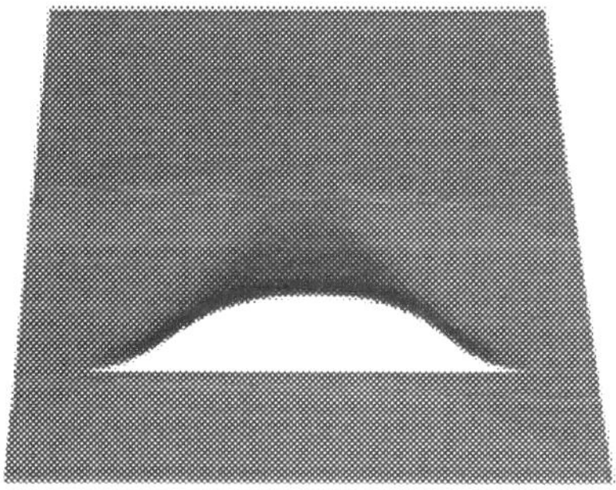

$N^I_{100}$ over Configuration 3

$N^I_{200}$ over Configuration 4

$N^I_{100}$ over Configuration 5

$N^I_{200}$ over Configuration 6

Figure 26: The quadratic normalized B-splines $N^I_{110}(\mathbf{u})$ and $N^I_{200}(\mathbf{u})$ over the six different knot configurations shown in the previous Figure.

## 6.2 Representing Piecewise Polynomials as Linear Combinations of B-Splines

We now wish to examine the spline space $\mathcal{S}_V^n$ as introduced in (6.10) a little bit closer. In particular, we will demonstrate that every piecewise polynomial $F$ over a given triangulation $\mathcal{T} = \{\Delta(I)|I \in \mathcal{I}\}$ can be represented as linear combination of the normalized B-splines $N_{ijk}^I(\mathbf{u})$, $I \in \mathcal{I}$, $i+j+k=n$, and we provide an explicit formula for this representation, using polar forms. This formula also implies that the normalized B-splines $N_{ijk}^I(\mathbf{u})$, do in fact form a partition of unity.

Starting point of our discussion is Equation (4.4) for B-patches which we would like to extend to B-splines. It is clear that every element

$$F(\mathbf{u}) = \sum_{I\in\mathcal{I}} \sum_{i+j+k=n} N_{ijk}^I(\mathbf{u})\mathbf{c}_{ijk}^I \tag{6.12}$$

of the spline space

$$\mathcal{S}_V^n = \operatorname{span}\{N_{ijk}^I(\mathbf{u})|I \in \mathcal{I}\} \tag{6.13}$$

on $\Omega_n^I$ reduces to $\sum_{i+j+k=n} N_{ijk}^I(\mathbf{u})\mathbf{c}_{ijk}^I$. In view of Theorem 6.4 it is also clear that any element of $\mathcal{S}_V^n$ can be evaluated efficiently by means of the de Boor Algorithm (Corollary 4.10) on any of the regions $\Omega_n^I$. The extension of (4.4) to B-splines is based upon studying to what extent this recursive evaluation persists for $\mathbf{u}$ outside the regions $\Omega_n^I$. The essential observation may be formulated as follows.

**Lemma 6.7 (Recursive Evaluation)** *Suppose that the sequence* $\{\mathbf{c}_{ijk}^I\}$, $I \in \mathcal{I}$, $i+j+k=n$ *has the following property. For any two adjacent triangles* $\Delta(I)$ *and* $\Delta(J)$ *whose common edge is* $[\mathbf{t}_{i_1,0}, \mathbf{t}_{i_2,0}]$ *one has*

$$\mathbf{c}_{0,j,k}^I = \mathbf{c}_{0,j,k}^J. \tag{6.14}$$

*Then*

$$\sum_{I\in\mathcal{I}} \sum_{i+j+k=n} N_{ijk}^I \ \mathbf{c}_{ijk}^I = \sum_{I\in\mathcal{I}} \sum_{i+j+k=n-1} N_{ijk}^I \ \mathbf{c}_{ijk}^{I\ (1)} \tag{6.15}$$

*where*

$$\mathbf{c}_{ijk}^{I\ (1)} = \lambda_{ijk,0}^I(\mathbf{u})\mathbf{c}_{i+1,j,k}^I + \lambda_{ijk,1}^I(\mathbf{u})\mathbf{c}_{i,j+1,k}^I + \lambda_{ijk,2}^I(\mathbf{u})\mathbf{c}_{i,j,k+1}^I. \tag{6.16}$$

**Proof:** The proof of Lemma 6.7 is based on studying how contributions of B-spline blending functions from different triangles in the recursion 5.2 cancel against each other. Although the proof is not really difficult, it is fairly involved, and is therefore omitted here, due to space limitations. The complete proof can be found in [17]. ♣

Using Lemma 6.7 we are then able to represent arbitrary piecewise polynomials as linear combinations of B-splines.

**Theorem 6.8 (Piecewise Polynomial Representation)** *Let $F$ be a piecewise polynomial of degree $n$ over a given triangulation $\mathcal{T}$, and let $F_I$ be the restriction of $F$ to the triangle $\Delta(I)$, $I \in \mathcal{I}$. Let $f_I$ be the polar form of $F_I$. Suppose that $F$ is $C^\nu$-continuous everywhere. Let $\mu = n - \nu$, and assume that every vertex $\mathbf{t}_i$ has multiplicity $\mu$, i.e., $\mathbf{t}_i = \mathbf{t}_{i,0} = \ldots = \mathbf{t}_{i,\mu-1}$. Then $F$ can be represented as linear combination of the normalized B-splines $N^I_{ijk}(\mathbf{u})$, and we get $F(\mathbf{u}) = \sum_{I\in\mathcal{I}} \sum_{i+j+k=n} N^I_{ijk}(\mathbf{u})\mathbf{c}^I_{ijk}$ with*

$$\mathbf{c}^I_{ijk} = f_I(\mathbf{t}_{i_0,0}\cdots\mathbf{t}_{i_0,i-1}\mathbf{t}_{i_1,0}\cdots\mathbf{t}_{i_1,j-1}\mathbf{t}_{i_2,0}\cdots\mathbf{t}_{i_2,k-1}). \tag{6.17}$$

**Proof:** We proceed by induction on $\nu$ and start with $\nu = -1$. In this situation we have to show that $C^{-1}$-continuous, i.e., discontinuous, piecewise polynomials of arbitrary degree $n$ can be represented as linear combinations of B-splines with multiple knots $\mathbf{t}_{i,0} = \ldots = \mathbf{t}_{i,n}$. However, in this situation the normalized B-splines $N^I_{ijk}(\mathbf{u}) = B^{\Delta(I)}_{ijk}(\mathbf{u})$ are equal to the Bernstein polynomials, and the assertion is trivial.

Suppose hat the statement holds for $\nu - 1$. Let $F$ be a $C^\nu$-continuous piecewise polynomial, and let $\mathbf{w} \in \mathbb{R}^2$ be arbitrary. We consider the piecewise polynomial $F_\mathbf{w}$ defined by $F_\mathbf{w}(\mathbf{u}) = f_I(\mathbf{u}^{n-1}\mathbf{w})$ for $\mathbf{u} \in \Delta(I)$. By definition, $F_\mathbf{w}$ is a piecewise polynomial of degree $n-1$ over $\mathcal{T}$. We claim that $F_\mathbf{w}$ is $C^{\nu-1}$-continuous: Consider a point $\mathbf{u}$ that lies on a common edge of the triangles $\Delta(I)$ and $\Delta(J)$. According to Theorem 2.10, the $C^\nu$-continuity of $F$ at $\mathbf{u}$ is equivalent to the identity $f_I(\mathbf{u}^{n-\nu}\mathbf{u}_1\cdots\mathbf{u}_\nu) = f_J(\mathbf{u}^{n-\nu}\mathbf{u}_1\cdots\mathbf{u}_\nu)$ for all $\mathbf{u}_i \in \mathbb{R}^2$. Let $f_{\mathbf{w}_I}$ be the polar form of the restriction of $F_\mathbf{w}$ to the triangle $\Delta(I)$. Then

$$\begin{aligned} f_{\mathbf{w}_I}(\mathbf{u}^{n-\nu}\mathbf{u}_1\cdots\mathbf{u}_{\nu-1}) &= f_I(\mathbf{u}^{n-\nu}\mathbf{u}_1\cdots\mathbf{u}_{\nu-1}\mathbf{w}) \\ &= f_J(\mathbf{u}^{n-\nu}\mathbf{u}_1\cdots\mathbf{u}_{\nu-1}\mathbf{w}) \\ &= f_{\mathbf{w}_J}(\mathbf{u}^{n-\nu}\mathbf{u}_1\cdots\mathbf{u}_{\nu-1}) \end{aligned}$$

for all $\mathbf{u}_1, \ldots, \mathbf{u}_{\nu-1} \in \mathbb{R}^2$, which in turn is equivalent to the $C^{\nu-1}$-continuity of $F_\mathbf{w}$ at $\mathbf{u}$. Using the induction hypothesis and the fact that the polar form $f_{\mathbf{w}_I}$ is affine in each component, we see that $F_\mathbf{w}$ satisfies the identity

$$\begin{aligned} F_\mathbf{w}(\mathbf{u}) &= \sum_I \sum_{i+j+k=n-1} N^I_{ijk}(\mathbf{u}) f_{\mathbf{w}_I}(\mathbf{t}_{i_0,0}\cdots\mathbf{t}_{i_0,i-1}\mathbf{t}_{i_1,0}\cdots\mathbf{t}_{i_1,j-1}\mathbf{t}_{i_2,0}\cdots\mathbf{t}_{i_2,k-1}) \\ &= \sum_I \sum_{i+j+k=n-1} N^I_{ijk}(\mathbf{u}) f_I(\mathbf{t}_{i_0,0}\cdots\mathbf{t}_{i_0,i-1}\mathbf{t}_{i_1,0}\cdots\mathbf{t}_{i_1,j-1}\mathbf{t}_{i_2,0}\cdots\mathbf{t}_{i_2,k-1}\mathbf{w}) \\ &= \sum_I \sum_{i+j+k=n-1} N^I_{ijk}(\mathbf{u})(\lambda_{ijk,0}(\mathbf{w})\mathbf{c}^I_{i+1,j,k} + \lambda_{ijk,1}(\mathbf{w})\mathbf{c}^I_{i,j+1,k} + \lambda_{ijk,2}(\mathbf{w})\mathbf{c}^I_{i,j,k+1}) \end{aligned}$$

for all $\mathbf{u} \in \mathbb{R}^2$. In order to complete the proof we use Lemma 6.7. Let $\mathbf{e} = [\mathbf{t}_{i_1,0}, \mathbf{t}_{i_2,0}]$ be a common edge of $\Delta(I)$ and $\Delta(J)$. Since every vertex $\mathbf{t}_{i,0} = \ldots = \mathbf{t}_{i,\mu-1}$ has multiplicity $\mu$, every sequence $(\mathbf{t}_{i_1,0}, \ldots, \mathbf{t}_{i_1,j-1}, \mathbf{t}_{i_2,0}, \ldots, \mathbf{t}_{i_2,k-1})$, $j + k = n$, contains at least $\mu$ many knots on $\mathbf{e}$. According to Theorem 2.10, the $C^\nu$-continuity of $F$ along $\mathbf{e}$ thus implies $f_I(\mathbf{t}_{i_1,0}\cdots\mathbf{t}_{i_1,j-1}\mathbf{t}_{i_2,0}\cdots\mathbf{t}_{i_2,k-1}) = f_J(\mathbf{t}_{i_1,0}\ldots\mathbf{t}_{i_1,j-1}\mathbf{t}_{i_2,0}\ldots\mathbf{t}_{i_2,k-1})$, and the points $c^I_{0jk} =$

$f_I(\mathbf{t}_{i_1,0}\cdots\mathbf{t}_{i_1,j-1}\mathbf{t}_{i_2,0}\cdots\mathbf{t}_{i_2,k-1})$ satisfy the assumption of Lemma 6.7. Using $F_{\mathbf{u}}(\mathbf{u}) = F(\mathbf{u})$ we therefore obtain

$$\begin{aligned}
F(\mathbf{u}) &= F_{\mathbf{u}}(\mathbf{u}) \\
&= \sum_{I\in\mathcal{I}}\sum_{i+j+k=n-1} N^I_{ijk}(\mathbf{u})(\lambda_{ijk,0}(\mathbf{u})\mathbf{c}^I_{i+1,j,k} + \lambda_{ijk,1}(\mathbf{w})\mathbf{c}^I_{i,j+1,k} + \lambda_{ijk,2}(\mathbf{w})\mathbf{c}^I_{i,j,k+1}) \\
&= \sum_{I\in\mathcal{I}}\sum_{i+j+k=n} N^I_{ijk}(\mathbf{u})\ \mathbf{c}_{ijk},
\end{aligned}$$

and the proof of Theorem 6.8 is complete. ♣

An application of Theorem 6.8 is shown in Figure 29. Theorem 6.8 also implies the following Corollary [17]:

**Corollary 6.9 (Polynomial Representation)** *Let $F$ be a polynomial of degree $n$ and let $f$ be its polar form. Then $F(\mathbf{u}) = \sum_{I\in\mathcal{I}}\sum_{i+j+k=n} N^{K_I}_{ijk}(\mathbf{u})\mathbf{c}^I_{ijk}$ with*

$$\mathbf{c}^I_{ijk} = f(\mathbf{t}_{i_0,0}\cdots\mathbf{t}_{i_0,i-1}\mathbf{t}_{i_1,0}\cdots\mathbf{t}_{i_1,j-1}\mathbf{t}_{i_2,0}\cdots\mathbf{t}_{i_2,k-1}). \tag{6.18}$$

*This means that the coefficients $\mathbf{c}_{ijk}$ for representing a given polynomial $F$ as linear combination of the normalized B-splines $N^I_{ijk}$, are obtained by evaluating the polar form $f$ on the knots.* ♣

Finally, application of Corollary 6.9 to the constant polynomial $F(\mathbf{u}) = 1$ yields

$$\sum_{I\in\mathcal{I}}\sum_{i+j+k=n} N^I_{ijk}(\mathbf{u}) = 1, \tag{6.19}$$

and the normalized B-splines $N^I_{ijk}(\mathbf{u})$ form a *partition of unity*. This ensures affine invariance of our surface scheme.

## 6.3 Properties of the New B-Spline Scheme

The new B-spline scheme that has been developed in this section inherits desirable properties from both simplex splines and B-patches. Locality, smoothness, and the modeling of discontinuities are inherited from simplex splines, while control points, affine invariance, and the representation of piecewise polynomials are derived via B-patches.

The main properties of the surface scheme are summarized in the following Theorem:

**Theorem 6.10 (Shape Control)** *Let* $F(\mathbf{u}) = \sum_{I \in \mathcal{I}} \sum_{i+j+k=n} N_{ijk}^{I}(\mathbf{u})\mathbf{c}_{ijk}^{I}$ *be a B-spline surface. Then the shape of the surface is related to the shape of its control net and to the location of the knots as follows:*

- **Piecewise Polynomial:** $F(\mathbf{u})$ *is a piecewise polynomial of degree* $n$.
- **Locality:** *Movement of a single control point* $\mathbf{c}_{ijk}^{I}$ *only influences the surface on the triangle* $\Delta(I)$ *and on the triangles directly surrounding* $\Delta(I)$.
- **Convex Hull Property:** $F(\mathbf{u})$ *lies inside the convex hull of its control net.*
- **Smoothness:** *If at most* $\mu$ *knots within each knot net* $\mathcal{K}_I$ *are collinear,* $2 \leq \mu \leq n+2$, *then* $F(\mathbf{u})$ *is* $C^{n+1-\mu}$*-continuous everywhere. Hence, if the knots are in general position (no three knots within the same knot net* $\mathcal{K}_I$ *are collinear), then* $F$ *is* $C^{n-1}$*-continuous everywhere.*
- **Affine Invariance:** *The relationship between the surface* $F$ *and its control net is affinely invariant: If* $\phi : \mathbb{R}^t \to \mathbb{R}^t$ *is an affine map, then the B-spline control points of the image surface* $\tilde{F} = F \circ \phi$ *are the images* $\phi(\mathbf{c}_{ijk}^{I})$ *of the B-spline control points* $\mathbf{c}_{ijk}^{I}$, *i.e., the diagram*

$$\begin{array}{ccc} (\mathbf{c}_{i,j,k}^{I})_{i+j+k=n} & \xrightarrow{\phi} & (\phi(\mathbf{c}_{i,j,k}^{I}))_{i+j+k=n} \\ \downarrow & & \downarrow \\ F & \xrightarrow{\phi} & \phi \circ F \end{array} \quad \clubsuit$$

*commutes.*

In addition to these properties, the B-spline scheme allows to represent polynomials and piecewise polynomials over arbitrary triangulations as linear combinations of B-splines. Piecewise polynomial representation is illustrated in Figure 27 and Figure 28, and an application to the polygonal hole problem is given in Figure 29.

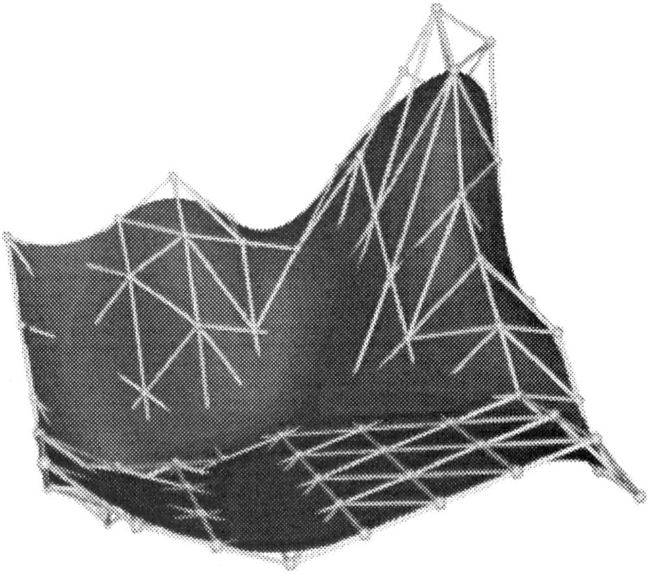

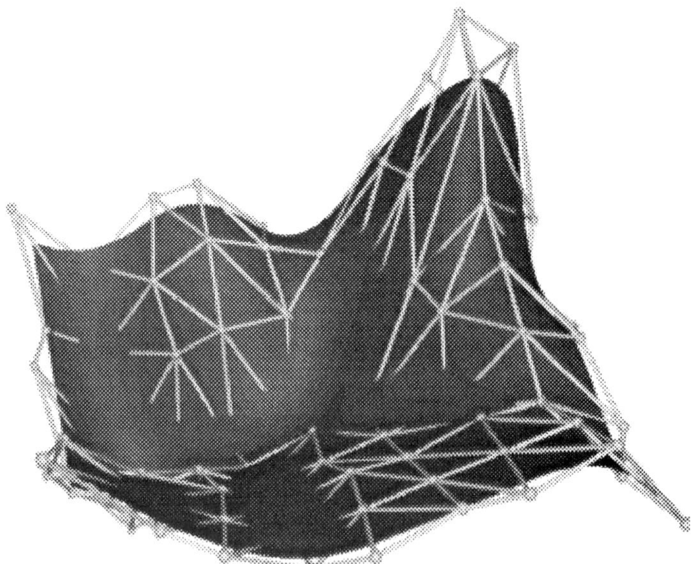

Figure 27: A quadratic $C^1$-continuous piecewise polynomial test surface with Bézier control net (top) and B-spline control net (bottom).

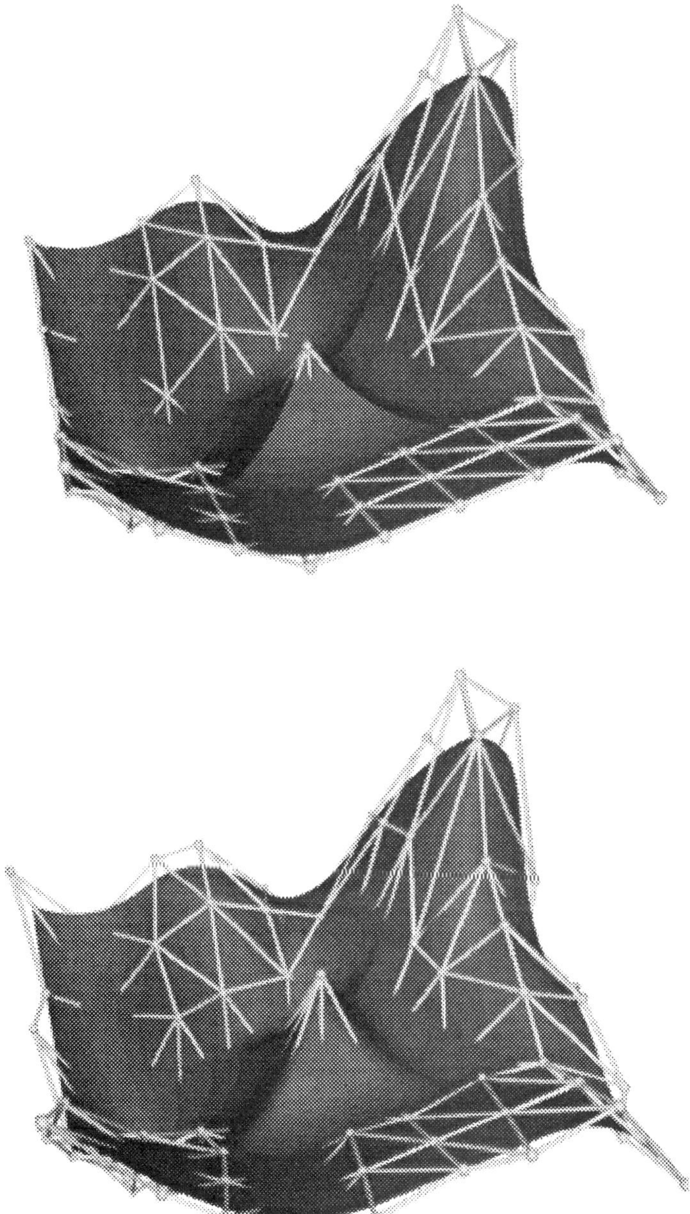

Figure 28: Influence of moving a single control point: If a Bézier control point is moved, the $C^1$-continuity of the surface is destroyed, and a sharp edge is introduced (top). If a B-spline control point is moved, the $C^1$-continuity of the surface is preserved (bottom).

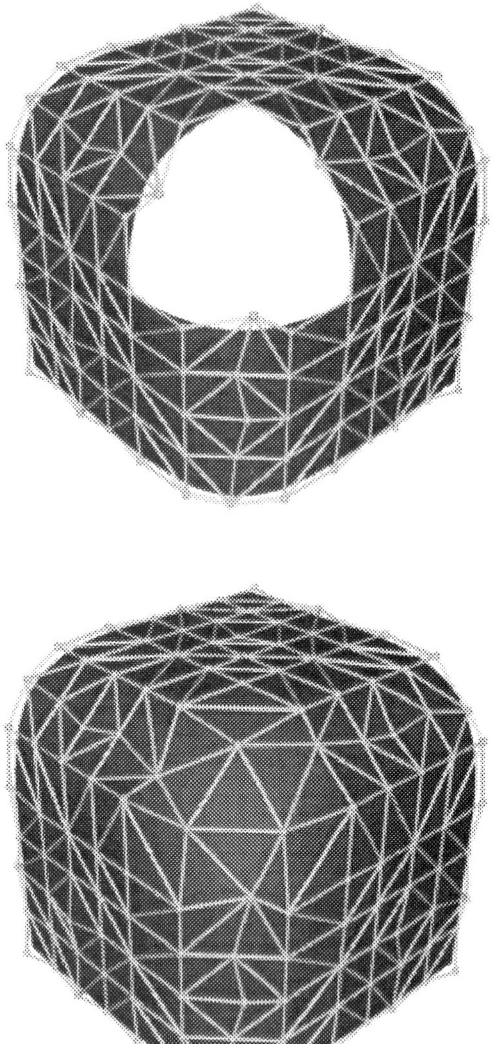

Figure 29: Solving the polygonal hole problem using triangular B-splines: First, the piecewise polynomial surface around the hole is represented as linear combination of B-splines (top). This B-spline surface can then be extended to produce an overall $C^{n-1}$-continuous fill of the hole (bottom). Note that this method can achieve $C^1$-continuity with piecewise quadratics, and $C^2$-continuity with piecewise cubics.

# References


[1] S. Auerbach, R.H.J. Gmelig Meyling, M. Neamtu, and H. Schaeben. Approximation and geometric modeling with simplex B-splines associated with irregular triangles. *Computer-Aided Geom. Design*, 8:67–87, 1991.

[2] R.E. Barnhill. Surfaces in computer aided geometric design. *Computer-Aided Geom. Design*, 2:1–17, 1985.

[3] R.H. Bartels, J.C. Beatty, and B.A. Barsky. *An Introduction to Splines for Use in Computer Graphics and Geometric Modeling*. Morgan Kaufmann Publishers, 1987.

[4] S. Bernstein. Démonstration du théorème de weierstrass, fondée sur le calcul des probabilités. *Comm. Soc. Math. Kharkov*, 13(2):1–2, 1912.

[5] P. Bézier . *Numerical Control, Mathematics and Applications*. John Wiley & Sons, London, 1972.

[6] W. Boehm. Inserting new knots into a B-spline curve. *Comput. Aided Design*, 12:50–62, 1980.

[7] W. Boehm. The de Boor algorithm for triangular splines. In R.E. Barnhill and W. Boehm, editors, *Surfaces in CAGD*, pages 109–120. North-Holland, 1983.

[8] W. Boehm and G. Farin. Letter to the editor. *Comput. Aided Design*, 15:260–261, 1983.

[9] W. Boehm, G. Farin, and J. Kahmann. A survey of curve and surface methods in CAGD. *Computer-Aided Geom. Design*, 1:1–60, 1984.

[10] E. Catmull and J. Clark. Recursively generated B-spline surfaces on arbitrary topological meshes. *Comput. Aided Design*, 10:350–355, 1978.

[11] G. Chang and P. Davis. The convexity of bernstein polynomials over triangles. *J. Approx. Th.*, 40:11–28, 1984.

[12] G. Chang and Y. Feng. An improved condition for the convexity of bernstein-Bézier surfaces over triangles. *Computer-Aided Geom. Design*, 2:279–283, 1985.

[13] W. Dahmen. Multivariate B-splines - recurrence relations and linear combinations of truncated powers. In W. Schempp and K. Zeller, editors, *Multivariate Approximation Theorey*, pages 64–82, Basel, 1979. Birkhäuser.

[14] W. Dahmen. Bernstein-Bézier representation for polynomial surfaces. In *Extension of B-Spline Curve Algorithms to Surfaces*. SIGGRAPH'86 Tutorial Notes, ACM SIGGRAPH, 1986.

[15] W. Dahmen and C.A. Micchelli. On the linear independence of multivariate B-splines I. Triangulations of simploids. *SIAM J. Numer. Anal.*, 19:993–1012, 1982.

[16] W. Dahmen and C.A. Micchelli. Recent progress in multivariate splines. In C.K. Chui, L.L. Schumaker, and J.D. Ward, editors, *Approximation Theory IV*, pages 27–121. Academic Press, 1983.

[17] W. Dahmen, C.A. Micchelli, and H.-P. Seidel. Blossoming begets B-Splines built better by B-Patches. Research Report RC 16261(#72182), IBM Research Division, Yorktown Heights, 1990. To appear in Math. Comp..

[18] C. de Boor. On calculating with B-splines. *J. Approx. Th.*, 6:50–62, 1972.

[19] C. de Boor. *A Practical Guide to Splines.* Springer, New York, 1978.

[20] C. de Boor. B-form basics. In G. Farin, editor, *Geometric Modeling, Algorithms and New Trends*, pages 131–148. SIAM, 1987.

[21] P. de Casteljau. Outillages méthodes calcul. Technical report, Andre Citroen, Paris, 1959.

[22] P. de Casteljau. Courbes et surfaces à pôles. Technical report, Andre Citroen, Paris, 1963.

[23] P. de Casteljau. *Formes à Pôles.* Hermes, Paris, 1985.

[24] P. de Casteljau. *Shape Mathematics and CAD.* Kogan Page Ltd, London, 1986.

[25] T. DeRose. *Geometric Continuity: A Parametrization Independent Measure of Continuity for Computer Aided Geometric Design.* PhD thesis, UC Berkeley, 1985.

[26] D. Doo and M.A. Sabin. Behaviour of recursive division surfaces near extraordinary points. *Comput. Aided Design*, 10:356–360, 1978.

[27] G.E. Farin. *Subsplines über Dreiecken.* PhD thesis, TU Braunschweig, Braunschweig, F.R.G., 1979.

[28] G.E. Farin. Bézier polynomials over triangles and the construction of piecewise $C^r$ polynomials. Technical Report TR/91, Dept. of Mathematics, Brunel University, Uxbridge, 1980.

[29] G.E. Farin. Triangular Bernstein-Bézier patches. *Computer-Aided Geom. Design*, 3:83–127, 1986.

[30] G.E. Farin. *Curves and Surfaces for Computer Aided Geometric Design.* Academic Press, 1988.

[31] G.E. Farin, B. Piper, and A. Worsey. The octant of a sphere as a nondegenerate triangular Bézier patch. *Computer-Aided Geom. Design*, 4(4):329–332, 1988.

[32] D. Filip. Adaptive subdivision algorithms for a set of Bézier triangles. *Comput. Aided Design*, 18(2):74–78, 1986.

[33] P. Fong. Shape control for B-splines over arbitrary triangulations. Master's thesis, University of Waterloo, Waterloo, 1992. in preparation.

[34] P. Fong and H.-P. Seidel. Control points for multivariate B-spline surfaces over arbitrary triangulations. *Computer Graphics Forum*, 1991. To appear.

[35] L. Frederickson. Triangular spline interpolation. Technical Report Math report 6/70, Lakehead University, Canada, 1970.

[36] L. Frederickson. Generalized triangular splines. Technical Report Math report 7/71, Lakehead University, Canada, 1971.

[37] R.H.J. Gmelig Meyling. *Polynomial spline approximation in two variables.* PhD thesis, University of Amsterdam, 1986.

[38] R.N. Goldman. Using degenerate Bézier triangles and tetrahedra to subdivide Bézier curves. *Comput. Aided Design*, 14:307–311, 1982.

[39] R.N. Goldman. Subdivision algorithms for Bézier triangles. *Comput. Aided Design*, 15:159–166, 1983.

[40] R.N. Goldman. Blossoming and knot insertion algorithms for B-spline curves. *Computer-Aided Geom. Design*, 7:69–81, 1990.

[41] W.J. Gordon and R.F. Riesenfeld. B-spline curves. In R.E. Barnhill and R.F. Riesenfeld, editors, *Computer Aided Geometric Design.* Academic Press, 1974.

[42] T.A. Grandine. *Computing with multivariate simples splines.* PhD thesis, University of Wisconsin, Madison, 1985.

[43] G. Herron. Techniques for visual continuity. In G.F. Farin, editor, *Geometric Modeling*, pages 163–174. SIAM, 1987.

[44] C.M. Hoffmann. *Geometric & Solid Modeling.* Morgan Kaufmann Publishers, San Mateo, California, 1989.

[45] K. Höllig. Multivariate splines. *SIAM J. Numer. Anal.*, 19:1013–1031, 1982.

[46] J. Hoschek and D. Lasser. *Grundlagen der geometrischen Datenverarbeitung.* Teubner, 1989.

[47] C. Loop and T.D. DeRose. A multisided generalization of Bézier surfaces. *ACM Trans. Graph.*, 8(3):204–234, 1989.

[48] C. Loop and T.D. DeRose. Generalized B-spline surfaces of arbitrary topology. In *Proc. SIGGRAPH'90*, pages 347–356. ACM SIGGRAPH, 1990.

[49] G.G. Lorentz. *Bernstein Polynomials*. University of Toronto Press, Toronto, 1953.

[50] C.A. Micchelli. On a numerically efficient method for computing with multivariate B-splines. In W. Schempp and K. Zeller, editors, *Multivariate Approximation Theory*, pages 211–248, Basel, 1979. Birkhäuser.

[51] C.A. Micchelli. A constructive approach to Kergin interpolation in $\mathbb{R}^k$, multivariate B-splines and lagrange interpolation. *Rocky Mt. J. Math.*, 10:485–497, 1980.

[52] J. Peters. *Fitting smooth parametric surfaces to 3D data*. PhD thesis, University of Wisconsin, Madison, 1990.

[53] B. Piper. Visually smooth interpolation with triangular Bézier patches. In G.E. Farin, editor, *Geometric Modeling*, pages 221–234. SIAM, 1987.

[54] W. Png. Adaptive bounding algorithms for the arclength and surface area of Bézier curves and surfaces. Master's thesis, University of Waterloo, Waterloo, 1991.

[55] M.J.D. Powell and M. Sabin. Piecewise quadratic approximations on triangles. *ACM Trans. Math. Software*, 3:316–325, 1977.

[56] H. Prautzsch. *Unterteilungsalgorithmen für multivariate Splines*. PhD thesis, TU Braunschweig, Braunschweig, F.R.G., 1984.

[57] L. Ramshaw. Blossoming: A connect-the-dots approach to splines. Technical report, Digital Systems Research Center, Palo Alto, 1987.

[58] L. Ramshaw. Béziers and B-splines as multiaffine maps. In *Theoretical Foundations of Computer Graphics and CAD*, pages 757–776. Springer, 1988.

[59] L. Ramshaw. Blossoms are polar forms. *Computer-Aided Geom. Design*, 6:323–358, 1989.

[60] R.F. Riesenfeld. *Applications of B-spline Approximation to Geometric Problems of Computer Aided Design*. PhD thesis, Syracuse University, 1973.

[61] M.A. Sabin. *The Use of Piecewise Forms for the Numerical Representation of Shape*. PhD thesis, Hungarian Academy of Sciences, Budapest, Hungary, 1976.

[62] M.A. Sabin. Non-rectangular surface patches suitable for inclusion in a B-spline surface. In *Proc. EUROGRAPHICS'83*, pages 57–69. North-Holland, 1983.

[63] P. Sablonniere. *Bases de Bernstein et approximants splines*. PhD thesis, University of Lille, Lille, France, 1982.

[64] P. Sablonniere. Bernstein-Bézier methods for the construction of bivariate spline approximants. *Computer-Aided Geom. Design*, 2:29–36, 1985.

[65] L.L. Schumaker. *Spline Functions: Basic Theory.* John Wiley & Sons, New York, 1981.

[66] H.-P. Seidel. A general subdivision theorem for Bézier triangles. In T. Lyche and L. L. Schumaker, editors, *Mathematical Methods in Computer Aided Geometric Design*, pages 573–581. Academic Press, 1989.

[67] H.-P. Seidel. A new multiaffine approach to B-splines. *Computer-Aided Geom. Design*, 6:23–32, 1989.

[68] H.-P. Seidel. *Polynome, Splines und symmetrische rekursive Algorithmen im Computer Aided Geometric Design.* Habilitationsschrift, Universität Tübingen, 1989.

[69] H.-P. Seidel. Algorithms for B-patches. In *Proc. Graphics Interface '91*, pages 8–15. Morgan Kaufmann Publishers, 1991.

[70] H.-P. Seidel. Polar forms and triangular B-Spline surfaces. In *Blossoming: The New Polar-Form Approach to Spline Curves and Surfaces, SIGGRAPH '91 Course Notes #26*, pages 8.1–8.52. ACM SIGGRAPH, 1991.

[71] H.-P. Seidel. Symmetric recursive algorithms for surfaces: B-patches and the de Boor algorithm for polynomials over triangles. *Constr. Approx.*, 7:257–279, 1991.

[72] H.-P. Seidel. Representing piecewise polynomials as linear combinations of multivariate B-splines. In T. Lyche and L. L. Schumaker, editors, *Curves and Surfaces.* Academic Press, 1992. To appear.

[73] L. Shirman and C. Sequin. Local surface interpolation with Bézier patches. *Computer-Aided Geom. Design*, 4(4):279–296, 1988.

[74] P.B. Slusallek. Flächenmodellierung mit Splines über Dreiecken. Master's thesis, University of Tübingen, Tübingen, F.R.G., 1989.

[75] P.B. Slusallek and H.-P. Seidel. TRIMO - a workstation-based interactive system for the generation, manipulation, and display of surfaces over arbitrary topological meshes. In *Proc. EUROGRAPHICS'90*, pages 147–158. North Holland, 1990.

[76] D.D. Stancu. De l'approximation, par des polynômes du type bernstein, des fonctions de deux variables. *Comm. Akad. R. P. Romine*, 9:773–777, 1959.

[77] E. Stärk. *Mehrfach differenzierbare Bézier -Kurven und Bézier -Flächen.* PhD thesis, TU Braunschweig, Braunschweig, F.R.G., 1976.

[78] S. Teller. Bézier surfaces that cover quadrics, and their equivalence to stereographic maps. Master's thesis, University of California, Berkeley, California, 1990.

[79] C.R. Traas. Practice of bivariate quadratic simplicial splines. In W. Dahmen, M. Gasca, and C.A. Micchelli, editors, *Computation of Curves and Surfaces*, pages 383–422, Dordrecht, 1990. NATO ASI Series, Kluwer Academic Publishers.

[80] B. L. van der Waerden. *Moderne Algebra*. Springer, 1930.

[81] J. Zhou. *The positivity and convexity of Bézier polynomials over triangles*. PhD thesis, Beijing University, Beijing, China, 1985.

*Computing in Euclidean Geometry*, pp. 287–385
Eds. D.-Z. Du and F. K. Hwang


# Computational Geometry and Topological Network Design

J. MACGREGOR SMITH

*Dept. of Industrial Engineering and Operations Research*

*University of Massachusetts Amherst, Massachusetts 01003 USA*

and

P. WINTER

*Dept. of Computer Science*

*University of Copenhagen DK-2100 Copenhagen Ø DENMARK*

**Abstract**

*Network Design* ($\mathcal{ND}$) is generally concerned with the design, analysis, and synthesis of communication, transportation and distribution networks. One of the fundamental issues when designing such networks is their physical topology. More specifically, *Topological Network Design* ($\mathcal{TND}$) concerns the identification, number, and location of nodes (relays, junctions, depots, etc.), and their interconnections (wires, roads, pipelines, etc.). In this survey, we discuss how the problems, paradigms, geometric data structures, and algorithms of *Computational Geometry* ($\mathcal{CG}$) can be successfully applied to such *geometric* $\mathcal{TND}$ problems.

Starting from the fundamental geometric data structures of $\mathcal{CG}$ (convex hulls, Voronoi diagrams and Delaunay triangulations, etc.), and paradigms (greedy, plane sweep, locus, divide and conquer, etc.), the advantages of a geometric, as opposed to an algebraic approach to geometric $\mathcal{TND}$ problems will be illustrated. Fundamental problems and algorithms dealing with the optimal location and clustering of nodes and/or optimal interconnections between nodes (possibly in the presence of obstacles) will be described. Both polynomially solvable problems and $\mathcal{NP}$-complete problems will be treated in order to emphasize the full power and beauty of the $\mathcal{CG}$ methodology.

In a final synthesis, geometric $\mathcal{TND}$ problems along with their geometric data structures and paradigms will be classified affording a complete and exhaustive morphology. This morphology should become an investigative tool for future research linking the fields of $\mathcal{TND}$ and $\mathcal{CG}$.

# 1. Introduction

## 1.1 Purpose and Focus

*Computational Geometry* ($\mathcal{CG}$) is concerned with the design and analysis of algorithms and heuristics for geometric problems or for problems which have a fundamental geometric nature. *Topological Network Design* ($\mathcal{TND}$) problems are concerned with the design and analysis of a network topology i.e. nodes, arcs and their flow relationships for structuring, analyzing, or characterizing a system and its constituent components. $\mathcal{CG}$ has developed into a significant and challenging discipline in the last decade and it is only natural that we explore its impact on the design and analysis of algorithms for geometric $\mathcal{TND}$ problems.

Some of the most critical issues addressed in this survey include: the nature, origins, fundamental paradigms and geometric data structures of $\mathcal{CG}$, the types of $\mathcal{TND}$ problems appropriate to $\mathcal{CG}$ analysis, and the crucial computational complexity relationships between $\mathcal{CG}$ and $\mathcal{TND}$. Our principal aim is to provide a classification system for $\mathcal{TND}$ so that future development of algorithms and heuristics for new as well as previously approached problems can be enhanced by paradigms and geometric data structures that have proved useful on related $\mathcal{CG}$ problems. We also hope that this survey will clearly illustrate the potential and beauty of $\mathcal{CG}$ for those network design researchers who are not familiar with this relatively novel field.

## 1.2 Overview of the Survey

In Section 2 we briefly discuss general characteristics of *Network Design* ($\mathcal{ND}$) problems. In particular, we argue that $\mathcal{ND}$ problems can be viewed as consisting of three closely related subproblems, one of them being the $\mathcal{TND}$ problem. Due to the inherent difficulty of $\mathcal{ND}$ problems, the subproblems are usually solved separately.

Section 3 gives a brief introduction to the field of $\mathcal{CG}$. The advantages of the $\mathcal{CG}$ approach over mathematical programming and combinatorial optimization approaches to $\mathcal{TND}$ problems are discussed.

Section 4 discusses the most important paradigms and geometric data structures of $\mathcal{CG}$. Examples of how they can be applied to some $\mathcal{TND}$ problems are given. These paradigms and geometric data structures by themselves and in concert often yield natural lower bounds on the problems and provide insights into effective algorithms and heuristics.

Section 5 is the heart of the survey. Our classification scheme for $\mathcal{TND}$ problems and the fundamental problem categories are explained in detail. Within each class, we review basic $\mathcal{TND}$ problems which have been approached by the $\mathcal{CG}$ methodology. In particular, we characterize these $\mathcal{TND}$ problems with respect to the $\mathcal{CG}$ paradigms and geometric data structures employed by the underlying algorithms and heuristics.

Given the paradigms, geometric data structures and geometric $\mathcal{TND}$ problem categories, we construct in Section 6 a 3-dimensional cube within which algorithms and heuristics for geometric $\mathcal{TND}$ problems will be classified. We will also examine the cube along its three dimensions: $\mathcal{TND}$ problems, $\mathcal{CG}$ Data Structures, $\mathcal{CG}$ Paradigms.

One of the major purposes of the cube and its decompositions is to suggest how one might design an algorithm or a heuristic to solve a geometric $\mathcal{TND}$ problem in a most efficient manner. Our understanding of the interrelationship between $\mathcal{CG}$ and geometric $\mathcal{TND}$ is only beginning to evolve. Consequently, we view the cube and its decompositions as a working tool, rather than a rigid set of "pigeon holes".

Section 6 also summarizes the methodological and algorithmic complexity results, discusses open problems, and provides an overall synthesis of the relationships between $\mathcal{CG}$ and $\mathcal{TND}$. Finally, Section 7 concludes the survey.

## 2. Network Design Problems

*Network design ($\mathcal{ND}$) is generally concerned with the design, analysis, and synthesis of communication, transportation and distribution networks.*

To illustrate the type of problems we are examining, consider the design of a *material handling system* (MHS) for a manufacturing plant consisting of an interconnected set of buildings with different levels and circulation elements. Typical $\mathcal{ND}$ problems within such a manufacturing plant entail the following three interrelated problems, see Figure 1:

- *Topological network design ($\mathcal{TND}$)*: is concerned with the number and location of *nodes* (e.g. workstations, warehouses, shipment and delivery points), identification of their interconnecting *edges* (e.g. corridors, conveyors, elevators, tow lines, escalators and lifts) and the general configuration of both components.
- *Routing network design ($\mathcal{RND}$)*: is concerned with the sequencing and routing of various units (e.g. products, vehicles, customers) across the network.
- *Capacitated network design ($\mathcal{CND}$)*: is concerned with the optimal sizing of nodes and edges to accommodate requisite flow volumes.

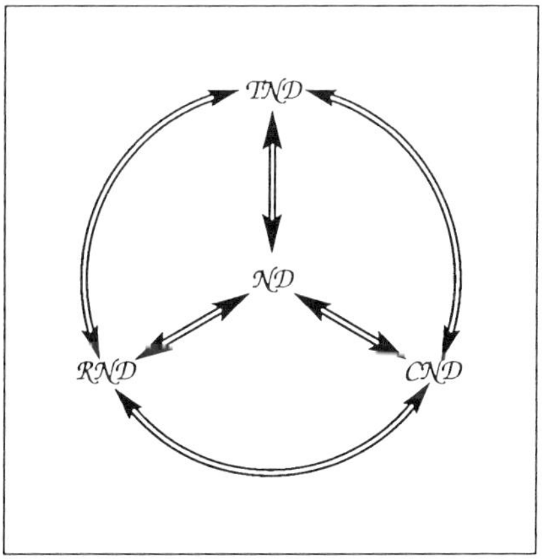


**Figure 1: Network Design Problems Medallion**

In Figure 2 the manufacturing/assembly plant and buildings are illustrated where the warehouse is represented by building 104 which has two interconnecting links to building 111 and a powered conveyor designed to transport material to the elevators and Connectors linking Buildings 111 and 102. Material must be moved from Building

104 through the MHS to the various work stations scattered on the different floor levels of Buildings 111 and 102. The material handling routes through the horizontal and vertical conveyors, elevators, and floor levels all must be topologically configured in a three-dimensional MHS network suitable for moving products throughout the overall production facility.

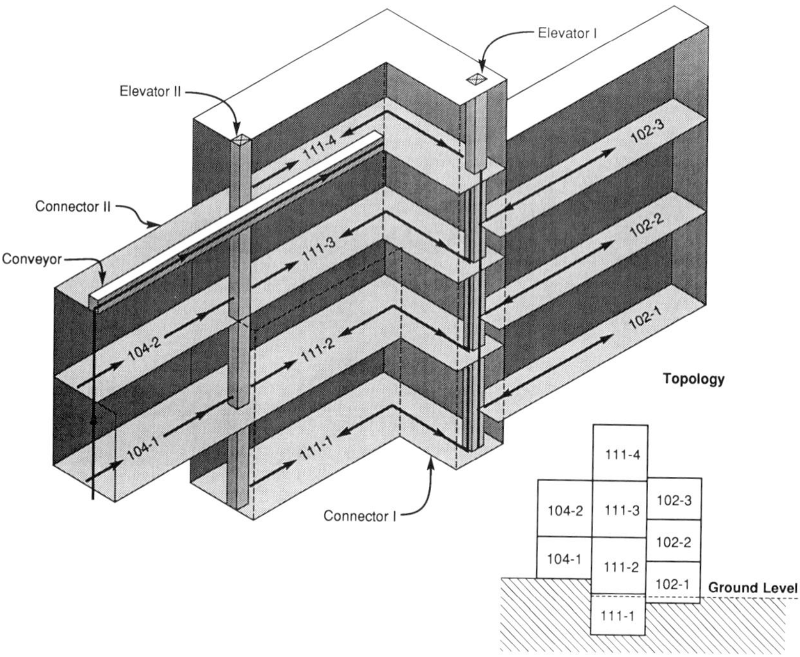


**Figure 2: Material Handling System**

As the above MHS example illustrates, typical $\mathcal{ND}$ problems comprise three separate but closely interrelated subproblems. We are not only concerned about the topology of the MHS. We must also take into account the routing of products and capacities of the edges and nodes which accommodate the product flows. Serious congestion problems can arise if the topology, routing, and capacitated resource allocation problems within the system are not properly resolved. These three subproblems will be briefly discussed in the following subsections.

## 2.1 Topological Network Design

*Topological Network Design ($\mathcal{TND}$) is concerned with the design and analysis of a physical system (topology) i.e. the optimal number and location of nodes, and how they should be interconnected.* The topology can be either in two or three dimensions and often follows certain patterns or configurations (e.g. star, grid, circuit, or backbone topology). The medallion in Figure 1 is comprised of a star and circuit topology.

Assume for our manufacturing example that a number of products in our MHS are to be manufactured or assembled. The products must flow through the MHS on their way from the originating warehouse to the final assembly or product shipment department. Therefore, we must have knowledge of the overall topology of the network before we begin to route the products, size the nodes and edges, and assess the resource requirements and staffing at the various workstations and loading/unloading points. Figure 3 illustrates three alternative MHS topologies for the facility. They were generated to solve the particular product flow problem (see [185] for further details).

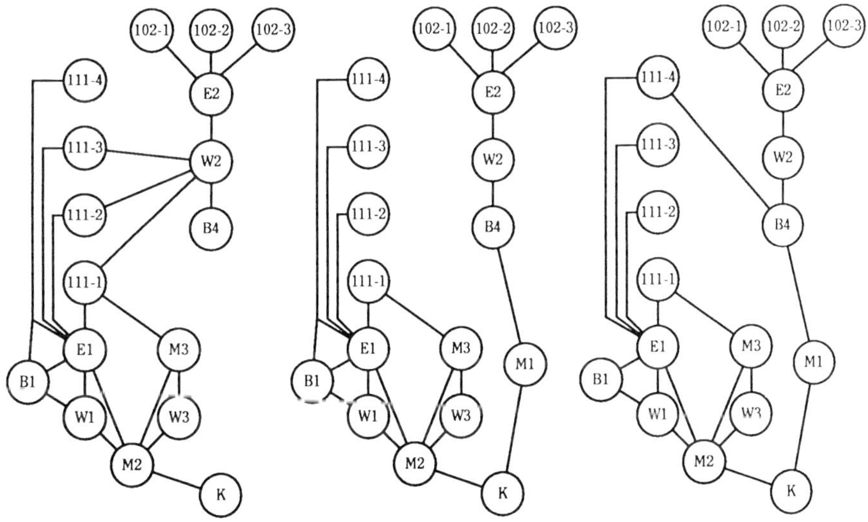


**Figure 3: Alternative Topologies for the MHS**

Usually *optimal* rather than *suboptimal* solutions are preferred. However, as will be demonstrated in this survey, optimal topologies can be rarely achieved.

Further classifications of $\mathcal{TND}$ problems can be made in terms of whether the network is being examined within a *stochastic* rather than a *deterministic* environment. For the most part, our survey concentrates on the *deterministic* $\mathcal{TND}$ problems.

## 2.2 Routing Network Design

*Given the network topology, Routing Network Design ($\mathcal{RND}$) is concerned with the sequencing and routing of such units as products, vehicles or customers across the network to extremize some objective function.*

Returning to the MHS example once again, products on their passage from one node to another, follow a pre-determined or random route depending on their machining and assembly requirements. This routing depends of course on the knowledge of the manufacturing network topology defined at the previous $\mathcal{TND}$ stage.

Routing problems tend to be classified as either steady-state or real-time depending on the information requirements or purpose of the routing decision at each node in the network.

## 2.3 Capacitated Network Design

*Finally, Capacitated Network Design ($\mathcal{CND}$) is concerned with the optimal sizing, shape and technical specifications (e.g. processing rates) of the nodes and edges so that the requisite flow volumes can be accommodated in order to extremize some objective function (e.g. cost, throughput, service utilization).*

Let us once again return to the MHS example. Usually some amount of time is spent processing the products. Various resources are required at some nodes and/or edges. We must decide how many staff, conveyors, or machines to allocate to these nodes and edges to handle the volume of traffic flowing throughout. Congestion may occur at each node or along the edges of the MHS. What is particularly irksome here is that if we know of the congestion beforehand, we can better design the topology of the network in the first place and we can forecast the traffic and plan the nodes and edges to better accommodate demand.

Unfortunately, this circularity brings us back to the $\mathcal{TND}$ problem. The topology needs to be reassessed in light of $\mathcal{RND}$ and $\mathcal{CND}$ decisions. While the three problems are highly interrelated, see Figure 1, one often treats them independently because each in their own right gives rise to a difficult optimization problem.

The focus of this survey is on geometric $\mathcal{TND}$ problems since the applicability of $\mathcal{CG}$ strategies and concepts is most significant for this class of problems. The applicability and significance of $\mathcal{CG}$ to $\mathcal{RND}$ and $\mathcal{CND}$ and to network analysis problems (concerned with the overall cost, performance or reliability of existing networks) would only lengthen an already extensive survey discussion for $\mathcal{TND}$ and $\mathcal{CG}$. However, several $\mathcal{TND}$ problems discussed in subsequent sections often arise in connection with the $\mathcal{RND}$ problems. This is for example the case for the *Euclidean traveling salesman problem* (Subsection 5.2.2) and the *shortest obstacle-avoiding path problem* (Subsection 5.4.2).

Below we present the important definitions, notation and assumptions relevant to our discussion.

## 2.4 Geometric Topological Network Design

$\mathcal{TND}$ problems often arise in practical applications where the points and edges of networks are embedded in some metric space. Such *geometric* $\mathcal{TND}$ problems can therefore be approached by the powerful tools of $\mathcal{CG}$. The main emphasis of this survey will be on geometric $\mathcal{TND}$ problems in the *Euclidean plane*. However, many of these problems, lower bounds and algorithms naturally generalize to higher dimensions and/or different metrics.

We have tried to be broad and inclusive in our categorization while at the same time identifying problems of most significance and which have been approached with $\mathcal{CG}$. We have also included many problems which so far remain open to the $\mathcal{CG}$ approach. We feel that there is an enormous potential in using $\mathcal{CG}$ in connection with these problems. The $\mathcal{CG}$ approach will undoubtedly provide new insights and algorithms to these problems traditionally solved by mathematical programming and combinatorial approaches.

## 2.5 $\mathcal{TND}$ Definitions and Assumptions

In order to provide a common background of network concepts for the survey, the following definitions are presented:

- We have a finite set of points $Z = \{z_1, z_2, ..., z_n\}$ given or randomly generated within a bounded region of the Euclidean plane $E^2$ with Cartesian coordinates $(x_i, y_i)$ for $i = 1, 2, ..., n$. The bounded region is often necessary for the probabilistic analysis of $\mathcal{CG}$ algorithms. In the Steiner network problems described later in Section 5, we will augment the set $Z$ with points from the set $S$ so that the entire point set for our $\mathcal{TND}$ problem may contain $V = Z \cup S$ points.

- When necessary, we assume that total cost of the network is a linear function of the flows in the network unless specified otherwise. In many problems to be described, we assume that the cost of an edge is proportional to its length.

- Also, where necessary, we assume a finite set of obstacles $\Omega = \{\omega_1, \omega_2, \ldots, \omega_k\}$ in $E^2$. Unless otherwise stated, the obstacles are assumed to be convex, polygonal and disjoint.

- We further assume that the distance function is an $L_p$ metric, where the distance between any two points $z_i$ and $z_j$ with coordinates $(x_i, y_i)$ and $(x_j, y_j)$ is given by:
$$d_p(z_i, z_j) = [| x_i - x_j |^p + | y_i - y_j |^p]^{1/p}$$
where $p$ is a real number $p \in [1, 2]$. In some problems such as constructing networks on the sphere or other surfaces, a *geodesic* distance norm is used.

- In certain problems and applications, $p$ will be set to an integral value. When $p = 2$, we have the Euclidean metric, and for $p = 1$, we have the rectilinear metric. When $p \in (0, 1)$, the distance function no longer remains convex. When

$p \in (2, \infty)$, its meaningfulness to real world applications has not yet been demonstrated. Also when $p$ ranges in the interval $[1, 2]$, it is a useful surrogate for measuring distance in the presence of obstacles while also being a well-behaved convex function [137]. There is also a close relationship between the $L_1$- and $L_\infty$-metrics.

- The triangle inequality does not hold in the strict sense for the rectilinear metric but does hold in the strict inequality sense for the $L_p$ metric, $p \in (1, 2)$.

Thus, we are interested in constructing a topology $G(Z, E, \Omega)$ to solve a given problem. As is typical for problems where only the coordinates of the point, edge, and obstacle sets $Z, E, \Omega$ are given, little structure about the relationships between the points is available unless there is some pre-processing involved. Additional information, however, is possible with a geometric approach, and it is this additional spatial information or *geometric data structure,* if you wish, which provides the impetus for the $\mathcal{CG}$ approach to $\mathcal{TND}$.

## 3. Computational Geometry

*Computational geometry* ($\mathcal{CG}$) is concerned with the design and analysis of algorithms and heuristics for geometric problems. The study of geometric problems has long and rich history which can be traced back to Euclid. However, it is only since 1975 that $\mathcal{CG}$ has been recognized as a field of theoretical computer science. The main emphasis of $\mathcal{CG}$ was initially on the development of asymptotically efficient algorithms for problems dealing with large numbers of simple geometric objects (e.g. points, lines, planes, polygons, and polyhedra). The theoretical interest in $\mathcal{CG}$ have been in recent years impressively supplemented by many practical applications (e.g. in computer graphics, computer-aided design, VLSI design, robotics, operations research). Apart from being asymptotically efficient, algorithms in such applications must perform well on average (or for special configurations of geometric objects). Furthermore, they must be relatively easy to implement. For many inherently difficult problems, optimal solutions are out of reach. These problems must be solved by fast heuristics yielding reasonable solutions. $\mathcal{CG}$ paradigms and geometric data structures proved extremely useful also in connection with the design of practical heuristics.

It is only recently that the powerful paradigms and geometric data structures of $\mathcal{CG}$ have been applied with striking success to some ($\mathcal{TND}$) problems of operations research. One such example is the *Euclidean Traveling Salesman Problem* (ETSP) where the $\mathcal{CG}$ approach made it possible to find near-optimal solutions to problem instances with hundred of thousands of points embedded in the plane [21]. Neither classical approaches of mathematical programming nor combinatorial optimization techniques traditionally applied to solve the ETSP can compete with the $\mathcal{CG}$ approach.

$\mathcal{CG}$ is generally regarded as coming of age circa 1975-1978 when M.I. Shamos in his Ph.D. dissertation [179] initiated an exhaustive study into this fascinating domain linking geometry and computer science. Since then several books providing a systematic coverage of $\mathcal{CG}$ appeared [145, 166, 76]. Several surveys providing an excellent overview of $\mathcal{CG}$ have also been published [131, 64, 214, 101].

There are essentially five basic problem areas of $\mathcal{CG}$ which have emerged in the field and which are of particular importance to $\mathcal{TND}$.

- **Searching & Sorting**: Point Location, Convex Inclusion, Star-Shaped Inclusion, General Polygon Inclusion, Inclusion in Planar Straight Line Graphs, Range Searching, ....
- **Convexity**: Convex Hulls in $E^2, E^3, E^d$, Statistical Estimation, Regression, Clustering, Bezier Curves, ....
- **Proximity**: General Location and Nearest Neighbor Search, Post Office Search, All Nearest Neighbor, Triangulation, Equation Solution, ...
- **Intersection**: Hidden Line Problems, Pattern Recognition, Wire and Component Layout, Linear Programming, ....
- **Optimization**: Network Design, Minimum Spanning Trees, Steiner Trees, Traveling Salesman, Chinese Postman, Linear and Integer Programming, Mini-Max and Maxi-Min Location Problems, ....

There are certainly other problem areas which have emerged over the years in $\mathcal{CG}$. One such area involves *arrangements* of lines, pseudo-lines, hyperplanes and other space dissection elements. Yet, these arrangement problems do not directly apply to the $\mathcal{TND}$ issues. For an informative survey on arrangements, see [76].

$\mathcal{CG}$ makes use of standard algorithm design techniques (e.g. sorting, searching, divide-and-conquer, graph searching and manipulation, dynamic programming). Some of these techniques (e.g. divide-and-conquer) have particular features when applied to geometric problems. In addition to the standard techniques, special geometric paradigms have been developed. They proved useful in connection with a wide range of $\mathcal{CG}$ problems. These paradigms will be discussed and exemplified in Section 4.1.

$\mathcal{CG}$ also makes use of standard data structures (e.g. lists, binary trees, heaps). In addition to these standard data structures, special *geometric* data structures have been developed (e.g. DCELs, segment trees, multidimensional binary search trees). Furthermore, such geometric objects as convex hulls, Voronoi diagrams and Delaunay triangulations play a crucial structuring role in connection with many $\mathcal{CG}$ problems. Consequently, also these geometric objects can be regarded as special purpose geometric data structures. Geometric data structures will be discussed and exemplified in Section 4.2.

In the remainder of this section, we argue why the classical mathematical programming and combinatorial optimization methods to solve $\mathcal{TND}$ problems are often inferior to the $\mathcal{CG}$ approach.

## 3.1 Mathematical Programming vs. Geometric Approaches

Let's explore the natural way of modelling geometric $\mathcal{TND}$ problems with classical mathematical programming methods by considering the *minisum Weber problem* (MSWP). This is a well-known location optimization problem in $E^2$ often denoted in the operations research literature as the *Weber problem in location economics.* We seek the location of a new facility $z_0 = (x_0, y_0)$ in relation to $n$ existing facilities $Z = \{z_1, z_2, ..., z_n\}$ in order to minimize the sum of their weighted distances to $z_0$. No constraints are imposed on the location of $z_0$. This problem can be formulated as a nonlinear convex programming problem:

$$\text{minimize } W = \sum_{i=1}^{n} \{w_i[(x_i - x_0)^2 + (y_i - y_0)^2]\}^{1/2}$$

Even when $n = 3$ and all weights are equal, classical methods to optimally solve the MSWP seem to be fraught with computational difficulties [83]. Furthermore, it has been recently shown by Bajaj [13] that an optimal solution to the MSWP cannot be determined by classical methods for $n \geq 5$. This is the case even if all weights are equal. This is a startling result. Until Bajaj's work, little was known about the algorithmic complexity of this problem, although Cockayne and Melzak [55] had indicated earlier of the difficulty of optimally solving these problems. In a recent paper, Bajaj further begins to classify these geometric optimization problems into the $D^p$ Completeness Class [14]. Of course by complicating matters with restricted solution regions and higher dimensional spaces, the results are aggravated even more. Thus, as Bajaj argues only numerical or symbolic approximation procedures can be reasonably expected to resolve the problem. In fact, only locally optimal solutions can be found by numerical algorithms which use first or second order information. As the number of existing facilities grows and $w_i \neq 1$, the number of locally optimal solutions explodes.

### 3.1.1 Steiner Minimal Trees

To illustrate the differences between the geometric and classical approaches, consider the MSWP for $n = 3$ and unit weights. Then it is equivalent to the *Euclidean Steiner minimal tree* (ESMT) problem (discussed in more detail in Subsection 5.3.1) where a shortest possible network in $E^2$ spanning $Z$ is sought. Figures 4a and 4b demonstrate two alternative geometric constructions for finding the Steiner point where the three edges from $z_1, z_2, z_3$ must meet in order to minimize the total length. We first determine third corners $\gamma_1, \gamma_2, \gamma_3$ of equilateral triangles with respectively $z_2z_3$, $z_3z_1$ and $z_1z_2$ as their sides. In the first construction, we draw three circles circumscribing the equilateral triangles. They will intersect in the sought Steiner point $s$. In the second construction, we draw line segments $z_1\gamma_1$, $z_2\gamma_2$, and $z_3\gamma_3$. They will also intersect in $s$. These constructions apply only if the interior angles of the triangle $\Delta z_1z_2z_3$ are less than 120 degrees. If the angle at $z_i$, $i \in \{1, 2, 3\}$ is 120 degrees or more, then the Steiner point coincides with $z_i$.

The MSWP for $n = 4$ and unit weights can also be solved by a very simple geometric construction. However, for $n \geq 5$, it is well-known [55] that the problem cannot be solved by geometric constructions. On the other hand, paradigms and geometric data structures which will be discussed in the next section will clearly illustrate that effective $O(n \log n)$ or even $O(n)$ heuristics can be designed using the $\mathcal{CG}$ approach.

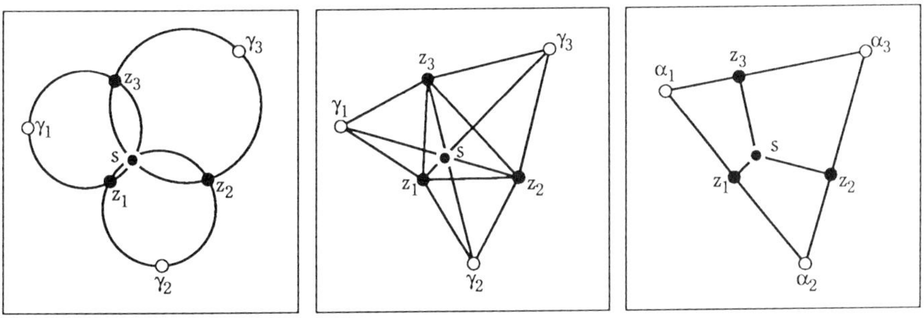


(a) (b) (c)

**Figure 4: Steiner Minimal Tree Constructions**

### 3.1.2 Weber Dual

Additional geometric insight can be gained from the MSWP for $n = 3$ and equal weights when one considers the mathematical programming dual problem. Considering the dual problem is often an invaluable formulation step in mathematical programming problems and in the Weber problem we fortunately have a dual solution which is actually a *geometric construct*.

We must find vectors $U_j, j = 1,2,3$ in order to:

$$\text{maximize } W = -\sum_{j=1}^{3} U_j z_j$$

subject to

$$\sum_{j=1}^{3} U_j = \bar{0}$$

$$||U_j|| \leq 1, j = 1,2,3$$

Fasbender proved in 1846 that there is a geometric construction which solves the dual and inadvertently also solves the primal problem (see Figure 4c). If the Steiner point $s$ does not overlap with any $Z$-point, then $\sum_j^3 d_2(s, z_j)$ equals the maximum altitude of the circumscribing equilateral triangle. This triangle has its sides perpendicular to the line segments $sz_1$, $sz_2$, $sz_3$.

## 3.2 Combinatorial Optimization vs. Geometric Approaches

Many geometric $\mathcal{TND}$ problems can also be formulated as combinatorial optimization problems on weighted graphs. Consequently, they can be solved by network optimization algorithms and heuristics. In particular, network flow formulations play a central role for $\mathcal{TND}$ problems formulated on weighted graphs [140, 152]. We feel that combinatorial and geometric approaches are complimentary rather than conflicting although some of the methods utilized in each approach are entirely different.

The most typical example of approaching geometric $\mathcal{TND}$ problems by combinatorial optimization methods is the *Euclidean minimum spanning tree* (EMST) problem. Let $Z$-points be represented by vertices of a complete weighted graph $K_n$. $K_n$ has edges between all pairs of vertices. The edge-weights are equal to distances between the end-vertices. The minimum spanning tree in $K_n$ can be determined in $O(n^2)$ time using Prim's algorithm [167]. The tree will also be the EMST for $Z$. As will be seen in Subsection 5.2.1, the EMST can be determined in $\Theta(n \log n)$ time if a $\mathcal{CG}$ approach is used to prune away most of the useless edges in $K_n$ before the minimum spanning tree is determined.

Another classical problem where the $\mathcal{CG}$ approach has been successfully used is the *Euclidean traveling salesman problem* (ETSP). This problem will be discussed in more detail in Subsection 5.2.2. Until recently, this problem was also reformulated to its counterpart in $K_n$. It has been recently demonstrated by Bentley [21] that geometric data structures of $\mathcal{CG}$ can be successfully utilized to speed up many heuristics for the ETSP. Consequently, considerably larger problem instances can now be solved than when using traditional combinatorial optimization [123].

The *rectilinear Steiner minimal tree* (RSMT) problem (which will be discussed in more detail in Subsection 5.3.2) can also be formulated as a problem on a weighted graph. Draw vertical and horizontal lines through the points of $Z$. Hakimi [103] proved that at least one optimal solution to the RSMT problem must follow these vertical and horizontal lines. Consequently, the RSMT problem can be reformulated as the Steiner minimal tree problem on a weighted grid graph induced by these lines. It is remarkable that so far no exact $\mathcal{CG}$ algorithm for the RSMT has been developed, although several heuristics based on $\mathcal{CG}$ concepts exist. Such an algorithm will undoubtedly be considerably faster than any of the algorithms for the Steiner tree problem in weighted graphs.

## 3.3 Easy vs. Difficult Problems

Our focus in this survey will primarily be on those $\mathcal{TND}$ problems which lend themselves to a $\mathcal{CG}$ approach. In this regard, we shall treat problems from both the classes $\mathcal{P}$ and $\mathcal{NP}$-complete. For many $\mathcal{NP}$-complete problems, a $\mathcal{CG}$ approach is instrumental because it provides a *unifying* approach both through its paradigms and geometric data structures so that effective and computationally efficient heuristics are achievable.

# 4. Paradigms and Geometric Data Structures

There are a number of $\mathcal{CG}$ paradigms and geometric data structures of major importance to geometric $\mathcal{TND}$. They will be briefly discussed in this section. We first discuss the paradigms as they often are used in algorithms generating the geometric data structures.

## 4.1 Paradigms

Several $\mathcal{CG}$ Paradigms reoccur in algorithms and heuristics for many different geometric $\mathcal{TND}$ problems. Consequently, they provide a natural and useful categorization tool for $\mathcal{TND}$ problems. Paradigms especially useful in connection with geometric $\mathcal{TND}$ problems are described in the following subsections. Each paradigm described below will be further exemplified by a representative algorithm for some geometric $\mathcal{TND}$ problem.

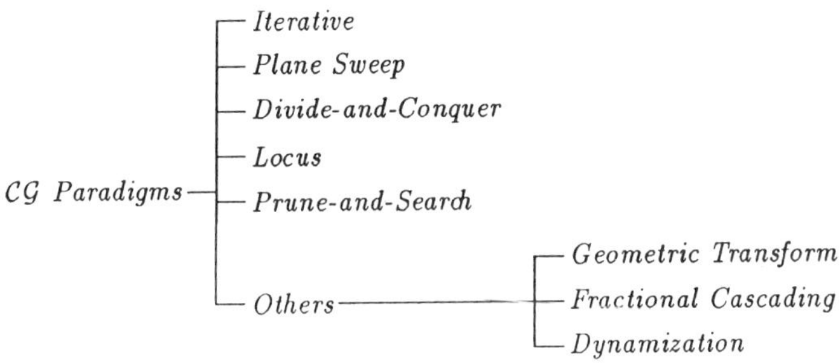


**Figure 5: Morphological Tree for $\mathcal{CG}$ Paradigms**

### 4.1.1 Iterative

Local properties of geometric objects characterizing a given problem (e.g. location of points, their mutual proximity) are utilized to iteratively expand (or improve) a solution. At each iteration several choices of how to expand a partial solution (or improve a feasible solution) are available. A *greedy* choice which gives the smallest cost increase (or largest cost reduction) is always selected.

The iterative paradigm is the most basic and intuitively appealing problem solving approach. It is direct and relatively easy to implement. When *expanding* partial solutions, the number of iterations is proportional to the size of the problem instance. For some polynomially solvable problems, this variant yields in fact optimal solutions. When applied to $\mathcal{NP}$-complete problems, however, it usually yields poor suboptimal solutions. When *improving* feasible solutions, the number of iterations can be substantial since improvements typically continue for as long as possible. However, when applied to $\mathcal{NP}$-complete problems, it usually yields better suboptimal solutions than the expansion variant.

As an example illustrating the (expanding) iterative paradigm, consider the *Euclidean minimum spanning tree* (EMST) problem: Given a set $Z$ of $n$ points in the plane, find a shortest interconnecting network (with junctions permitted only at $Z$-points). This EMST problem can be solved in $O(n^2)$ by a straightforward adaptation of Prim's minimum spanning tree algorithm in weighted graphs [167] (Figure 6). Prim's algorithm is one of the most popular algorithms for the EMST. It requires relatively simple and local proximity information. For each $Z$-point not yet in the partial solution $T$, we merely need to know its nearest neighbour in $T$. This information is readily available from the $n \times n$ distance matrix. More sophisticated geometric data structures are necessary to achieve the optimal $\Theta(n \log n)$ algorithm (see Subsection 5.2.1 for more details).

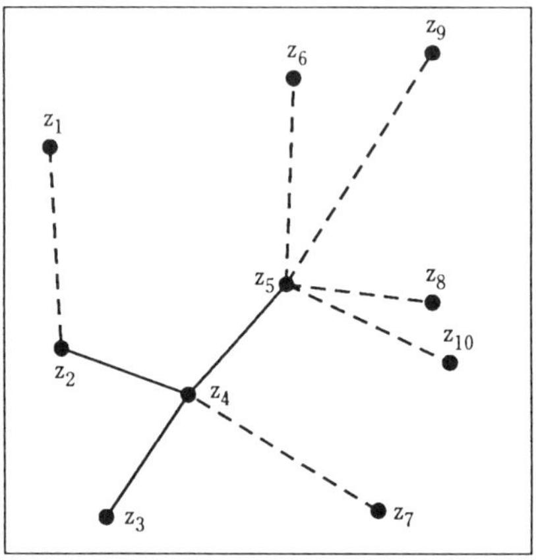


*Pre-Processing:*

**Step 1:** Read in a set $Z$ of $n$ points in the plane.

**Step 2:** Construct a distance matrix between all pairs of these $Z$-points.

*Algorithm:*

**Step 1:** Select any $Z$-point as the initial tree $T$.

**Step 2:** **STOP** if $T$ spans all $Z$-points.

**Step 3:** Find a $Z$-point $z \notin T$ closest to a $Z$-point already in $T$.

**Step 4:** Add $z$ and the shortest interconnecting line segment to $T$.

**Figure 6: A Stage of the Iterative EMST Algorithm**

### 4.1.2 Plane Sweep

Move a *sweep line* $S$ across the plane where some configuration of geometric objects (e.g. points, line segments, rectangles) is embedded. Collect appropriate information about the configuration at the cross-sections by updating the *status* of $S$. Typically, the status changes only at finitely many strategic positions of $S$. These strategic positions are collected on an *event line* $E$ perpendicular to $S$. Events can be identified before the sweep starts. They also can be added or removed dynamically as $S$ moves across the plane.

The plane sweep paradigm is a purely geometrical paradigm. It is often very effective because it transforms a higher dimensional problem into a sequence of lower dimensional problems (one at each event).

The plane sweep paradigm is often used to construct geometric data structures in a preprocessing phase for multiple query problems. This is the case in connection with some point location problems where a subdivision of the plane is constructed gradually as $S$ moves across the plane. However, it also occurs when answering queries. This is also the case in connection with the problem of finding all intersections between horizontal and vertical line segments (Figure 7). In this problem all events are known before the sweep begins. If line segments are not restricted to horizontal and vertical directions, a more elaborate plane sweep algorithm is required. The status of $S$ indicates the ordering of line segments currently intersected by $S$. Initially, projections of end-points of line segments are the only events on $E$. When $S$ reaches the lower end-point of a line segment, this line segment is placed in a (sorted) status list. If the line segment intersects its predecessor (respectively successor) on the status list, projections of these intersections are placed on $E$. When $S$ reaches an upper end-point of a line segment, this line segment is removed from $E$. If its predecessor and successor on the status list intersect above $S$, the projection of this intersection is added to $E$.

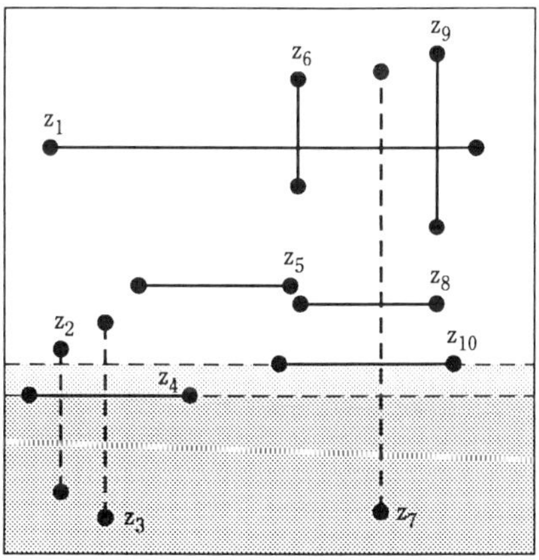


*Algorithm:*

**Step 1:** Project end-points of line segments onto the event line $E$. Make all vertical line segments *inactive*. Place a horizontal sweep line $S$ below all events.

**Step 2:** **STOP** if there is no events above $S$. Otherwise, move $S$ upward to the next event $P$.

**Step 3:** If $P$ corresponds to a lower (respectively upper) end-point of a vertical line segment $L$, make $L$ *active* (respectively *inactive*).

**Step 4:** If $P$ corresponds to a horizontal line segment $L$, determine all intersections of $L$ with active line segments. Go to **Step 2**.

**Figure 7: A Stage of the Plane Sweep Intersection Algorithm**

### 4.1.3 Divide-and-Conquer

*Divide-and-conquer* is a very general paradigm which has been applied to a wide variety of problems. Not surprisingly, it also plays a very important role when designing algorithms for geometric problems.

If a problem instance is small (i.e. its size is bounded by some appropriately chosen constant), the solution is found by some direct (usually trivial) method. Otherwise, the problem instance is subdivided into two instances, each at least a constant fraction smaller than the original instance. The smaller instances are solved recursively. The solutions are then *merged* together yielding a solution to the original problem instance. The merging step is crucial in all divide-and-conquer algorithms. In geometric divide-and-conquer algorithms, it is often possible to use geometric aspects of the problem so that the merging step requires $O(n)$ time. This in turn ensures the overall $O(n \log n)$ bound.

In many divide-and-conquer algorithms, the subdivision of a problem instance into two smaller instances is trivial (e.g. arbitrary partition of $Z$-points into two evenly sized subsets) or based on standard techniques (e.g. partition of $Z$-points into two evenly sized subsets separated by a vertical median line which can be found without sorting in $O(n)$ time using Blum's et al. median algorithm [28]). However, special separation algorithms for $\mathcal{CG}$ problems have also been developed. Perhaps the most significant algorithm of this kind is the polygon cutting algorithm due to Chazelle [37]. This algorithm finds a diagonal cut of a simple polygon in $O(n)$ time such that one of the resulting subpolygons has roughly not more than twice as many vertices as the other subpolygon.

The divide-and-conquer paradigm has been used in connection with several algorithms for the convex hull of a point set $Z$. One of these algorithms is given in Figure 8. Both the subdivision and merging can be done in $O(n)$ time.

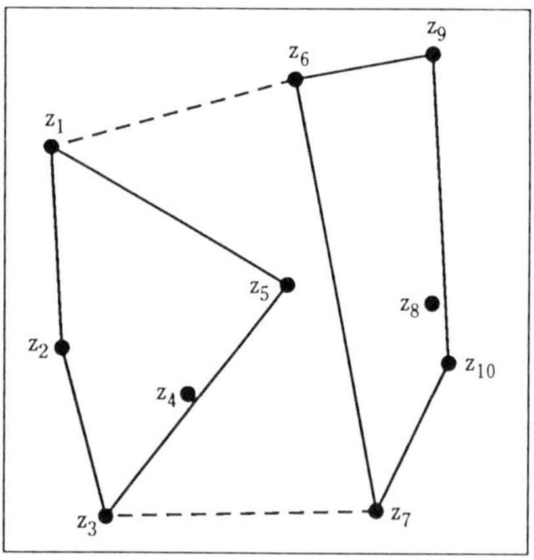


*Algorithm:*

**Step 1:** If $|Z| \leq 2$, then solve directly and **RETURN**. Otherwise, partition $Z$ into two evenly sized subsets $Z_1$ and $Z_2$ such that $x$-coordinate of any point in $Z_1$ is less than or equal to $x$-coordinate of any point in $Z_2$.

**Step 2:** Construct $CH(Z_1)$ and $CH(Z_2)$ recursively.

**Step 3:** Merge $CH(Z_1)$ and $CH(Z_2)$ by identifying the upper and lower *bridges* and by removing *internal* edges.

**Figure 8: A Stage of the Divide-and-Conquer Convex Hull Algorithm**

### 4.1.4 Locus

The *locus* paradigm is extremely useful in connection with multiple point queries (e.g. when determining nearest neighbors to arriving query points). The space is partitioned into cells during the preprocessing phase. The partition is chosen in such a way that answers are invariant for all query points falling in the same cell.

Answering a query amounts to the identification of a cell within which the query point falls. $\Omega(\log n)$ is a natural lower bound for the query time. It follows from the fact that searching for a number in a sorted list with $n$ elements is linear-time transformable to the location of a query point in one of $n$ intervals on the $x$-axis. Searching is known to require $\Omega(\log n)$ time.

The locus paradigm is illustrated in Figure 9 where it is used to determine efficiently the number of points falling within a rectangular query window with horizontal and vertical sides. This example also illustrates that a query does not need to be a point. The query rectangle is *transformed* into four point queries. Query transformations make the locus paradigm applicable to non-geometric problems. In particular, data base queries can be efficiently answered using the locus approach (by appropriate transformation of records and queries to points in multidimensional spaces).

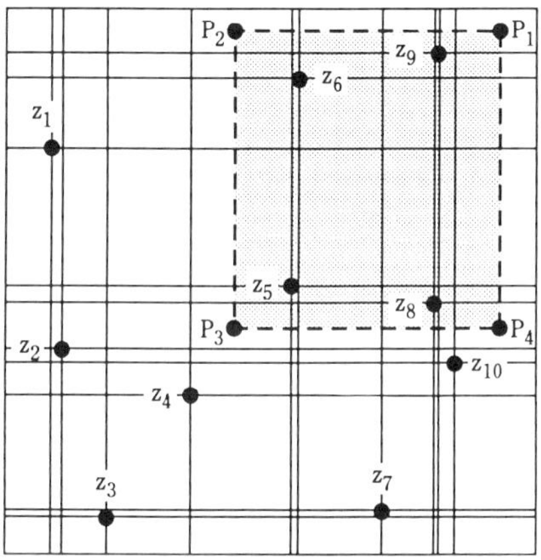


*Preprocessing:*

**Step 1:** Draw a horizontal and vertical line through each point.

**Step 2:** Within each cell $P$ determine the number $Q(P)$ of points to the left and below the bottom left corner of $P$.

*Query:*

**Step 1:** Identify cells $P_1, P_2, P_3, P_4$ containing respectively top right, bottom right, top left and bottom left corners of $R$.

**Step 2:** $N(R) = Q(P_1) - Q(P_2) - Q(P_4) + Q(P_3)$

**Figure 9: Answering Rectangle Queries Using Locus Paradigm**

In the example (Figure 9), the partition has a very regular structure. Answering a query amounts to 8 binary searches, each requiring $\Theta(\log n)$ time. If a partition has an irregular structure, queries cannot be answered quickly unless a suitable data structure to represent the partition is available. In order to facilitate $\Theta(\log n)$ query time, such data structures must be variants of balanced binary trees. Each non-leaf node $v$ represents a certain part of the partition. Its two sons represent a subdivision of this part into two roughly evenly sized subparts. In addition, $v$ has associated sufficient information which makes it possible to decide in constant time whether the search for a region containing a query point (or search for points falling within a query window) should continue through the left or right son (or both). We will return to this kind of data structures in Subsection 4.2.5.

### 4.1.5 Prune-and-Search

The *prune-and-search* paradigm applies to problems where the solution depends on only a few geometric objects (e.g. points, line segments, rectangles). Irrelevant geometric objects are gradually pruned away while searching for the solution. If some fixed fraction of the objects can be pruned away in linear time during each iteration until the problem can be solved in linear time by some other method, the overall worst-case time complexity of the algorithm will be $O(n)$.

The prune-and-search paradigm was first used by Megiddo and Dyer [146, 73] to solve two and three-dimensional linear programming problems. Subsequently, it yielded efficient algorithms for several $\mathcal{CG}$ problems. The paradigm is illustrated in Figure 10 in connection with the problem of finding the common upper tangent of two evenly sized subsets of points separated by a vertical line. As will be seen in Subsection 4.2.1, this problem arises in connection with one of the convex hull algorithms.

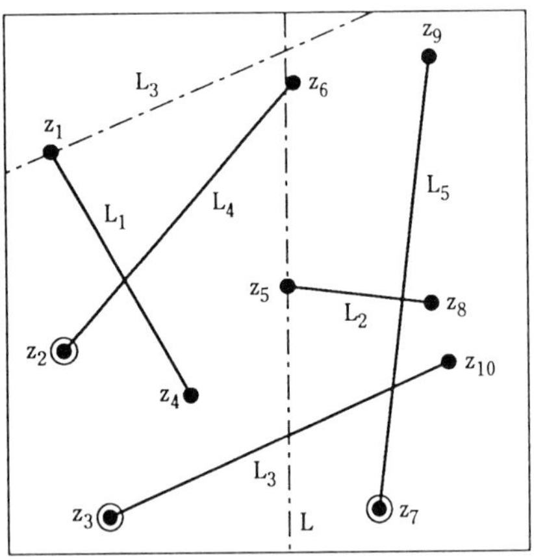


*Pruning Step:*

**Step 1:** Pair up the points in an arbitrary way yielding $\lfloor n/2 \rfloor$ lines.

**Step 2:** Find among these $\lfloor n/2 \rfloor$ lines one (denoted by $L_i$) with median slope $\alpha_i$.

**Step 3:** Find points $z_k, z_m \in Z$, both on the line $L'_i$ with slope $\alpha_i$, no $Z$-points above $L'_i$, and such that $z_k$ (resp. $z_m$) has the smallest (resp. largest) $x$-coordinate.

**Step 4:** If $z_k$ and $z_m$ are respectively to the left and to the right of $L$, then $z_k z_m$ is the sought bridge. If both $z_k$ and $z_m$ are to the left (resp. right) of $L$, then the leftmost (resp. rightmost) $Z$-points of $\lfloor n/4 \rfloor$ line segments with slope greater (resp. smaller) than or equal to $\alpha_i$ cannot be end-points of the upper bridge across $L$.

**Figure 10: Pruning Step of the Common Upper Tangent Algorithm**

### 4.1.6 Other Paradigms

Another important $\mathcal{CG}$ paradigm (which is especially useful in higher dimensions) is the *geometric transformation* paradigm where a given problem is transformed into another problem whose solution can be transformed back into a solution of the original problem. Several types of geometric transformations have been use in $\mathcal{CG}$ (see [214,

101] for further references). The usefulness of geometric transformations to geometric $\mathcal{TND}$ has so far been rather limited. However, as will be seen in Subsection 4.2.3, it has been used to construct some geometric data structures which play an important role in geometric $\mathcal{TND}$.

In various $\mathcal{CG}$ problems one often faces the task of searching through a sequence of different but related ordered sets. A typical example arises when each set is an order-preserving subset of its predecessor. *Fractional cascading* is a technique which replaces searches in each set by a single search in the first set followed by a traversal of appropriate pointers *bridging* consecutive sets. This idea has been applied to several $\mathcal{CG}$ problems [204, 139, 200, 112, 57]. Its generality and formalization was demonstrated by Chazelle and Guibas [42]. Two examples of fractional cascading will be mentioned in Subsections 4.2.5 and 4.2.6.

*Dynamization* involves a transformation of a good geometric data structure for *static* problems (where the geometric objects characterizing the problem remain unchanged) into a *dynamic* data structure (which can accommodate insertions and deletions of geometric objects). Applications of dynamization to geometric $\mathcal{TND}$ problems will most likely play a very important role in future. In this survey we will confine ourselves to static geometric data structures. For more on dynamization, the reader is referred to [131, 166, 76].

## 4.2 Geometric Data Structures

In the past, geometric $\mathcal{TND}$ problems were solved with simple geometric information (e.g. distance matrices). One of the key contributions of $\mathcal{CG}$ has been to introduce new geometric objects (e.g. convex hulls, Voronoi diagrams, Delaunay triangulations, planar decompositions) and design efficient algorithms for their construction. Construction of such geometric objects is fairly complex. It involves use of various $\mathcal{CG}$ paradigms and elaborate data structures. However, the informational and computational benefits far outweigh this disadvantage. As will be demonstrated in Section 5, these geometric objects provide critical information leading to efficient and effective algorithms and heuristics for geometric $\mathcal{TND}$ problems.

Once determined, the geometric objects can be considered as *geometric data structures*. It is these geometric data structures that are the focus of this subsection. As they are described below, we will also discuss how they can be constructed and what data structures are needed in order to determine and represent them.

This section naturally concentrates on geometric data structures of particular importance to geometric $\mathcal{TND}$ problems. However, we will also discuss geometric data structures which currently are not applied to geometric $\mathcal{TND}$ problems. Some of these structures will most likely prove useful in the future. Furthermore, techniques for their generation and representation can provide useful guidelines for new developments of particular importance to geometric $\mathcal{TND}$.

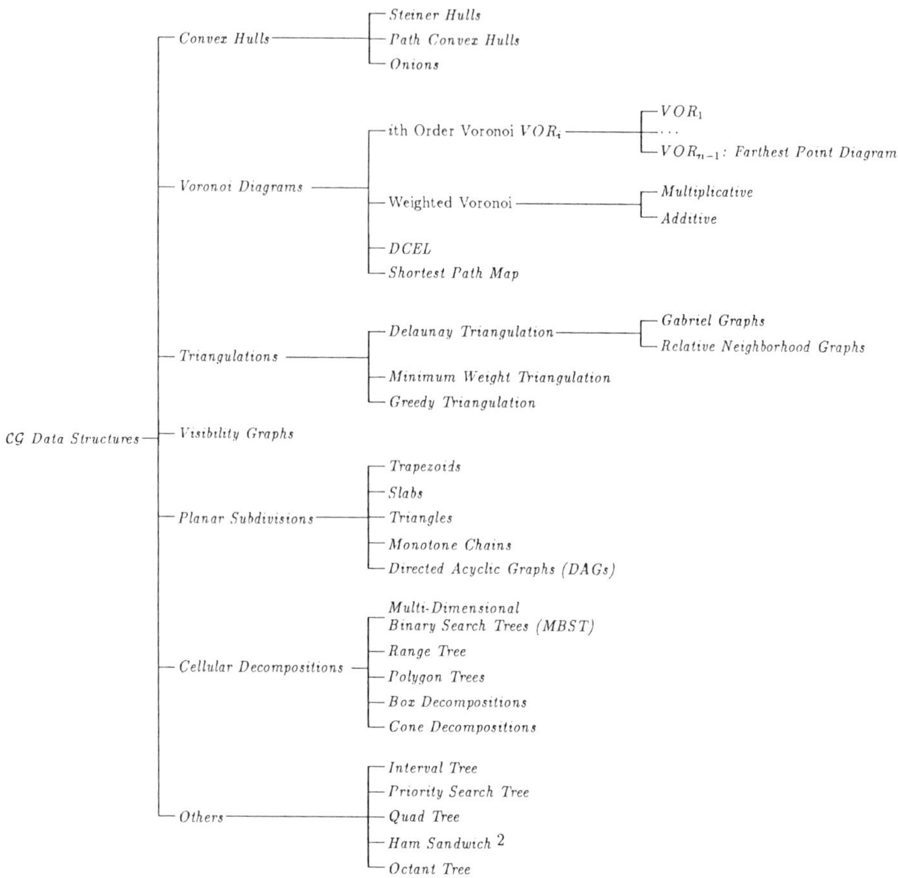


**Figure 11: Morphological Tree for $\mathcal{CG}$ Data Structures**

### 4.2.1 Convex Hulls

A *convex hull* $CH(Z)$ is the smallest convex set containing $Z$ (Figure 12a). $CH(Z)$ is usually represented by the sequence of its extreme points. Hence, more or less elaborate lists (e.g., doubly-linked circular lists) are adequate as the underlying data structures.

Several $CH(Z)$ algorithms have been suggested in the literature [166]. In particular, $\Theta(n \log n)$ algorithms are available. This bound is optimal since sorting of $n$ numbers is linear-time transformable to the convex hull problem. Sorting is known to require $\Omega(n \log n)$ time.

$CH(Z)$ can be determined in $\Theta(n \log n)$ time by an iterative *Graham's scan* algorithm [96]. $Z$-points are sorted by their polar angles around an interior point. This interior point can be found in $O(n)$ time. Consecutive $Z$-points are then connected by edges. The resulting simple polygon is then linearly scanned. $Z$-points with reflex interior angle are thrown out one by one. Its two neighbors on the polygon are connected by an edge. If the scan is done carefully, $O(n)$ time is needed to obtain a convex polygon which by necessity is the boundary of $CH(Z)$.

An $O(hn)$ iterative *Jarvis' march* algorithm (where $h$ is the number of extreme points on $CH(Z)$) has also been suggested [113]. It sequentially finds the edges of $CH(Z)$. Given an edge of $CH(Z)$, the next edge has one end-point in common with it, and the angle formed by these two edges must be the largest possible. Hence, the second end-point can be found in linear time. Although this algorithm is not optimal, it is easy to implement, performs very good on average, and can be generalized to higher dimensions [34].

One $\Theta(n \log n)$ divide-and-conquer algorithm developed by Preparata and Hong [165] has already been discussed in Subsection 4.1.3. Shamos [179] suggested another divide-and-conquer algorithm where the initial sorting of $Z$-points can be avoided. $Z$-points are partitioned arbitrarily into two evenly sized subsets $Z_1$ and $Z_2$. $CH(Z_1)$ and $CH(Z_2)$ are determined recursively. Extreme points of $CH(Z_1)$ and $CH(Z_2)$ are in sorted order. They can be merged together in $O(n)$ time into a simple polygon. Graham's scan is then used to throw out points with reflex angles.

Kirkpatrick and Seidel [120] suggested a $\Theta(n \log h)$ *marriage-before-conquest* algorithm which combines divide-and-conquer and prune-and-search paradigms. The upper and lower part of $CH(Z)$ are determined separately. When determining the upper part, the upper bridge is found first as already explained in Subsection 4.1.5. All points below this upper bridge are pruned away. Points to the left and to the right of the upper bridge are then dealt with recursively. The lower part is determined in analogous way.

In higher dimensions, Chand and Kapur [34] suggested a *gift-wrapping* algorithm which is conceptually related to the Jarvis' march algorithm in $E^2$. The gift-wrapping algorithm requires $O(n^{\lfloor d/2 \rfloor + 1})$ time [27]. Kallay [114] suggested another algorithm, usually referred to as the *beneath-beyond* algorithm. It has the same time complexity as the gift-wrapping algorithm but is in addition on-line (i.e. $Z$-points need not to be known beforehand). Seidel [178] developed another on-line algorithm which requires $O(n^{\lfloor d/2+1 \rfloor})$ time. This is optimal for even $d$. Preparata and Hong [165] suggested a $\Theta(n \log n)$ divide-and-conquer algorithm for $CH(Z)$ in $E^3$.

Cockayne [54] introduced Steiner hulls which proved useful in connection with the Euclidean Steiner minimal tree problem (see Subsection 5.3.1). A *Steiner hull* $SH(Z)$ is obtained from $CH(Z)$ in the following iterative way. Initially $SH(Z) = CH(Z)$. Let $z_i$ and $z_j$ denote a pair of consecutive $Z$-points on the boundary of $SH(Z)$. If a triangle $\triangle z_i z_j z_k$ for some $z_k \in Z \setminus \{z_i, z_j\}$ has the interior angle at $z_k$ of $120^o$ or more, and it contains no $Z$-points, then the boundary edge $z_i z_j$ is replaced by $z_i z_k$

and $z_k z_j$. This is continued until no replacements are possible. It can be shown that $SH(Z)$ is unique and can be obtained independent of the ordering in which boundary edges are processed. Furthermore, it can be shown that $SH(Z)$ contains an optimal solution to the Euclidean Steiner minimal tree problem.

Provan [169] introduced the notion of a *path-convex hull* $PCH(Z, \Omega)$ for a set of $Z$-points and a set $\Omega$ of polygonal obstacles. It is a polygonal region containing all $Z$-points, its boundary avoids the obstacle and has shortest possible length. $PCH(Z, \Omega)$ can be constructed in $O(N^3)$ time where $N$ is the number of $Z$-points and vertices of obstacles. It can be shown that the optimal solution to the Euclidean Steiner minimal tree problem with obstacles (see Subsection 5.4.3) is completely within $PCH(Z, \Omega)$.

A nested collection of convex hulls, sometimes called an *onion* [76], proved very useful in certain applications such as range searching [43] and Euclidean Steiner minimal trees [188]. The onion, denoted by $CH^*(Z)$, is a sequence of convex hulls $CH(Z_1), CH(Z_2), ..., CH(Z_q)$ where $Z_1 = Z$, $Z_i$, $2 \leq i \leq q$, is obtained by removing all extreme points of $CH(Z_{i-1})$ from $Z_{i-1}$, and $CH(Z_q)$ has no interior $Z$-points (Figure 12a). The onion can be constructed in $O(n^2 \log n)$ time by successive computations of convex hulls for $Z_1, Z_2, ..., Z_q$. A straightforward modification of Jarvis' march algorithm yields an $O(n^2)$ algorithm [179]. If the $CH(Z)$ is appropriately represented [158, 166] it is possible to determine $CH(Z \setminus \{z\})$ for any $z \in Z$ in $O(\log^2 n)$. Consequently, the onion can be determined in $O(n \log^2 n)$. Finally, an optimal $\Theta(n \log n)$ time onion algorithm has been developed by Chazelle [38].

### 4.2.2 Voronoi Diagrams

As indicated in Subsection 3.1.1, some geometric $\mathcal{TND}$ problems are concerned with the proximity relations among $Z$-points (or other geometric objects). Another class of geometric $\mathcal{TND}$ problems is concerned with the proximity relations of arriving query points to $Z$-points. Furthermore, proximity issues of this type often arise as subproblems when solving geometric $\mathcal{TND}$ problems in other classes. Geometric data structures which partition the plane into appropriate regions so that proximity relations can be quickly resolved are of paramount importance.

Let $z_i$ and $z_j$ denote two distinct $Z$-points. Furthermore, let $H(z_i, z_j)$ denote the set of points not farther from $z_i$ than from $z_j$. $H(z_i, z_j)$ is a half-plane. Let

$$V(z_i) = \bigcap_{z_j \in Z \setminus \{z_i\}} H(z_i, z_j)$$

Hence, $V(z_i)$ is convex and its interior is the locus of points closer to $z_i$ than to any other $Z$-point (Figure 12b). Let $P(z_i)$ denote the boundary of $V(z_i)$. The union of these boundaries forms the *Voronoi diagram* for $Z$ (Figure 12c), denoted by $VOR(Z)$. Its edges are called *Voronoi edges*. Points where Voronoi edges meet are called *Voronoi points*. Voronoi diagrams form the foundation of many $\mathcal{CG}$ algorithms in geometric $\mathcal{TND}$. For a detailed survey on Voronoi diagrams, the reader is referred to Boots [30] and Aurenhamer [10].

$Z$-points within the unbounded faces of $VOR(Z)$ are precisely the extreme points of $CH(Z)$. Consequently, any $VOR(Z)$ algorithm will require $\Omega(n \log n)$ time.

A $\Theta(n \log n)$ divide-and-conquer Voronoi diagram algorithm is available [180]. $Z$-points are bisected by a vertical line into two subsets $Z_1$ and $Z_2$. $VOR(Z_1)$ and $VOR(Z_2)$ are determined recursively. Subsequent merging is based on the fact that Voronoi edges of $VOR(Z)$ separating $V(z_1)$, $z_1 \in Z_1$ from $V(z_2)$, $z_2 \in Z_2$, form a single downward-going chain $\sigma$. Line segments of $\sigma$ can be determined in $O(n)$ time. Portions of $VOR(Z_1)$ and $VOR(Z_2)$ respectively to the right and left of $\sigma$ are deleted yielding $VOR(Z)$.

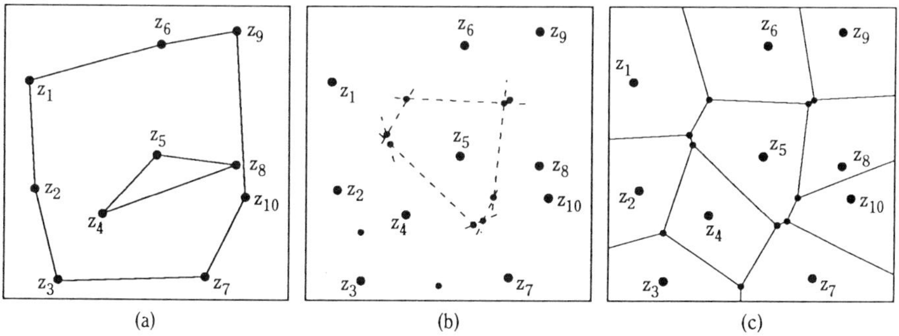


**Figure 12:** $CH^*(Z), V(z_5), VOR(Z)$

Fortune [82] suggested a $\Theta(n \log n)$ plane sweep algorithm. An excellent description of this algorithm can be found in [101].

In higher dimensions, $VOR(Z)$ can be obtained by exploiting duality between $VOR(Z)$ in a $d$-dimensional space and $CH(Z)$ in a $(d+1)$-dimensional space [31, 178].

Due to the utmost importance of Voronoi diagrams in $\mathcal{CG}$, many variants and generalizations have been suggested in the literature. Voronoi diagrams with other metrics (and with the $L_1$-metric in particular), higher-order Voronoi diagrams, weighted Voronoi diagrams andVoronoi diagrams for line segments and circular arcs are just few examples. Here we will only briefly mention higher-order Voronoi diagrams and weighted Voronoi diagrams.

The *i-th order Voronoi diagram* $VOR_i(Z)$, $1 \leq i < n$, is a partition of the plane into regions $V(Z_i)$, $Z_i \subseteq Z$, $|Z_i| = i$. The interior of $V(Z_i)$ is the locus of points closer to every point in $Z_i$ than to any point in $Z \setminus Z_i$. Note that some $V(Z_i)$ may be empty. $VOR_1(Z) = VOR(Z)$. $VOR_{n-1}(Z)$ is usually called the *farthest point Voronoi diagram.*

Edelsbrunner et al. [79] developed an $O(n^{d+1})$ time and space algorithm which constructs *all* $VOR_i(Z)$, $i = 1, 2, ..., n-1$, in a $d$-dimensional space. Lee [128] developed an algorithm for the construction of $VOR_i(Z)$ in $O(i^2 n \log n)$ time using $O(i^2(n-i))$ space. It is based on successive modifications of $VOR_1(Z)$, $VOR_2(Z)$,..., $VOR_{i-1}(Z)$. These bounds were improved to $O(i(n-i)\sqrt{n} \log n)$ time and $\Theta(i(n-i))$ space by Edelsbrunner [75]. Chazelle and Edelsbrunner [41] suggested yet two improvements leading to $O(n^2 \log n + i(n-i) \log^2 n)$ time, $\Theta(i(n-i))$ space algorithm and $O(n^2 + i(n-i) \log^2 n)$ time and $O(n^2)$ space algorithm. Aurenhammer [11] discovered a new duality between $VOR_i(Z)$ in $E^d$ and $CH(Z)$ in $E^{d+1}$. This resulted in a different and simpler algorithm for $VOR_i(Z)$ which requires $O(i^2 n \log n)$ time and $\Theta(i(n-i))$ space. Contrary to the Lee's algorithm [128], this simpler algorithm generalizes to higher dimensions. $VOR_{n-1}(Z)$ can also be constructed directly in $\Theta(n \log n)$ time by a divide-and-conquer algorithm similar to the algorithm for $VOR(Z)$ [179, 126].

Suppose that a positive weight $w_i$ is associated with each $Z$-point. A *weighted Voronoi diagram* is a partition of the plane into regions $V(z_1), V(z_2), ..., V(z_n)$ such that $q \in V(z_i)$ iff the weighted distance from $q$ to $z_i$, $z_i \in Z$, is not larger than the weighted distance from $q$ to any other $Z$-point. There are quite a few weighted distance functions possible. Here we mention just two which have proved most useful in connection with geometric $\mathcal{TND}$ problems.

- *Multiplicative* distance function:

  $$V(z_i) = \{q | w_i d(q, z_i) \leq w_j d(q, z_j) \text{ for all } z_j \in Z \setminus \{z_i\}\}$$

  The boundary of $V(z_i)$ can be a mixture of line segments and circular arcs. The resulting Voronoi diagram appears as a 2-dimensional soap bubble. It can be constructed in $O(n^2)$ time [12].

- *Additive* distance function:

  $$V(z_i) = \{q | d(q, z_i) - w_i \leq d(q, z_j) - w_j \text{ for all } z_j \in Z \setminus \{z_i\}\}$$

  The boundary of $V(z_i)$ can be a mixture of line segments and hyperbolic arcs. This Voronoi diagram can be constructed in $\Theta(n \log n)$ time by a sweep plane algorithm which is a modification of Fortune's algorithm for $VOR(Z)$ [82].

Voronoi diagrams (as well as other higher-level geometric data structures which will be discussed below) are plane straight-line graphs. A *doubly connected edge list* (DCEL) is a data structure particularly well-suited to represent such graphs. A DCEL is a list of records, each representing one edge of the graph. The following information is associated with each record:

- $V_1$ and $V_2$: End-points of the edge.

- $F_1$ and $F_2$: Faces on respectively left and right side of the edge when looking from $V_1$ toward $V_2$.

- $P_1$ and $P_2$: Pointers to records containing the first edge encountered after the edge $(V_1, V_2)$ when moving counterclockwise around $V_1$ (resp. $V_2$).

- $S_1$ and $S_2$: Pointers to records containing the first edge encountered after the edge $(V_1, V_2)$ when moving clockwise around $V_1$ (resp. $V_2$).

A DCEL can be constructed in $\Theta(n)$ time and space (provided that the edges incident to each vertex of the graph are given in their angular order). Once constructed, the edges incident to a given vertex or the edges enclosing a given face can be found efficiently. Also such operations as deletion and addition of an edge can be carried out efficiently.

Another geometric data structure which subdivides the plane into regions and which is useful in connection with obstacle-avoiding shortest path problems (discussed in detail in Subsection 5.4.2) is the *shortest path map* $SPM(z, \Omega)$ for a $Z$-point $z$ and a set $\Omega = \{\omega_1, \omega_2, ..., \omega_k\}$ of disjoint polygonal obstacles with total $c$ extreme points. $SPM(z, \Omega)$ is a subdivision of plane into regions such that two points $p$ and $q$ belong to the same region iff obstacle-avoiding shortest paths from $z$ to $p$ and from $z$ to $q$ go through the same sequence of extreme points of obstacles. Given a query point $v$, the shortest obstacle-avoiding path from $z$ to $v$ can be quickly determined by identifying the region of $SPM(z, \Omega)$ containing $v$. $O(\log c)$ techniques to identify the region containing $v$ are available [6].

### 4.2.3 Delaunay Triangulations

A *triangulation* of $Z$-points is a plane straight-line graph with $Z$ as its vertex set and with maximal number of edges. It follows that every interior face of a triangulation is a triangle. In some applications, triangulations are *constrained* in the sense that some of its edges are beforehand forced to be included.

Triangulations play a very important role in $\mathcal{CG}$. First of all, triangulations provide a decomposition of the plane into manageable regions. Consequently, problems involving location of query points in any straight-line subdivision can be simplified if the subdivision is triangulated in the preprocessing phase. We will return to this kind of triangulations in Subsection 4.2.5. Secondly, triangulations can be constructed so that they in one way or another capture the proximity relations between $Z$-points. Then, they can be used in geometric $\mathcal{TND}$ problems as a pruning tool to remove irrelevant connections. Since triangulations are planar graphs, they have $O(n)$ edges. Complete graphs which represent all possible connections between $Z$-points have $O(n^2)$ edges. We will see the application of such pruning in connection with the *Euclidean minimum spanning tree* problem discussed in detail in Subsection 5.2.1. Triangulations which capture proximity relations also provide a tool to cluster groups of three (or more) $Z$-points. We will see the application of such clustering in connection with the *Euclidean Steiner minimal tree* problem discussed in Subsections 5.3.1 and 5.4.3.

Several kinds of triangulations which capture proximity relations have been suggested in the literature. One of the most important triangulations of this type is the *Delaunay triangulation* (Figure 13a), denoted by $DT(Z)$. It is defined as the unique triangulation such that circumcircle of each triangle does not contain any other $Z$-point in its interior. Triangles of $DT(Z)$ tend to be as "equilateral" as possible in the sense that the minimum angle of its triangles is maximum over all triangulations.

The straight-line dual of $VOR(Z)$ obtained by connecting a pair $z_i$ and $z_j$ of $Z$-points if and only if $P(z_i)$ and $P(z_j)$ share a common side yields in fact $DT(Z)$. Hence, $DT(Z)$ can be determined in $\Theta(n \log n)$ time. The optimality of this algorithm follows from the fact that sorting of $n$ numbers is linear-time transformable to the problem of finding a triangulation of $Z$-points.

$DT(Z)$ can also be determined via a geometric transformation of $Z$-points onto a paraboloid of revolution $z = x^2 + y^2$ via the mapping $(x, y) \rightarrow (x, y, x^2 + y^2)$. Faces of the 3-dimensional convex hull visible from the $x, y$-plane will form $DT(Z)$ when mapped back onto the plane. Furthermore, faces not visible from the $x, y$-plane form a geometric dual of $VOR_{n-1}(Z)$ when mapped back onto the plane [100].

$\Theta(n \log n)$ divide-and-conquer algorithms determining $DT(Z)$ directly are available [127, 100]. These algorithms are in particular useful for the $L_1$-metric where the straight-line dual of $VOR(Z)$ does not in general yields $DT(Z)$.

There are two other geometric data structures closely related to $DT(Z)$. They also capture proximity relations (although they are not necessarily triangulations). They have so far only been of marginal importance to geometric $\mathcal{TND}$.

- Let $z_i$ and $z_j$ denote two distinct $Z$-points. Let $L(z_i, z_j)$ be a lune obtained as an intersection of two discs with radius $|z_i z_j|$ and centered at $z_i$ and $z_j$ respectively. A *relative neighborhood graph* $RNG(Z)$ has $Z$ as its vertex set. A pair of $Z$-points $z_i$ and $z_j$ is adjacent iff $L(z_i, z_j)$ contains no other $Z$-point [194] (Figure 13b). A $\Theta(n \log n)$ plane sweep algorithm for the construction of $RNG(Z)$ has been suggested by Supowit [191].

- Let $z_i$ and $z_j$ denote two distinct $Z$-points. Let $D(z_i, z_j)$ denote a disc with $z_i z_j$ as its diameter. A *Gabriel graph* $GG(Z)$ has $Z$ as its vertex set. A pair of $Z$-points $z_i$ and $z_j$ is adjacent iff $D(z_i, z_j)$ contains no other $Z$-point (Figure 13c). $GG(Z)$ can be constructed in $\Theta(n \log n)$ time by removing from $DT(Z)$ each edge not intersecting its dual Voronoi edge [143].

It can be shown that $RNG(Z)$ is a subgraph of $GG(Z)$ which in turn is a subgraph of $DT(Z)$. It can also be shown that $RNG(Z)$ contains at least one Euclidean minimum spanning tree for $Z$ as a subgraph. The implications of this fact will be discussed in Subsection 5.2.1.

Another triangulation which indirectly captures the proximity of $Z$-points is the *minimum weight triangulation* where the total weight of edges is to be minimized. No polynomial algorithm for this problem is available. On the other hand, it is unsettled whether the problem of determining the minimum weight triangulation is $\mathcal{NP}$-complete. Minimum weight triangulations of the interior of a simple polygon can be found in $O(n^3)$ time [92].

The *greedy triangulation* heuristic determines a low weight triangulation in the following way. Line segments between all pairs of $Z$-points are sorted in nondecreasing order of their length. Line segments are then added one by one. A line segment is added only if it does not intersect any of the edges already in the triangulation. Sorting of the line segments requires $O(n^2 \log n)$ time. Intersection checks require in total $O(n^3)$ time if done in a trivial way. This can be reduced to $O(n^2 \log n)$ using a more elaborate geometric data structure introduced below. Note that the greedy triangulation also works in constrained situations.

A *segment tree* was originally introduced by Bentley in 1977 (see [166]). It can be used to represent various arrangements of line segments. The segment tree $T$ is a balanced binary tree with each node representing an appropriate subinterval of $[1, n[$.

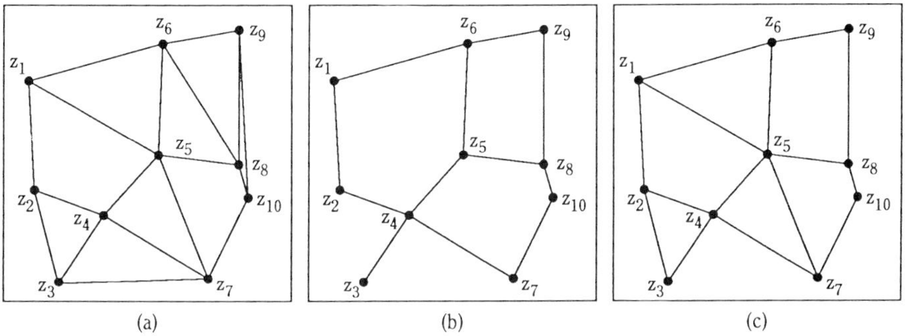


**Figure 13:** $DT(Z), RNG(Z), GG(Z)$

Any secondary data structure can be associated with each node of the segment tree. If a node $v$ in $T$ represents an interval $[a, b[$, then the left son of $v$ represents the interval $[a, \lfloor (a+b)/2 \rfloor[$, and the right son of $v$ represents the interval $[\lfloor (a+b)/2 \rfloor, b[$. $T$ has $n$ leaves and has height $O(\log n)$. It can be constructed in $\Theta(n)$ time and space.

Let us now return to the greedy triangulation. For each $Z$-point $z$, the remaining $Z$-points are numbered from 1 to $n-1$ according to their clockwise order of appearance around $z$. A segment tree $ST(z)$ with the root representing the interval $[1, n[$ is associated with every $Z$-point.

When a line segment $z_i z_j$ is to be added to the greedy triangulation, each $ST(z)$, $z \neq z_i, z_j$, is updated in the following manner. Let $e$ (respectively $f$) denote the position of $z_i$ (respectively $z_j$) in the circular list around $z$. Assume that $1 \leq e < f$ (other possibilities can be dealt with in similar manner). $ST(z)$ is traversed beginning from the root. When traversing a node $v$ representing an interval $[a, b[$, the following action is taken.

- If there is a line segment already allocated at $v$ and it is closer to $z$ than $z_i z_j$, then the traversal stops. $ST(z)$ remains unchanged.
- Otherwise:
    1. If $e \leq a$ and $b \leq f$, then $z_i z_j$ is allocated at $v$. The line segment previously allocated at $v$ (if any) is removed.
    2. If $e < \lfloor (a+b)/2 \rfloor$, then the traversal continues through the left son.
    3. If $\lfloor (a+b)/2 \rfloor < f$, then the traversal continues through the right son.

Note that the second and third case are not mutually exclusive. It can be shown that the interval $[e, f[$ will be associated with at most $O(\log n)$ nodes. Union of the intervals of these nodes will be precisely $[e, f[$. Insertion of a line segment into $n-2$ segment trees requires $O(n \log n)$ time. Since triangulations are planar, $O(n)$ line segments will be inserted. Consequently, insertions require in total $O(n^2 \log n)$ time.

In order to check whether a line segment $z_i z_j$ intersects any of the line segments already in the triangulation, we only need to search through $ST(z_i)$. Let $f$ denote the position of $z_j$ in the circular list around $z_i$. Traverse the path from the root through nodes representing intervals containing $f$. At each such node, check whether the associated line segment (if any) separates $z_j$ from $z_i$. If this is the case, the traversal stops since $z_i z_j$ intersects one of line segments already in the triangulation. Each intersection check requires $O(\log n)$ time. There are $O(n^2)$ line segments to check in total. Consequently, intersection checks require in total $O(n^2 \log n)$ time.

### 4.2.4 Visibility Graphs

Let $Z = \{z_1, z_2, \ldots, z_n\}$ be a set of points in $E^2$. Furthermore, let $\Omega = \{\omega_1, \omega_2, \ldots, \omega_k\}$ denote a collection of disjoint polygonal obstacles with $C$ denoting the set of all extreme points. Let $c = |C|$ and $N = n + c$. A *visibility graph* $VG(Z, \Omega)$ for $Z$ and $\Omega$ is a plane straight-line graph with $Z \cup C$ as its vertex set. Two vertices are adjacent iff the line segment connecting them avoids all obstacles.

$\Theta(N^2)$ algorithms for the determination of $VG(Z, \Omega)$ are available [202, 8]. The optimality follows from the fact that $VG(Z, \Omega)$ can have up to $O(N^2)$ edges. Ghosh and Mount [90] gave an optimal output-sensitive $\Theta(m + N \log N)$ algorithm where $m$ is the number of edges in $VG(Z, \Omega)$. As will be seen in Subsection 5.4.2, visibility graphs play a crucial role in connection with the determination of obstacle avoiding shortest paths.

### 4.2.5 Planar Subdivisions

Many $\mathcal{CG}$ problems involve identification of a region in a prespecified planar subdivision within which a query point falls. In the following we assume that boundaries of regions are straight line segments intersecting at $Z$-points only. Usually there will be many query points whose location has to be established. Consequently, it is necessary to further subdivide the regions into simpler structures (e.g. triangles, rectangles, trapezoids, convex polygons) in the preprocessing phase. It is also necessary to organize these structures in such a way that subsequent queries can be answered efficiently (and preferable in $\Theta(\log n)$ time).

As we already mentioned in connection with the locus paradigm (Subsection 4.1.4), efficient algorithms to many $\mathcal{CG}$ problems often involve some kind of subdivision of the plane into suitable regions. Onions, Voronoi diagrams, Delaunay and greedy triangulations discussed earlier in this section are examples of such subdivisions. They depend in one way or another on proximity relations between $Z$-points, and as such are not artificial. They require more overhead and impede efficiency in higher dimensions.

In this subsection we discuss *artificial* subdivisions which only loosely depend on distances between geometric objects. As a consequence, they are much easier to construct and can be generalized to higher dimensions. Furthermore, they are much more suitable in dynamic scenarios where geometric objects arrive and disappear from the scene. See [118] for a survey on planar subdivisions (including Steiner-like subdivisions where additional intersections can be introduced).

Artificial subdivisions arise not only in location problems. Many problems involving polygons are solved by decomposing them into simpler structures, applying efficient algorithms to each structure, and combining partial solutions. This approach is for example used to find a shortest path between two points within a simple polygon (Subsection 5.4.2).

Suppose that we are given a planar subdivision $G$ which is a plane straight-line graph with $Z$-points as its vertices. We would like to further partition $G$ so that all interior faces are triangles and the exterior face is convex. $G$ is first *regularized* so that each $Z$-point except the top and bottom one has at least one line segment coming from above and at least one line segment coming from below. Such regularization can be achieved in $O(n \log n)$ time by two plane sweeps. In the first (bottom-to-top) sweep, new line segments are added to $G$ so that every $Z$-point (except the top one) has at least one line segment coming from above. In the second (top-to-bottom) sweep, new line segments are added to $G$ so that every $Z$-point (except the bottom one) has at least one line segment coming from below. In both sweeps the event list consists of $y$-coordinates of $Z$-points. The status of the horizontal sweep line $S$ is a list of line segments intersected by $S$ (ordered from left to right). In addition, during the bottom-to-top sweep, $S$ keeps a $Z$-point with highest ordinate below $S$ for each interval between two intersections with consecutive edges. When arriving at some event (i.e. a $Z$-point $z$) during the bottom-to-top sweep, the interval on $S$ containing $z$ is identified. Let $z_i$ denote the $Z$-point associated with this interval. Line segments terminating at $z$ are deleted from the status list. This in general causes two edges on $S$ to become consecutive with $z$ being a $Z$-point with the highest ordinate between them. If no line segment goes up from $z$, the line segment $zz_i$ is added to $G$ If there are line segments going up from $z$, they are added to $S$ with $z$ as the $Z$-point with the highest ordinate between every pair of added consecutive edges, and the sweep continues. The top-to-bottom sweep is done in analogous way. If the status list is represented by any data structure which supports insertions and deletions of elements in $\Theta(n \log n)$ time, the entire regularization can be done in $O(n \log n)$ time [166].

A polygon is *monotone* if any horizontal line between its top and bottom $Z$-point intersects the polygon in exactly two places. All polygons in a regularized planar subdivision $G$ are monotone [166]. In order to triangulate a regular subdivision $G$, we only need to triangulate its monotone polygons. A monotone polygon can be triangulated in linear time by an elegant iterative algorithm due to Garey et al. [89] which traverses both sides of the polygon in synchronized fashion. Since the number of line segments in a planar partition is $O(n)$ and each line segment occurs in at most two monotone polygons, it follows that all monotone polygons of $G$ can be triangulated in $O(n)$ time. Consequently, triangulation of an arbitrary planar subdivision can be done in $\Theta(n \log n)$.

A very important type of triangulation arises when the original subdivision is a simple polygon. The prespecified line segments are no longer arbitrary, so that the above approach is not necessarily optimal. In fact, an $O(n \log \log n)$ algorithm for triangulation of simple polygons has been developed by Tarjan and van Wyk [193]. Recently, Chazelle [40] gave an optimal $\Theta(n)$ algorithm for this problem.

The detailed discussion of slabs, triangles, separating chains and layered DAGS in relation to point location clustering problems is reserved for discussion in Subsection 5.1.5 where these data structures are most applicable.

### 4.2.6 Cellular Decompositions

Many $\mathcal{CG}$ problems are approached by some decomposition of the geometric objects involved. Usually such decompositions are governed by subdivisions of the plane into (possible hierarchical) cells capturing the proximity relations of the objects. Consequently, such decompositions are closely related to the theme of the previous subsection.

One of the most important among cellular decompositions is a *multidimensional binary search tree* (MBST) introduced by Bentley [19]. It is obtained in the following manner.

- Partition $Z$ into two subsets $Z_1$ and $Z_2$ by means of a vertical bisecting line such that roughly half of $Z$-points is in each subset.

- Partition $Z_1$ into two subsets $Z_{11}$ and $Z_{12}$ by means of a horizontal bisecting line such that roughly half of $Z_1$-points is in each subset. Partition $Z_2$ into subsets $Z_{21}$ and $Z_{22}$ in similar way.

- Keep partitioning generated subsets by alternating vertical and horizontal bisecting lines. Stop when subsets with some small prespecified number of $Z$-points emerge.

The resulting hierarchy can be conveniently represented by a balanced binary tree. $Z$ is associated with the root, $Z_1$ (respectively $Z_2$) is associated with its left (respectively right) son, etc. Furthermore, rectangular regions containing these subsets (and determined by up to four bisecting lines) are also associated with nodes. Note that these rectangular regions in fact determine a planar subdivision.

The generalization of MBSTs to $d$-dimensional spaces, $d \geq 2$, is straightforward. The space is partitioned by hyperplanes orthogonal to coordinate axes. Orientations of hyperplanes at each level changes according to the cyclic order of coordinate indices. MBSTs can be built in $\Theta(dn \log n)$ time using $\Theta(dn)$ space.

MBSTs are extremely useful in connection with *range searching*. Consider a rectangle query window $R$ with its sides parallel to coordinate axes. Suppose that we are interested in determining all Z-points within $R$. Search through the tree beginning at the root. When visiting a node $v$ of the tree the following action is taken.

- If the rectangular region associated with $v$ is completely covered by $R$, all $Z$-points associated with $v$ are in $R$. *Backtrack.*

- Otherwise, if the rectangular region associated with $v$ and $R$ are disjoint, no $Z$-point associated with $v$ is in $R$. *Backtrack.*

- Otherwise search through both the left and right son of $v$.

Lee and Wong [133] proved that range searching in MBSTs requires $O(dn^{1-1/d}+k)$ time where $k$ is the number of points which eventually are found to be within $R$.

MBSTs are also useful when solving proximity problems. Suppose that we want to find a nearest neighbor to a query point $q$. Follow a path in the MBST from the root to the leaf whose rectangular region contains $q$. Find a $Z$-point $z$ inside this region, and nearest to $q$. Let $r=|qz|$. Let $R$ be a $2r \times 2r$ square centered at $q$. Go up the path in the MBST. At each node, visit its other son and carry out a range search with $R$ as a query window. If any $Z$-point $z'$ falls within $R$, check its distance to $q$. If it is less than $r$, reduce the size of $R$ accordingly.

We have previously indicated that Voronoi diagrams provide a planar subdivision of the plane so that the above problem can be solved in $\Theta(\log n)$ time using one of the point location methods described in Subsection 4.2.5. However, Voronoi diagrams are difficult to construct in higher dimensions. Furthermore, in some 2-dimensional problems (e.g. in the *Euclidean traveling salesman problem* described in detail in Subsection 5.2.2), the nearest neighbors have to be found many times for dynamically changing set of $Z$-points. In particular, $Z$-points are deleted and added. Each deletion and addition requires substantial modifications of the Voronoi diagram. On the other hand, MBSTs remain basically unchanged. It suffices to mark $Z$-points as deleted. When they subsequently reenter $Z$, the mark is simply removed.

Another subdivision of $Z$-points which is useful in connection with range search problems is a *range tree* suggested by Bentley [23]. Suppose that the $x$-coordinates of $Z$-points are normalized to integers $1,2,..,n$ (preserving their order). Consider a segment tree representing subintervals of $[1,n[$. We already introduced segment trees in connection with the greedy triangulations (Subsection 4.2.3). An interval node $v$ representing an interval $[a,b[$ has associated with it $Z$-points with $x$-coordinates $a,a+1,...,b-1$. These $Z$-points are sorted w.r.t. their $y$-coordinate and stored in a *threaded binary search tree* $T_v$. $T_v$ is a balanced binary tree with leaves representing $Z$-points. $Z$-points in the left subtree of any node in $T_v$ have $y$-coordinates smaller than $y$-coordinates of $Z$-points in the right subtree. Furthermore, leaves are threaded to form a list reflecting the ordering of $Z$-points. Note that node $v$ can be considered as representing a vertical slab with vertical lines through $a$ and $b$ as its sides.

In order to identify $Z$-points within a (normalized) rectangular query window $R$, we only need to identify $O(\log n)$ nodes of the segment tree whose intervals partition the interval between the vertical sides of $R$. For each such interval, $Z$-points in its threaded binary tree between the horizontal sides of $R$ can be identified in $O(\log n+k)$ time where $k$ is the number of $Z$-points in $R$.

The $O((\log n)^2+k)$ query time can be reduced to $O(\log n)$ using fractional cascading as suggested independently by Willard [204] and Lueker [139]. Fractional cascading replaces $O(\log n)$ searches through threaded binary search trees by one search at the root followed by a simple traversal through an appropriate data structure. Note that $Z$-points associated with any non-leaf node $v$ of the segment tree are partitioned into two subsets associated with its two sons $v_L$ and $v_R$. Furthermore, this partitioning is order-preserving. Let $z$ be any $Z$-point associated with $v$. Associate two pointers $p_L$ and $p_R$ with the copy of $z$ in $v$. $p_L$ points to the first $Z$-point in $v_L$ not below $z$ (if $z \in v_L$ then $p_L = z$). $p_R$ points to the first $Z$-point in $v_R$ not below $z$ (if $z \in v_R$ then $p_R = z$).

In order to identify $Z$-points within a normalized rectangular query window $R$, we again need to identify $O(\log n)$ nodes of the segment tree whose intervals partition the interval between the vertical sides of $R$. However, now the threaded binary search tree associated with the root is searched in $O(\log n)$ time for the first $Z$-point above the lower horizontal side of $R$. When going down the segment tree, we in addition follow left and/or right pointers to identify first $Z$-points above the lower horizontal side of $R$ in son nodes. When arriving at an interval completely within the normalized vertical sides of $R$, the first $Z$-point above the lower horizontal side of $R$ is given. Remaining $Z$-points below the upper horizontal side of $R$ can be identified by a sequential scan. Since the search through threaded binary search trees is done only at the root, all other threaded binary search trees can be replaced by simple lists.

The method generalizes in a natural way to arbitrary $d, d \geq 2$. It can be shown that $O((n \log n)^{d-1})$ time and space is needed to construct a range-tree. Furthermore, a query takes $O((\log n)^{d-1} + k)$ time.

*Polygon trees* [205] are closely related to MBSTs. $Z$-points are partitioned into four subsets, each with at most $n/4$ $Z$-points, by a vertical and horizontal bisecting line. Smaller subsets are partitioned recursively. Polygon trees proved useful in connection with line queries (i.e. identification of all $Z$-points on a query line or in one of the half-planes defined by the query line).

Cellular decompositions described so far are generated in such a way that the *number* of $Z$-points in cells on the same level is more or less the same. Another class of cellular decompositions aims to keep the *size* of the cells more or less the same on the same level. Perhaps the most straightforward decomposition of this type is the *box decomposition*. In the plane boxes are squares. The top-level square (containing all $Z$-points) is partitioned into four equal size squares. Each of them is partitioned recursively. Partitioning stops at some prespecified level or when a square contains a fixed number of $Z$-points. This kind of decomposition generalizes in a natural way to higher dimensions. We will see applications of box decomposition in Subsections 5.1.4 and 5.2.1.

In some cases the plane is partitioned into wedges rooted at $Z$-points. Each $Z$-point is a root of a fixed number of wedges. If wedges are sufficiently narrow, the search for nearest neighbor to a given $Z$-point $z$ can be confined to a constant number of $Z$-points within each wedge rooted at $z$. This decomposition generalizes in a natural way to higher dimensions (with cones rooted at $Z$-points.) We will see applications of this kind of *cone decompositions* in Sections 5.1 and Subsection 5.2.1.

### 4.2.7 Other Geometric Data Structures

An *interval tree* introduced by Edelsbrunner [74] is a data structure suitable to represent parallel line segments. It is in particular well-suited in connection with the plane sweep algorithm for finding all intersecting pairs of $n$ rectangles with sides parallel to the coordinate axes. For a detailed description of this data structure, the reader is referred to [166]. An alternative data structure supporting representation of parallel line segment is a *priority search tree* suggested by McCreight [144]. Recall that segment trees introduced in Subsection 4.2.3 are also suitable to represent parallel line segments.

A *quadtree* is a data structure which can be used to organize a rectangular grid of $n \times m$ cells determined by $n+1$ horizontal lines and $m+1$ vertical lines. It can be considered as a 2-dimensional generalization of segment trees. Assume for simplicity that $n = m = 2^k$ for some integer $k$. The quad-tree $T$ is a quaternary tree where the entire grid is associated with the root (which is on level 0). On level $i$, $0 \leq i < k-1$, the grid associated with a node of $T$ on level $i$ is subdivided into four subgrids determined by horizontal and vertical bisecting lines. Since $T$ has $O(n^2)$ leaves, its construction takes $O(n^2)$ time and space.

*Ham-sandwich trees* [80] are a refinement of polygon trees by taking advantage of the fact that bisecting lines do not need to be orthogonal. In addition to answering line queries, ham-sandwich trees can be used to answer triangular range queries (see also [76]).

*Octant trees* [215] are generalizations of ham-sandwich trees to three dimensions. $Z$-points are partitioned using 3 planes into 8 subsets each containing at most $n/8$ $Z$-points. Smaller subsets are partitioned recursively.

## 5. Geometric $\mathcal{TND}$ Problems

In this section we will examine the geometric $\mathcal{TND}$ problems according to the following tree decomposition categories:

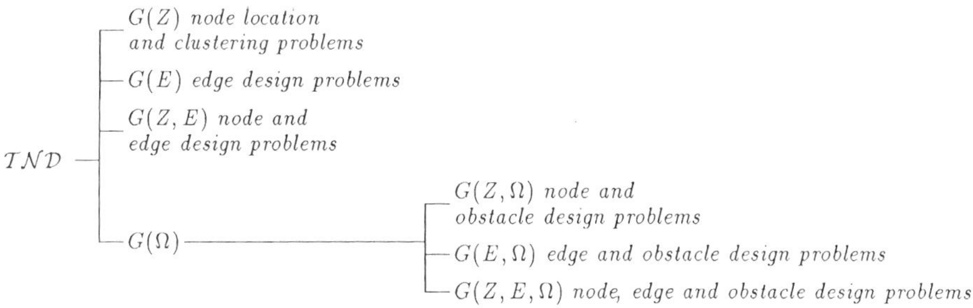


**Figure 14: Morphological Tree for Geometric $\mathcal{TND}$ Problems**

The three sub-categories of the obstacle problems $G(\Omega)$ will be addressed in a single section (5.4) since the number of research efforts has been limited. We will discuss the most important problems in each category. We will identify their lower bounds and describe their algorithms and heuristics. In particular, we will point out what $\mathcal{CG}$ paradigms and geometric data structures proved useful in connection with these algorithms. We will also discuss the worst-case time complexity of these algorithms. We will conclude the discussion of each category by providing a classification table illustrating the importance of particular paradigms and geometric data structures. Further classification will be given in Section 6.

## 5.1 $G(Z)$ Problems

$G(Z)$ problems are considered part of $\mathcal{TND}$ since once a new facility is located, it begins to establish road, communication and utility network relationships critical to the other $Z$−points in the plane. Thus optimal location and clustering of $Z$−points is central to $\mathcal{TND}$. In fact, $G(Z)$ problems seem to fall into two main subcategories:

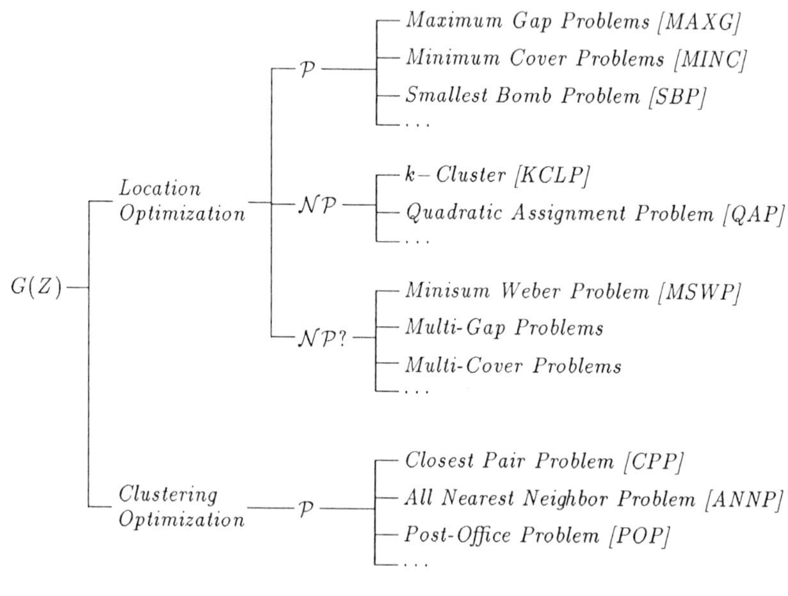


**Figure 15: Morphological Tree for $G(Z)$ Problems**

- *Location Optimization Problems* where we wish to locate one or several new points so that some objective function involving distances to $Z$-points is minimized (*cover problems*) or maximized (*gap problems*). New points must usually satisfy some side constraints. A typical side constraint which arises in connection with gap problems is that the new points must be located within some bounded region enclosing all $Z$-points.

The simplest cover (respectively gap) problem is to find a new point which serves as a center of a smallest (respectively largest) possible disc (or other regular geometric object such as a cone, diamond or ellipse) containing all (respectively none) $Z$-points. In the MHS discussed in Section 2, we might for example wish to locate a new elevator to service a cluster of workstations located on different floors of the manufacturing facility.

The objective function of location optimization problems is usually nonlinear. This normally leads one to consider mathematical programming or numerical approaches to solve these problems. As we shall argue below, $\mathcal{CG}$ offers novel and powerful alternative approaches.

- *Cluster Optimization Problems* where we wish to geometrically characterize the set of $Z$-points so that subsequent queries about proximity of $Z$-points to other $Z$-points or to new points can be answered efficiently. For instance, when solving the *Euclidean traveling salesman problem*, we often need to identify the closest pair of $Z$-points or the closest neighbor for every $Z$-point. Alternatively, when new families (new points) move into an urban area, they must be assigned to schools, medical services, or other public utilities ($Z$-points).

  A geometrical characterization of $Z$ with respect to a particular type of queries is usually obtained by an appropriate subdivision of the space into regions. Consequently, clustering optimization problems are closely related to the geometric data structures which can represent such subdivisions. Furthermore, these geometric data structures must be relatively easy to construct, require little space, and facilitate quick answers to multiple queries.

The morphological tree shown in Figure 15 indicates some of the most important $G(Z)$ problems. Both location and cluster optimization problems are further subdivided into polynomial and $\mathcal{NP}$-complete problems.

Within this subsection, we discuss two representative location problems which are the *maximum gap* (MAXG) problem and the *minimum cover* (MINC) problem. Following this we discuss three representative clustering problems: *Closest Pair Problem* (CPP), *All Nearest Neighbors Problem* (ANNP) and the *Post Office Problem* (POFP).

### 5.1.1 Maximum Gap Problem

Consider the following problem which in the operations research literature is usually referred to as the *maxi-min* problem, Figure 16.

The simplest form of the maxi-min problem is the *Maximum Gap* (MAXG) problem which arises when all weighting factors are equal. This problem has an obvious geometric formulation when $p = 2$. It asks for the largest circle with center $z$ in $R$, and with no $Z$-point inside (Figure 16).

- **Given:** $Z = \{z_1, z_2, ..., z_n\} \in E^2$ with the $L_p$-metric, non-negative weighting factors $\{w_1, w_2, ..., w_n\}$, and a bounded region $R$ containing all $Z$-points.

- **Find:** $z = (x, y) \in R$ such that:

$$\min_{1 \leq i \leq n} \{w_i[|x_i - x|^p + |y_i - y|^p]^{1/p}\}$$

is maximized.

- **Data Structure:** The underlying Voronoi Diagram is illustrated along with the center of the largest circle corresponding to a Voronoi point of the point set. For this example we assumed all $w_i$ were equal and $p = 2$.

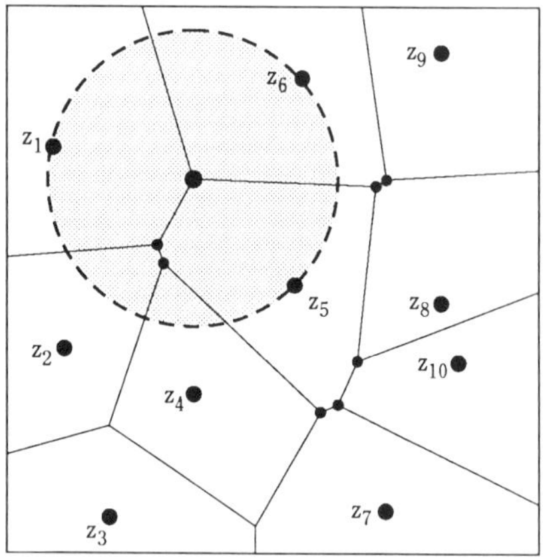


**Figure 16: Maximum Gap Solution**

Unequal weighting factors represent the relative incompatibility between the existing $Z$-points and the new undesireable point $z$. The general maxi-min problem is much more difficult than the equally weighted MAXG problem.

The bounded region $R$ is often identical with $CH(Z)$. Although the $L_2$-metric is the most prevalent setting, other metrics ($L_1$- and $L_\infty$-metrics in particular) have also been considered in the literature.

**Applications.** There are many applications for the maxi-min and MAXG problems. The following are just a few of the most illustrative examples.

- Location of nuclear power plants and other noxious public facilities within a large geographic region.

- Location of commercial enterprises to minimize competition from similar commercial outlets.

- Location of military facilities, telecommunication equipment, or other equipment for communication or defensive reasons.

- Location of equipment or processes in a facility to minimize possible hazards (i.e. melting metal, radiant energy, flying debris), stresses (i.e. noise, vibration, heat) or contaminents [147].

**Lower Bounds.** The MAXG problem for the special case with all $Z$-points on the $x$-axis has been shown to require $\Omega(n \log n)$ time in the linear decision-tree model [141] and in the algebraic decision-tree model [134] (see [166] for the definition of these models). However, this problem can be solved in $\Theta(n)$ time if *floor functions* $\lfloor$ and $\rfloor$ are permitted.

**Algorithms.** When all $Z$-points are on the $x$-axis, a $\Theta(n)$ algorithm for the MAXG problem was developed by Gonzalez [95]. First, $n-1$ equally-sized buckets dividing the interval between the leftmost and rightmost $Z$-points are created. The remaining $n-2$ $Z$-points are placed in their appropriate buckets (this involves floor functions). The leftmost and rightmost $Z$-point in each bucket $i$, $i = 1, 2, ..., n-1$, is retained in $LOW[i]$ and $HIGH[i]$ respectively. Since $n-2$ $Z$-points are placed in $n-1$ buckets, at least one bucket remains empty. Consequently, the largest gap cannot occur between two $Z$-points in the same bucket. A linear time scan of $LOW$ and $HIGH$ reveals the largest gap. Further algorithmic details occur in [166]. This algorithm is unfortunately not generalizeable to two dimensions.

An $O(n \log n)$ algorithm for the MAXG problem in $E^2$ with $R = CH(Z)$ was developed by Shamos [179]. The algorithm is based on the observation that $z$ is either a Voronoi point of $VOR(Z)$ or it occurs at an intersection of $VOR(Z)$ with $CH(Z)$. There is $O(n)$ Voronoi points and $O(n)$ intersections of $VOR(Z)$ with $CH(Z)$. Furthermore, these intersections can be determined in $O(n)$ time if $VOR(Z)$ is represented by a DCEL.

There are many classical mathematical programming approaches to the maxi-min problem. Melachrinoudis developed $O(n^4)$ algorithms for the general weighted case [147, 149] where he employs the Kuhn-Tucker conditions of mathematical programming. An $O(n^3)$ algorithm for the MAXG problem was developed by Dasarthy and White [58, 59].

There exists a $\Theta(n^2)$ algorithm for constructing the weighted (with multiplicative distance function) Voronoi diagram in $E^2$ (see Subsection 4.2.2). Its application to the location problems has not been reported upon. Given this weighted Voronoi diagram, one should be able to achieve an improved algorithm over Melachrinoudis' $O(n^4)$ algorithms utilizing a $\mathcal{CG}$ approach.

**Heuristics.** Melachrinoudis and Cullinane [147, 148] discussed some $\mathcal{CG}$ heuristics and extensions of the maxi-min problem. In particular, they developed a heuristic for the maxi-min problem which performs on average $O(n^2)$. No formal proof of its asymptotic complexity either in the average or worst-case exists.

- **Step 1:** The feasible region $R$ is partitioned into narrow vertical strips and the points inside each strip are sorted along the $y$-axis.

- **Step 2:** For each of the vertical strips, every $Z$-point is combined with any two other $Z$-points which are within a specified threshold distance $r$. For each acceptable triple of $Z$-points, the intersection of conical equations derived from the Kuhn-Tucker conditions is found and the same tests for optimality as in the exact algorithm are applied. All local maxima in each strip are calculated and stored in a list.

- **Step 3:** A search for local maxima along the boundary of $R$ is undertaken.
- **Step 4:** The best solution found in **Step 2** and **Step 3** is selected.

Further details for the size and number of the strips and the determination of $r$ occur in [147, 148]. The solutions obtained by this heuristic seem to be very close to optimal solutions.

It is worth mentioning at this point that some improvements in the above heuristic could reduce its complexity while yielding better solutions. Construction of the weighted Voronoi diagram for $Z$ (see Subsection 4.2.2) would make the vertical strips unnecessary. Furthermore, triples determined in **Step 2** could be more easily identified from the locus information available from the weighted Voronoi diagram.

### 5.1.2 Minimum Cover Problem

Consider the following problem, Figure 17, which in the operations research literature is usually referred to as the *mini-max* problem. It can be considered as dual to the maxi-min problem.

- **Given:** $Z = \{z_1, z_2, ..., z_n\} \in E^2$ with the $L_p$-metric, non-negative weighting factors $\{w_1, w_2, ..., w_n\}$,
- **Find:** $z = (x, y)$ such that:

$$\max_{1 \le i \le n} \{w_i[|x_i - x|^p + |y_i - y|^p]^{1/p}\}$$

  is minimized.

- **Data Structure:** We illustrate the convex hull data structure which is utilized in an algorithm designed by Shamos to solve the MINC problem. The optimal solution actually corresponds to the midpoint of the diameter of the point set after constructing the $CH(Z)$.

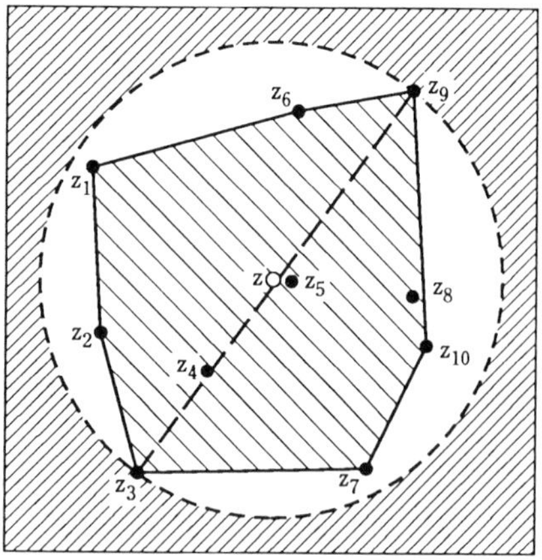


**Figure 17: Minimum Covering Solution**

The *minimum cover* (MINC) problem arises when all weighting factors are equal to 1. This problem has then an obvious geometric formulation when $p = 2$. It asks for the smallest circle with all $Z$-points inside Figure 17.

**Lower Bounds.** $\Omega(n)$ is a trivial lower bound for the MINC problem. As will be seen below, there exists a $\Theta(n)$ algorithm for this problem.

**Algorithms.** A straightforward $O(n^4)$ algorithm for the MINC problem is to construct a circle for every triple of $Z$-points and check whether the remaining $n-3$ $Z$-points are inside. Elzinga and Hearn [81] improved this bound to an $O(n^2)$ algorithm by reducing the number of triples that need to be checked. We will not describe this algorithm as more efficient approaches are available.

The smallest circle enclosing all $Z$-points is unique. Furthermore, it goes either through at least three $Z$-points or through two $Z$-points defining the diameter. Shamos [179] used this fact to obtain an $O(n \log n)$ algorithm.

- **Step 1:** Determine $CH(Z)$ in $\Theta(n \log n)$ time (see Subsection 4.2.1).
- **Step 2:** Determine two farthest apart extreme points $z_i$ and $z_j$ in $CH(Z)$. This can be done in $\Theta(n)$ time [166]. If the circle with $z_i z_j$ as its diameter contains all $Z$-points, then $z$ occurs at the midpoint of $z_i z_j$. **STOP.**
- **Step 3:** Otherwise, construct $VOR_{n-1}(Z)$ in $\Theta(n \log n)$ time (see Subsection 4.2.2).
- **Step 4:** Let $z$ be a Voronoi point in $VOR_{n-1}(Z)$ such that its three determining $Z$-points are farthest away from $z$. Clearly, $z$ can be found in $O(n)$ time. **STOP.**

The **MINC** problem can be solved in $\Theta(n)$ time by transforming it into a special variant of the three-variable quadratic programming problem. Particular characteristics of this problem permit us to solve it by the prune-and-search technique applied to two- and three-variable linear programming problems. Before discussing the algorithm for the MINC problem, we will briefly discuss two- and three-variable linear programming problems.

One way to solve a two-variable (respectively three-variable) linear programming problem with $n$ constraints is two determine the common intersection of $n$ half-planes (3-dimensional half-spaces) determined by the constraints. This can be done in $O(n \log n)$ time [166]. The optimal solution will occur in one of the extreme points of this convex intersection. It can be located in $O(\log n)$ time. Hence, this algorithm requires $O(n \log n)$ time. Note that once the convex intersection is known, any linear programming problem with the same set of constraints (but varying objective function) can be solved in $O(\log n)$ time.

As shown independently by Megiddo [146] and Dyer [73], two- and three-variable linear programming problems can be solved without the construction of the convex intersection of $n$ half-planes (3-dimensional half-spaces). The optimum solution occurs at the intersection of two half-planes (three 3-dimensional half-spaces) defined by two (three) constraints. The remaining constraints are redundant and can be pruned away. The prune-and-search algorithms for two-variable (three-variable) linear programming problems consist of a number of iterations. At each iteration, at least a constant fraction $\alpha$ of remaining constraints are identified as redundant. In fact, the value $\alpha = 1/4$ can be achieved. Consequently, the number of constraints at the beginning of each iteration is at most $(1-\alpha)^{i-1}n$. Identification of redundant constraints during the $i$-th iteration requires $O((1-\alpha)^{i-1}n)$ time using a technique closely related to the technique in the marriage-before-conquest algorithm for convex hulls (see Subsection 4.2.1). It can be easily verified that after at most $\log_{1/(1-\alpha)} n$ iterations all redundant constraints will be pruned away and the optimal solution can be found in constant time. It can also be easily verified that the total time used by the pruning process is $O(n)$. It follows that a two-variable (three-variable) linear programming problem can be solved in $\Theta(n)$ time. Note however, that if the objective function is changed while the original constraints remain the same, $O(n)$ time is needed. The intersection algorithm required only $O(\log n)$ time. For further details on the linear programming variant, the reader is referred to original papers of Megiddo or Dyer, or to the excellent description in [166].

We are now in position to return to the MINC problem. The transformation is obtained as follows. By introducing a new variable $\zeta$, the MINC problem can be reformulated to:

$$\text{minimize } \zeta$$

subject to

$$\zeta \geq (x_i - x)^2 + (y_i - y)^2 \quad \forall\, z_i \in Z$$

Quadratic constrains can be rewritten to:

$$\zeta \geq -2x_i x - 2y_i y + (x_i^2 + y_i^2) + (x^2 + y^2) \quad \forall\, z_i \in Z$$

If the quadratic term $(x^2 + y^2)$ is removed from each constraint, the set of resulting linear constraints defines a polyhedron, whose boundary is described by $\zeta = f(x, y)$ with $f$ a downward-convex function. Since $(x^2 + y^2)$ is also a downward-convex function, so is $f(x,y) + (x^2 + y^2)$. The surface determined by $f(x,y) + (x^2 + y^2)$ resembles a polyhedral surface. More specifically, its "faces" are portions of surfaces of paraboloids, and its vertices are intersections of triples of paraboloids. In order to locate the minimum, we only need to remove redundant constraints (paraboloids). The $\Theta(n)$ prune-and-search algorithm for the three-variable linear programming problem extends in a straightforward manner to this special case of a three-variable quadratic programming problem. When three non-redundant constraints have been identified, the vertex in which they intersect may not be the minimum (since "faces" are portions of surfaces of paraboloids). However, it suffices to search the "edges" incident with this vertex (formed by pairwise intersections of paraboloids) in order to find the minimum.

**Related Problems.** Shamos [179] gave an $O(n^4)$ time algorithm for the *smallest bomb problem* (SMBP):

- **Given:** $Z = \{z_1, z_2, \ldots, z_n\} \in E^2$,
- **Find:** Smallest circle enclosing least $k$ of $n$ given $Z$-points.

and discussed the design of algorithms and heuristics for the $k$-Cluster location problem using $\mathcal{CG}$. The MAXG, MINC, Weber, Mini-Sum, Maxi-Sum, and Steiner problems (see Subsection 5.3.1) are closely related because of the complexity nature of the unconstrained objective function in the mathematical programming formulation of the problems [13, 137]. Also, the body of literature relating mini-max, maxi-min and related location problems on networks is of interest [94].

### 5.1.3 Closest Pair Problem

The first clustering problem which will be discussed is the *closest pair problem* (CPP). It is formulated as follows in Figure 18.

- **Given:** $Z = \{z_1, z_2, ..., z_n\} \in E^2$,
- **Find:** A pair of $Z$-points which is closest to each other.
- **Data Structure:** Here we illustrate the first step of the divide-and-conquer approach (described below) for resolving the CPP. The closest pair of points is also indicated in the figure.

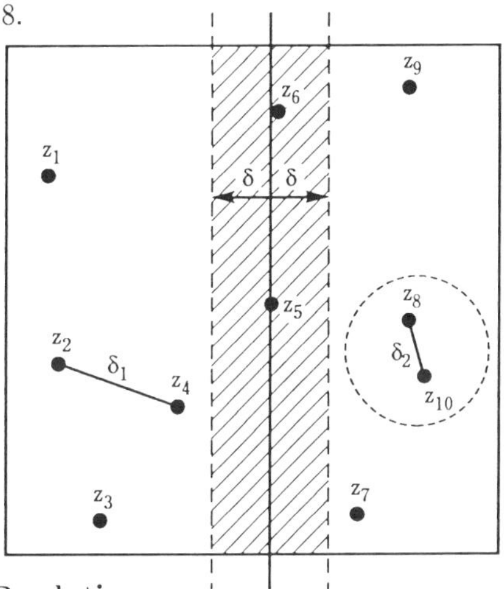


Figure 18: CPP solution

**Applications.** This is a critical problem in certain real-time applications such as determining the closest pair of airplanes in an air traffic control display.

**Lower Bounds.** The problem of deciding whether a set of $n$ real numbers contains two identical numbers is linear-time transformable to the CPP. Map the $n$ numbers onto the $x$-axis and regard them as $Z$-points (with $y$-coordinates equal to 0). Solve the CPP. If the distance between the closest pair is nonzero, all numbers are distinct. It is well-known that the *element uniqueness problem* requires $\Omega(n \log n)$ time [63]. Hence, this is also the case for the CPP.

**Algorithms.** The CPP can be solved in $O(n^2)$ time by a straightforward algorithm which compares distances of all pairs of $Z$-points.

A $\Theta(n \log n)$ divide-and-conquer algorithm for the CPP was suggested by Bentley and Shamos [24] and Bentley [20]. The algorithm is as follows.

- **Step 1:** Partition $Z$ into two approximately equal-size subsets $Z_1$ and $Z_2$ by a vertical median line $L$. This can be done in $\Theta(n)$ time.
- **Step 2:** Solve the CPP recursively for $Z_1$ and $Z_2$. Let $\delta = \min\{\delta_1, \delta_2\}$ where $\delta_1$ and $\delta_2$ denote the distance between the closest pair in $Z_1$ and in $Z_2$ respectively.
- **Step 3:** Remove all $Z$-points at distance $\delta$ or more from $L$. Let $Z_1'$ and $Z_2'$ denote the remaining $Z$-points sorted by their $y$-coordinate. This can be done in $\Theta(n)$ time assuming that entire $Z$ is initially presorted by $y$-coordinate in $\Theta(n \log n)$ time.
- **Step 4:** Scan $Z_1'$. For each $z_i = (x_i, y_i) \in Z_1'$, inspect the points of $Z_2'$ with their $y$-coordinate in the interval $]y_i - \delta, y_i + \delta[$. There can be at most 3 such $Z$-points in $Z_2'$. Since both $Z_1'$ and $Z_2'$ are presorted by $y$-coordinate, finding the closest pair with one $Z$-point belonging to $Z_1$ and the other $Z$-point belonging to $Z_2$ can be done in $O(n)$ time. Let $\delta'$ denote the distance between such closest pair.
- **Step 5:** Select the pair with distance $\min\{\delta, \delta'\}$ as the solution.

Bentley and Shamos [24] and Bentley [20] proved that the divide-and-conquer algorithm can be generalized to solve the CPP in $d$-dimensional space, $d \geq 2$, in $\Theta(n \log n)$ time.

The line segment connecting the closest pair of $Z$-points always belongs to $DT(Z)$. Since $DT(Z)$ can be constructed in $\Theta(n \log n)$ time and contains $O(n)$ line segments, another $\Theta(n \log n)$ algorithm for the CPP thus follows.

### 5.1.4 All Nearest Neighbors Problem

The *all nearest neighbor problem* (ANNP) can be formulated as follows, Figure 19.

**Applications.** Among some of the applications of the **ANNP** are: ecological mapping of the territoriality of species, geography, marketing science, urban planning, particle and solid state physics.

**Lower Bounds.** The ANNP is a generalization of the CPP discussed in the previous subsection. Consequently, the lower bound $\Omega(n \log n)$ for the CPP carries over to the ANNP.

**Algorithms.** The ANNP can be solved in $O(n^2)$ time by a straightforward algorithm which compares distances of each $Z$-point to the remaining $n - 1$ $Z$-points.

The line segments connecting each $Z$-point to its nearest neighbor form an *all nearest neighbor graph*, denoted by $ANNG(Z)$. The $ANNG(Z)$ is a subgraph of $DT(Z)$.

- **Given:** $Z = \{z_1, z_2, ..., z_n\} \in E^2$,
- **Find:** Nearest $Z$-point to each $Z$-point where $z_j$ is a nearest neighbor of $z_i$ if:

$$d_2(z_i, z_j) = min_{z_k \in Z - z_i}\{d_2(z_i, z_k)\}$$

- **Data Structure:** The cubical box data structure is illustrated for solving the problem based on Vaidya's algorithm. Notice that the ANNP solution illustrated by the curved segments is not symmetric. Another alternative data structure useful here is DT(Z).

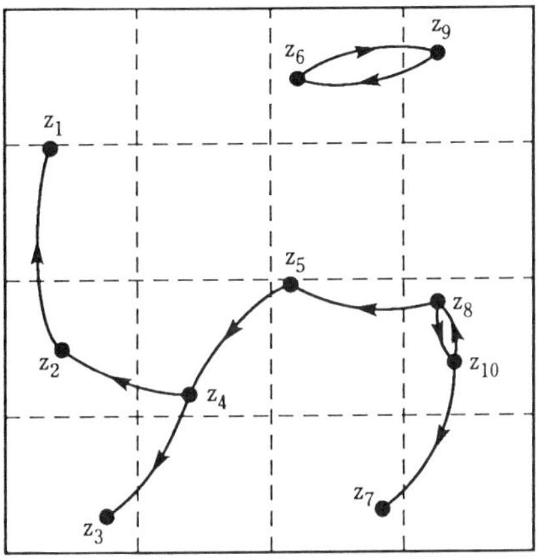


**Figure 19: ANNP Solution**

Since $DT(Z)$ has $O(n)$ edges, it follows immediately that the ANNP can be solved in $\Theta(n \log n)$ time. This algorithm was originally suggested by Shamos and Hoey [180].

Bentley and Shamos [24] and Bentley [20] utilized multi-dimensional divide-and-conquer to develop an $O(n(\log n)^{d-1})$ algorithm for the ANNP in a $d$-dimensional space, $d \geq 2$.

Vaidya [196, 199] developed a $\Theta(n \log n)$ algorithm for the ANNP in a $d$-dimensional space, $d \geq 2$. It partitions $Z$-points into $d$-dimensional cubes. These are further subdivided with neighborhood information around each $d$-cube. The partitioning process is repeated until at most one $Z$-point lies inside each $d$-cube. The nearest neighbors are obtained from the neighborhood information for each $d-$cube. This approach is also generalizable to the all $k$-th nearest neighbor problem yielding an $O(kn \log n)$ algorithm

Similar ideas and techniques have been developed by Clarkson [50] and Gabow et al. [85]. They suggested $O(n \log \delta)$ algorithms where $\delta$ is the ratio of the maximum and minimum distance between pairs of $Z$-points.

### 5.1.5 Post Office Problem

While the above clustering problems allocate $Z$-points to other $Z$-points, the *post office problem* (POFP) allocates a priori unknown query points to $Z$-points, Figure 20.

In many applications, the Euclidean distance between query points and $Z$-points is overly simplistic to capture the proximity relations. Other distance criteria and additional constraints are often needed to model more realistic situations.

**Applications.** There have been numerous applications of the POFP. Examples include information retrieval (database queries) and pattern recognition (classification problems). The POFP also arises as a subproblem in many other $\mathcal{CG}$ problems.

- **Given:** $Z = \{z_1, z_2, ..., z_n\} \in E^2$, and a query point $q$,
- **Find:** $Z$-point nearest to $q$.
- **Data Structure:** The Slab data structure of Dobkin and Lipton is illustrated for the POFP problem along with the underlying Voronoi diagram. The slab data structure is constructed on the point set of the VOR(Z).

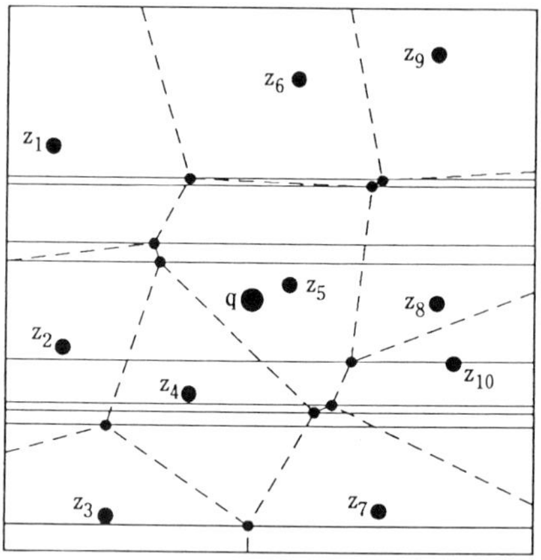


**Figure 20: POFP Solution**

**Lower Bounds.** If the number of queries is limited, it usually does not pay to preprocess the $Z$-points so that the queries can be answered more efficiently. Hence, $\Omega(n)$ is a trivial lower bound.

Typically, there will be many query points. In order to allocate them efficiently, the $Z$-points have to be preprocessed in some way. This preprocessing usually results in some kind of a geometric data structure. The query time is naturally dependent on how much time and in particular how much space is needed to construct the geometric data structure during the preprocessing phase. Since the geometric data structure must capture the location of all $Z$-points, it is reasonable to assume that it must require $\Omega(n)$ space.

Consider the POFP in 1-dimensional space (i.e. $Z$-points are all located on the $x$-axis). Independent on what kind of preprocessing is applied to the $Z$-points, finding nearest $Z$-point to a query point will require $\Omega(\log n)$ time. Hence, $\Omega(\log n)$ is a lower bound for the query time in any dimension.

**Algorithms.** Algorithms for the POFP subdivide the plane into regions such that any query point falling into the same subregion has the same nearest $Z$-point. As already discussed in Subsection 4.2.2, Voronoi diagrams provide this kind of subdivision. They can be constructed in $\Theta(n \log n)$ time using $\Theta(n)$ space. Once $VOR(Z)$ is given, the subregion containing a query point can be identified in $\Theta(\log n)$ time using any of the point location methods described below. Some of these point location methods may require that $VOR(Z)$ is further subdivided into smaller regions.

When other distance criteria and additional constraints are to be taken into account, variants of Voronoi diagrams (see Subsection 4.2.2) can be used. Point location in these variants can usually also be carried out in $\Theta(\log n)$ time.

Several approaches to point location problems based on various kinds of planar subdivisions have been suggested in the literature. They will be briefly discussed here. Detailed descriptions can be found in [166].

- The *slab* method developed by Dobkin and Lipton [62] further partitions a given planar subdivision $G$ into trapezoids in the preprocessing phase. Draw horizontal lines through $Z$-points. Sides of trapezoids within each slab (i.e. pair of neighboring horizontal lines) do not intersect and can be ordered from left to right. In order to find a region within which a query point $q$ falls, it suffices to find a trapezoid containing $q$. This can be done in $\Theta(\log n)$ time by two binary searches:

  1. *Find the slab containing $q$.*
  2. *Find the side of a trapezoid with $q$ immediately to its right.*

  The weakness of this intuitively simple approach is that there can be $O(n^2)$ trapezoids. Both preprocessing and space requirements are of this order.

- The *trapezoid* method developed by Preparata [164] is a refinement of the slab method. It reduces the preprocessing and space requirements to $\Theta(n \log n)$ while preserving the $\Theta(\log n)$ query time. The basic idea is to avoid cutting long line segments of $G$ into small pieces as is the case in the slab method. Assume w.l.o.g. that the exterior face of $G$ is a trapezoid. Let $L$ be a horizontal line through a $Z$-point $z$ such that roughly half of the $Z$-points is above $L$. Any line segment intersected by $L$ is divided into two line segments by introducing a new point at the intersection. The upper part of $G$ is then scanned horizontally (the same is subsequently done with the lower part of $G$). Every line segment spanning the upper part of $G$ from the top to $L$ is identified. These spanning line segments form a sequence of trapezoids. Each trapezoid can be empty or it can contain some $Z$-points. In the latter case, the trapezoid is recursively subdivided. The entire decomposition can be represented by a binary tree with internal nodes corresponding to splitting horizontal lines and spanning line segments, and with leaves corresponding to empty trapezoids. Furthermore, the binary tree can be balanced in such a way that subsequent queries will involve $O(\log n)$ discriminations against the splitting horizontal lines and spanning line segments.

- The *triangulation* method was developed by Kirkpatrick [119]. Assume w.l.o.g. that the exterior face of $G$ is a triangle. Let $\mathcal{T}_1$ denote a triangulation of $G$. It is always possible to find in $\mathcal{T}_1$ a subset of at least $n/24$ independent (i.e. mutually nonadjacent) interior $Z$-points of degree less than 12. These $Z$-points and their incident line segments are removed. Non-triangle faces are triangulated yielding a coarser triangulation $\mathcal{T}_2$. This process of generating coarser and coarser triangulations is repeated until a triangulation $\mathcal{T}_k$ consisting of three exterior $Z$-points and a single triangle is obtained. At least a constant fraction of $Z$-points is deleted at each iteration. Consequently, $k = O(\log n)$. Given $\mathcal{T}_1, \mathcal{T}_2, ..., \mathcal{T}_k$, a directed acyclic graph $D$ with $k$ levels is generated. Vertices on level $i$, $1 \leq i \leq k$, represent triangles of $\mathcal{T}_{k-i+1}$. There is an arc from a vertex on level $i$, $1 \leq i \leq k-1$, to a vertex on level $i+1$ iff the intersection of the corresponding triangles is nonempty. Each vertex has at most 10 outgoing arcs. $D$ can be constructed in $O(n \log n)$ time using $O(n)$ space.

Given $D$, point location starts at level 1. If the query point $q$ is outside of the exterior triangle, we are done; $q$ belongs to the exterior face of $G$. Otherwise, descendants on level 2 are examined in $O(1)$ time to identify the unique triangle containing $q$. This process continues until a triangle on level $k$ containing $q$ is reached. It follows that each query requires $\Theta(\log n)$ time.

- The *separating chain* method was developed by Lee and Preparata [130]. $G$ is regularized first. Regular subdivisions can be covered by a set of $k = O(n)$ *monotone* chains (i.e. downward going chains from the top point to the bottom point) which do not intersect but can share some of the line segments. Chains are placed on a balanced binary *chain tree* with the middle chain at the root. Chains to the left (respectively right) of the middle chain are stored in the left (respectively right) subtree. A balanced binary tree is associated with each node of the chain tree. It contains line segments of the chain not covered by any chain higher up in the chain tree. With each line segment we also associate the index of the leftmost and rightmost chain (below in the chain tree) containing it. This representation of chains requires $O(n)$ space and can be obtained in $O(n \log n)$ time.

  In order to locate a query point $q$, it is first discriminated against the root chain. $O(\log n)$ time is needed to decide whether $q$ is to its left or right. As a byproduct of this discrimination, we also obtain the line segment $e$ of the chain intersected by the horizontal line $L$ through $q$. Suppose that $q$ is to the left of the chain. Next discrimination should be against the left-son chain. However, if the leftmost chain through $e$ has index less than or equal to the index of the left-son chain, this discrimination is superfluous, and we can jump directly to the left-grandson chain. This principle is used throughout the search down the chain tree until a leaf node is reached. The query point will be either immediately to its left or right. At most $O(\log n)$ discriminations is required. Each discrimination takes $O(\log n)$ time. Hence, the query time is $O((\log n)^2)$.

  The query time can be further reduced to $\Theta(\log n)$ time using fractional cascading as suggested by Edelsbrunner et al. [77]. Recall that during the first discrimination, the line segment $e$ intersected by a horizontal line $L$ through $q$ is identified. If appropriate *horizontal bridges* between each chain and its left-son and right-son chains are maintained, the line segment of the left-son (or right-son) chain intersected by $L$ can be identified in $O(1)$ time. We will not illustrate fractional cascading in detail here, since the *layered DAG* data structure supporting it is rather complex (see also [76]). A simpler example of fractional cascading has been described in connection with range searching problems in Sub-section 4.2.6.

The POFP in $d$-dimensional spaces, $d \geq 2$, is much more difficult. Dobkin and Lipton [62] and Yao [213] suggested techniques which preserve the $\Theta(\log n)$ query time but require $O(n^{2^{d+1}})$ space and preprocessing time. These bounds remain valid for $L_1$- and $L_\infty$-metrics. For $d = 3$, Chazelle [39] reduced the space and preprocessing time to $O(n^2)$ with $O((\log n)^2)$ query time.

As already indicated in Subsection 4.2.6, multidimensional binary search trees proved very useful in connection with various proximity problems. In particular, they can be used to solve the POFP.

### 5.1.6 Classification of $G(Z)$ Problems

Table 5.1.6 illustrates the range of $\mathcal{CG}$ paradigms and geometric data structures which have been utilized in approaches to the $G(Z)$ problems. Each entry in the table corresponds to an algorithm or a heuristic for a $G(Z)$ problem (indicated by its acronym). Acronyms in italics indicate heuristics or special case algorithms. Algorithms and heuristics are placed according to their most important paradigm and geometric data structure used.

With each entry we also associated its worst-case time complexity. If an algorithm is exponential, we write $O(exp)$. If the complexity of an algorithm or heuristic is unknown, we write $O(?)$. $O_e$ denotes expected running times. $O_o$ denotes empirically observed average running times. We use the notation $O(p(n), s(n), q(n))$ in connection with algorithms where preprocessing is required to solve query problems. Functions $p(n)$, $s(n)$ and $q(n)$ denote worst-case preprocessing time, space and query time respectively. We have also identified the key references associated with these algorithms. This notation will be used throughout all classification tables in the sequel.

## 5.2 $G(E)$ Problems

We are given a set of edges $E = \{e_1, e_2, ..., e_m\}$ representing potential connections between the $Z$-points. We wish to select a subset of edges yielding a network extremizing some objective function subject to side constraints pertaining to the topology of the network. Unless otherwise specified, we will assume that the objective function is the sum of lengths of selected edges. Usually, we are interested in constructing one or several (possibly disjoint) paths, trees, circuits. Determination of survivable networks with respect to various survivability criteria also belongs to this class. The most significant problem of this type is that of determining least cost network with prespecified vertex or edge connectivities between $Z$-points.

Referring back to our MHS (Section 2), we might wish to find the shortest circuit for dispatching a pallet truck from the warehouse to a workstation and its return. In

| | $\mathcal{CG}$ Paradigms | | | | |
|---|---|---|---|---|---|
| **G(Z)** | *ITER* | *PSWE* | *D&C* | *LOCUS* | *P&S* |
| $CH(Z)$ | | | | | |
| $VOR(Z)$ | MAXG: $O(n \log n)$ [179] | | | POFP: $O(n \log n, n, \log n)$ [179, 166, 76] | MINC: $O(n \log n)$ [179] |
| $DT(Z)$ | | | | | ANNP: $\Theta(n \log n)$ [180] |
| $VG(Z)$ | | | | | |
| $CD(Z)$ | MAXG: $\Theta(n)$ [95] | | ANNP: $O(n(\log n)^{d-1})$ [24, 20] | POFP: $O(n^2, n^2, (\log n)^2)$ [39]<br>POFP: $O(n^{2^{d+1}}, n^{2^{d+1}}, \log n)$ [62, 213] | *MAXG*: $O_o(n^2)$ [147, 148]<br>ANNP: $\Theta(n \log n)$ [196, 199]<br>ANNP: $O(n \log \delta)$ [50, 85] |
| $OTHERS$ | MINC: $O(n^2)$ [81]<br>SMBO: $O(n^4)$ [179] | | CPP: $\Theta(n \log n)$ [24, 20] | | MINC: $\Theta(n)$ [146, 73] |

**Table 1: $G(Z)$ Classification Table**

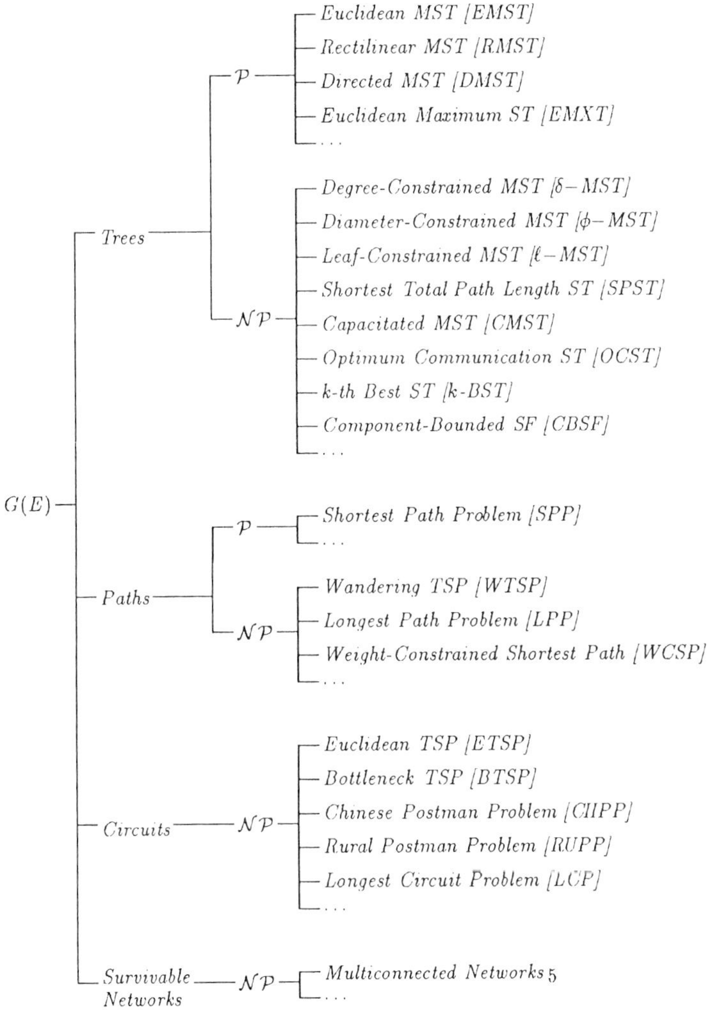


**Figure 21: Morphological Tree for $G(E)$ Problems**

fact, this decision might have to be repeated a number of times during the day, so we wish to ensure an efficient solution to such multiple queries.

The morphological tree shown in Figure 21 indicates some of the problems in the $G(E)$ category. The *Euclidean minimum spanning tree problem* (EMST) and *Euclidean traveling salesman problem* (ETSP) are perhaps the most well-known among them. Most of these problems (except for polynomially solvable tree problems) remain open to the $\mathcal{CG}$ approach. For example, effective heuristics for the ETSP (discussed in detail in Subsection 5.2.2) incorporating $\mathcal{CG}$ paradigms and geometric data structures are only just beginning to be adequately designed and analyzed.

In this section we discuss how the $\mathcal{CG}$ methodology has been applied to some of the most important geometric $\mathcal{TND}$ problems of the class $G(E)$. In Subsection 5.2.1 the *Euclidean minimum spanning tree problem* is discussed. Subsection 5.2.2 deals with the *Euclidean travelling salesman problem*. The *Euclidean maximum spanning tree problem* is discussed in Subsection 5.2.3. Some other problems belonging to the $G(E)$ class are covered in Subsection 5.2.4.

### 5.2.1 Euclidean Minimum Spanning Tree Problem

The *Euclidean minimum spanning tree* (EMST) problem can be formulated as follows, Figure 22.

- **Given:** $Z = \{z_1, z_2, ..., z_n\} \in E^2$,
- **Find:** Minimum length connected network $T(Z)$ spanning $Z$.
- **Data Structure:** The Delaunay Triangulation DT(Z) along with the Voronoi Diagram (its dual) are illustrated and the EMST which is a subgraph of the DT(Z) is clearly demarcated.

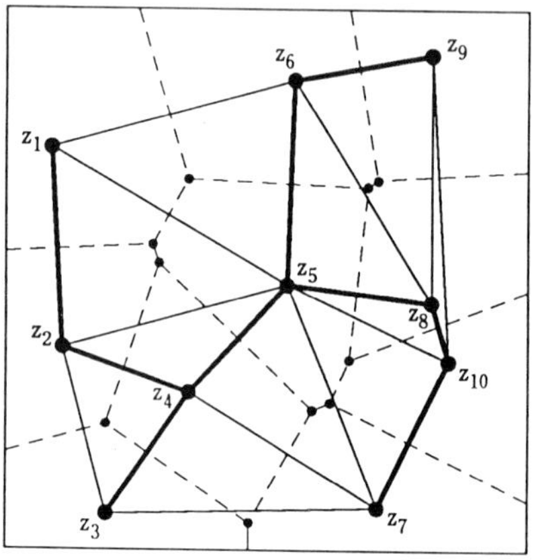


**Figure 22: EMST Solution**

The EMST problem is illustrative of CG approaches to TND problems because full advantage of the geometric nature of the problem has been taken into account in some of the algorithms. This is in particular the case for the EMST problem in higher dimensions.

In many applications it is more appropriate to use the rectilinear $L_1$-metric. Considerable amount of research has been devoted to this *rectilinear minimum spanning tree* (RMST) problem. It is closely related to the $L_\infty$*-metric minimum spanning tree* ($L_\infty$-MST) problem. In particular, there is a distance preserving isometry between $L_1$- and $L_\infty$-metrics [125]. Consequently, existence of polynomial time algorithms for the rectilinear case implies the existence of polynomial time algorithms for the $L_\infty$-metric and vice versa. This observation applies also to other "distance-based" problems such as the travelling salesman problem (Subsection 5.2.2) and Steiner minimal tree problem (Subsection 5.3.1). Some work has also been done on the EMST problem with $L_p$-metrics, $p \in ]1, 2[$.

**Applications.** There is a vast number of different applications of the EMST problem. It is a common component in many applications involving construction of communication, transportation and distribution networks. It is used in pattern recognition (e.g. two and three-dimensional clustering). It also plays a very important role in VLSI design (rectilinear case in particular),

**Lower Bounds.** Sorting is transformable to the EMST problem in $E^2$ in linear time [166]. Hence $\Omega(n \log n)$ is a lower bound for the EMST problem. This lower bound in fact applies to all $L_p$-MST problems, $p \in [1, \infty]$ in every dimension $d$, $d \geq 2$ [125].

**Algorithms.** An $\Theta(n \log n)$ algorithm for the EMST problem in $E^2$ is based on the observation that at least one $T(Z)$ is a subgraph of the Delaunay triangulation $DT(Z)$ [179]. Various $\Theta(n \log n)$ algorithms for the construction of $DT(Z)$ (either directly or via dualization of Voronoi diagrams) are available (see Subsection 4.2.3). Since $DT(Z)$ is planar, $T(Z)$ can be found in $\Theta(n)$ time once $DT(Z)$ is given [45]. As already mentioned in Subsection 4.2.3, the Gabriel graph $GG(Z)$ and the relative neighborhood graph $RNG(Z)$ are subgraphs of $DT(Z)$. Both contain at least one $T(Z)$. Hence, also these geometric data structures lead to $\Theta(n \log n)$ algorithms for the EMST problem in $E^2$.

Chang and Lee [35] proposed a divide-and-conquer approach to the EMST problem in $E^2$ which constructs $DT(Z)$ and identifies the edges of $T(Z)$ in one go. The original $\Theta(n \log n)$ worst-case time complexity claim was subsequently disproved by Stojmenovic and Langston [190]; the algorithm requires in fact $O(n^2 \log n)$ time. Consequently, the existence of a $\Theta(n \log n)$ algorithm for the EMST in $E^2$ which does not employ graph algorithms is an open problem.

It has been shown [125, 132] that the RMST, and more generally $L_p$-MST, $p \in [1, \infty]$, can be found by constructing a geometric dual to the rectilinear Voronoi diagram. It is planar and has at least one rectilinear $T(Z)$ as a subgraph. Hence, $T(Z)$ can be found in the dual in $O(n)$ time. Although the construction of rectilinear Voronoi diagrams is more complicated than the construction of Euclidean Voronoi diagrams, it can still be done in $\Theta(n \log n)$ time.

While the EMST and RMST problems in $E^2$ can be considered as solved, constructions of $T(Z)$ in higher dimensions can still be considered as not completely understood. In particular, pruning approaches based on constructions of Delaunay triangulations, Gabriel graphs, or relative neighbourhood graphs cannot be generalized to higher dimensions.

Bentley and Friedman [22] presented an EMST algorithm in arbitrary $E^d$, $d \geq 2$. A simplified version of this algorithm was given by Nevalainen et al. [156].

Katajainen [117] proved that the worst-case time complexity of this algorithm is $O(n^2 \log n)$. He also demonstrated that there are problem instances for which this bound is reached. The technique is as follows. Initially, the solution consists of $n$-subtrees, each being an isolated point. A priority queue (heap) is associated with each subtree. It stores the closest "foreign" subtree for each $Z$-point (in nondecreasing order of their distances). At each iteration, the shortest subtree is joined to its nearest neighbour. As a consequence two priority queues have to be merged. Furthermore, some of the entries in the merged priority queue must be updated. To facilitate the updating, a multidimensional binary search tree of $Z$-points is maintained. The strategy of selecting the shortest subtree at each iteration proved to keep the number of necessary priority queue updates relatively low.

Yao [213] developed another algorithm for the EMST problem in $E^d$, $d \geq 2$. It prunes all but $O(n)$ line segments away such that the remaining graph contains at least one $T(Z)$. The algorithm is also applicable to the $L_1$-MST and $L_\infty$-MST problems. The approach is as follows. A family of cones $C_z$ rooted at $z$ is generated for every $Z$-point $z$. A nearest neighbour to $z$ is determined within each cone of $C_z$. Let $E_z$ denote the set of line segments connecting $z$ with its nearest neighbour in each cone. If cones are appropriately chosen (for $d = 2$ it suffices to take $45^o$ wedges), it can be shown that $T(Z) \subseteq \cup_{z \in Z} E_z$. The algorithm requires $O(n^{2-\alpha}(\log n)^{1-\alpha})$ time for any fixed $d \geq 2$ where $\alpha = 2^{-d-1}$. For $d = 3$ this bound can be further improved to $O((n \log n)^{1.5})$ due to a better algorithm for the post-office (nearest neighbour query) problem in $E^3$ given by Chazelle [39].

Vaidya [197] suggested another "safe" pruning algorithm. The space enclosing $Z$-points (e.g. unit $d$-cube) is partitioned into $2^{di}$ identical cubical boxes with side-length $l_i$. This is done for $i = 1, 2, ..., \delta$, where $\delta$ is an appropriately chosen integer. If two boxes $b_i^1$ and $b_i^2$ on level $i$ satisfy the following conditions:

- no pair of $Z$-points (one in each box) is less than $l_i/3$ and more than $l_i$ apart,
- there is no $Z$-point $z$ such that its distance to farthest $Z$-point in respectively $b_i^1$ and $b_i^2$ is less than the distance between the closest pair of $Z$-points, one in $b_i^1$ and one in $b_i^2$,

then $Z$-points in $b_i^1$ and $b_i^2$ are made pairwise adjacent. Furthermore, $Z$-points which belong to level $\delta$ boxes $b_\delta^1$ and $b_\delta^2$ with the closest pair being no more than $l_\delta$ apart are made adjacent. All other edges are pruned away. If $\delta$ and $l_i$ are appropriately chosen, it can be shown that the remaining edges contain at least one $T(Z)$. If $Z$-points are random, uniformly and independently distributed in the unit $d$-cube $[0, 1]^d$, then the probability that the pruned graph is sparse can be shown to be high. The expected running time of this minimum spanning tree algorithm is $O_e(n\alpha(cn, n))$ where $c$ is a constant dependent on $d$ and $\alpha$ is the inverse Ackermann's function.

Agarwal et al. [4] suggested yet another approach to the EMST problem by establishing its relation to the *bichromatic closest pairs* (BCP) problem: Given a set of $n_r$ red points and a set of $n_b$ blue points in $E^d$, $d \geq 2$, find a red point $z_r$ and a blue point $z_b$ such that the distance between $z_r$ and $z_b$ is minimum among all red-blue pairs. The algorithm for the EMST subdivides the edges into $O(n \log^{d-1} n)$ groups, each group forming a complete bipartite subgraph on two subsets $Z_r$ and $Z_b$ of $Z$. These groups are chosen such that only a closest pair between $Z_r$ and $Z_b$ can be an edge of any $T(Z)$. In order to solve the BCP problem for given $Z_r$ and $Z_b$, a random sample of one of the point sets can be used to decompose to original problem into many small subproblems. Each of these small subproblems can be solved using any algorithm for the post-office problem. This approach leads to an $O((n_r n_b \log n_r \log n_b)^{2/3} + n_r \log^2 n_b + n_b \log^2 n_r)$ expected time algorithm for the BCP problem in $E^3$. This reduces to $O((n \log n)^{4/3})$ for $d = 3$.

**Heuristics.** Vaidya [197] also suggested a heuristic closely related to his algorithm discussed above. The space enclosing the $Z$-points is partitioned as in the exact algorithm. For each non-empty box $b_i$ in each partition $P_i$, a representative $Z$-point is chosen. Furthermore, partitions are nested, and a representative $z_i$ of a larger box $b_i$ must also be chosen as the representative for each of the nested boxes containing $z_i$. If two representatives on the same level $i$ are at most $l_i$ apart, the interconnecting edge is not pruned. $Z$-points which are not representatives on any level are made adjacent to their representatives on the lowest level $\delta$. If $\delta$ and $l_i$ are appropriately chosen, it can be shown that not pruned edges must contain a spanning tree which is at most $(1+\varepsilon)$ times the length of $T(Z)$. Furthermore, pruning requires $O(\varepsilon^{-d} n \log n)$ time. The heuristic generalizes in a straightforward manner to $d$ dimensional spaces with any $L_p$-metric, $p \in [1, \infty]$.

A less general (for $d = 3$) but similar prune-and-search heuristic was developed by Clarkson [51]. It finds a spanning tree with the length at most $(1+\varepsilon)$ times the length of $T(Z)$. It requires $O(n(\log n + (1/\varepsilon) \log \delta))$ time and $O(n \log \delta)$ space, where $\delta$ is the ratio of the distance between the farthest pair of $Z$-points to the distance between the closest pair of $Z$-points. Thus, contrary to Vaidya's heuristic, the running time is dependent on a particular $Z$-point distribution. Furthermore, in view of the fact that the (expected time) subquadratic exact algorithms are available for $d = 3$, the applicability of this heuristic is rather limited.

### 5.2.2 Euclidean Traveling Salesman Problem

The *Euclidean traveling salesman problem* (ETSP) can be formulated as follows, Figure 23.

**Applications.** Most applications of the ETSP arise in connection with various kinds of routing and scheduling problems. However, the problem arises also in connection with network design problems. Application examples include single and multiple depot vehicle routing, delivery, automated wiring of printed circuit boards, scheduling of jobs with machine set-up costs.

- **Given:** $Z = \{z_1, z_2, ..., z_n\} \in E^2$,
- **Find:** Shortest closed tour through all $Z$-points.
- **Data Structure:** The Multi Dimensional Binary Search Tree (MBST) data structure utilized by Bentley described below is illustrated for this point set along with the optimal ETSP solution. Once we separate $Z$ into $Z_1, Z_2$ with the vertical line, we alternate between horizontal and vertical lines through appropriate $z$−points to further subdivide the sets.

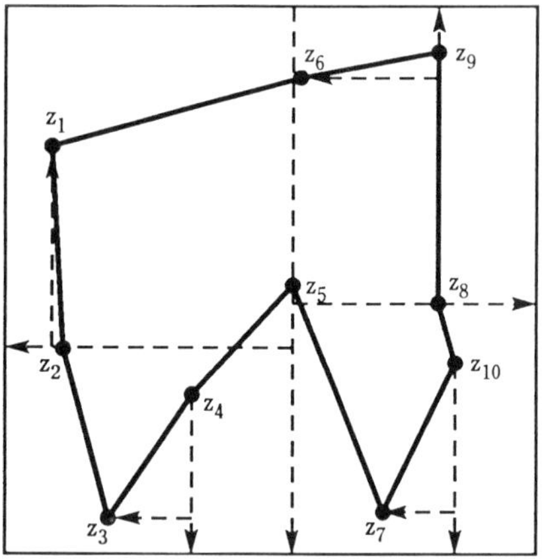


**Figure 23: ETSP Solution**

**Lower Bounds.** The problem is known to be $\mathcal{NP}$-complete in the strong sense when the discretized Euclidean metric is used [86, 162, 159]. It remains $\mathcal{NP}$-hard in the strong sense if the nondiscretized Euclidean metric is used, but it is then not known to be in $\mathcal{NP}$ [86].

**Algorithms.** Various exponential enumeration algorithms have been developed. They will not be discussed here since they in general do not exploit the geometric structure of the problem. The interested reader is referred to [123] for an exhaustive survey. Smith [183] developed an $2^{O(\sqrt{n})}$ algorithm for the ETSP. It will not be discussed here.

**Heuristics.** Most of the heuristics for the ETSP have been developed for the more general problem in the graph-theoretic setting. It asks for a minimum length tour through all vertices of a complete graph with edge-costs satisfying the triangle inequality. The interest in this problem has been stimulated by the fact that the traveling salesman problem in an arbitrary graph can always be reduced to it.

Until recently, the application of CG techniques in connection with these heuristics was surprisingly limited. However, an extensive study of using multidimensional binary search trees (MBSTs) (see Subsection 4.2.6) has been recently carried out by Bentley [21]. It gives a striking example of how powerful CG approaches can be used in solving difficult geometric TND problems.

The *nearest neighbor* heuristic [18] begins with a single $Z$-point as an initial path.

During each iteration a $Z$-point closest to the end of the path is added to the partial solution. When all $Z$-points are on the path, the closing edge is added. This heuristic has been shown to yield solutions which are at most $(\lceil \log n \rceil + 1)/2$ the optimal [176]. One way to improve it is to use several or all $Z$-points as starting points. Another improvement possibility is to allow the path to be expanded at both ends. In the graph setting, the nearest neighbour heuristic requires $O(n^2)$. In the geometric setting, Bentley [21] suggested a variant of MBSTs to quickly identify nearest $Z$-points to the growing path. Extensive experiments indicate $O(n \log n)$ expected running time for this implementation (for various distributions over the unit square). No worst-case time complexity for the heuristic using MBSTs is available.

The *multiple fragment* heuristic starts with all $Z$-points as isolated points in the partial solution. During each iteration, the shortest line segment which does not close a cycle with less than $n$ $Z$-points, and which does not introduce $Z$-points of degree more than 2, is added to the partial solution. In the graph setting, the worst-case time complexity of this heuristic is $O(n^2 \log n)$. In the geometric setting, Bentley [21] indicated that if a variant of MBSTs is used, the heuristic runs in $O(n \log n)$ expected time (for various distributions over the unit square). No worst-case time complexity for this implementation is available.

The *nearest insertion* heuristic begins with a single $Z$-point as an initial tour. At each iteration a $Z$-point $z$ closest to the tour is found. It is inserted into the tour by replacing one of the edges incident to $z$, denoted by $e$, by the two edges from $z$ to the endpoints of $e$. The edge $e$ is usually chosen such that the length increase of the tour is as small as possible. More sophisticated rules for the identification of the outgoing edge have been suggested [21]. In the *farthest insertion* heuristic, the $Z$-point $z$ is chosen as the farthest $Z$-point from the tour. In the *random insertion* heuristic, $z$ is chosen at random. In the graph setting all these variants can be implemented so that their worst-case time complexity is $O(n^2)$. In the geometric setting, computational experience indicates that use of MBSTs yields expected running times which are not more than $O(n \log n)$ [21]. As for the other heuristics, no worst-case time complexity bound is available.

Rather than choosing a $Z$-point closest to the tour, one could settle on a $Z$-point $z$ which minimizes the length of the enlarged tour [157]. This modified insertion heuristic has been implemented so that it requires $O(n^2 \log n)$ time in the graph setting. It can be shown to yield solutions at most twice the optimal, and the bound is tight. We are not aware whether this heuristic was implemented in the geometric setting. Yet another variation of the insertion heuristic [176] is to select $z$ such that the difference between the length of some edge $(z_i, z_j)$ in the tour and the sum of lengths of edges $(z_i, z)$ and $(z_j, z)$ is minimized. If the tour consists of a single $Z$-point, the closest $Z$-point is selected as $z$. It can be shown that the solutions obtained by this heuristic are at most $\lceil \log n \rceil + 1$ times the optimal. Once again, we are not aware of any geometric implementation of this variant.

The *minimum spanning tree* heuristic is based on the construction of the EMST $T(Z)$. A walk around $T(Z)$ yields a tour where each edge is traversed twice. Furthermore, if a $Z$-point is visited more than once, shortcuts are available. It can be shown that this heuristic yields solutions at most twice the optimal [175]. In addition, this bound is tight. In the graph setting, this heuristic requires $O(n^2)$ time in the worst-case. In the geometric setting, it reduces to $O(n \log n)$ time as this is the complexity of constructing $T(Z)$. The same technique applies to the RTSP [125].

The *matching* heuristic was developed by Christofides [47]. First the EMST $T(Z)$ is determined in $\Theta(n \log n)$ time. Next, the minimum weight matching $M$ for odd-degree vertices in $T(Z)$ is found. In the Euclidean setting, this can be done in $O(n^{2.5}(\log n)^4)$ time using dynamic nearest-neighbour query in a weighted Voronoi diagram [198]. $T(Z)$ together with the matching forms an Eulerian graph, and therefore yields a tour through all $Z$-points. If a $Z$-point is visited more than once, shortcuts are available. It can be shown that this $O(n^{2.5}(\log n)^4)$ heuristic (which requires $O(n^3)$ in the graph setting) yields solutions at most $\frac{3}{2}$ the optimal, and this bound is tight. Bentley [21] discusses a heuristic for matching which uses MBSTs for the nearest neighbour identification. The observed running times are $O(n \log n)$ for various distribution over the unit square.

Another class of heuristics for the ETSP begins with some tour and then perturbs it as long as cost reductions are possible. One approach of this kind was suggested by Lin and Karnighan [136]. A *k-edge-exchange*, $2 \leq k \leq n$, of a tour consists of the replacement of $k$ edges in a tour by $k$ other edges provided that another tour is obtained. A tour is said to be *k-optimal* if no k-edge-exchange can be used to obtain a shorter tour. In the graph setting, identification of applicable $k$-edge-exchanges requires $O(n^k)$ time. Good empirical results were obtained for $k = 3$. Experiments with MBSTs have been carried out by Bentley [21].

The $O(n \log n)$ divide-and-conquer heuristic for the ETSP has been developed by Karp [116]. The region enclosing all $Z$-points is divided into appropriate subregions. This process continues until subregions with sufficiently few $Z$-points are obtained. The ETSP is solved exactly for $Z$-points in these subregions. Solutions obtained are then patched together yielding larger and larger solutions. Use of MBSTs to subdivide point sets and then to find nearest neighbours have been discussed in [21]. This approach can cause problems due to particular region partition. A bottom-up method in which the actual distribution of the vertices is taken into account has been suggested [98]. Karp's heuristic has been claimed to produce solutions within 1 or 2 percent of the optimal tour whenever the points are independently, uniformly distributed in the unit square.

Another polynomial heuristic was developed by Stewart [189]. It uses the boundary of the convex hull of $Z$ as the initial subtour. The quality of this heuristic has been reported to be good for a variety of test problems [93].

One of the most dominating methods for the determination of the EMSTs and RMSTs is based on pruning of useless connections by means of duals of Delaunay triangulations. Could such methods be applied to the ETSP and RTSP? Unfortunately, this is not the case in general [61, 115]. Nor is it the case when the minimum weight triangulation is used. Furthermore, the travelling salesman problem is NP-hard in planar graphs. In spite of these negative results, it would perhaps be worthwhile to devise specialized heuristics for the TSP in planar graphs embedded in the plane, and with the Euclidean or rectilinear distances as edge-costs.

### 5.2.3 Euclidean Maximum Spanning Tree Problem

An interesting approach to the *Euclidean maximum spanning tree* (EMXT) problem was suggested by Monma et al. [154]. Its applications arise in cluster analysis where one often needs to partition a set of entities into well-separated and homogeneous classes. In particular, the problem of separating $n$ points into two classes such that the largest diameter is minimized can be solved in $O(n \log n)$ time given the maximum spanning tree.

In the graph setting, the maximum spanning tree problem is the same as the minimum spanning tree problem: it suffices to replace edge-lengths by their negative values. However, in the Euclidean setting, the use of Delaunay triangulations to prune away edges does not work for obvious reasons. One could hope that there is a relation between the EMXT problem and the furthest-point Voronoi diagram (and its dual). Unfortunately, neither is this the case.

The algorithm suggested by Monma et al. [154] requires $O(n \log h)$ time and $O(n)$ space where $h$ is the number of $Z$-points on the convex hull of $Z$. First, for each $Z$-point $z$, its farthest mate is determined. Then, a directed *farthest neighbour network*, denoted by $F(Z)$, is constructed in the following way. Its vertices are $Z$-points, its directed arcs connect each $Z$-point to its farthest mate. Every directed path in $F(Z)$ consists of a sequence of arcs (of strictly increasing length) until a two-arc cycle is reached. All edges corresponding to arcs in $F(Z)$ belong to the maximum spanning tree.

Let $C_0, C_1, ..., C_{k-1}$ denote the weak components of $F(Z)$. $Z$-points belonging to $C_i$, $0 \leq i \leq k-1$, and on the boundary of $CH(Z)$ fall into two clusters $A_i$ and $B_i$. $A_i$ contains first, third, fifth,... $Z$-point on the directed path. $B_i$ contains second, fourth, sixth,... $Z$-point on the directed path. It can be shown that $A_i \cap CH(Z)$ and $B_i \cap CH(Z)$ are consecutive points on $CH(Z)$ for every $i$, $0 \leq i \leq k-1$. Furthermore (after possible relabeling), $A_i$ and $B_i$, $0 \leq i \leq k-1$, the portions of clusters will appear on $CH(Z)$ in the following order: $A_0 \cap CH(Z), A_1 \cap CH(Z), ..., A_{k-1} \cap CH(Z), B_0 \cap CH(Z), B_1 \cap CH(Z), ..., B_{k-1} \cap CH(Z)$.

Any edge of the maximum spanning tree not in $F(Z)$ must have at least one end-point in $CH(Z)$. Furthermore, such an edge must join consecutive components (i.e. any edge from $C_i$ in the maximum spanning tree must go either to $C_{i-1}$ or $C_{i+1}$, addition modulo $k$). So we only need to determine the longest edge between every pair of consecutive components (restricting the search even further by using the fact that at least one end-point is in $CH(Z)$). This will yield $k$ edges which connect the $k$ components $C_0, C_1, ..., C_{k-1}$ into a cycle. The maximum spanning tree for $Z$ can now be obtained by deleting the shortest of these $k$ edges.

The time complexity of this algorithm is dominated by the time needed to determine $CH(Z)$. The best known algorithm requires $\Theta(n \log h)$ time (see Subsection

4.2.1). Furthermore, it is shown in [154] that the maximum spanning tree algorithm is in fact $\Theta(n \log h)$. If all $Z$-points are on the boundary of $CH(Z)$, the time complexity of this algorithm reduces to $O(n)$. The algorithm generalizes to the $d$-dimensional case. Its time complexity is then $O((n \log n)^{2-\alpha(d)})$ where $\alpha(d) = 2^{-(d+1)}$. For $d = 3$, the algorithm can be further improved (using Chazelle's technique for neighbour search [39]) to run in $O(n^{1.5}(\log n)^{2.5})$ time.

### 5.2.4 Euclidean Spanning Tree with Low Stabbing Number

The problem of finding spanning trees with low *stabbing number* is to find $k = O(\log n)$ spanning trees $T_1, T_2, ..., T_k$ of $Z$ with the property that for any line $L$, there is a tree $T_i$ such that $L$ intersects it in at most $O(\sqrt{n})$ edges. Chazelle and Welzl [44] proved that for any $Z \subset E^d$, there exists a spanning tree with the stabbing number $\Theta(n^{1-1/d})$. Algorithms for this problem were given in [78, 142]. Recently, Agarwal [1, 2, 3] presented an $O(n^{3/2} \log^c n)$ algorithm (where $c$ is a constant). With additional $O(n \log n)$ preprocessing time and $O(n)$ space, the tree $T_i$ corresponding to the query line $L$ can be determined in $O(\log n)$ time.

### 5.2.5 Classification of $G(E)$ Problems

Table 5.2.5 illustrates the range of $\mathcal{CG}$ methods and geometric data structures which have been utilized in approaches to the $G(E)$ problems.

Algorithms and heuristics are placed according to their most important strategy and geometric data structure used. Few comments seem in order. The EMST algorithms which use Delaunay triangulations are placed in the prune-and-search column although the construction of $DT(Z)$ is based on the divide-and-conquer, plane-sweep or geometric transformation (see Subsection 4.2.3). We consider the construction of $DT(Z)$ as a pruning phase for the subsequent determination of a minimum spanning tree in a planar network. The same applies to the matching heuristic for the ETSP.

Greedy heuristics for the ETSP have been all collected in one entry ($CD(Z)$ row and $ITER$ column) with $O_o(n \log n)$ empirically observed average complexity. Worst-case bounds for these heuristics in the Euclidean setting are unknown. It is interesting to note that plane sweep seems to be of no importance to $G(E)$ problems.

## 5.3 $G(Z, E)$ Problems

We are given a set of $Z$-points. We wish to select a subset of edges forming a network extremizing some objective function subject to some side constraints pertaining to the topology of the network. In contrast with the $G(E)$ problems, we are here allowed to include some additional *Steiner points*. This increases the number of potential connections and will usually lead to lower cost networks. Steiner points can be taken from an a priori given set of points $S$. Alternatively, the number and location of Steiner points can be unknown. When designing the MHS discussed in Section 2, we might for example wish to determine a minimal length circuit for an

| | $\mathcal{CG}$ Paradigms | | | | |
|---|---|---|---|---|---|
| **G(E)** | *ITER* | *PSWE* | *D&C* | *LOCUS* | *P&S* |
| $CH(Z)$ | *ETSP*: $O(?)$ [189] | | | | EMXT: $\Theta(n \log h)$ [154] |
| $VOR(Z)$ | | | | | |
| $DT(Z)$ | | | | | EMST: $\Theta(n \log n)$ [179]<br>RMST: $\Theta(n \log n)$ [125, 132] |
| $VG(Z)$ | | | | | |
| $CD(Z)$ | *ETSP*: $O_o(n \log n)$ [21] | | *ETSP*: $O(n \log n)$ [116] | | EMST: $O((n \log n)^{1.5})$ [213, 39]<br>EMST: $O_e((n \log n)^{4/3})$ [4]<br>EMST: $O_e(\alpha(cn, n)n)$ [197]<br>*EMST*: $O(\epsilon^{-d} n \log n)$[197]<br>*EMST*: $O(n(\log n + (1/\varepsilon) \log \delta))$ [51] |
| $OTHERS$ | EMST: $O(n^2 \log n)$ [22, 156<br>ETSP: $O(exp)$ | | EMST: $O(n^2 \log n)$ [35] | | *ETSP*: $O(n^{2.5}(\log n)^4)$ [47] |

**Table 2:** $G(E)$ **Classification Table**

automated guided vehicle allowing for additional stops and junctures so that the total circuit length between workstations is minimized. This problem can be formulated as the *rectilinear Steiner minimal circuit problem.* However, the most well-known $G(Z,E)$ problems are the *Euclidean* and *rectilinear Steiner minimal tree problems.*

Figure 24 illustrates the breakdown of geometric $\mathcal{TND}$ problems in the class $G(Z,E)$. We only indicated problems which received most attention in the literature. However, practically all problems in $G(E)$ can be formulated in the more general $G(Z,E)$ setting. Furthermore, approaches to the Steiner path and Steiner circuit problems can be more or less directly derived from their $G(E)$ counterparts. Whereas some tree problems in $G(E)$ are in the class $\mathcal{P}$, Steiner tree problems seem to be inherently difficult.

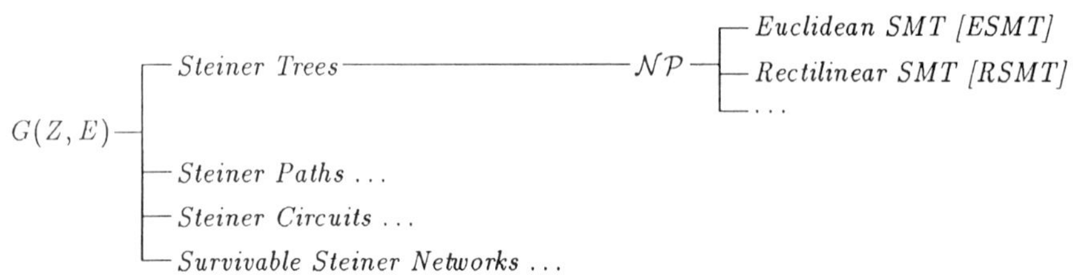


**Figure 24: Morphological Tree for $G(Z,E)$ Problems**

In this section we discuss CG approaches to geometric TND problems where location of some new points together with a suitable set of edges is required. The most well-known problems of this kind are undoubtedly the Euclidean and rectilinear Steiner minimal tree problems. We have chosen these two problems to illustrate the range of CG algorithms and heuristics for this problem class. Not only are the Euclidean and rectilinear aspects of CG important here, but these two problems have a number of applications and the number of recent contributions to their solution has been mushrooming over the past few years.

### 5.3.1 Euclidean Steiner Minimal Tree Problem

The *Euclidean Steiner minimal tree* (ESMT) problem can be formulated as follows, Figure 25.

**Lower Bounds.** The ESMT problem is known to be $\mathcal{NP}$-hard [87]. It is not known to be in $\mathcal{NP}$. The discretized version of the ESMT problem is $\mathcal{NP}$-complete.

**Algorithms.** The first finite algorithm for the ESMT problem in $E^2$ was due to Melzak [151]. In his approach each subset of $Z$-points is considered separately.

- **Given:** $Z = \{z_1, z_2, ..., z_n\} \in E^2$
- **Find:** Shortest tree $T(Z)$ spanning $Z$. Contrary to the EMST problem, a set $S$ of additional so-called *Steiner points* can appear anywhere in the plane if they reduce the total length.
- **Data Structure:** Here the Delaunay Triangulation not only provides the underlying bound for the EMST, it acts as a geometric data structure for local Steiner triangle solutions. The heuristic found in [187] utilizes the DT(Z) and VOR(Z) to find local 3 and 4-point Steiner solutions and eventually concatenates these into a solution for the entire point set.

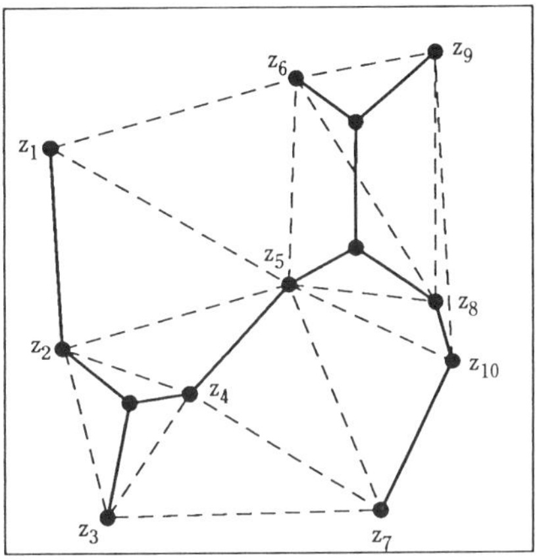


**Figure 25: ESMT Solution**

- Given a subset of $p$ $Z$-points, $2 \leq p \leq n$, there is a finite (but superexponential) number of *full Steiner topologies* in which all $Z$-points have degree 1 and all $p - 2$ Steiner points have degree 3. The number of full Steiner topologies can be reduced to an exponential number by examining particular locations of $Z$-points.
- Given a full Steiner topology, Hwang [108] gave a linear time geometric algorithm which either yields the (unique) *full Steiner tree* with this topology, and with all edges meeting at Steiner points at $120^o$, or decides that no such tree exists. The algorithm consists of two phases, each involving $p-2$ iterations. In the first phase, the algorithm replaces two suitably chosen $Z$-points and their common Steiner point $s$ by their appropriately located *equilateral point* (acting as a $Z$-point) which is made adjacent to the third neighbor of $s$. This process continues until a full Steiner topology with two points is obtained. In the second phase, the algorithm works backwards and determines the location of one Steiner point at a time. Many pruning techniques for quick elimination of full Steiner topologies for which a full Steiner tree does not exist or cannot be a part of the overall optimal solution, have been suggested [54, 207]. Some of them make use of convex hulls while others rely on the proximity information of $Z$-points. In other words, geometric properties of $Z$-points play a crucial role in this elimination process.
- Given all full Steiner trees, an exhaustive concatenation of them is carried out. The shortest concatenation spanning all $Z$-points is $T(Z)$. In order to

- simplify the concatenation process, various decomposition techniques have been suggested [91, 54, 109].

Hwang and Weng et al. [111] suggested that the two-phase geometric algorithm for the determination of the locations of Steiner points for a given full Steiner topology can be replaced by an algebraic one-phase algorithm involving solving a system of linear equations.

The bottleneck of the algorithms for the ESMT problem seems to be the concatenation of full Steiner trees. The recent results suggest that it does not need to be so (although no implementations have been reported in the literature as yet). Let $D_S(T)$ denote all topologies that can be obtained from a full Steiner topology $T$ by any sequence of edge shrinking (i.e. deleting edges and collapsing the end-points) where one of the end-points belongs to $Z$ (considering the collapsed point as a $Z$-point after the shrinking). Trietsch and Hwang [195] observed that the ESMT for $Z$ has a (not necessarily full) topology which belongs to at least one $D_S(T)$ for at least one full Steiner topology $T$ on all $n$ $Z$-points. They also suggested a *negative edge* algorithm which finds the shortest among all trees with topologies belonging to $D_S(T)$ (if it exists). Unfortunately, there is no proof that this algorithm terminates in polynomial time. Recently, Hwang and Weng [110] suggested a *luminary* algorithm which solves this problem in $O(n^2)$ time. Both the negative edge and the luminary algorithms are quite complex and will not be described here.

The ESMT in $E^d$, $d > 2$, seems to be much harder to solve than in the 2-dimensional case. It also has received considerably less attention. Some basic properties (similar to the properties in $E^2$) of ESMTs in $E^d$ were given by Gilbert and Pollak [91]. Smith [183] developed an enumeration algorithm which in its overall structure follows the approach for the 2-dimensional case (in fact it can be used in $E^2$ as well). Each full Steiner topology on all $n$ $Z$-points (and $n-2$ Steiner points) is generated. Although no exact algorithm for the determination of optimal locations of Steiner points for a given full Steiner topology is known for $d \geq 3$, the length of such a tree is a strict convex function of the locations of Steiner points. Consequently, numerical methods can be applied. As noted by Smith, setting partial derivatives to zero and applying standard numerical methods is ineffective since the length function has nonsmooth, complicated behaviour near the global minimum. It is better to iteratively update the location of each Steiner point by solving exactly the ESMT for its three neighbours. Unfortunately, the convergence can be very slow. Smith proved that it is possible to update all Steiner points simultaneously and compatibly by solving at the $(i+1)$-st iteration a system of $2n-4$ linear equations

$$(x_k^{i+1}, y_k^{i+1}) = \left(\sum_{e_{kj}^i} \frac{x_j^{i+1}}{|e_{kj}^i|} \Big/ \sum_{e_{kj}^i} \frac{1}{|e_{kj}^i|}, \sum_{e_{kj}^i} \frac{y_j^{i+1}}{|e_{kj}^i|} \Big/ \sum_{e_{kj}^i} \frac{1}{|e_{kj}^i|}\right), k = 1, 2, ..., n-2$$

where each summation is over the three edges adjacent to the $k$-th Steiner point. If the initial locations $(x_k^0, y_k^0)$ of Steiner points form a set of non-zero measure in

$R^{2n-4}$, then the solutions converge to the unique global minimum. Furthermore, the sequence of solutions is of decreasing length. The system of linear equations can be solved in $O(n)$ time using floating point operations and Gaussian elimination method. Experimental results indicate that the rate of convergence is geometric on average.

**Heuristics.** The EMST for $Z$ is a straightforward suboptimal solution to the ESMT problem. One of the most intriguing (and until recently open) problems in TND and CG was the lower bound on the ratio between the lengths of ESMT and EMST for any configuration of $Z$-points. It has been a long-standing conjecture [91] that no problem instance can have this ratio less than $\sqrt{3}/2 = 0.8666....$ Several papers which either proved this conjecture for small $n = 3,4,5,6$ [91, 163, 72, 67, 71, 84, 29, 177] or gave not tight lower bounds [91, 97, 48, 66, 49] appeared. Du and Hwang [68] recently closed this sequence of papers by proving the conjecture true in the general case.

Several heuristics for the ESMT problem are based on the idea of local improvements of the current tree $T$ by the introduction and/or replacement of appropriately located Steiner points [36, 122].

An $O(n \log n)$ locus heuristic based on the concatenation of appropriately chosen ESMTs for 2, 3 and 4 $Z$-points was given by Smith et al. [187]. The identification of 2, 3 and 4 $Z$-point sets is based on the information available from $DT(Z)$ and the $VOR(Z)$ along with the EMST for $Z$. More specifically, only the following subsets of size $k = 2,3,4$ are considered during the concatenation phase.

- $k = 2$: *pairs of $Z$-points connected by an edge of the EMST,*
- $k = 3$; *triples of $Z$-points on a common triangle of $DT(Z)$ and with two of its edges in the EMST,*
- $k = 4$: *quadruples of $Z$-points based on minimizing the Voronoi edge distance of two adjacent edge-sharing triangles of the $DT(Z)$ which are also connected by EMST edges.*

The heuristic, in fact, first attempts to construct the $k = 4$ clusters based on the Voronoi information, then the $k = 3$, and finally the $k = 2$ edges.

Beasley [17] suggested a heuristic which constructs ESMTs for all 3 and 4 $Z$-points sets. These trees are then sorted according to the ratio with their EMST, and concatenated in a greedy fashion to obtain a tree spanning all $Z$-points. His heuristic does not utilize the locus information provided by the $VOR(Z)$ or $DT(Z)$ nor has convergence been proved although his reductions are very good.

**Special Cases.** When $Z$-points are on a *circle* with radius $r$, the ESMT for $Z$ is the perimeter of the $n$-gon minus the longest side provided that it is longer than $0.557r$ [67]. When $Z$-points are vertices of a *regular $n$-gon*, then for $n = 3,4,5$, the ESMT for $Z$ has full topology, and for $n \geq 6$, the ESMT for $Z$ is the perimeter of the $n$-gon minus one side [70]. When $Z$-points form a *ladder* (i.e. are placed on the longer sides of an rectangle (and in its corners) such that the distance between two consecutive $Z$-points is equal to the length of the shorter side). Complete characterization of this type of ESMTs has been given in [48]. Characterization of ESMTs for $Z$-points forming *zig-zag lines* was given in [69]. Characterization of ESMTs in *splitting trees* was given in [111].

### 5.3.2 Rectilinear Steiner Minimal Tree Problem

The *rectilinear Steiner minimal tree* (RSMT) problem is defined as the ESMT problem but with the $L_1$-metric, Figure 26.

- **Given:** $Z = \{z_1, z_2, ..., z_n\} \in E^2$
- **Find:** Shortest tree $T(Z)$ spanning $Z$. Contrary to the EMST problem, a set $S$ of additional so-called *Steiner points* can appear anywhere in the plane if they reduce the total length.
- **Data Structure:** The $L_1$ Voronoi Diagram is illustrated along with the $L_1$ RSMT solution. The locus information provided by the $L_1 VOR(Z)$ is crucial to the identification of the 4-point clusters in this solution generated by the heuristic found in [186].

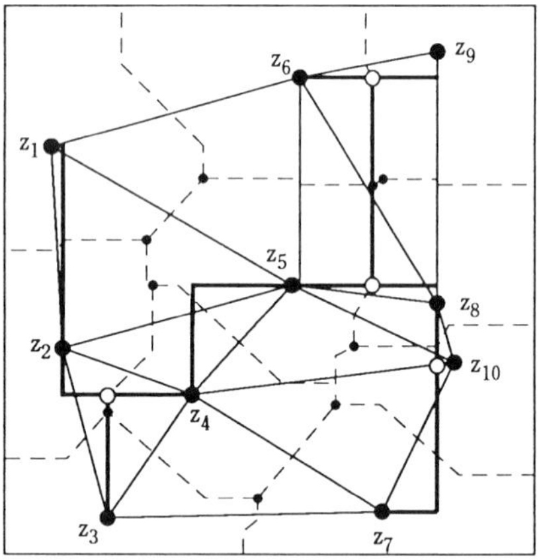


**Figure 26: RSMT Solution**

**Lower Bounds.** The RSMT is known to be NP-hard [88]. It is not known to be in NP. The discretized version of the RSMT is NP-complete.

**Algorithms.** Hanan [104] proved that at least one solution to the RSMT problem must follow horizontal and vertical lines through $Z$-points. Consequently, the RSMT can be found by any of the exact algorithms for the Steiner problem in graphs. For a survey of methods on the graph-theoretic version of the Steiner minimal tree problem see [208]. As a consequence, very little attention has been given to developing exact algorithms exploiting particular properties of the $L_1$-metric. The only exact algorithm developed especially for the RSMT problem is a branch-and-bound algorithm due to Yang and Wing [212].

**Heuristics.** The RMST for $Z$ is a straightforward suboptimal solution to the RSMT. It has been proved [106] that the ratio between the lengths of the RSMT and the RMST for any configuration of $Z$-points cannot be less than 2/3. Furthermore, this bound is tight. It was conjectured [91] that the ratio is $\frac{d-1}{d}$ in an arbitrary $d$-dimensional rectilinear space.

Hwang [107] developed a Prim-like $O(n \log n)$ heuristic. It is based on the $O(n^2)$ approach suggested in [124]. Instead of simply connecting the nearest not included $Z$-point, these heuristics use *three Z-points connection schemes.*

An $O(n^4)$ local improvement heuristic based on the onion data structure was first suggested by Smith and Liebman [188]. An $O(n \log n)$ locus heuristic based on the $L_1$ $DT(Z)$ and $VOR(Z)$ was given in [186] where 3 and 4-point local improvement solutions are concatenated into a sub-optimal solution for the entire point set. In a related heuristic using 3 and 4-point local improvements without the Delaunay triangulation was also extended by Beasley to yield even closer optimal solutions, yet at the tradeoff of increased algorithmic running time [17].

Richards [172] developed an $O(n \log n)$ plane-sweep heuristic. $Z$-points are ordered by non-decreasing $y$-coordinates. The heuristic begins with a tree $T$ containing the $Z$-point with the smallest $y$-coordinate. Subsequent $Z$-points are added to $T$ in their predefined order by shortest paths. The complexity of identifying these shortest paths to $T$ is reduced since $Z$-points in lower parts of $T$ are hidden and cannot be endpoints of such shortest paths.

Bern and Carvalho [25] developed an $O(n^3)$ Kruskal-like heuristic which begins with $n$ isolated $Z$-points. At each iteration two closest trees are interconnected.

Komlos and Shing [121] developed a divide-and-conquer heuristic based on alternating vertical and horizontal partitioning of the plane with $Z$-points. The initial enclosing rectangle is partitioned and represented by a multidimensional binary search tree with at most $t$ $Z$-points in each leaf. RSMTs in each rectangle are determined. These solutions are then combined into an overall suboptimal solution. The heuristic requires $O(n \log n)$ time if $t = O(\log \log n)$.

**Special Cases.** An $O(n^3)$ dynamic programming algorithm has been developed by Aho, Garey and Hwang [5] for the case when $Z$-points are on the *boundary of a rectangle.* Cohoon et al. [56] recently developed an $O(n)$ algorithm. An $O(n^6)$ algorithms were independently developed by Bern [26] and Provan [168] for the case when the $Z$-points are on the *boundary of the rectilinear convex hull.*

### 5.3.3 Classification of $G(Z, E)$ Problems

Table 5.3.3 illustrates the time complexities for the algorithms and heuristics for the ESMT and RSMT problems. It also indicates CG paradigms and geometric data structures used.

| | $\mathcal{CG}$ Paradigms | | | | |
|---|---|---|---|---|---|
| **G(Z,E)** | *ITER* | *PSWE* | *D&C* | *LOCUS* | *P&S* |
| $CH(Z)$ | *RSMT*: $O(n^4)$ [186]<br>*RSMT*: $O(n^6)$ [26, 168] | *RSMT*: $O(n^3)$ [5]<br>*RSMT*: $O(n)$ [56] | | | |
| $VOR(Z)$ | | | | | |
| $DT(Z)$ | | | | *RSMT*: $O(n \log n)$ [186]<br>*ESMT*: $O(n \log n)$ [187] | |
| $VG(Z)$ | | | | | |
| $CD(Z)$ | | | *RSMT*: $O(n \log n)$ [121] | | |
| $OTHERS$ | *ESMT*: $O(?)$ [36, 122]<br>*ESMT*: $O(n^3)$ [17]<br>*RSMT*: $O(n^3)$[17, 25]<br>*RSMT*: $O(n \log n)$ [107] | *RMST*: $O(n \log n)$ [172] | | | ESMT: $O(exp)$ [151, 183] |

Table 3: $G(Z, E)$ Classification Table

## 5.4 $G(\Omega)$ Problems

Our last class of problems is characterized by the presence of a finite set $\Omega = \{\omega_1, \omega_2, ..., \omega_k\}$ of obstacles in some metric space. Unless otherwise specified, the obstacles will be convex, polygonal and disjoint. Each problem in the previous three classes can be generalized to one with obstacles. Consequently, $G(\Omega)$ problems can be further subdivided into the following three subcategories:

- $G(Z, \Omega)$ problems: When locating new point(s) with respect to $Z$-points, obstacles usually indicate forbidden regions. Furthermore, geodesic distances are typically used. In the clustering problems, the subdivision of the space in connection with subsequent multiple queries becomes more complicated as the obstacles have to be taken into account.

- $G(E, \Omega)$ problems: The edges which will occur in the constructed networks must avoid the obstacles. Consequently, they will no longer be straight line segments but rather polygonal chains.

- $G(Z, E, \Omega)$ problems: Both edges (polygonal chains) and additional Steiner points which will occur in constructed networks must avoid the obstacles.

$G(\Omega)$ problems are usually very difficult since we have discontinuities in the objective functions and non-convex solution spaces. These problems are of tremendous practical importance. For example, when designing an MHS within a manufacturing facility, obstacles and impediments (e.g. walls, columns, machine centers) are prevalent. They must be accounted for in the optimal design of paths, trees, circuits and other types of networks.

As already mentioned, all problems in the previous three classes can be formulated as problems with obstacles. However, there are some issues unique to obstacle problems which deserve special attention. In the morphological tree depicted in Figure 27 we only indicated those problems which received the most attention in the literature and which will be discussed in detail below.

$G(\Omega)$ problems are usually much more difficult than their obstacle-free counterparts. Consequently, these problems are typically approached by heuristic methods. Until recently, $\mathcal{CG}$ was mainly oriented toward development of efficient, polynomial algorithms for geometric problems. However, due to many important applications, interesting approaches to solve obstacle problems are beginning to emerge.

We have chosen to illustrate approaches to $G(\Omega)$ problems by examining one problem from each of its subclasses $G(Z, \Omega)$, $G(E, \Omega)$ and $G(Z, E, \Omega)$.

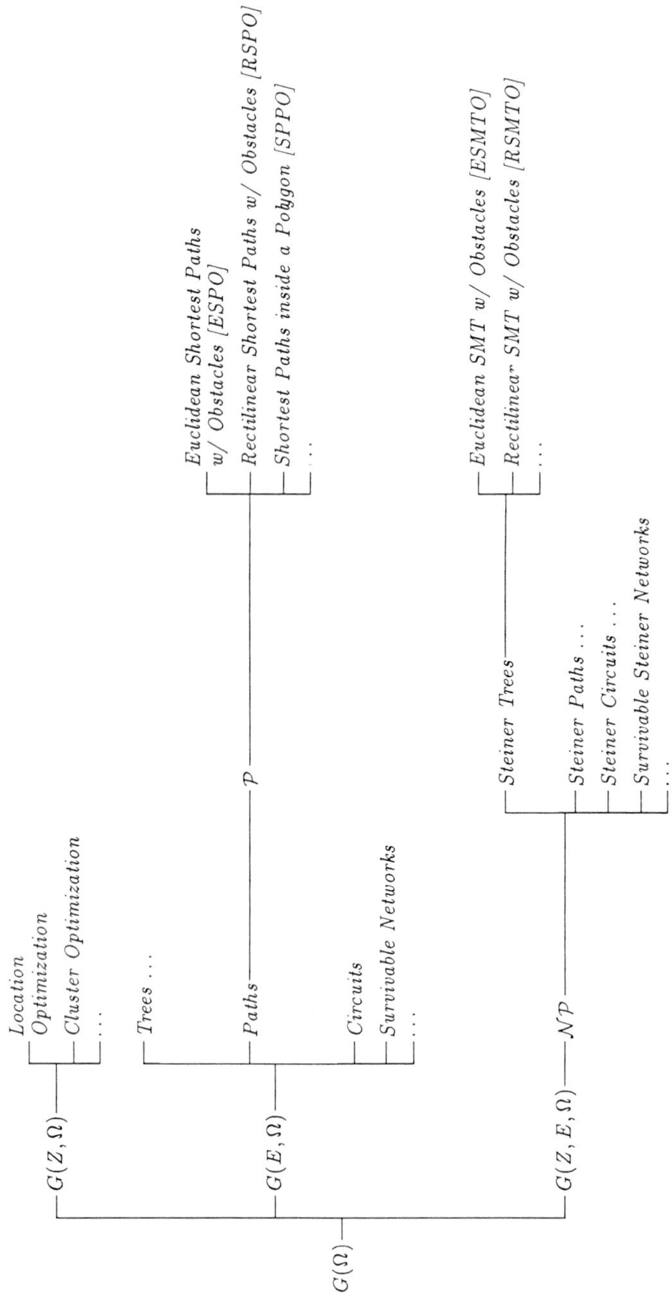


**Figure 27: Morphological Tree for $G(\Omega)$ Problems**

### 5.4.1 Maximum Gap Problem with Obstacles

The *maximum gap problem with obstacles* (MAXG($\Omega$)) belongs to $G(Z, \Omega)$ and can be formulated as follows, Figure 28.

- **Given:** A set $\Omega = \{\omega_1, \omega_2, ..., \omega_n\}$ of obstacles in $E^2$, and a bounded region $R$ containing all obstacles.
- **Find:** $z = (x, y) \in R$ such that its shortest distance to any obstacle is maximized. If obstacles are circles with radius $r_i$ and centered at $z_i = (x_i, y_i)$, $i = 1, 2, ..., n$,

$$\text{maximize } \zeta$$
$$s.t.\ \zeta^2 \leq (x - x_i)^2 + (y - y_i)^2\ \forall i$$
$$r_i^2 \leq (x - x_i)^2 + (y - y_i)^2\ \forall i$$
$$(x, y) \in R$$

- **Data Structure** We have overlayed the Voronoi Diagram in this problem instance to illustrate the relationship between the circular forbidden regions and the Voronoi points.

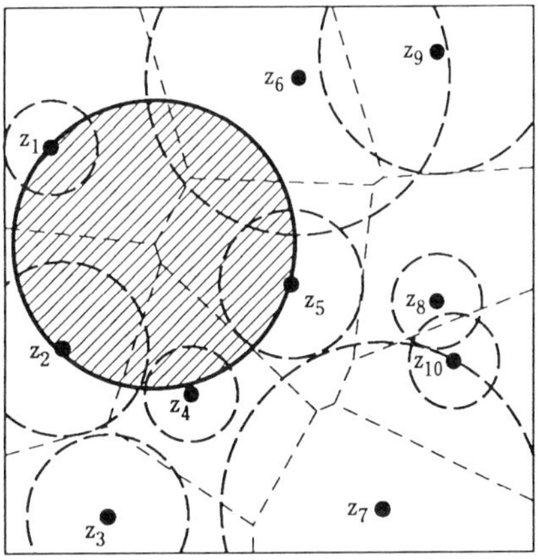


**Figure 28: MAXG($\Omega$) Solution**

**Algorithms.** Melachrinoudis and Cullinane [149, 150] suggested the following heuristic.

- **Step 1:** Solve the unconstrained problem by relaxing the forbidden circle constraints (see also Subsection 5.1.1).
- **Step 2:** Check whether all local maxima satisfy the forbidden circle constraints. If the best local maximum satisfies the circle constraints, then **Stop**.
- **Step 3:** Find all violated constraints and resolve the problem utilizing the conical equations of the Kuhn-Tucker conditions along with the algebraic information on the forbidden circles to generate new local maxima and return to **Step 2**.

This approach is not a pure $\mathcal{CG}$ one. Nevertheless, it is based on cellular decomposition. It also employs the iterative strategy which first ignores the obstacles and then gradually readjusts the solution. This approach is very common in many algorithms for obstacle problems.

### 5.4.2 Euclidean Shortest Path Problem with Obstacles

The *Euclidean shortest path problem with obstacles* (ESPO) can be formulated as follows, Figure 29.

- **Given:** Two points $s$ and $t$ in $E^2$ and a set $\Omega = \{\omega_1, \omega_2, ..., \omega_k\}$ of polygonal obstacles.
- **Find:** Shortest path from $s$ to $t$ avoiding $\Omega$.
- **Data Structure** Here the Visibility Graph with $O(n^2)$ segments is critical to the solution. The density of the line segments clearly indicates the price of the $O(n^2)$ algorithm.

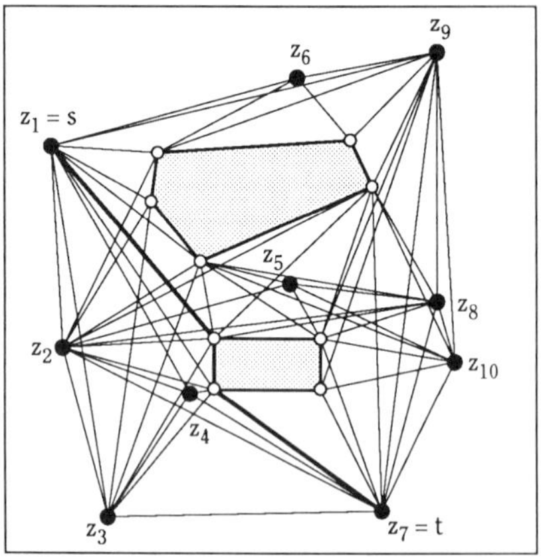


Figure 29: SPWO Solution

**Applications.** There are a vast number of applications for the SPWO both in $\mathcal{TND}$ and other fields such as robotics, motion planning and computer graphics.

**Lower Bounds.** As will be seen below, the ESPO in $E^2$ is solvable in polynomial time. Canny and Reif [33] proved that the problem of determining the length of the shortest path in $E^3$ with polyhedral obstacles is NP-hard. Hence this is also the case for the ESPO where not merely the length but the path has to be determined.

**Algorithms.** The shortest path from $s$ to $t$ must be a polygonal chain whose vertices are extreme points of polygonal obstacles. As first observed by Lozano-Perez and Wesley [138], the problem consequently reduces to the shortest path problem in the visibility graph with $n+2$ vertices, where $n$ is the total number of extreme points of the obstacles. Using Dijkstra's shortest path algorithm [60] together with the $\Theta(n^2)$ visibility graph algorithm (see Subsection 4.2.4), the ESPO can be solved in $O(n^2)$. If the visibility graph is sparse, it can be determined in $O(m + n \log n)$ time where $m$ is the number of edges in the visibility graph [90]. This bound can be preserved for the shortest path computations if Fibonacci heaps are used as data structures in Dijkstra's algorithm [192].

Reif and Storer [171] used the shortest path map $SPM(s, \Omega)$ (see Subsection 4.2.4) to solve the ESPO for *arbitrary polygonal obstacles* in $O(n(k + \log n))$ time. If $SPM(\Omega, s)$ is given, the region containing $t$ can be determined in $O(\log n)$ time. The path to $t$ can be determined in $O(b+\log n)$ time where $b$ is the number of line segments on the path.

Using visibility graphs, Welzl [202] and Asano et al. [8] gave an $O(n^2)$ algorithm for the ESPO in $E^2$ with $n$ disjoint (except perhaps at their endpoints) *line segments.* This algorithm can be easily modified to deal with disjoint polygonal obstacles (with the same time complexity). Note that the bound $O(n^2)$ cannot be improved upon if the visibility graph approach is used.

An $O(n^2)$ algorithm for the shortest path between two polygonal objects (rather than points) amidst polygonal obstacles is also known [9]. Lee and Preparata [129] developed a $\Theta(n \log n)$ plane sweep algorithm when $\Omega$ is a set of *parallel line segments.* The line segments are swept across and a shortest path map (see Subsection 4.2.2) is constructed up to the sweep line. No $O(n^2)$ time algorithm for the shortest path in $E^2$ avoiding $n$ *arbitrary intersecting line segments* is known.

Rohnert [173] gave an $O(n \log n + k^2)$ time algorithm for the shortest path in $E^2$ avoiding $k$ *convex disjoint obstacles.* This time bound is achieved provided that the obstacles are preprocessed in $O(k^2 \log n + n)$ time and $O(n + k^2)$ space. Preprocessing involves construction of a subgraph of the visibility graph.

Rohnert [174] also gave an $O(kn \log n)$ time and $O(n)$ space algorithm for the shortest path in $E^2$ avoiding $k$ *convex disjoint obstacles.* Rather than preprocessing the obstacles, visibility is computed on-line by exploiting the particular nature of Dijkstra's shortest path algorithm.

Rohnert [174] suggested an $O(n \log n + k^2)$ time algorithm for finding the shortest path between two arbitrary points in $E^2$ with $k$ *convex obstacles which boundaries may intersect pairwise at most twice.* This time bound is achieved provided that the obstacles are preprocessed in $O(n \log n + k^3)$ time and $O(n + k^2)$ space.

Sharir and Schorr [182] presented an $O(n^n)$ algorithm for the ESPO in $E^3$ with *polyhedral obstacles.* This algorithm was improved to $O(n^{k^r})$ (where $r$ is a constant) by Reif and Storer [171]. An even more efficient but still exponential time algorithm was recently given by Canny [32]. The algorithm is more general as it is applicable to obstacles which need not to be polyhedral. Furthermore, the object to be moved needs not to be a point but essentially can be of arbitrary shape.

If the problem is restricted to finding the shortest path between two points $s$ and $t$ in $E^3$ avoiding a single convex polyhedral obstacle, an $O(n^2 \log n)$ algorithm has been suggested by Mount [155] and Sharir and Shorr [182].

When the number of convex polyhedral obstacles is increased to two, a polynomial time algorithm (almost proportional to $O(n^3 \log n)$) has been obtained by Baltsan and Sharir [16]. Sharir [181] shows the existence of polynomial time algorithms for shortest paths amidst any *fixed* number of convex polyhedral obstacles.

**Heuristics.** In view of the inherent difficulty in solving the ESPO in $E^3$, approximative methods are of interest. Papadimitriou [160] developed a polynomial approximation algorithm which finds a path of length at most $(1+\varepsilon)$ times the length of the optimal path. Another approximative algorithm was developed by Clarkson [52] which is faster for larger $\varepsilon$.

**Generalizations and Special Cases.** There are numerous problems related to the ESPO. Only a few will be mentioned below.

- **Motion Planning.** One interesting generalization of the ESPO arises when the path to be found is required to have some minimum width so that an object can be moved from $s$ to $t$. Most approaches to this problem are closely related to the visibility graph approaches to the ESPO. In particular, Baker [15] and Chew [46] obtained $O(n^2 \log n)$ algorithms for a disk moving among polygonal obstacles using visibility graphs. Reif and Storer [170] obtained an $O(n^2)$ algorithm for moving a disk among polygonal obstacles using the shortest path map approach.

  The visibility graph approach was generalized by Hershberger and Guibas [105]. They obtained an $O(n^2)$ algorithm for the movement of arbitrary convex objects among polygonal obstacles under the assumption that the motion consists of a sequence of translations but no rotations.

  The problem becomes very complex when both translations and rotations are allowed. Papadimitriou and Silverberg [161] developed an $O(n^4 \log n)$ algorithm for the convex polygonal object with a fixed reference point which must move along the edges of the visibility graph.

- **Shortest Path Inside a Polygon.** An interesting special case arises when there is only one obstacle which is the region *outside* a simple polygon. Lee and Preparata [129] suggested an $O(n \log\log n)$ algorithm for this problem. The interior of the polygon is triangulated. A uniquely determined sequence of triangles between the triangle containing $s$ and the triangle containing $t$ is then identified. It contains the shortest path from $s$ to $t$ which can be found in $O(n)$ time. A preprocessed version of this algorithm has been discussed by Guibas and Hershberger [99].

- **Weighted Region Problem.** Another generalization was suggested by Mitchell and Papadimitriou [153]. The plane is partitioned into polygonal regions. A weight is assigned to each region. The cost of a path is defined as the sum of the weights times the length of line segments of the path passing through different regions. The problem is to find the least cost path between $s$ and $t$. The ESPO is a special case of this *weighted region problem*: Assign weight $\infty$ to obstacle regions, and weight 1 to the rest of the plane. This problem has many practical applications. Mitchell and Papadimitriou showed that the least cost path consists of straight line segments within the region. They presented an $O(n^7 L)$ algorithm where $L$ is the number of bits to represent the input and the output.

- **Moving Obstacles.** Yet another generalization of tremendous practical importance arises when the obstacles are permitted to move [170].

- **Rectilinear Case.** The *rectilinear shortest path problem with obstacles* (RSPO) problem is closely related to the ESPO. We will not review the RSPO here.

Widmayer [203] surveys the literature for the RSPO from the $\mathcal{CG}$ point of view and provides some new algorithms.

### 5.4.3 Steiner Minimal Tree Problem with Obstacles

The *Euclidean Steiner minimal tree problem with obstacles* (ESMTO) can be formulated as follows, Figure 30.

- **Given:** $Z = \{z_1, z_2, ..., z_n\} \in E^2$ and a set $\Omega = \{\omega_1, \omega_2, ..., \omega_k\}$ of polygonal obstacles.
- **Find:** Shortest network spanning $Z$ and avoiding all obstacles.
- **Data Structure:** Here the triangulation of the extreme points of the obstacles and the point set illustrates one useful data structure for this problem. The solution generated for this example was based on a heuristic described in [184].

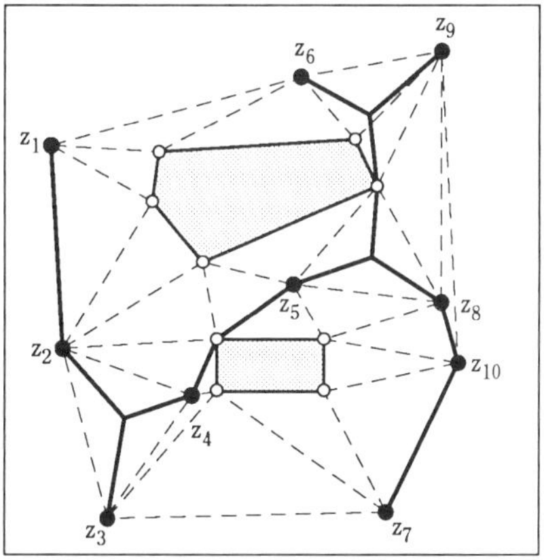


**Figure 30: SMTO Solution**

**Applications.** There are many application areas for this problem ranging from the optimal design of oil and natural gas pipeline systems in regions with obstacles such as lakes, mountains and other natural phenomena to the routing of mechanical and electrical systems around the architectural and structural elements within buildings [188].

**Lower Bounds.** The ESMTO is a generalization of the ESMT problem discussed in Subsection 5.3.1. Hence the ESMTO is also $\mathcal{NP}$-hard.

**Algorithms.** No general exact algorithm for the ESMTO problem is available. Basic properties for routing Steiner minimal trees around obstacles were given by Smith [184]. In particular, an $O(N \log N)$ algorithm for three $Z$-points in the presence of a single convex polygonal obstacle was given ($N$ denotes the total number of $Z$-points and obstacle vertices). Winter and Smith [211] improved this to a $\Theta(N)$ algorithm. If the obstacle is preprocessed in an appropriate way in $O(N)$ time, the ESMTO can be solved in $O(\log N)$ time for any triple of $Z$-points. A straightforward $O(N^2)$ algorithm for four $Z$-points and one convex polygonal obstacle was suggested in [209]. No $\Theta(N)$ algorithm for this case is currently available.

**Heuristics.** Provan [169] developed an $\epsilon$-approximation scheme for the ESMTO problem. It operates in three stages.

- **Step 1:** The path-convex-hull $PCH(Z,\Omega)$ (see Subsection 4.2.1) is constructed in $O(N^3)$ time. Obstacles not in $PCH(Z,\Omega)$ are pruned away since an optimal solution to the ESMTO is bound to be inside.

- **Step 2:** A set $S$ of additional grid points (whose density depends on $\epsilon$) is placed within the $PCH(Z,\Omega)$ (outside of the obstacles). The visibility graph $VG(Z \cup S,\Omega)$ is constructed.

- **Step 3:** Any algorithm [208] or heuristic [210] for the Steiner minimal tree problem in graphs is used to find Steiner minimal tree for $Z$ in $VG(Z \cup S,\Omega)$.

An $O(N^3)$ heuristic for the ESMTO problem was proposed by Armillotta and Mummolo [7]. It consists of four steps.

- **Step 1:** Construct the visibility graph $VG(Z,\Omega)$.

- **Step 2:** Find a low-cost tree $T$ in $VG(Z,\Omega)$ spanning all $Z$-points.

- **Step 3:** Improve $T$ by replacing pairs of incident edges by Euclidean Steiner minimal trees for their three end-points (disregarding the obstacles).

- **Step 4:** Readjust Steiner points if some of their incident edges penetrate obstacles. **Stop** if no readjustments were required. Otherwise go to **Step 3**.

Winter and Smith [209] outlined another heuristic which is conceptually closely related to the $O(n \log n)$ heuristic for the ESMT problem [187] (see Subsection 5.3.1). It involves constrained triangulation of $Z$-points as well as solving the ESMTO for three and four $Z$-points with one or few convex polygonal obstacles.

**Related Problems.** There are a number of extensions possible depending on how one characterizes the obstacles, metrics, dimensions and their interrelationships.

- **Rectilinear Steiner Minimal Tree Problem with Obstacles.** This problem is of major importance to VLSI design and routing of building services within facilities. Relatively little work has been done on this problem. Liestman [135] derived some properties of optimal solutions and discussed solution methods for small problem instances. Watanabe and Sugiyama [201] suggested a parallel heuristic.

- **Euclidean Steiner Minimal Tree Problem with Curvilinear Obstacles.** Constructing Euclidean Steiner minimal trees in geographic regions will often encounter lakes and other natural phenomenon which have natural curvilinear edges and boundaries. In principle, the solutions for navigating these curvilinear obstacles are closely related to the ESMTO solutions for polygonal obstacles.

- **Euclidean Steiner Minimal Tree Problem on Riemann Surfaces.** The construction of Steiner minimal trees on the sphere, torus, and other three-dimensional objects or Riemann surfaces is of interest to many terrestrial as well as extra-terrestrial engineering applications. An $O(n \log n)$ heuristic for the ESMT on a sphere was developed by Dolan et al. [65].

- **Three-Dimensional Euclidean and Rectilinear Steiner Minimal Tree Problems with Obstacles.** There are various applications of the Euclidean version. One important example is piping and mechanical system design within facilities. The rectilinear version is also important in this connection. Furthermore, it is of critical importance in VLSI design, where the interconnections between transistors and their multi-level planes is necessarily three-dimensional.

It is perhaps hard not to wax about these problems because if one looks around his or her environment, one soon sees that the SMTO problems are omnipresent. They represent one of the most exciting directions for future $\mathcal{CG}$ and geometric $\mathcal{TND}$ research.

### 5.4.4 Classification of $G(\Omega)$ Problems

Table 4 illustrates the key $\mathcal{CG}$ approaches and their worst-case running time algorithms. It is perhaps not surprising that the predominant algorithms developed for the $G(\Omega)$ problems are iterative in nature and primarily depend on the visibility graphs. We have included many algorithms which are special cases. This is only natural considering the complexity of the obstacle avoidance problems.

The wide open nature of the problems in this class together with the frugal utilization of some of the other paradigms and geometric data structures represents perhaps the most fruitful avenue for future research.

## 5.5 $\mathcal{C}^{\mathcal{CG}}_{\mathcal{TND}}$ Cube

Given our $\mathcal{CG}$ Data Structures, Strategies and geometric $\mathcal{TND}$ Problem categories, we have constructed a three-dimensional cube within which the various algorithms developed for geometric $\mathcal{TND}$ problems can be classified. In fact we will refer to it as a composite "$\mathcal{C}^{\mathcal{CG}}_{\mathcal{TND}}$" cube. This type of morphological analysis is very useful in categorizing the $\mathcal{CG}$ algorithms as well as indicating where new algorithms might be developed or discovered. One of the major purposes of this type of classification schema is that it suggests how and with what concepts one might design an algorithm or a heuristic to solve a geometric $\mathcal{TND}$ problem in a most efficient way considering time and space tradeoffs and its underlying geometric nature. Figure 31 illustrates this three-dimensional cube.

In the following subsections, we will examine the $\mathcal{C}^{\mathcal{CG}}_{\mathcal{TND}}$ cube with three different graphical and visual tools providing a comprehensive picture of the way $\mathcal{CG}$ and $\mathcal{TND}$ problems currently interact. The three tools we shall utilize are respectively: *1) Composite Matrix, 2) Planar Graph Diagram, 3) Cube Decomposition.*

| | $\mathcal{CG}$ Paradigms | | | | |
|---|---|---|---|---|---|
| **G**($\Omega$) | *ITER* | *PSWE* | *D&C* | *LOCUS* | *P&S* |
| $CH(Z)$ | **ESMTO:** $O(?)$ **[169]** | | | | |
| $VOR(Z)$ | **ESPO:** $O(n(k + \log n))$ **[171]** | **ESPO:** $O(n \log n)$ **[129]** | | | |
| $DT(Z)$ | **ESMTO:** $O(?)$ **[209]** | | | | |
| $VG(Z)$ | **ESPO:** $O(n^2)$ **[138, 9, 202, 8]** | | | | |
| | **ESPO:** $O(k^2 \log n, n + k^2, n \log n + k^2)$ **[173]** | | | | |
| | **ESPO:** $O(n \log n + K^3, n + k^2, n \log n + k^2)$ **[174]** | | | | |
| | **ESPO:** $O(n^2 \log n)$ **[155, 182]** | | | | |
| | **ESPO:** $O(n \log \log n)$ **[129]** | | | | |
| | **ESPO:** $O(n^7 L)$ **[153]** | | | | |
| | **ESPO:** $O(n^3 \log n)$ **[16]** | | | | |
| | **ESPO:** $O(n^4 \log n)$ **[161]** | | | | |
| | **ESPO:** $O(n^2 \log n)$ **[15, 46]** | | | | |
| | **ESPO:** $O(n^2)$ **[170]** | | | | |
| | **ESPO:** $O(?)$ **[160, 52]** | | | | |
| | **ESPO:** $O(n^{k^r})$ **[171]** | | | | |
| | **ESPO:** $O(n^n)$ **[182, 32]** | | | | |
| $CD(Z)$ | **ESMTO:** $O(?)$ **[7]** | | | | |
| | **MAXG**($\Omega$)**:** $O(?)$ **[149, 150]** | | | | |
| $OTHERS(Z)$ | **ESMTO:** $\Theta(N)$ **[211]** | | | | |

**Table 4:** $G(\Omega)$ **Classification Table**

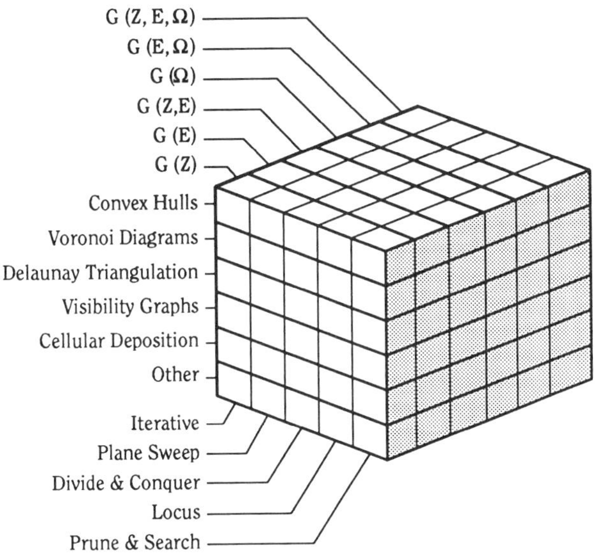


**Figure 31:** $\mathcal{C}^{\mathcal{CG}}_{\mathcal{TND}}$ **Cube**

**Composite Matrix** We first combine all the classification matrices previously arrayed in Tables 1–4 into one large two-dimensional matrix labelled Table 5. This enables us to summarize and categorize all methods and data structures across $\mathcal{TND}$ problem classes as viewed from the point of view held throughout Section 5.

**Planar Graph Diagram** Also, a *Planar Graph Diagram* is utilized to provide a two-dimensional representation of the pairwise entities in the composite matrix. The Planar Graph Diagram re-interprets the composite matrix in a graphical format in order to show more clearly the pairwise relationships in Table 5.

Within the Planar Graph Diagram, the heavier weight lines indicate the strength of relationship or importance between the pairwise relationships embedded in the composite matrix. The numbers within the edges refer to the number of contributions that we have identified in Table 5 for each pairwise relationship.

$\mathcal{C}^{\mathcal{CG}}_{\mathcal{TND}}$ **Cube Decomposition** After we have examined the Composite Matrix, we present a three-dimensional decomposition of the $\mathcal{C}^{\mathcal{CG}}_{\mathcal{TND}}$ cube. This three-dimensional decomposition is essential in visualizing the three-way relationships embodied in the two-dimensional Composite Matrix and Planar Graph diagram.

When examining the decomposed $\mathcal{C}^{\mathcal{CG}}_{\mathcal{TND}}$ we have identified the number of algorithms within a particular cell with a single dot entry to simplify discussion of the three-dimensional link between the dimensions $\mathcal{CG}$ Data Structure, Paradigms and $\mathcal{TND}$ Problem.

| | $\mathcal{CG}$ Paradigms | | | | |
|---|---|---|---|---|---|
| $G(Z,E,\Omega)$ | *ITER* | *PSWE* | *D&C* | *LOCUS* | *P&S* |
| $CH(Z)$ | *ETSP*: $O(?)$ [189]<br>*RSMT*: $O(n^4)$ [186]<br>*RSMT*: $O(n^6)$ [26, 168]<br>ESMTO: $O(?)$ [169] | *RSMT*: $O(n^3)$ [5]<br>*RSMT*: $O(n)$ [56] | | | EMXT: $\Theta(n \log h)$ [154] |
| $VD(Z)$ | MAXG: $O(n \log n)$ [179]<br>ESPO: $O(n(k + \log n))$ [171] | ESPO: $O(n \log n)$ [129] | | POFP: $O(n \log n, n, \log n)$ [179, 166, 76] | MINC: $O(n \log n)$ [179] |
| $DT(Z)$ | ESMTO: $O(?)$ [209] | | | *RSMT*: $O(n \log n)$ [186]<br>*ESMT*: $O(n \log n)$ [187] | ANNP: $\Theta(n \log n)$ [180]<br>EMST: $\Theta(n \log n)$ [179]<br>RMST: $\Theta(n \log n)$ [125, 132] |
| $VG(Z)$ | ESPO: $O(n^2)$ [138, 9, 202, 8]<br>ESPO: $O(k^2 \log n, n + k^2, n \log n + k^2)$ [173]<br>ESPO: $O(n \log n + K^3, n + k^2, n \log n + k^2)$ [174]<br>ESPO: $O(n^2 \log n)$ [155, 182]<br>ESPO: $O(n \log \log n)$ [129]<br>ESPO: $O(n^7 L)$ [153]<br>ESPO: $O(n^3 \log n)$ [16]<br>ESPO: $O(n^4 \log n)$ [161]<br>ESPO: $O(n^2 \log n)$ [15, 46]<br>ESPO: $O(n^2)$ [170]<br>ESPO: $O(?)$ [160, 52]<br>ESPO: $O(n^{k^r})$ [171]<br>ESPO: $O(n^n)$ [182, 32] | | | | |
| $CD(Z)$ | MAXG: $\Theta(n)$ [95]<br>*ETSP*: $O_a(n \log n)$ [21]<br>ESMTO: $O(?)$ [7]<br>MAXG($\Omega$): $O(?)$ [149, 150] | | ANNP: $O(n(\log n)^{d-1})$ [24, 20]<br>*ETSP*: $O(n \log n)$ [116]<br>*RSMT*: $O(n \log n)$ [121] | POFP: $O(n^{2^{d+1}}, n^{2^{d+1}}, \log n)$ [62, 213]<br>POFP: $O(n^2, n^2, (\log n)^2)$ [39] | *MAXG*: $O_a(n^2)$ [147, 148]<br>ANNP: $\Theta(n \log n)$ [196, 199]<br>ANNP: $O(n \log \delta)$ [50, 85]<br>EMST:$O(n \log n)^{1.5}$ [213, 39]<br>EMST:$O_e((n \log n)^{4/3})$ [4]<br>EMST $O_e(\alpha(cn, n)n)$ [197]<br>*EMST*: $O(\epsilon^{-d} n \log n)$[197]<br>*EMST*: $O(n(\log n + (1/\varepsilon) \log \delta))$ [51] |
| *OTHERS* | MINC: $O(n^2)$ [81]<br>SMBO: $O(n^4)$ [179]<br>EMST: $O(n^2 \log n)$ [22, 156]<br>ETSP: $O(exp)$<br>*ESMT*: $O(?)$ [36, 122]<br>*ESMT*: $O(n^3)$ [17]<br>*RSMT*: $O(n^3)$[17, 25]<br>*RSMT*: $O(n \log n)$ [107]<br>ESMTO: $\Theta(N)$ [211] | *RSMT*: $O(n \log n)$ [172] | CPP: $\Theta(n \log n)$ [24, 20]<br>EMST: $O(n^2 \log n)$ [35] | | MINC: $\Theta(n)$ [146, 73]<br>*ETSP*: $O(n^{2.5} (\log n)^4)$ [47]<br>ESMT: $O(exp)$ [151, 183] |

Table 5: $G(Z, E, \Omega)$ Classification Table

In the best of all worlds, we would have a holographic color gradation display capable of providing the reader with a dynamic three-dimensional means of examining the relationship between $\mathcal{TND}$ and $\mathcal{CG}$ through the $\mathcal{C}^{\mathcal{CG}}_{\mathcal{TND}}$. Our **3-d** decomposition affords a simplified focal point since the decomposed matrix representations allow one a means to visualize the interaction between $\mathcal{TND}$ and $\mathcal{CG}$.

## 5.6 Composite Matrix

Table 5 represents the composite matrix of Tables 1–4 which appeared throughout Section 5. We examine Table 5 firstly from a columnar $\mathcal{CG}$ Paradigmatic perspective, then secondly a row-wise $\mathcal{CG}$ Data Structure perspective.

$\mathcal{CG}$ **Paradigms Columnar Perspective** :

- If we review the matrix Table 5 via the columns, *Iterative and Prune-and-Search* paradigms are the most powerful in solving $\mathcal{CG}$. Roughly speaking there are 33 entries in the *Iterative* category, 16 in *Prune-and-Search* while the rest are equally spread among *Plane Sweep, Divide-and-Conquer and Locus.* The *Iterative* paradigm is dominant probably due to its straightforward implementation.
- It is surprising that *Plane Sweep, Divide-and-Conquer and Locus* are not as critical in solving geometric $\mathcal{TND}$ problems as much as they are in constructing the $\mathcal{CG}$ geometric objects themselves.
- Certainly other strategies such as *Geometric Transform* and *Dynamization* are appropriate, however their exploitation in $\mathcal{CG}$ and $\mathcal{TND}$ remains to be fully developed.

$\mathcal{CG}$ **Data Structures Row-wise Perspective** :

- Examining the row entries in Table 5, *Cellular Decomposition* has 17 entries, *Others* has 15 and *Visibility Graphs* has 13 entries. These three data structures dominate whereas *Convex Hulls* has 7 entries, *Delaunay Triangulation* 6 and *Voronoi Diagram* 5.
- Finally, roughly 40% of the cells in the matrix remain unfilled which would indicate that many potential $\mathcal{CG}$ data structure and paradigmatic approaches remain open for exploration. For example, there is great potential for the use of *Visibility Graphs* along with the other $\mathcal{CG}$ strategies in $\mathcal{TND}$ obstacle problems. This potential is further amplified through open cells in the cube $\mathcal{C}^{\mathcal{CG}}_{\mathcal{TND}}$ decomposition, see Figure 34.

## 5.7 Planar Graph Diagram

- The Planar Graph Diagram clearly reinforces the comments about the importance of the *Iterative* and *Prune-and-Search* strategies. These two strategies are central in the diagram because the vertex degree of these two strategies is the

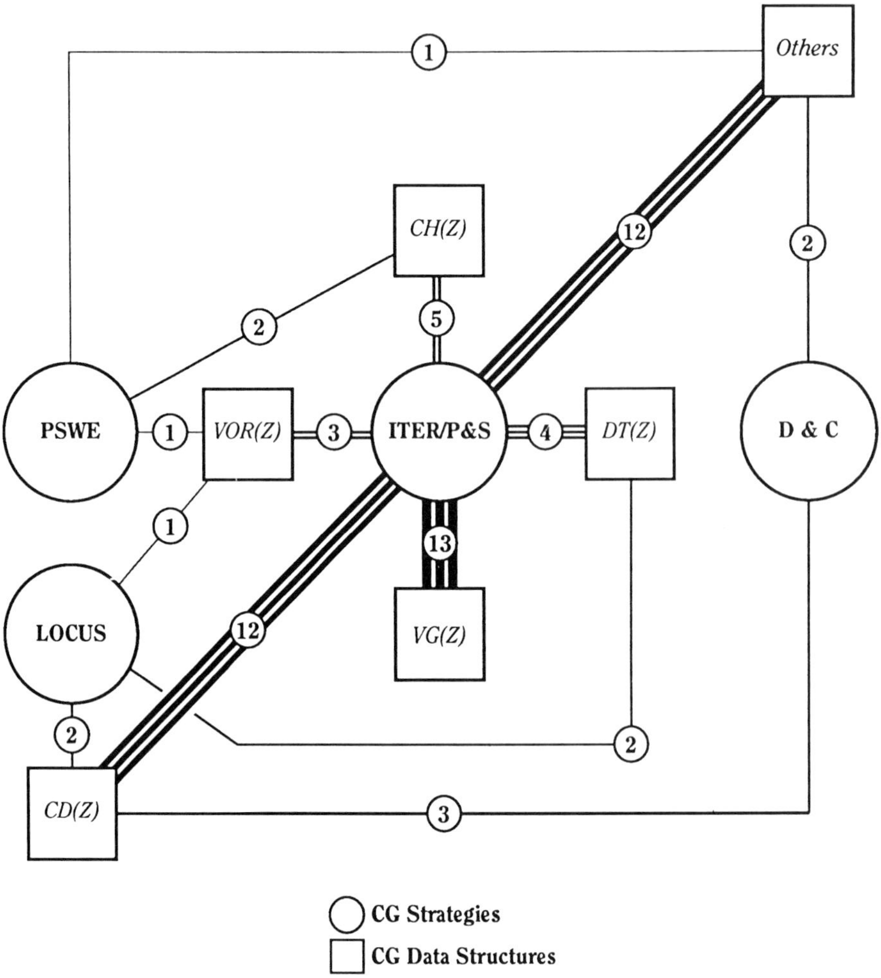


Figure 32: Planar Graph Diagram of CG Geometric Data Structures and Paradigms

highest of all strategies and in order to keep the graph planar, they should be centrally located in the diagram. *Divide-and-Conquer, Plane Sweep, and Locus* are surprisingly less utilized and not centrally located in the diagram.

- What is most striking about the Planar Graph diagram is that it clearly shows that in solving $G(Z, E, \Omega)$ problems, people have taken Occam's Razor to heart: apply the simplest data structure along with the simplest strategy in order to solve a $\mathcal{TND}$ problem.
- Elegance in solving $\mathcal{TND}$ problems seems lacking. For example, one wonders why there has been no efforts to link *Visibility Graphs* and the *Locus* strategy especially for obstacle problems rather than relying on purely *Iterative* methods.

## 5.8 $\mathcal{C}^{\mathcal{CG}}_{\mathcal{TND}}$ Decomposition

In our final synthesis, we examine the $\mathcal{C}^{\mathcal{CG}}_{\mathcal{TND}}$ Cube three-dimensionally and discuss the conclusions we can make about each dimension and their three-way relationships.

**$\mathcal{TND}$ Problem Category:** In evaluating the most active $\mathcal{TND}$ problem categories the following breakdown is recorded:

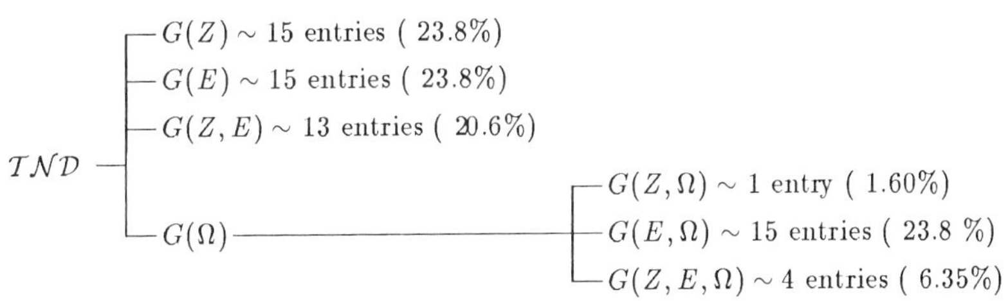


**Figure 33: Morphological Tree for $G(Z, E, \Omega)$ Problems**

Thus, as is clearly shown in the tree diagram, most research efforts have been concentrated on the $G(Z), G(E), G(Z, E)$ problem categories with the exception of the $G(E, \Omega)$ shortest path problem. The latter effort is largely motivated by applications to vehicle routing and robotics.

**CG Paradigms:** Upon inspecting the cube decomposition, it is obvious that *Iterative* is the most general and robust paradigm across all $\mathcal{TND}$ problems while *Prune-and-Search* have largely been applied to $\mathcal{TND}$ problems without obstacles. The three middle paradigms *Plane Sweep, Divide-and-Conquer* and *Locus* have had spotty performance and no clear trends occur. This again is surprising in that the three middle paradigms are central to the construction of $\mathcal{CG}$ objects yet do not re-appear in $\mathcal{TND}$ problem solving except incidentally.

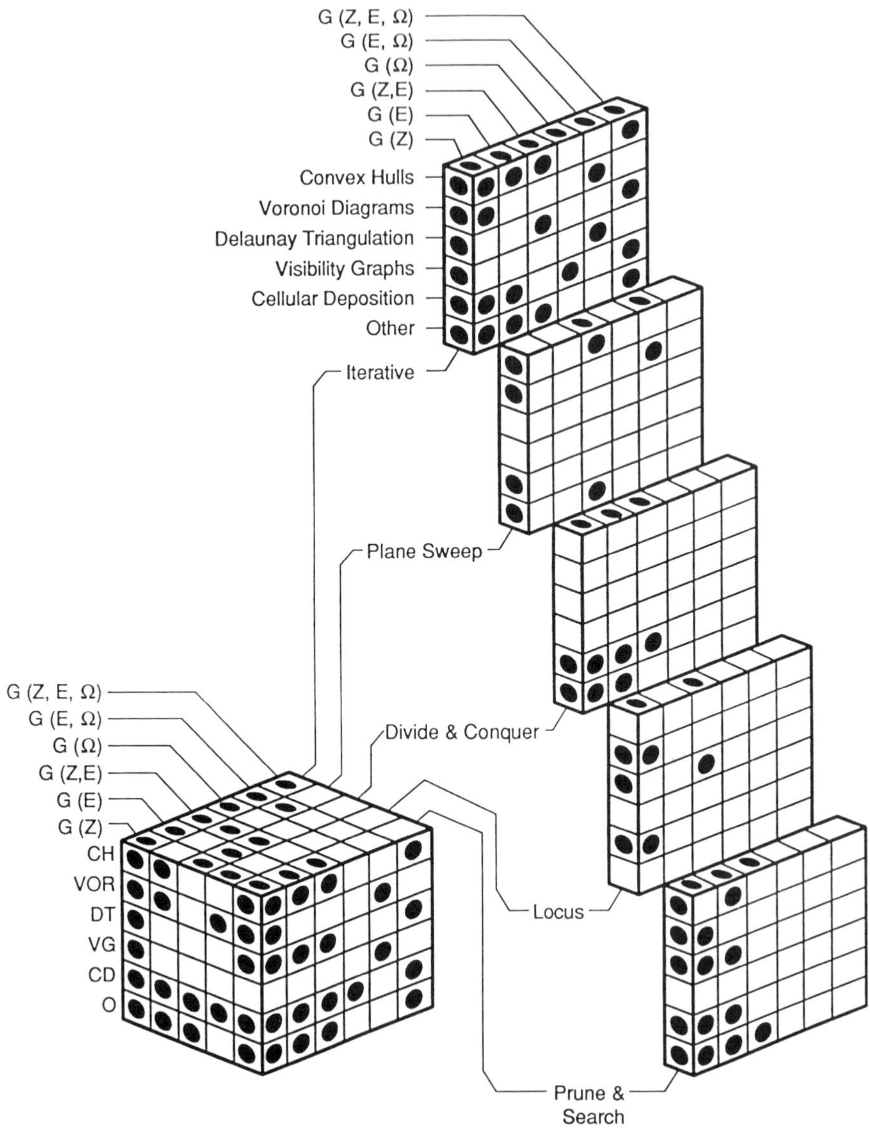


**Figure 34:** $\mathcal{C}^{\mathcal{CG}}_{\mathcal{TND}}$ **Cube Decomposition**

$\mathcal{CG}$ **Data Structures:** Again to summarize what we have said already, *Cellular Decomposition* and *Other* data structures dominate. *Convex Hulls, Voronoi Diagrams* and *Delaunay Triangulations* are indirectly used to construct the geometric objects rather than directly used as data structures in solving $\mathcal{TND}$ problems. There are exceptions of course, but their general usefulness is surprising in light of the geometric nature of some of the $\mathcal{TND}$ problems. The focus of *Visibility Graphs* only on one $\mathcal{TND}$ problem category has limited its generality.

In conclusion, the mainstay $\mathcal{CG}$ Paradigms and Data Structures have not been explored to their fullest potential in the design of algorithms for geometric $\mathcal{TND}$ problems. This is surprising because most $\mathcal{TND}$ problems are in two and three dimensions where the $\mathcal{CG}$ Data Structures have low order complexity. Rather, it is the simpler $\mathcal{CG}$ Paradigms and Data Structures which dominate. Perhaps elegance awaits in the wings.

## 5.9 Open Problems

One might argue that perhaps we should have provided more details and more room for other $\mathcal{CG}$ Data Structures, $\mathcal{CG}$ Paradigms, and perhaps even additional $\mathcal{TND}$ problem types. This is certainly possible and we encourage others to criticize and propose additions and modifications to the $\mathcal{C}^{\mathcal{CG}}_{\mathcal{TND}}$. In fact, there are probably three cubes total because a cube for the routing and capacitated resource problems should be developed: $\mathcal{C}^{\mathcal{CG}}_{\mathcal{RND}}; \mathcal{C}^{\mathcal{CG}}_{\mathcal{CND}}$. Better yet, one cube with all three problem types should be created: $\mathcal{C}^{\mathcal{CG}}_{\mathcal{ND}}$.

To say that much more needs to be done and that there are many open problems is probably too obvious a statement to make, yet the morphological classification schema avaliable with the cubes $\mathcal{C}^{\mathcal{CG}}_{\mathcal{TND}}$ demonstrates great potential in recognizing and stimulating new algorithmic approaches since we can begin to conjoin the $\mathcal{TND}$ problems with $\mathcal{CG}$ Data Structures and $\mathcal{CG}$ Strategies, where none have been attempted before.

One definition of creativity centers around the apparent joining of two unrelated concepts or ideas, where none have been joined before. We hope that our work sparks the imagination of some researchers to make the leap that will yield new ideas, algorithms and a better understanding of the relationship between $\mathcal{TND}$ and $\mathcal{CG}$.

## 6. Summary and Conclusions

When we set out to do this survey, we had little idea of the Herculean adventure on which we were embarking. We have attempted in this survey to provide an overview and an in-depth understanding of the relationship between $\mathcal{TND}$ Problems and $\mathcal{CG}$ Data Structures and Paradigms. As we have argued, this is a rich tapestry and while we have tried to weave together three different perspectives through the cubes, matrices and graph diagrams, we have only begun to hint at what is possible for these fields.

The two and three-dimensional tables, matrices, and cubical decompositions illustrate not only what we know already but indicate what work lies in the future for weaving together these concepts.

While we have indicated some of the key relationships, there is a great deal of material upon which we have not begun to elaborate since this survey is already rather detailed and lengthy. Given more time, we would like to have addressed parallel algorithms and stochastic network analysis which are very important to the general $\mathcal{ND}$ problems we discussed at the beginning of the paper as well as its many applications. Also, the issues of $\mathcal{TND}$ problem/model integration especially in relation to the many algorithms available for $\mathcal{TND}$, $\mathcal{RND}$, and $\mathcal{CND}$ problems (see Figure 1).

This entire area interrelating $\mathcal{TND}$ and $\mathcal{CG}$ is one which should grow and flourish over the years and we trust that this survey will inspire some others to further explore the richness and beauty of the tapestry of these disciplines.

## REFERENCES


[1] P.K. Agarwal, *"Partitioning Arrangements of Lines I: An Efficient Deterministic Algorithm,"* *Discrete and Computational Geometry* **5** (1990) 449-483.

[2] P.K. Agarwal, *"Partitioning Arrangements of Lines II: Applications,"* *Discrete and Computational Geometry* **5** (1990) 533-573.

[3] P.K. Agarwal, *"Ray Shooting and Other Applications of Spanning Trees with Low Stabbing Number,"* *to appear in SIAM J. Comput.* (1990).

[4] P.K. Agarwal, H. Edelsbrunner, O. Schwarzkopf and E. Welzl, *"Euclidean Minimum Spanning Trees and Bichromatic Closest Pairs,"* DIMACS Tech. Report 90-54 (1990).

[5] A.V. Aho, M.R. Garey and F.K. Hwang, *"Rectilinear Steiner Trees: Efficient Special-Case Algorithms,"* *Networks* **7** (1977) 37-58.

[6] H. Alt and E. Welzl, *"Visibility Graphs and Obstacle Avoiding Shortest Paths,"* *ZOR* **32** (1988) 145-164.

[7] A. Armillotta and G. Mummolo, *"A Heuristic for the Steiner Tree Problem with Obstacles,"* Technical Report, Department of Industrial Design and Production, University of Bari, Italy (1988).

[8] Te. Asano, Ta. Asano, L.G. Guibas, J. Hershberger and H. Imai, *"Visibility of Disjoint Polygons,"* *Algorithmica* **1** (1986) 49-63.

[9] Ta. Asano, Te. Asano and H. Imai, *"Shortest Path Between Two Simple Polygons,"* *IPL* **24** (1987) 285-288.

[10] F. Aurenhammer, *"Voronoi Diagrams - A Survey of a Fundamental Geometric Data Structure,"* Technical Report B-90-09, Department of Mathematics, Free University of Berlin (1990).

[11] F. Aurenhammer, *"A New Duality Result Concerning Voronoi Diagrams,"* *Discrete and Computational Geometry* **5** (1990) 243-254.

[12] F. Aurenhammer and H. Edelsbrunner, *"An Optimal Algorithm for Constructing the Weighted Voronoi Diagram in the Plane,"* *Pattern Recognition* **17** (1984) 251-257.

[13] C. Bajaj, *"The Algebraic Degree of Geometric Optimization Problems,"* *Discrete and Computational Geometry* **3** (1988) 177-191.

[14] C. Bajaj, *"Geometric Optimization and $D^p$ Completeness,"* *Discrete and Computational Geometry* **4** (1989) 3-13.

[15] B. Baker, *"Shortest Paths with Unit Clearance among Polygonal Obstacles," presented at the SIAM Conference on Geometrical Modelling and Robotics, 1985*

[16] A. Baltsan and M. Sharir, *"On Shortest Paths between Two Convex Polyhedra,"* Technical Report 180, Computer Science Dept., Courant Institute, New York (1985).

[17] J.E. Beasley, *"A Greedy Heuristic for the Euclidean and Rectilinear Steiner Problem,"* Working Paper, Dept. of Management Science, Imperial Collage, London (1989).

[18] M. Bellmore and L. Nemhauser, *"The Traveling Salesman Problem: A Survey," Operations Research* **16** (1968) 538-558.

[19] J.L. Bentley, *"Multidimensional Search Trees Used for Associative Searching," Comm. ACM* **18** (1975) 509-517.

[20] J.L. Bentley, *"Multidimensional Divide and Conquer," Comm. ACM* **23** (1980) 214-229.

[21] J.L. Bentley, *"Experiments on Geometric Traveling Salesman Heuristics"* Tech. Report, AT & T Bell Laboratories, Murray Hill (1990).

[22] J.L. Bentley and J.H. Friedman, *"Fast Algorithms for Constructing Spanning Trees in Coordinate Spaces" IEEE Transactions on Computers* **C-27** (1978) 97-105.

[23] J.L. Bentley and H.A. Maurer, *"A Note on Euclidean Near Neighbor Searching in the Plane," IPL* **8** (1979) 133-136.

[24] J.L. Bentley and M. Shamos, *"Divide and Conquer in Multidimensional Space," Proc. of the 8-th ACM Ann. Symp. on the Theory of Computation* (1976) 220-230.

[25] M.W. Bern and M. de Carvaho, *"A Greedy Heuristic for the Rectilinear Steiner Problem,"* Technical Report, Computer Science Division, Univ. of California at Berkeley (1985).

[26] M.W. Bern, *"Network Design Problems: Steiner Trees and Spanning k-Trees,"* Ph.D. Thesis, Computer Science Dept., Univ. of California at Berkeley (1987).

[27] B. Bhattacharya, *"Worst-Case Analysis of a Convex Hull Algorithm,"* Unpublished manuscript, Dept. of Computer Science, Simon Fraser University (1982).

[28] M. Blum, R.W. Floyd, V. Pratt, R. Rivest and R.E. Tarjan, *"Time Bounds for Selection," J. of Computer and Systems Science* **7** (1973) 448-461.

[29] R.S. Booth, *"The Steiner Ratio for Five Points," (to appear) Annals of Operations Research.*

[30] B.N. Boots, *"Voronoi (Thiessen) Polygons"* Norwich, England: Geo Books, W.H. Hutchins & Sons (1986).

[31] K.Q. Brown, *"Voronoi Diagrams from Convex Hulls,"* *IPL* **9** (1979) 223-228.

[32] J. Canny, *"A New Algebraic Methods for Robot Motion Planning and Real Geometry,"* *Proc. of the 28-th IEEE Symp. on the Foundations of Computer Science* (1987) 39-48.

[33] J. Canny and J. Reif, *"New Lower Bound Techniques for Robot Motion Planning Problems,"* *Proc. of the 28-th IEEE Symp. on the Foundations of Computer Science* (1987) 49-60.

[34] D.R. Chand and S.S. Kapur, *"An Algorithm for Convex Polytopes,"* *J. ACM* **17** (1970) 78-86.

[35] R.C. Chang and R.C.T. Lee, *"An $O(n \log n)$ Minimal Spanning Tree Algorithm for n Points in the Plane,"* *BIT* **26** (1986) 7-16.

[36] S.K. Chang, *"The Generation of Minimal Trees with a Steiner Topology,"* *J. ACM* **19** (1972) 699-711.

[37] B.M. Chazelle, *"A Theorem on Polygon Cutting with Applications,"* *Proc. of 23-th Ann. IEEE Symp. On Foundations of Computer Science* (1982) 339-349.

[38] B.M. Chazelle, *"On the Convex Layers of a Convex Set,"* *IEEE Transactions on Information Theory* **IT-31** (1985) 509-517.

[39] B.M. Chazelle, *"How to Search in History,"* *Information and Control* **64** (1985) 77-99.

[40] B.M. Chazelle, *"Triangulating a Simple Polygon in Linear Time, "* Technical Report CS-TR-264-90, Princeton Univ., Princeton, NJ (1990).

[41] B.M. Chazelle and H. Edelsbrunner, *"An Improved Algorithm for Constructing k-th Order Voronoi Diagrams,"* *IEEE Transactions on Computers* **36** (1987) 1349-1354.

[42] B.M. Chazelle and L.J. Guibas, *"Fractional Cascading: I. A Data Structuring Technique,"* *Algorithmica* **1** (1986) 133-162.

[43] B.M. Chazelle, L.J. Guibas and D.T. Lee, *"The Power of Geometric Duality,"* *BIT* **25** (1985) 76-90.

[44] B.M. Chazelle and E. Welzl, *"Quasi Optimal Range Searching in Spaces with Finite VC-Dimensions,"* *Discrete and Computational Geometry* **4** (1989) 467-489.

[45] D. Cheriton and R.E. Tarjan, *"Finding Minimum Spanning Trees,"* *SIAM J. Comput.* **5** (1976) 724-742.

[46] L.P. Chew, *"Planning the Shortest Path for a Disk in $O(n^2 \log n)$ Time,"* *Proc. of the 1-st Ann. Symp. on Computational Geometry* (1985) 214-220.

[47] N. Christofides, *"Worst-Case Analysis of a New Heuristic for the Travelling Salesman Problem," Proc. Symp. on New Directions and Recent Results in Algorithms and Complexity*, Carnegie-Mellon Univ., Pittsburgh (1976).

[48] F.R.K. Chung and R.L. Graham, *"Steiner Trees for Ladders," Annals of Discrete Mathematics* **2** (1978) 173-200.

[49] F.R.K. Chung and R.L. Graham, *"A New Bound for Euclidean Steiner Minimal Trees", Ann. N.Y. Acad. Sci.* **440** (1985) 328-346.

[50] K.L. Clarkson, *"Fast Algorithms for the All-Nearest-Neighbours Problem," Proc. of the 24-th Ann. Symp. on the Foundations of Computer Science* (1983) 226-232.

[51] K.L. Clarkson, *"Faster Expected-Time and Approximation Algorithms for the Geometric Minimum Spanning Trees," Proc. of the 16-th Ann. ACM Symp. on the Theory of Computing* (1984) 342-348.

[52] K.L. Clarkson, *"Approximation Algorithms for Shortest Path Motion Planning," Proc. of the 19th Ann. ACM Symp. on the Theory of Computing* (1987) 56-65.

[53] K.L. Clarkson, K. Kapoor and P. Vaidya, *"Rectilinear Shortest Paths through Polygonal Obstacles in $O(n \log^2 n)$," Proc. of the 3-rd Ann. ACM Symp. on Computational Geometry*, Waterloo, Ontario (1987) 251-257.

[54] E.J. Cockayne, *"On the Efficiency of the Algorithm for Steiner Minimal Trees", SIAM J. Appl. Math.* **18** (1970) 150-159.

[55] E.J. Cockayne and Z.A. Melzak, *"Euclidean Constructability in Graph-Minimization Problems," Mathematics Magazine* **42** (1969) 206-208.

[56] J.P. Cohoon, D.S. Richards, and J.S. Salowe, *"A Linear Time Steiner Tree Routing Algorithm for Terminals on the Boundary of a Rectangle," Proc. of the IEEE CAD Conference* (1988) 402-405.

[57] R. Cole, *"Searching and Storing Similar Lists," J. of Algorithms* **7** (1986) 202-220.

[58] B. Dasarthy and L.J. White, *"On Some Maximin Location and Classifier Problems, "* Unpublished Lecture, Computer Science Conference, Washington D.C. (1975).

[59] B. Dasarthy and L.J. White, *"A Maximin Location Problem," Operations Research* **28** (1980) 1385-1401.

[60] E.W. Dijkstra, *"A Note on Two Problems in Connection with Graphs," Numerical Mathematics* **1** (1959) 269-271.

[61] M. Dillencourt, *"Travelling Salesman Cycles are not Always Subgraphs of Delaunay Triangulations or of Minimum Weight Triangulations," IPL* **24** (1987) 339-342.

[62] D.P. Dobkin and R.J. Lipton, *"Multidimensional Searching Problems," SIAM J. Comput.* **5** (1976) 181-186.

[63] D.P. Dobkin and R.J. Lipton, *"On the Complexity of Computations under Varying Set of Primitives," J. of Computer and Systems Sciences* **18** (1979) 86-91.

[64] D.P. Dobkin and D.L. Souvaine, *"Computational Geometry - A User's Guide,"* Technical Report 334, Dept. of EE and CS, Princeton Univ. (1985).

[65] J. Dolan, R. Weiss, and J. MacGregor Smith, *"Minimal Length Tree Networks on the Unit Sphere," (to appear in) Annals of Operations Research.*

[66] D.Z. Du and F.K. Hwang, *"A New Bound for the Steiner Ratio," Amer. Math. Soc.* **278** (1983) 137-148.

[67] D.Z. Du and F.K. Hwang, *"Steiner Minimal Trees for Bar Waves," Chinese J. of Mathematics* (1987)

[68] D.Z. Du and F.K. Hwang, *"The Steiner Ratio Conjecture of Gilbert and Pollak is True," Proc. Nat. Acad. Sci.* **87** (1990) 9464-9466.

[69] D.Z. Du, F.K. Hwang, and J.F. Weng, *"Steiner Minimal Trees on Zig-Zag Lines," Trans. Amer. Math. Soc.* **278** (1983) 149-156.

[70] D.Z. Du, F.K. Hwang and J.F. Weng, *"Steiner Minimal Trees for Regular Polygons" Discrete and Computational Geometry* **2** (1987) 65-87.

[71] D.Z. Du, F.K. Hwang, and E.Y. Yao, *"The Steiner Ratio Conjecture Is True for Five Points," J. Comb. Theory, Ser. A* **38** (1985) 230-240.

[72] D.Z. Du, E.Y. Yao, and F.K. Hwang, *"A Short Proof of the Result of Pollak on Steiner Minimal Trees," J. Comb. Theory, Ser. A* **32** (1982) 396-400.

[73] M.E. Dyer, *"Linear Time Algorithms for Two- and Three-Variable Linear Programs," SIAM J. Comput.* **13** (1984) 31-45.

[74] H. Edelsbrunner, *"Dynamic Rectangle Intersection Searching, "* Technical Report F47, Inst. für Informationsverarbeitung, Tech. Univ., Graz, Austria (1980).

[75] H. Edelsbrunner, *"Edge-Skeletons in Arrangements with Applications," Algorithmica* **1** (1986) 93-109.

[76] H. Edelsbrunner, *"Algorithms in Combinatorial Geometry,"* Springer-Verlag (1987).

[77] H. Edelsbrunner, L.J. Guibas, and J. Stolfi, *"Optimal Point Location in a Monotone Subdivision," SIAM J. Comput.* **15** (1986) 317-340.

[78] H. Edelsbrunner, H.A. Maurer, F.P. Preparata, A.L. Rosenberg, E. Welzl, D. Wood, *"Stabbing Line Segments," BIT* **22** (1982) 274-281.

[79] H. Edelsbrunner, J. O'Rourke, and R. Seidel, *"Constructing Arrangements of Lines and Hyperplanes with Applications," SIAM J. Comput.* **15** (1986) 341-363.

[80] H. Edelsbrunner and E. Welzl, *"Halfplanar Range Search in Linear Space and $O(n^{0.695})$ Query Time," IPL* **23** (1986) 289-293.

[81] J. Elzinga and D.W. Hearn, *"Geometrical Solutions for Some Minimax Location Problems," Transportation Science* **6** (1971) 379-394.

[82] S. Fortune, *"A Sweepline Algorithm for Voronoi Diagrams," Algorithmica* **2** (1987) 153-174.

[83] R.L. Francis and J.A. White, *"Facility Layout and Location: An Analytical Approach,"* Englewood Cliffs NJ, Prentice-Hall (1974).

[84] J. Friedel and P. Wiedmayer, *"A Simple Proof of the Steiner Ratio Conjecture for Five Points," SIAM J. Appl. Math.* **49** (1989) 960-967.

[85] H.N. Gabow, J.L. Bentley, and R.E. Tarjan, *"Scaling and Related Techniques for Geometry Problems," Proc. of the 16-th Ann. ACM Symp. on Theory of Computing* (1984) 135-143.

[86] M.R. Garey, R.L. Graham, D.S. Johnson, *"Some NP-Complete Geometric Problems," Proc. of the 8-th Ann. ACM Symp. on Theory of Computing* (1976) 10-22.

[87] M.R. Garey, L.R. Graham, and D.S. Johnson, *"The Complexity of Computing Steiner Minimal Trees," SIAM J. Appl. Math.* **32** (1977) 835-859.

[88] M.R. Garey and D.S. Johnson, *"The Rectilinear Steiner Tree Problem Is NP-Complete," SIAM J. Appl. Math.* **32** (1977) 826-834.

[89] M.R. Garey, D.S. Johnson, F.P. Preparata and R.E. Tarjan, *"Triangulating a Simple Polygon," IPL* **7** (1978) 175-180.

[90] S.K. Ghosh and D.M. Mount, *"An Output Sensitive Algorithm for Computing Visibility Graphs," Proc. of the 28-th Ann. Symp. on Foundations of Computer Science* (1987) 11-19.

[91] E.N. Gilbert and H.O. Pollak, *"Steiner Minimal Trees," SIAM J. Appl. Math.* **16** (1968) 1-29.

[92] P.N. Gilbert, *"New Results on Planar Triangulations,"* Technical Report ACT-15, Coord. Sci. Lab., Univ. of Illinois at Urbana (1979).

[93] B. Golden, L. Bodin, T. Doyle, and W. Stewart Jr., *"Approximate Traveling Salesman Algorithms," Operations Research* **28** (1980) 694-711.

[94] A.J. Goldman and P.M. Dearing, *"Concepts of Optimal Location for Partially Noxious Facilities,"* Paper presented at the ORSA/TIMS National Meeting, Chicago, Il. (1975).

[95] T. Gonzalez, *"Algorithms on Sets and Related Problems,"* Technical Report, Dept. of Computer Science, University of Oklahoma (1975).

[96] R.L. Graham, *"An Efficient Algorithm for Determining the Convex Hull of a Finite Planar Set," IPL* **1** (1972) 32-33.

[97] R.L. Graham and F.K. Hwang, *"Remarks on Steiner Minimal Trees," Bull. Inst. Math. Acad. Sinica* **4** (1976) 177-182.

[98] V.M. Grout, P.W. Sanders, and C.T. Stockel, *"Reduction Techniques Providing Initial Groupings for Euclidean Travelling Salesman Patching Algorithm," Applied Mathematical Modelling* **13** (1989) 110-115.

[99] L.J. Guibas and J. Hershberger, *"Optimal Shortest Path Queries in a Simple Polygon," Proc. of the 3-th Ann. Symp. on Computational Geometry* (1987) 50-63.

[100] L.J. Guibas and J. Stolfi, *"Primitives for the Manipulation of General Subdivisions and the Computation of Voronoi Diagrams," ACM Transactions on Graphics* **4** (1985) 74-123.

[101] L.J. Guibas and J. Stolfi, *"Ruler, Compass and Computer: The Design and Analysis of Geometric Algorithms,"* manuscript (undated).

[102] F.O. Hadlock, *"A Shortest Path Algorithm for Grid Graphs," Networks* **7** (1977) 323-334.

[103] S.L. Hakimi, *"Steiner's Problem in Graphs and Its Implications," Networks* **1** (1971) 113-133.

[104] M. Hanan, *"On Steiner's Problem with Rectilinear Distance," J. SIAM Appl. Math.* **14** (1966) 255-265.

[105] J. Hershberger and L.J. Guibas, *"An $O(n^2)$ Shortest Path Algorithm for a Non-Rotating Convex Body,"* Technical Report, DEC Systems Research Center (1986).

[106] F.K. Hwang, *"On Steiner Minimal Trees with Rectilinear Distance," SIAM J. Appl. Math.* **30** (1976) 104-114.

[107] F.K. Hwang, *"An $O(n \log n)$ Algorithm for Suboptimal Rectilinear Steiner Trees," IEEE Transactions on Circuits and Systems* **CAS-26** (1979) 75-77.

[108] F.K. Hwang, *"A Linear Time Algorithm for Full Steiner Trees," Operations Research Letters* **5** (1986) 235-237.

[109] Hwang, F.K., G.D. Song, G.Y. Ting and D.Z. Du, *"A Decomposition Theorem on Euclidean Steiner Minimal Trees," Discrete and Computational Geometry* **3** (1988) 367-382.

[110] F.K. Hwang and J.F. Weng, *"The Shortest Network Under a Given Topology,"* to appear.

[111] F.K. Hwang, J.F. Weng, and D.Z. Du, *"A Class of Full Steiner Minimal Trees," Discrete Mathematics* **45** (1983) 107-112.

[112] T. Imai and T. Asano, *"Dynamic Segment Intersection with Applications," Proc. of the 28-th Ann. Symp. on Foundations of Computer Science* (1984) 393-402.

[113] R.A. Jarvis, *"On the Identification of the Convex Hull of a Finite Set of Points in the Plane," IPL* **2** (1973) 18-21.

[114] M. Kallay, *"Convex Hull Algorithms in Higher Dimensions, "* unpublished manuscript, Dept of Mathematics, Univ. of Oklahoma, Norman, Oklahoma (1981).

[115] V. Kantabutra, *"Travelling Salesman Cycles Are Not Always Subgraphs of Voronoi Dual," IPL* **16** (1983) 11-12.

[116] R.M. Karp, *"The Probabilistic Analysis of Some Combinatorial Search Algorithms, "* in J.F. Traub (ed.) *"Algorithms and Complexity: New Directions and Recent Results,"* Academic Press, New York (1976) 1-19.

[117] J. Katajainen, *"On the Worst Case of a Minimal Spanning Tree Algorithm for Euclidean Space," BIT* **23** (1983) 2-8.

[118] J.M. Keil and J.R. Sack, *"Minimum Decomposition of Polygonal Objects, "* in G.T. Toussaint (ed.), *"Computational Geometry,"* North-Holland, Amsterdam (1985).

[119] D.G. Kirkpatrick, *"Optimal Search in Planar Subdivisions," SIAM J. Comput.* **12** (1983) 28-35.

[120] D.G. Kirkpatrick and R. Seidel, *"The Ultimate Planar Convex Hull Algorithm?," SIAM J. Comput.* **15** (1986) 287-299.

[121] J. Komlos and M.T. Shing, *"Probabilistic Partitioning Algorithms for the Rectilinear Steiner Problem," Networks* **15** (1985) 413-423.

[122] P. Korhonen, *"An Algorithm for Transforming a Spanning Tree into a Steiner Tree," Proc. of 9-th Mathematical Programming Symp.* **2** (1979) 349-357.

[123] E.L. Lawler, J.K. Lenstra, A.H.G. Rinnooy Kan and D. Shmoys (eds.), *"The Traveling Salesman Problem,"* John Wiley and Sons (1985).

[124] J.H. Lee, N.K. Bose, and F.K. Hwang, *"Use of Steiner's Problem in Sub-Optimal Routing in Rectilinear Metric," IEEE Transactions on Circuits and Systems,* **CAS-23** (1976) 470-476.

[125] D.T. Lee, *"Two Dimensional Voronoi Diagram in the $L_p$-metric," J. ACM* **27** (1980) 604-618.

[126] D.T. Lee, *"Furthest Neighbor Voronoi Diagrams and Applications,"* Tech. Rep. No. 80-11-FC-04, Dept. EE/CS, Northwestern Univ. (1980).

[127] D.T. Lee, *"Two Algorithms for Constructing Delaunay Triangulations," Int. J. Comput. and Info. Sci.* **9** (1980) 219-242.

[128] D.T. Lee, *"On k-Nearest Neighbor Voronoi Diagrams in the Plane," IEEE Transactions on Computers* **C-31** (1982) 478-487.

[129] D.T. Lee and F.P. Preparata, *"Euclidean Shortest Paths in the Presence of Rectilinear Barriers," Networks* **14** (1984) 393-410.

[130] D.T. Lee and F.P. Preparata, *"Location of a Point in a Planar Subdivision and Its Applications," SIAM J. Comput.* **6** (1977) 594-606.

[131] D.T. Lee and F.P. Preparata, *"Computational Geometry - A Survey," IEEE Transactions on Computers* **C-33** (1984) 1072-1101.

[132] D.T. Lee and C.K. Wong, *"Voronoi Diagrams in $L_1$ ($L_\infty$) Metrics with 2-Dimensional Storage Applications," SIAM J. Comput.* **9** (1980) 200-211.

[133] D.T. Lee and C.K. Wong, *"Worst-Case Analysis for Region and Partial Region Searches in Multidimensional Binary Search Trees and Balanced Quad Trees," Acta Informatica* **9** (1977) 23-29.

[134] D.T. Lee and Y.F. Wu, *"Geometric Complexity of some Location Problems," Algorithmica* **1** (1986) 193-211.

[135] A. Liestman, *"Construction of Steiner Trees with Obstacles in the Plane,"* M.Sc. Thesis, Dept. of Computer Science, University of Illinois, Urbana, Illinois (1978).

[136] S. Lin and B.W. Kernigham, *"An Effective Heuristic Algorithm for the Travelling Salesman Problem," Operations Research* **21** (1973) 498-516.

[137] R.F. Love, J.G. Morris, and G.O. Wesolowsky, *"Facilities Location: Models and Methods,"* North-Holland (1988).

[138] T. Lozano-Perez and M.A. Wesley, *"An Algorithm for Planning Collision-Free Paths among Polyhedral Obstacles," Comm. ACM* **22** (1979) 560-570.

[139] G.S. Lueker, *"A Data Structure for Orthogonal Range Queries," Proc. of the 19-th Ann. Symp. On Foundations of Computer Science* (1978) 28-34.

[140] T.L. Magnanti and R.T. Wong, *"Network Design and Transportation Planning: Models and Algorithms," Transportation Science* **18** (1984) 1-55.

[141] U. Manber and M. Tompa, *"The Complexity of Problems on Probabilistic, Nondeterministic and Alternating Decision Trees," J. ACM* **32** (1985) 720-732.

[142] J. Matousek, *"Construction of Epsilon Nets," Discrete and Computational Geometry* **5** (1990) 427-448.

[143] D.W. Matula and R.R. Sokal, *"Properties of Gabriel Graphs Relevant to Geographic Variation Research and the Clustering of Points in the Plane," Geographical Analysis* **12** (1980) 205-222.

[144] E.M. McCreight, *"Priority Search Trees," SIAM J. Comput.* **14** (1985) 257-276.

[145] K. Mehlhorn, *"Data Structures and Algorithms: 3. Multi-Dimensional Searching and Computational Geometry,"* Springer-Verlag (1984).

[146] N. Megiddo, *"Linear Time Algorithm for Linear Programming in* $R^3$ *and Related Problems," SIAM J. Comput.* **12** (1983) 759-776.

[147] E. Melachrinoudis, *"The Maximin Single Facility Location Problem Using a Euclidean Metric, "* Ph.D. Thesis, Dept. of Industrial Engineering and Operations Research, University of Massachusetts, Amherst, MA (1980).

[148] E. Melachrinoudis and T.P. Cullinane, *"A Heuristic Approach to the Single Facility Maximin Location Problem," Int. J. Prod. Res.* **23** (1985) 523-532.

[149] E. Melachrinoudis and T.P. Cullinane, *"Locating an Undesireable Facility within a Geographic Region Using the MAXIMIN Criterion," J. Reg. Sci.* **25** (1985) 115-127.

[150] E. Melachrinoudis and T.P. Cullinane, *"Locating an Obnoxious Facility within a Polygonal Region," Annals of Operations Research* **6** (1986) 137-145.

[151] Z.A. Melzak, *"On the Problem of Steiner," Canad. Math. Bull.* **4** (1961) 143-148.

[152] M. Minoux, *"Network Synthesis and Optimum Network Design Problems: Models, Solution Methods and Applications," Networks* **19** (1989) 313-360.

[153] J.S.B. Mitchell and C.H. Papapdimitriou, *"The Weighted Region Problem," Proc. of the 3-rd Ann. Symp. on Computational Geometry* (1987) 30-38.

[154] C. Monma, M. Paterson, S. Suri, and F. Yao, *"Computing Euclidean Maximum Spanning Trees," Algorithmica* **5** (1990) 407-419.

[155] D.M. Mount, *"On Finding Shortest Paths on Convex Polyhedra,"* Technical Report, Computer Science Dept., Univ. of Maryland, College Park, MD (1984).

[156] O. Nevalainen, J. Ernvall, and J. Katajainen, *"Finding Minimal Spanning Tree in a Euclidean Coordinate Space," BIT* **21** (1981) 46-54.

[157] T.A.J. Nicholson, *"A Sequential Method for Discrete Optimization Problems and Its Application to the Assignment, Traveling Salesman, and Three Machine Scheduling Problems," J. Inst. Math. Appl.* **3** (1967) 362-375.

[158] M.H. Overmars and J. van Leeuven, *"Maintenance of Configurations in the Plane," J. Comput. Syst. Sci.* **23** (1981) 166-207.

[159] C.H. Papadimitriou, *"The Euclidean Traveling Salesman Problem Is NP-Complete," Theoretical Computer Science* **4** (1977) 237-244.

[160] C.H. Papadimitriou, *"An Algorithm for Shortest Path Motion in Three Dimensions," IPL* **20** (1985) 259-263.

[161] C.H. Papadimitriou and E.B. Silverberg, *"Optimal Piecewise Linear Motion of an Object among Obstacles,"* Technical Report, Dept. of OR, Stanford University (1986).

[162] C.H. Papadimitriou and K. Steiglitz, *"Some Complexity Results for the Traveling Salesman Problem," Proc. of the 8-th Ann. Symp. on Theory of Computing* (1976) 1-9.

[163] H.O. Pollak, *"Some Remarks on the Steiner Problem," J. Comb. Theory, Ser. A* **24** (1978) 278-295.

[164] F.P. Preparata, *"A New Approach to Planar Point Location," SIAM J. Comput.* **10** (1981) 473-482.

[165] F.P. Preparata and S.J. Hong, *"Convex Hulls of Finite Sets of Points in Two and Three Dimensions," Comm. ACM* **20** (1977) 87-93.

[166] F.P. Preparata and M.I. Shamos, *"Computational Geometry: An Introduction,"* Springer-Verlag, New York (1985).

[167] R.C. Prim, *"Shortest Connecting Networks and Some Generalizations," Bell Syst. Tech. J.* **36** (1957) 1389-1401.

[168] J.S. Provan, *"Convexity and the Steiner Tree Problem," Networks* **18** (1988) 55-72.

[169] J.S. Provan, *"An Approximation Scheme For Finding Steiner Trees with Obstacles," SIAM J. Comput.* **17** (1988) 920-934.

[170] J. Reif and J.A. Storer, *"Motion Planning in the Presence of Moving Obstacles," Proc. of the 26-th Symp. on the Foundations of Computer Science* (1985) 144-154.

[171] J. Reif and J.A. Storer, *"Shortest Paths in Euclidean Space with Polyhedral Obstacles,"* Technical Report CS-85-121, Computer Science Department, Brandeis University, Waltham, MA (1985).

[172] D. Richards, *"Fast Heuristic Algorithms for Rectilinear Steiner Trees,"* Technical Report TR-86-28, Dept. of Computer Science, Univ. of Virginia (1986).

[173] H. Rohnert, *"Shortest Paths in the Plane with Convex Polygonal Obstacles," IPL* **23** (1986) 71-76.

[174] H. Rohnert, *"Time and Space Efficient Algorithms for Shortest Paths between Convex Polygons," IPL* **27** (1988) 175-179.

[175] Rosenkrantz, D.J., R.E. Stearns and P.M. Lewis, *"Approximate Algorithms for the Traveling Salesperson Problem," 15-th Ann. IEEE Symp. on Switching and Automata Theory* (1974) 33-42.

[176] D.J. Rosenkrantz, R.E. Stearns, and P.M. Lewis, *"An Analysis of Several Heuristics for the Traveling Salesman," SIAM J. Comput.* **6** (1977) 563-581.

[177] J.H. Rubinstein and D.A. Thomas, *"The Steiner Ratio Conjecture for Six Points," J. Comb. Theory, Ser. A*, to appear.

[178] R. Seidel, *"A Convex Hull Algorithm Optimal for Point-Sets in Even Dimensions,"* M.Sc. Thesis, Technical Report 81-14, Dept. of Computer Science, Univ. of British Colombia, Vancouver, Canada (1981).

[179] M.I. Shamos, *"Computational Geometry", Ph.D. Thesis*, Dept. of Computer Science, Yale University (1978).

[180] M.I. Shamos and D. Hoey, *"Closest-Point Problems," Proc. 16-th IEEE Ann. Symp. on the Foundations of Computer Science* (1975) 151-162.

[181] M. Sharir, *"On Shortest Paths amidst Convex Polyhedra," SIAM J. Comput.* **16** (1987) 561-572.

[182] M. Sharir and A. Schorr, *"On Shortest Paths in Polyhedral Spaces," SIAM J. Comput.* **15** (1986) 193-215.

[183] W.D. Smith, *"Studies in Discrete and Computational Geometry,"* PhD. Thesis, Program in Applied and Computational Math., Princeton Univ. (1988).

[184] J.MacGregor Smith, *"Steiner Minimal Trees with Obstacles," Paper Presented at the ORSA/TIMS Meeting Detroit, Michigan* (1982).

[185] J.MacGregor Smith, R.J. Graves, and L. Kerbache, *"QNET: An Open Queueing Network Model for Material Handling System Analysis," Material Flow* **3** (1986) 225-242.

[186] J. MacGregor Smith, D.T. Lee and J.S. Liebman, *"An $O(n \log n)$ Heuristic Algorithm for the Rectilinear Steiner Minimal Tree Problem," Engineering Optimization* **4** (1980) 179-192.

[187] J. MacGregor Smith, D.T. Lee and J. S. Liebman, *"An $O(n \log n)$ Heuristic for Steiner Minimal Tree Problems on the Euclidean Metric," Networks* **11** (1981) 23-29.

[188] J. MacGregor Smith and J.S. Liebman, *"Steiner Trees, Steiner Circuits, and the Interference Problem in Building Design," Engineering Design* **4** (1979) 15-36.

[189] J. Stewart Jr., *"A Computationally Efficient Heuristic for the Traveling Salesman Problem," Proc. 13-th Ann. Meeting of S.E. TIMS* (1977) 75-85.

[190] I. Stojmenovic and M.A. Langston, *"On a Proposed Divide-and-Conquer Minimal Spanning Tree Algorithm," BIT* **28** (1988) 785-791.

[191] K.J. Supowit, *"The Relative Neighborhood Graph with an Application to Minimum Spanning Trees," J. ACM* **30** (1983) 428-447.

[192] R.E. Tarjan and M.L. Fredman, *"Fibonacci Heaps and Their Uses in Improved Network Optimization Algorithms," Proc. of 25-th Ann. IEEE Symp. on Foundations of Computer Science* 338-346 (1980).

[193] R.E. Tarjan and C.J. van Wyk, *"An $O(n \log \log n)$-time Algorithm for Triangulating Simple Polygons," SIAM J. Comput.* **17** (1988) 143-178.

[194] G.T. Toussaint, *"The Relative Neighborhood Graph of a Finite Planar Set," Pattern Recognition* **12** (1980) 261-268.

[195] D. Trietsch and F.K. Hwang, *"An Improved Algorithm for Steiner Trees," SIAM J. Appl. Math.* **50** (1990) 244-263.

[196] P.M. Vaidya, *"An Optimal Algorithm for the All-Nearest-Neighbors Problem," Proc. of the 27-th Ann. IEEE Symp. on the Foundations of Computer Science* (1986) 117-122.

[197] P.M. Vaidya, *"Minimum Spanning Trees in k-Dimensional Space," SIAM J. Comput.* **17** (1988) 572-582.

[198] P.M. Vaidya, *"Geometry Helps in Matching," Proc. of the 20-th Ann. ACM Symp. on Theory of Computing* (1988) 422-425.

[199] P.M. Vaidya, *"An $O(n \log n)$ Algorithm for the All-Nearest Neighbors Problem," Discrete and Computational Geometry* **4** (1989) 101-115.

[200] V. Vaishnavi and D. Wood, *"Rectilinear Line Segment Intersection, Layered Segment Trees, and Dynamization," J. of Algorithms* **3** (1982) 160-176.

[201] T. Watanabe and Y. Sugiyama, *"A New Routing Algorithm and Its Hardware Implementation," Proc. of the 23-th ACM/IEEE Design Automation Conf.* (1986) 574-580.

[202] E. Welzl, *"Constructing the Visibility Graph for n Line Segments in $O(n^2)$ Time," IPL* **20** (1985) 167-171.

[203] P. Widmayer *"On Graphs Preserving Rectilinear Shortest Paths in the Presence of Obstacles," (to appear) Annals of Operations Research.*

[204] D.E. Willard, *"Predicate-Oriented Database Search Algorithms, "* Ph.D. Thesis, Report TR-20, Aiken Computation Lab., Harvard Univ., Cambridge, MA (1978).

[205] D.E. Willard, *"Polygon Retrieval," SIAM J. Comput.* **11** (1982) 149-165.

[206] H.P. Williams, *"Optimization is Best, "* Inaugural Lecture, Univ. of Southampton, Faculty of Mathematical Studies (1986).

[207] P. Winter, *"An Algorithm for the Steiner Problem in the Euclidean Plane," Networks* **15** (1985) 323-345.

[208] P. Winter, *"Steiner Problem in Networks: A Survey," Networks* **17** (1987) 129-167.

[209] P. Winter and J.MacGregor Smith, *"Steiner Minimal Trees with Obstacles: 3 and 4 Point Case' Paper Presented at the NATO/ARW on Topological Network Design* Schaeffergarden, Denmark (1989).

[210] P. Winter and J. MacGregor Smith, *"Path-Distance Heuristics for the Steiner Problem in Undirected Networks," to appear in Algorithmica.*

[211] P. Winter and J. MacGregor Smith, *"Steiner Minimal Trees for Three Points with One Convex Polygonal Obstacle," to appear in Annals of Operations Research*

[212] Y.Y. Yang and O. Wing, *"Optimal and Suboptimal Solution Algorithms for the Wiring Problem," Proc. of the IEEE Int. Symp. on Circuit Theory* (1972) 154-158.

[213] A.C. Yao, *"On Constructing Minimum Spanning Trees in k-Dimensional Space and Related Problems," SIAM J. Comput.* **11** (1982) 721-736.

[214] F. Yao, *"Computational Geometry, "* in J. van Leeuven (ed.) *"Handbook of Theoretical Computer Science,"* Elsevier Science Publishers B.V. (1990) 344-389.

[215] Y.Y. Yao, D.P. Dobkin, H. Edelsbrunner and M.S. Paterson, *"Partitioning Space for Range Queries," SIAM J. Comput.* **18** (1989) 371-384.